THE STEAM-ENGINE
AND
OTHER HEAT-ENGINES

THE STEAM-ENGINE

AND

OTHER HEAT-ENGINES

BY

SIR J. ALFRED EWING, K.C.B.

LL.D., D.Sc., F.R.S., M.Inst.C.E.

PRESIDENT OF THE ROYAL SOCIETY OF EDINBURGH
PRINCIPAL AND VICE-CHANCELLOR OF THE UNIVERSITY OF EDINBURGH
HONORARY FELLOW OF KING'S COLLEGE, CAMBRIDGE
FORMERLY PROFESSOR OF MECHANISM AND APPLIED MECHANICS IN THE UNIVERSITY OF CAMBRIDGE
SOMETIME DIRECTOR OF NAVAL EDUCATION

FOURTH EDITION
REVISED AND ENLARGED

CAMBRIDGE
AT THE UNIVERSITY PRESS
1926

CAMBRIDGE UNIVERSITY PRESS
Cambridge, New York, Melbourne, Madrid, Cape Town,
Singapore, São Paulo, Delhi, Mexico City

Cambridge University Press
The Edinburgh Building, Cambridge CB2 8RU, UK

Published in the United States of America by Cambridge University Press, New York

www.cambridge.org
Information on this title: www.cambridge.org/9781107615632

First Edition 1894
Second Edition, revised 1897
Reprinted 1899, 1902, 1906
Third Edition, revised and enlarged 1910
Reprinted 1914, 1920
Fourth Edition, revised and enlarged 1926
First published 1926
First paperback edition 2013

A catalogue record for this publication is available from the British Library

ISBN 978-1-107-61563-2 Paperback

PREFACE

TO THE FOURTH EDITION

THE purpose of this book is to present the subject of heat-engines, in their mechanical as well as their thermodynamical aspects, with sufficient fulness for the ordinary needs of University students of engineering.

Since 1910, when a Third and enlarged edition was published, there have been reprints involving little alteration.

In this—the Fourth—edition the book has been revised throughout and there are many additions. Recent developments have made it desirable to re-write large parts of the chapters on steam turbines, steam boilers, and internal-combustion engines. Not only is much of the descriptive matter new, but the treatment of theory has undergone some change. In chapters relating to the properties of steam advantage has been taken of the data supplied by Callendar in his Steam Tables, first published in 1915. A selection of these is included as an appendix, along with a short account of the formulas on which the Tables are based. The discussion of thermodynamic principles in their application to engines has been brought more closely into correspondence with the methods adopted in my *Thermodynamics for Engineers*, published in 1920. The conception of heat-drop, now familiar through its engineering uses, especially in relation to turbine design, has been given a prominent place.

Particulars of modern practice, with materials for illustration, have been kindly furnished by various firms whose names are noted below. I would also thank several friends for helpful information or suggestions, notably Sir Charles Parsons, Sir Dugald Clerk, Sir Henry Fowler, Professors Callendar, Dalby, Inglis, Jenkin, and Sir Thomas Hudson Beare. To Mr Stanley S. Cook I am specially grateful, not only for much information, but for reading in proof the sections which treat of steam turbines.

For particulars of turbines thanks are due to Messrs C. A. Parsons and Co., the Parsons Marine Steam Turbine Co., the English Electric Co., the Metropolitan-Vickers Co., the Brush Co., the British Thomson-Houston Co., Messrs Brown Boveri, Messrs Escher Wyss and Co., and the International General Electric Co. of New York.

For information about boilers and furnaces, to Messrs Galloways, Messrs William Beardmore and Co., Messrs Babcock and Wilcox, the Stirling Boiler Co., Messrs Yarrow and Co., Messrs John I. Thornycroft and Co., Messrs J. Samuel White and Co., Messrs Cochran and Co., John Thompson (Wolverhampton), the British Niclausse Co., the Howden-Ljungström Co., and International Combustion Ltd.

For information about gas and oil-engines, to the National Co., Crossley Brothers, the Premier Co., Messrs Ruston and Hornsby, Messrs Galloways, Messrs Petters, Messrs Mirrlees Bickerton and Day, Messrs Harland and Wolff, Messrs William Doxford and Sons, and Messrs Worthington-Simpson. Particulars of engines of special types and of various measuring appliances have been given by Messrs Hathorn Davey and Co., Sulzer Brothers, Messrs Belliss and Morcom, Messrs Browett Lindley and Co., the Baldwin Locomotive Works, Siemens Brothers, Messrs Dobbie McInnes and Clyde, the Crosby Steam Valve Co., and the Cambridge Instrument Co.

The Institutions of Naval Architects, of Civil Engineers, and of Mechanical Engineers have kindly allowed illustrations to be reproduced.

J. A. EWING

The University, Edinburgh
July 1926

CONTENTS

FOLDING PLATES

Images available for download from www.cambridge.org/9781107615632

CHAPTER I

THE EARLY HISTORY OF THE STEAM-ENGINE

1. Heat-Engines in general. In the scientific treatment of the steam-engine we have in the first place, and mainly, to regard it as a heat-engine—that is, a machine in which heat is employed to do mechanical work. Other aspects of the steam-engine will present themselves when we come to examine the action of the mechanism in detail, but the foremost place must be given to thermodynamic considerations. From the thermodynamic point of view the function of a heat-engine is to get as much work as possible from a given supply of heat, or (to go a step further back) from the combustion of a given quantity of fuel. Hence a large part of our subject is the discussion of what is called the *efficiency* of the engine, which is the ratio of the work done to the heat supplied. We have to consider on what conditions efficiency depends, how its value is limited in theory and how nearly the limiting value may be attained in practice. We have to describe means of testing the efficiency of engines, and the results which such tests have given in actual cases. Much of what has to be said in regard to efficiency is applicable to all heat-engines, whatever be the character of the substance which is made use of as the means of doing work within the engine. In all practical heat-engines work is done through the expansion by heat of a fluid which exerts pressure and overcomes resistance as it expands. Thus in steam-engines the working substance is water and water-vapour, and work is done by the pressure which the substance exerts while its volume is undergoing change. This is true of engines of the turbine type as well as those in which the steam presses against a moving piston: in the former the work is done by setting in motion a part of the working substance itself, and causing its momentum to exert force upon a moving part of the machine. In air-engines the working substance is atmospheric air; in gas-engines and oil or petrol-engines it is a mixture of air with gas or with the vapour of an inflammable liquid and with gaseous products of combustion. Engines of this last type are called *internal-combustion* engines because the heat which it is the function of the engine to convert into power is developed by combustion within the working substance itself, instead of reaching the substance from an external source. Thus in an engine of the internal-combustion type the heat does not have to pass into the working substance by conduction through the wall of a containing vessel, such as a boiler, strong enough to bear the pressure of the fluid within. This secures advantages in respect of lightness and efficiency which have led to a greatly extended application of the internal-combustion principle, especially for small power installations and for the driving of road vehicles, aircraft, and ships. Notwithstanding, however, the multiplied use of internal-combustion engines during the first quarter of the twentieth century, and their recent application to the driving of large ships, the steam-engine (including, of course, its turbine form) continues to be the chief means of producing power on a large scale out of the potential energy of fuel.

As a preliminary to the study of the modern engine it will be useful to review, if only very briefly, some of the stages through which it has passed in its development. In any such historical sketch the largest share of attention necessarily falls to the work of Watt, whose inventions were as remarkable for their scientific interest as for their industrial importance. But it should be borne in mind that a process of evolution had been going on before the time of Watt which prepared the steam-engine for the immense improvements it received at his hands. The labours of Watt stand in a natural sequence to those of Newcomen, and Newcomen's to those of Papin and Savery. Savery's engine, again, was the reduction to practical form of a contrivance which had long before been known as a scientific toy.

2. Hero of Alexandria. The earliest notices of heat-engines are found in the *Pneumatics* of Hero of Alexandria, which dates from the second century before Christ. One of the contrivances mentioned there is the æolipile, a steam reaction-turbine consisting of a spherical vessel pivoted on a central axis and supplied with steam through one of the pivots. The steam escapes by bent pipes facing tangentially in opposite directions, at opposite ends of a diameter perpendicular to the axis. The globe revolves by reaction from the escaping steam, just as a Barker's mill is driven by escaping water. Another apparatus described by Hero (fig. 1)[1] is

[1] From Greenwood's translation of Hero's *Pneumatics*, edited by B. Woodcroft, 1851.

interesting as the prototype of a class of engines which long afterwards became practically important. A hollow altar containing air is heated by a fire kindled on it; the air in expanding drives some of the water contained in a spherical vessel beneath the altar into a bucket, which descends and opens the temple doors above by pulling round a pair of vertical posts to which the doors are fixed. When the fire is extinguished the air cools, the water leaves the bucket, and the doors close. In another device a jet of water driven out by expanding air is turned to account as a fountain. Several other philosophical toys or pieces of conjuring apparatus of the like kind are also described, but there is no suggestion that the methods they illustrate could be applied on a large scale or turned to any useful account.

Fig. 1. Apparatus described by Hero.

3. Della Porta and De Caus. From the time of Hero to the seventeenth century there is no progress to record, though here and there we find evidence that appliances like those described by Hero were used for trivial purposes, such as organ-blowing and the turning of spits. The next distinct step was the publication in 1606 of a treatise on pneumatics by Giovanni Battista Della Porta, in which he shows an apparatus similar to Hero's fountain, but with steam instead of air as the displacing fluid. Steam generated in a separate vessel passed into a closed chamber containing water, and drove the water out through a pipe which opened near the bottom of the vessel. He also points out that the condensation of steam in the closed chamber may be used to produce a vacuum and suck up water from a lower level. In fact, his suggestions anticipate very fully the principle which a century later was applied by Savery in the earliest commercially successful steam-engine. In 1615 Salomon De Caus gives a plan of forcing up water by a steam-fountain which differs from Porta's only in having one vessel serve both as boiler and as displacement-chamber, the hot water being itself raised.

4. Branca's Steam Turbine. Another line of invention was taken by Giovanni Branca (1629), who designed an engine shaped like a water-wheel, to be driven by the impact of a jet of steam on its vanes, and, in its turn, to drive other mechanism for various useful purposes—what we should now call an impulse turbine. But Branca's suggestion was unproductive, and we find the course of invention revert to the line followed by Porta and De Caus.

5. Marquis of Worcester. The next contributor is one whose place is not easily assigned. To Edward Somerset, second marquis of Worcester, appears to be due the credit of proposing, if not of making, the first useful steam-engine. Its object was to raise water, and it worked probably like Porta's model, but with a pair of displacement-chambers, from each of which alternately water was forced by steam from an independent boiler, or perhaps by applying heat to the chamber itself, while the other vessel was allowed to refill. The only description of the engine is found in Art. 68 of Worcester's *Century of Inventions* (1663). There are no drawings, and the notice is so obscure that it is difficult to say whether there were any distinctly novel features except the double action. The inventor's account leaves much to the imagination. It is entitled "A Fire Water-work," and runs thus:—

An admirable and most forcible way to drive up water by fire, not by drawing or sucking it upwards, for that must be as the Philosopher calleth it, *Intra sphaeram activitatis*, which is but at such a distance. But this way hath no Bounder, if the Vessels be strong enough; for I have taken a piece of a whole Cannon, whereof the end was burst, and filled it three-quarters full of water, stopping and scruing up the broken end; as also the Touch-hole; and making a constant fire under it, within 24 hours it burst and made a great crack. So that having a way to make my Vessels, so that they are strengthened by the force within them, and the one to fill after the other, I have seen the water run like a constant Fountaine-stream forty foot high; one Vessel of water rarified by fire driveth up forty of cold water. And a man that tends the work is but to turn two Cocks, that one Vessel of water being consumed, another begins to force and re-fill with cold water, and so successively, the fire being tended and kept constant, which the self-same Person may likewise abundantly perform in the interim between the necessity of turning the said Cocks.

Later articles in the *Century of Inventions* contain notices of a device which, under the name of a "Water-commanding Engine," received protection by Act of Parliament and was experimented on by Worcester on a large scale at Vauxhall. But there is nothing to show distinctly that the Water-commanding Engine was a heat-engine at all, and the meagre accounts that have been given of it rather point to the conclusion that it was a form of "Perpetual Motion." In any case the experiments led to no practical result.

6. Savery. The steam-engine became commercially successful in the hands of Thomas Savery, who in 1698 obtained a patent for a water-raising engine, shown in fig. 2. Steam is admitted to one of the oval vessels *A*, displacing water, which it drives up through the check-valve *B*. When the vessel *A* is emptied of water, the supply of steam is stopped, and the steam already there is condensed by allowing a jet of cold water from a cistern above to stream over the outer surface of the vessel. This produces a vacuum and causes water to be sucked up through the pipe *C* and the valve *D*. Meanwhile, steam has been displacing water from the other vessel, and is ready to be condensed there. The valves *B* and *D* open only upwards. The supplementary boiler and furnace *E* are for feeding water to the main boiler; *E* is filled while cold and a fire is lighted under it; it then acts like the vessel of De Caus in forcing a supply of feed-water into the main boiler *F*. The gauge-cocks *G*, *G* for testing the level of the water in the boiler are an interesting feature of detail. Another form of Savery's engine had only one displacement-chamber and worked intermittently. In the use of

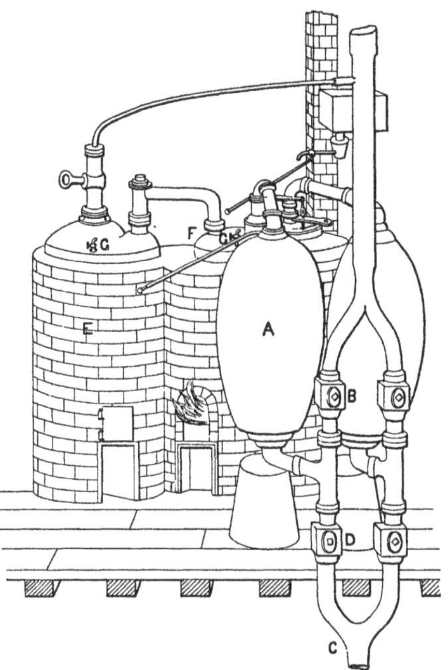

FIG. 2. Savery's Pumping Engine, 1698.

artificial means to condense the steam, and in the application of the vacuum so formed to raise water by suction from a level lower than that of the engine, the action used by Savery was probably an advance on that proposed or used by Worcester; in any case Savery's was the first engine to take a really practical shape. It found considerable employment in pumping mines and in raising water to supply houses and towns, and even to drive water-wheels. A serious difficulty which prevented its general use in mines was the fact that the height through which it would lift water was limited by the pressure the boiler and vessels could bear. Pressures as high as 8 or 10 atmospheres were employed—and that, too, without a safety-valve. But Savery found it no easy matter to deal with high-pressure steam; he complains that it melted his common solder, and forced him, as Desaguliers tells us, "to be at the pains and charge to have all his joints soldered with spelter." Apart from this drawback the waste of fuel was enormous, from the condensation of steam which took place on the surface of the water and on the sides of the displacement-chamber at each stroke; the consumption of coal was, in proportion to the work done, some twenty times greater than it is in a good modern steam-engine. In a tract called *The Miner's Friend,* Savery alludes thus to the alternate heating and cooling of the water-vessel: "On the outside of the vessel you may see how the water goes out as well as if the vessel were transparent, for so far as the steam continues within the vessel so far is the vessel dry without, and so very hot as scarce to endure the least touch of the hand. But as far as the water is, the said vessel will be cold and wet where any water has fallen on it; which cold and moisture vanishes as fast as the steam in its descent takes place of the water." Before Savery's engine was entirely displaced by its successor, Newcomen's, it was improved by Desaguliers, who applied to it the safety-valve (invented by Papin), and substituted condensation by a jet of cold water within the vessel for the surface condensation used by Savery.

To Savery is ascribed the first use of the familiar term "horse-power" as a measure of the performance of an engine.

7. Gunpowder Engines. Some twenty years before the date of Savery's patent, proposals had been made by several inventors to raise water by means of the explosive power of gunpowder.

One scheme was to explode the powder in a closed vessel furnished with valves which opened outwards and allowed a great part of the air and burnt gases to escape when the explosion took place. As the gas that remained became cool a partial vacuum was formed in the vessel, and this was used to draw up water from a lower level. It does not appear that these schemes were ever put in practice except experimentally. The most interesting of the gun-powder engines was that of Huygens (1680), who for the first time introduced the piston and cylinder as constituent parts of a heat-engine. In Huygens' engine the piston was set at the top of a vertical cylinder and a charge of powder was exploded below it. This expelled part of the gaseous contents through valves which opened outwards, and then the cooling of the remainder caused the piston to descend under atmospheric pressure. The piston in descending did work by raising a weight through the medium of a cord and pulley.

8. Papin. In 1690 Denis Papin, who ten years before had invented the safety-valve as an adjunct to his "digester," suggested that the condensation of steam should be employed to make a vacuum under a piston which had been previously raised by the expansion of the steam. Papin had been associated with Huygens in his experiments on the production of a vacuum under a piston by means of gunpowder, and had described Huygens' machine to the Royal Society. Noticing that after the explosion enough gas remained in the cylinder to fill about one-fifth of its volume, after cooling, he cast about for some means of obtaining a better vacuum. "By another way, therefore, I endeavoured to attain the same end, and since it is a property of water that a small quantity of it, converted into steam by heat, has an elastic force like that of air, but when cold supervenes, is again resolved into water so that no trace of the said elastic force remains, I saw that machines might be constructed wherein water, by means of no very intense heat and at small cost, might produce that perfect vacuum which had failed to be obtained by the use of gunpowder." He goes on to describe what was unquestionably the earliest cylinder and piston steam-engine, and his plan of using steam was that which afterwards took practical shape in the atmospheric engine of Newcomen. But his scheme was made unworkable by the fact that he proposed to use but one vessel as both boiler and cylinder. A small quantity of water was placed at the bottom of a cylinder and heat was applied. When the piston had risen the fire was removed, the steam was allowed to cool, and the piston did work in its down-stroke under the pressure of the atmosphere.

After hearing of Savery's engine in 1705 Papin turned his attention to improving it, and devised a modified form, shown in fig. 3, in which the displacement-chamber *A* was a cylinder, with a floating diaphragm or piston on the top of the water to keep the water and steam from direct contact with one another. The water was delivered into a closed air-vessel *B*, from which it issued in a continuous stream against the vanes of a water-wheel. After the steam had done its work in the displacement-chamber it was

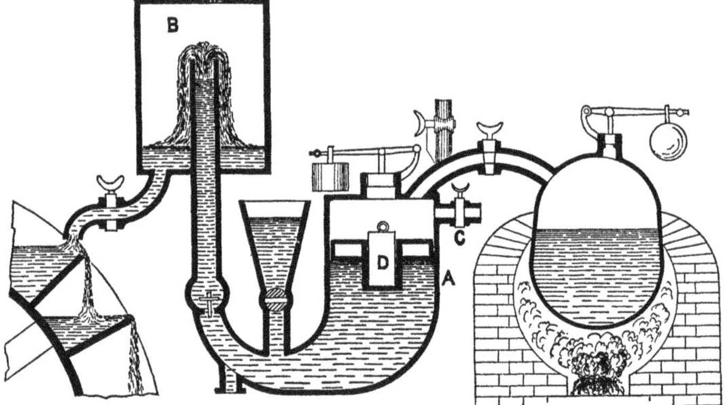

FIG. 3. Papin's modification of Savery's Engine, 1705.

allowed to escape by the stop-cock *C* instead of being condensed. This second engine of Papin's was in fact a non-condensing single-acting steam-pump, with steam-cylinder and pump-cylinder in one. A curious feature of it was the heater *D*, a mass of hot metal placed in the diaphragm for the purpose of keeping the steam dry. Among the many inventions of Papin was a boiler with an internal fire-box—the earliest example of a construction that is now almost universal[1].

9. Newcomen's "Atmospheric" Engine. While Papin was thus going back from his first notion of a piston engine to Savery's cruder type, a new inventor had appeared who made the piston

[1] For an account of Papin's inventions, see his *Life, and Correspondence with Leibnitz and Huygens*, by Dr E. Gerland, Berlin, 1881. See also Muirhead's *Life of Watt.*

engine a practical success by separating the boiler from the cylinder and by using (as Savery had done) artificial means to condense the steam. This was Newcomen, who in 1705, in conjunction with Savery and with Cawley, gave the steam-engine the form shown in fig. 4. The piston was connected by a chain with one end of an overhead beam. Steam admitted from the boiler to the cylinder allowed the piston to be raised by a heavy counterpoise hanging

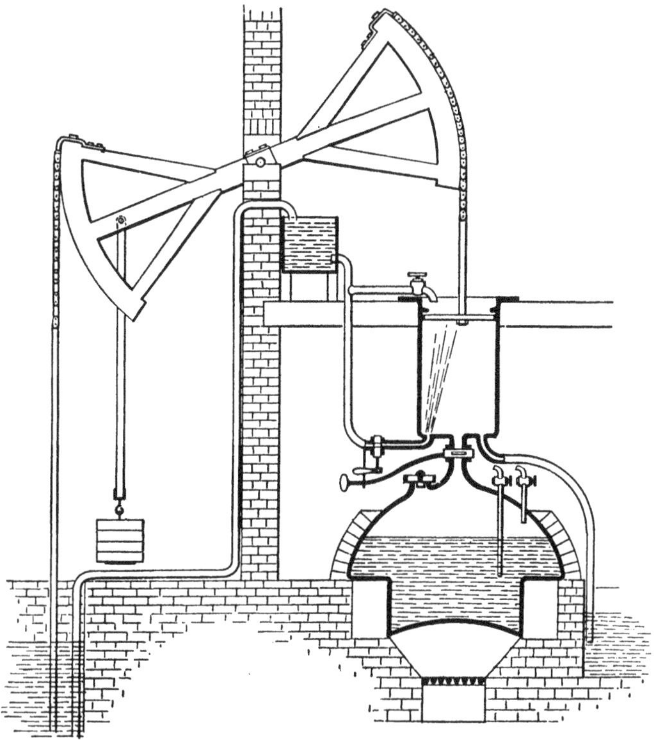

FIG. 4. Newcomen's Atmospheric Engine, 1705.

from the beam near the other end. Then the steam-valve was shut and a jet of cold water entered the cylinder and condensed the steam. The piston was consequently forced down by the pressure of the atmosphere and did work on the pump through the medium of a long rod which hung from the other end of the beam. The next entry of steam expelled the condensed water from the cylinder through an escape valve. The piston was kept tight by a layer of water on its upper surface. Condensation was at first effected by cooling the outside of the cylinder, but an accidental leakage of the packing water past the piston showed the advantage of condensing by a jet of injection water, and this plan took the place of surface condensation. The engine used steam which had a pressure little if at all greater than that of the atmosphere; sometimes indeed it was worked with the manhole-lid off the boiler. The function of the steam was merely to allow the piston to be raised, by making the pressure on the under side equal or nearly equal to the pressure on the top, and then to produce a vacuum by being condensed. Newcomen's engine was essentially the cylinder and piston of Papin combined with the separate boiler of Savery.

About 1711 Newcomen's engine began to be introduced for pumping mines. It is doubtful whether the engine was originally automatic in its action or depended on the periodical turning of taps by an attendant. An old print of an engine erected by Newcomen in 1712 near Dudley Castle shows a species of automatic gear. The common story is that in 1713 a boy named Humphrey Potter, whose duty it was to open and shut the valves of an engine he attended, made the engine self-acting by causing the beam itself to open and close the valves by means of cords and catches. This rude device was simplified in 1718 by Henry Beighton, who suspended from the beam a rod called the plug-tree, which worked the valves by means of tappets. By 1725 the engine was in common use in collieries, and it held its place without material change for about three-quarters of a century in all. Near the close of its career the atmospheric engine was much improved in its mechanical details by Smeaton, who built many large engines of this type about the year 1770, just after the great step which was to make Newcomen's engine obsolete had been taken by James Watt.

Like Savery's engine, Newcomen's was put to no other use than to pump water—in some instances for the purpose of turning water-wheels to drive other machinery. Compared with Savery's it had the great advantage that the intensity of pressure in the pump was not in any way limited by the pressure of the steam, but could be made as great as might be desired by reducing the area of the pump plunger. It shared with Savery's, in a scarcely less degree, the defect already pointed out, that steam was wasted by the alternate heating and cooling of the vessel into which it was led. Even contemporary writers complain of its "vast consumption of fuel," which appears to have been scarcely smaller than that of the engine of Savery.

10. James Watt. In 1763 James Watt, an instrument maker in Glasgow, while engaged by the University in repairing a model of Newcomen's engine, was struck with the waste of steam to which the alternate chilling and heating of the cylinder gave rise. He saw that the remedy, in his own words, would lie in keeping the cylinder as hot as the steam that entered it. With this view he added to the engine a new organ—namely, the *condenser*—a vessel separate from the cylinder, into which the steam should be allowed to escape from the cylinder, to be condensed there by the application of cold water either outside or as a jet. To preserve the vacuum in his condenser he added a pump, called the air-pump, whose function was to pump from it the condensed steam and water of condensation, as well as the air which would otherwise accumulate by leakage inwards or by being brought in with the steam or with the injection water. Then as the cylinder was itself no longer used as the chamber in which the steam was condensed he was able to keep it continuously hot by clothing it with non-conducting bodies, and in particular by the use of a *steam-jacket*, or layer of hot steam between the cylinder and an external casing. Further, and still with the same object, he covered in the top of the cylinder, taking the piston-rod out through a steam-tight stuffing-box, and allowed steam instead of air to press upon the piston's upper surface. The idea of using a separate condenser had no sooner occurred to Watt than he put it to the test by constructing the apparatus shown in fig. 5. There *A* is the cylinder, *B* a condenser (of the type now distinguished as a surface-condenser) and *C* is the air-pump. The cylinder was filled with steam above the piston, and a vacuum was formed in the surface-condenser *B*. On opening the stop-cock *D* the steam rushed over from the cylinder and was condensed, while the piston rose and lifted a weight. A fuller account of this experiment will be found in Watt's narrative, below.

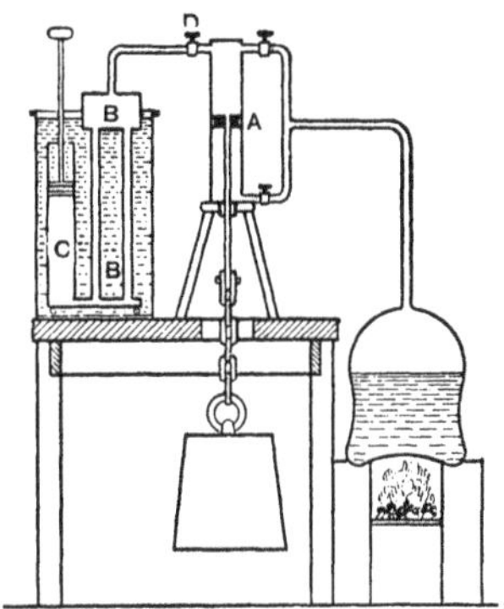

FIG. 5. Watt's Experimental Apparatus.

After several trials Watt patented his improvements in 1769; they are described in his specification in the following words, which, apart from their immense historical interest, deserve careful study as a statement of principles which to this day guide the scientific development of the steam-engine:—

My method of lessening the consumption of steam, and consequently fuel, in fire-engines, consists of the following principles:—

First, That vessel in which the powers of steam are to be employed to work the engine, which is called the cylinder in common fire-engines, and which I call the steam-vessel, must, during the whole time the engine is at work, be kept as hot as the steam that enters it; first by enclosing it in a case of wood, or any other materials that transmit heat slowly; secondly, by surrounding it with steam, or other heated bodies; and, thirdly, by suffering neither water nor any other substance colder than the steam to enter or touch it during that time.

Secondly, In engines that are to be worked wholly or partially by condensation of steam, the steam is to be condensed in vessels distinct from the steam-vessels or cylinders, although occasionally communicating with them; these vessels I call condensers; and, whilst the engines are working, these condensers ought at least to be kept as cold as the air in the neighbourhood of the engines, by application of water, or other cold bodies.

Thirdly, Whatever air or other elastic vapour is not condensed by the cold of the condenser, and may impede the working of the engine, is to be drawn out of the steam-vessels or condensers by means of pumps, wrought by the engines themselves, or otherwise.

Fourthly, I intend in many cases to employ the expansive force of steam to press on the pistons, or whatever may be used instead of them, in the same manner as the pressure of the atmosphere is now employed in common fire-engines. In cases where cold water cannot be had in plenty, the engines may be wrought by this force of steam only, by discharging the steam into the open air after it has done its office.

Sixthly, I intend in some cases to apply a degree of cold not capable of reducing the steam to water, but of contracting it considerably, so that the engines shall be worked by the alternate expansion and contraction of the steam.

Lastly, Instead of using water to render the piston or other parts of the engines air and steam-tight, I employ oils, wax, resinous bodies, fat of animals, quicksilver, and other metals, in their fluid state.

The fifth claim was for a rotary engine, and need not be quoted here.

The "common fire-engine" alluded to was the steam-engine, or, as it was more generally called, the "atmospheric" engine, of Newcomen. Enormously important as Watt's first patent was, it resulted for a time in the production of nothing more than a greatly improved engine of the Newcomen type, much less wasteful of fuel, able to make faster strokes, but still only suitable for pumping, still single-acting, with steam admitted during the whole stroke, the piston still pulling the beam by a chain working on a circular arc. The condenser was generally kept cool by the injection of cold water, but Watt has left a model of a surface-condenser made up of small tubes, in every essential respect like the condensers now used in marine engines. He also used, as we have seen, a surface-condenser in the experimental apparatus by which the practicability of condensation in a separate vessel was first demonstrated.

11. Watt's pumping-engine of 1769. Fig. 6 is an example of the Watt pumping-engine of this period. It should be noticed that, although the top of the cylinder is closed and steam has access to the upper side of the piston, this is done only to keep the cylinder and piston warm. The engine is still single-acting; the steam on the upper side merely plays the part which was played in Newcomen's engine by the atmosphere; and it is the lower end of the cylinder alone that is ever put in communication with the condenser. There are three valves—the "steam" valve *a*, the "equilibrium" valve *b*, and the "exhaust" valve *c*. At the beginning of the down-stroke *c* is opened to produce a vacuum below the piston and *a* is opened to admit steam above it. At the end of the down-stroke *a* and *c* are shut and *b* is opened. This puts the two sides of the piston in equilibrium, and allows the piston to be pulled up by the pump-rod *P*, which is heavy enough to serve as a counterpoise. *C* is the condenser, and *A* the air-pump, which discharges into the hot-well *H*, whence the supply of the feed-pump *F* is drawn.

12. Watt's narrative of his invention. In a note appended to the article "Steam-Engine" in Robison's *System of Mechanical Philosophy*[1] (1822) Watt has given the following account of the experiments and reflections which led up to his first patent. This narrative is of so particular interest that no apology need be made for reproducing it in full.

My attention was first directed in the year 1759 to the subject of steam-engines, by the late Dr Robison himself, then a student in the University of Glasgow, and nearly of my own age. He at that time

[1] John Robison, Professor of Natural Philosophy, University of Edinburgh, 1774–1805.

threw out an idea of applying the power of the steam-engine to the moving of wheel-carriages, and to other purposes, but the scheme was not matured, and was soon abandoned on his going abroad.

About the year 1761 or 1762, I tried some experiments on the force of steam, in a Papin's digester, and formed a species of steam-engine by

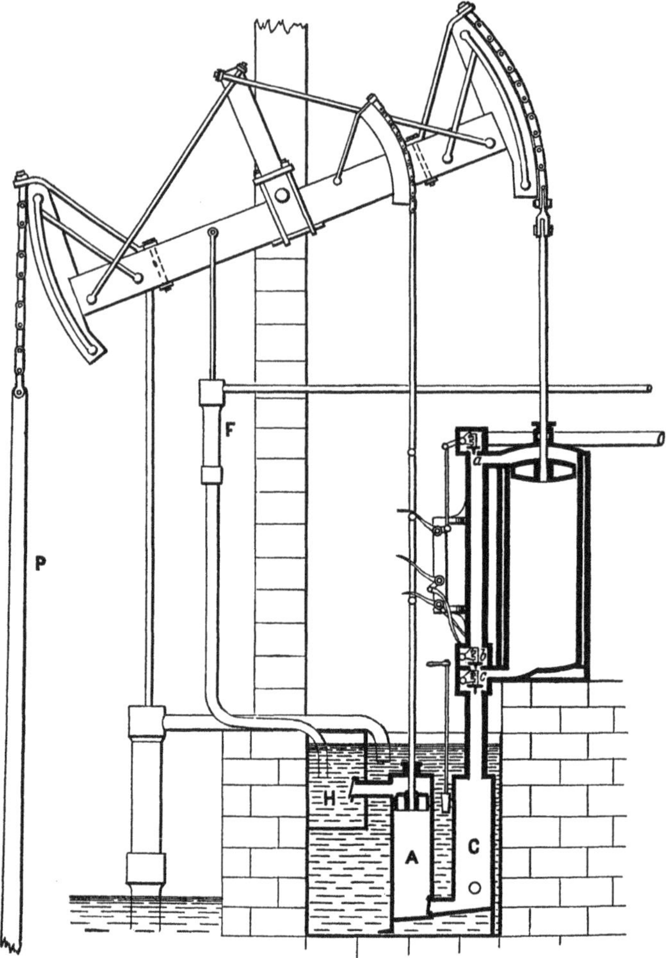

FIG. 6. Watt's Single-acting Engine, 1769.

fixing upon it a syringe one-third of an inch diameter, with a solid piston, and furnished also with a cock to admit the steam from the digester, or shut it off at pleasure, as well as to open a communication from the inside of the syringe to the open air, by which the steam contained in the syringe might escape. When the communication between the digester and syringe was opened, the steam entered the syringe, and by its action upon the piston raised a considerable weight (15 lb.) with which it was loaded. When this was raised as high as was thought proper, the communication with the digester was shut, and that with the atmosphere opened; the steam then made its escape, and the weight descended. The operations were repeated, and though in this experiment the cock was turned by hand, it was easy to see how it could be done by the machine itself, and to make it work with perfect regularity. But I soon relinquished the idea of constructing an engine upon this principle, from being sensible it would be liable to some of the objections against Savery's engine, viz. the danger of bursting the boiler, and the difficulty of making the joints tight, and also that a great part of the power of the steam would be lost, because no vacuum was formed to assist the descent of the piston. [I, however, described this engine in the fourth article of the specification of my patent of 1769; and again in the specification of another patent in the year 1784, together with a mode of applying it to the moving of wheel-carriages.]

The attention necessary to the avocations of business prevented me from then prosecuting the subject farther; but in the winter of 1763–4, having occasion to repair a model of Newcomen's engine belonging to the natural philosophy class of the University of Glasgow, my mind was again directed to it. At that period, my knowledge was derived principally from Desaguliers, and partly from Belidor. I set about repairing it as a mere mechanician, and when that was done and it was set to work, I was surprised to find that its boiler could not supply it with steam, though apparently quite large enough (the cylinder of the model being two inches in diameter and six inches stroke, and the boiler about nine inches diameter). By blowing the fire it was made to take a few strokes, but required an enormous quantity of injection water, though it was very lightly loaded by the column of water in the pump. It soon occurred that this was caused by the little cylinder exposing a greater surface to condense the steam than the cylinders of larger engines did in proportion to their respective contents. It was found that by shortening the column of water in the pump, the boiler could supply the cylinder with steam, and that the engine would work regularly with a moderate quantity of injection. It now appeared that the cylinder of the model being of brass, would conduct heat much better than the cast-iron cylinders of larger engines (generally covered on the inside with a stony crust), and that considerable advantage could be gained by making the cylinders of some substance that would receive and give out heat slowly: of these, wood seemed to be the most likely, provided it should prove sufficiently durable.

A small engine was therefore constructed with a cylinder six inches diameter, and twelve inches stroke, made of wood, soaked in linseed oil, and baked to dryness. With this engine many experiments were made; but it was soon found that the wooden cylinder was not likely to prove durable, and that the steam condensed in filling it still exceeded the proportion of that required for large engines according to the statements of Desaguliers. It was also found, that all attempts to produce a better exhaustion by throwing in more injection, caused a disproportionate waste of steam. On reflection, the cause of this seemed to be the boiling of water in vacuo at low heats, a discovery lately made by Dr Cullen, and some other philosophers (below 100°, as I was then informed), and, consequently, at greater heats, the water in the cylinder would produce a steam which would, in part, resist the pressure of the atmosphere.

By experiments which I then tried upon the heats at which water boils under several pressures greater than that of the atmosphere, it appeared, that when the heats proceeded in an arithmetical, the elasticities proceeded in some geometrical ratio; and by laying down a curve from my data, I ascertained the particular one near enough for my purpose. It also appeared, that any approach to a vacuum could only be obtained by throwing in large quantities of injection, which would cool the cylinder so much as to require quantities of steam to heat it again, out of proportion to the power gained by the more perfect vacuum; and that the old engineers had acted wisely in contenting themselves with loading the engine with only six or seven pounds on each square inch of the area of the piston.

It being evident that there was a great error in Dr Desaguliers' calculations of Mr Beighton's experiments on the bulk of steam, a Florence flask, capable of containing about a pound of water, had about one ounce of distilled water put into it; a glass tube was fitted into its mouth, and the joining made tight by lapping that part of the tube with packthread covered with glazier's putty. When the flask was set upright, the tube reached down near to the surface of the water, and in that position the whole was placed in a tin reflecting oven before a fire, until the water was wholly evaporated, which happened in about an hour, and might have been done sooner had I not wished the heat not much to exceed that of boiling water. As the air in the flask was heavier than the steam, the latter ascended to the top, and expelled the air through the tube. When the water was all evaporated, the oven and flask were removed from the fire, and a blast of cold air was directed against one side of the flask, to collect the condensed steam in one place. When all was cold, the tube was removed, the flask and its contents were weighed with care; and the flask being made hot, it was dried by blowing into it by bellows, and when weighed again, was found to have lost rather more than four grains, estimated at 4⅓ grains. When the flask was filled with water, it was found to contain about 17⅛ ounces avoirdupois of that fluid, which gave about 1800 for the expansion of water converted into steam of the heat of boiling water.

This experiment was repeated with nearly the same result; and in order to ascertain whether the flask had been wholly filled with steam, a similar quantity of water was for the third time evaporated; and, while the flask was still cold, it was placed inverted, with its mouth (contracted by the tube) immersed in a vessel of water, which it sucked in as it cooled, until in the temperature of the atmosphere it was filled to within half-an-ounce measure of water. [In the contrivance of this experiment I was assisted by Dr Black. In Dr Robison's edition of

Dr Black's lectures, vol. I, p. 147, the latter hints at some experiments upon this subject as made by him; but I have no knowledge of any except those which I made myself.]

In repetitions of this experiment at a later date, I simplified the apparatus by omitting the tube, and laying the flask upon its side in the oven, partly closing its mouth by a cork having a notch on one side, and otherwise proceeding as has been mentioned. I do not consider these experiments as extremely accurate, the only scale-beam of a proper size which I had then at my command not being very sensible, and the bulk of the steam being liable to be influenced by the heat to which it is exposed, which, in the way described, is not easily regulated or ascertained; but, from my experience in actual practice, I esteem the expansion to be rather more than I have computed.

A boiler was constructed which showed, by inspection, the quantity of water evaporated in any given time, and thereby ascertained the quantity of steam used in every stroke by the engine, which I found to be several times the full of the cylinder. Astonished at the quantity of water required for the injection, and the great heat it had acquired from the small quantity of water in the form of steam which had been used in filling the cylinder, and thinking I had made some mistake, the following experiment was tried:—A glass tube was bent at right angles, one end was inserted horizontally into the spout of a tea-kettle, and the other part was immersed perpendicularly in well-water contained in a cylindric glass vessel, and steam was made to pass through it until it ceased to be condensed, and the water in the glass vessel was become nearly boiling hot. The water in the glass vessel was then found to have gained an addition of about one-sixth part from the condensed steam. Consequently, water converted into steam can heat about six times its own weight of well-water to 212°, or till it can condense no more steam. Being struck with this remarkable fact, and not understanding the reason of it, I mentioned it to my friend Dr Black, who then explained to me his doctrine of latent heat, which he had taught for some time before this period (summer 1764), but having myself been occupied with the pursuits of business, if I had heard of it I had not attended to it, when I thus stumbled upon one of the material facts by which that beautiful theory is supported.

On reflecting further, I perceived, that in order to make the best use of steam, it was necessary, first, that the cylinder should be maintained always as hot as the steam which entered it; and secondly, that when the steam was condensed, the water of which it was composed, and the injection itself, should be cooled down to 100°, or lower, where that was possible. The means of accomplishing these points did not immediately present themselves; but early in 1765 it occurred to me, that if a communication were opened between a cylinder containing steam, and another vessel which was exhausted of air and other fluids, the steam, as an elastic fluid, would immediately rush into the empty vessel, and continue so to do until it had established an equilibrium; and if that vessel were kept very cool by an injection, or otherwise, more steam would continue to enter until the whole was condensed. But both the vessels being exhausted, or nearly so, how was the injection water, the air which would enter with it, and the condensed steam, to be got out? This I proposed, in my own mind, to perform in two ways. One was by adapting to the second vessel a pipe reaching downwards more than 34 feet, by which the water would descend (a column of that length overbalancing the atmosphere), and by extracting the air by means of a pump.

The second method was by employing a pump, or pumps, to extract both the air and the water, which would be applicable in all places, and essential in those cases where there was no well or pit.

This latter method was the one I then preferred, and is the only one I afterwards continued to use.

In Newcomen's engine, the piston is kept tight by water, which could not be applicable in this new method; as, if any of it entered into a partially exhausted and hot cylinder, it would boil and prevent the production of a vacuum, and would also cool the cylinder, by its evaporation during the descent of the piston. I proposed to remedy this defect by employing wax, tallow, or other grease, to lubricate and keep the piston tight. It next occurred to me, that the mouth of the cylinder being open, the air which entered to act on the piston would cool the cylinder, and condense some steam on again filling it, I therefore proposed to put an air-tight cover upon the cylinder, with a hole and stuffing-box for the piston-rod to slide through and to admit steam above the piston to act upon it instead of the atmosphere. [N.B. The piston-rod sliding through a stuffing-box was new in steam-engines; it was not necessary in Newcomen's engine, as the mouth of the cylinder was open, and the piston stem was square and very clumsy. The fitting the piston-rod to the piston by a cone was an after improvement of mine (about 1774).] There still remained another source of the destruction of steam, the cooling of the cylinder by the external air, which would produce an internal condensation whenever steam entered it, and which would be repeated every stroke; this I proposed to remedy by an external cylinder containing steam, surrounded by another of wood, or of some other substance which would conduct heat slowly.

When once the idea of the separate condensation was started, all these improvements followed as corollaries in quick succession, so that in the course of one or two days, the invention was thus far complete in my mind, and I immediately set about an experiment to verify it practically. I took a large brass syringe, $1\frac{3}{4}$ inches diameter, and 10 inches long, made a cover and bottom to it of tin-plate, with a pipe to convey steam to both ends of the cylinder from the boiler; another pipe to convey steam from the upper end to the condenser (for, to save apparatus, I inverted the cylinder). I drilled a hole longitudinally through the axis of the stem of the piston, and fixed a valve at its lower end, to permit the water which was produced by the condensed steam on first filling the cylinder, to issue. The condenser used upon this occasion consisted of two pipes of thin tin-plate, ten or twelve inches long, and about one-sixth inch diameter, standing perpendicular, and communicating at top with a short horizontal pipe of large diameter, having an aperture on its upper side which was shut by a valve opening upwards. These pipes were joined at bottom to another perpendicular pipe of about an inch diameter, which served for the air and water-pump; and both the condensing pipes and the air-pump were placed in a small cistern filled with cold water. [N.B. This construction of the condenser was employed from knowing that heat penetrated thin plates of metal very quickly, and considering that if no injection was thrown into an exhausted vessel, there would be only the water of which the steam had been composed, and the air which entered with the steam, or through the leaks, to extract.]

The steam-pipe was adjusted to a small boiler. When steam was produced, it was admitted into the cylinder, and soon issued through the perforation of the rod, and at the valve of the condenser. When it was judged that the air was expelled, the steam-cock was shut, and the air-pump piston-rod was drawn up, which leaving the small pipes of the condenser in a state of vacuum, the steam entered them and was condensed. The piston of the cylinder immediately rose and lifted a weight of about 18 lb., which was hung to the lower end of the piston-rod. The exhaustion cock was shut, the steam was readmitted into the cylinder, and the operation was repeated, the quantity of steam consumed, and the weights it could raise were observed, and, excepting the non-application of the steam-case and external covering, the invention was complete, in so far as regarded the savings of steam and fuel.

A large model, with an outer cylinder and wooden case, was immediately constructed, and the experiments made with it served to verify the expectations I had formed, and to place the advantage of the invention beyond the reach of doubt. It was found convenient afterwards to change the pipe-condenser for an empty vessel, generally of a cylindrical form, into which an injection played, and in consequence of there being more water and air to extract, to enlarge the air-pump.

The change was made, because, in order to procure a surface sufficiently extensive to condense the steam of a large engine, the pipe-condenser would require to be very voluminous, and because the bad water with which engines are frequently supplied would crust over the thin plates, and prevent their conveying the heat sufficiently quick. The cylinders were also placed with their mouths upwards, and furnished with a working-beam, and other apparatus, as was usual in the ancient engines; the inversion of the cylinder, or rather of the piston-rod, in the model, being only an expedient to try more easily the new invention, and being subject to many objections in large engines.

13. Development of Watt's Engine: the rotative type. In a second patent (1781) Watt describes the "sun-and-planet" wheels and other methods of making the engine give continuous revolving motion to a shaft provided with a fly-wheel. He had intended to use the crank and connecting-rod, for this purpose (a mechanical device familiar even at that time from its use in the common foot-lathe), and had even made a model of it, but the application of the crank to the steam-engine had meanwhile been patented by one Pickard, and Watt, rather than make terms with Pickard, made use of his sun-and-planet motion until the patent for the application of the crank expired. The reciprocating motion of earlier forms had served only for pumping, but by this invention Watt opened up for the steam-engine many other channels of usefulness. The engine was still single-acting; the connecting-rod was attached to the far end of the beam, and a counterpoise was provided which served to raise the piston when steam was admitted both below and above it.

14. Further improvements by Watt. In 1782 Watt patented two further improvements of the first importance, both of which he had invented some years before. One was the use of double action, that is to say, the application of steam and vacuum to each side of the piston alternately. The other, which had been invented as early as 1769, was the use of steam expansively, in other words, the plan (now used in all engines that aim at economy of fuel) of stopping the admission of steam when the piston had made only a part of its stroke, and allowing the rest of the stroke to be performed by the expansion of the steam already in the cylinder. To let the piston push as well as pull the end of the beam Watt devised his so-called parallel motion, an arrangement of links connecting the piston-rod head with the beam in such a way as to guide the rod to move in a very nearly straight line. He further added the throttle-valve, for regulating the rate of admission of steam, and the centrifugal governor, a double conical pendulum, which controlled the speed by acting on the throttle-valve. The stage of development reached at this time is illustrated by the engine of fig. 7 (from Stuart's *History of the Steam-Engine*), which shows the parallel motion *pp*, the governor *g*, the throttle-valve *t*, and a pair of steam and exhaust-valves at each end of the cylinder.

Among other inventions of Watt were the "indicator," by which diagrams showing the relation of the steam pressure in the cylinder to the movement of the piston are automatically drawn; a steam tilt-hammer; and also a steam locomotive for ordinary roads—but this invention was not prosecuted. As an inventor Watt was skilfully seconded by his assistant Murdoch, to whose ingenuity, he says, are due "many improvements"—amongst them, the introduction of the slide-valve as a means of controlling the admission and release of steam.

In partnership with Matthew Boulton, Watt carried on in Birmingham the manufacture and sale of his engines with the

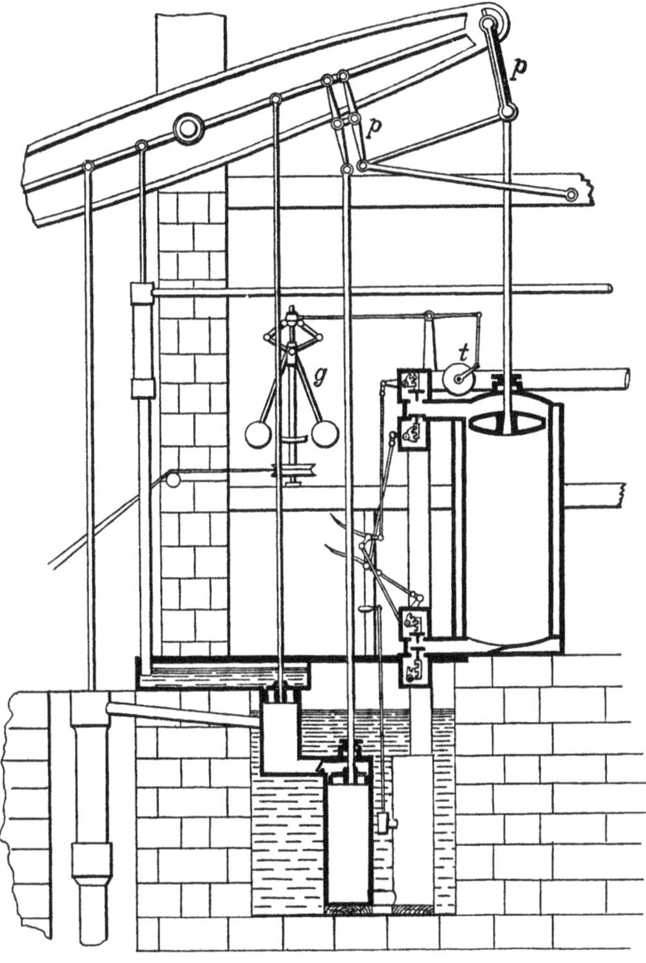

FIG. 7. Watt's Double-acting Engine, 1782.

utmost success, and held the field against all rivals in spite of severe assaults on the validity of his patents. A special Act of Parliament was obtained which extended the patent monopoly for a term of twenty-five years from 1775. Notwithstanding Watt's knowledge of the advantage to be gained by using steam expansively he continued to employ only low pressures—seldom more than 7 lb. per sq. inch over that of the atmosphere. His boilers were fed, as Newcomen's had been, through an open pipe which rose high enough to let the column of water in it balance the pressure of the steam. Following Savery, he adopted the term "horse-power" as a mode of rating engines and gave it a particular meaning, by defining one horse-power as the rate at which work is done when 33,000 lb. are raised one foot in one minute. This estimate was based on trials of the work done by horses; it is excessive as a statement of what an average horse can do in working continuously for any long time, but Watt purposely made it excessive in order that his customers might have no reason to complain on this score.

15. Non-condensing Steam-Engines. In the fourth claim in Watt's first patent, the second sentence describes a non-condensing engine, which would have required steam of a considerably higher pressure than served in the condensing engine. His narrative also shows that he had made experiments in this direction before devising the separate condenser. This, however, was a line of invention which Watt did not follow up, perhaps because so early as 1725 a non-condensing engine had been described by Leupold in his *Theatrum Machinarum*. Leupold's proposed engine (for the main features of which he professes himself indebted to Papin) is shown in fig. 8, which makes its action sufficiently clear. Watt's aversion to high-pressure steam was strong, and its influence on steam-engine practice long survived the expiry of his patents. So much indeed was this the case that the terms "high-pressure" and "non-condensing" were for many years synonymous, in contradistinction to the "low-pressure" or condensing engines of Watt. This nomenclature no longer holds good; in modern practice condensing engines use pressures quite as high as non-condensing engines, and by doing so are able to take advantage of Watt's great invention of expansive working to a degree which was impossible in his own practice.

16. Use of comparatively high-pressure steam. The introduction of the non-condensing and, at that time, relatively high-pressure engine was effected in England by Trevithick and in America by Oliver Evans about 1800. Both Evans and Trevithick applied their engines to propel carriages on roads, and both used for boiler a cylindrical vessel with a cylindrical flue inside—the construction now known as the Cornish boiler. In partnership with Bull, who had been a workman in the employment of Boulton and Watt, Trevithick had previously made direct-acting pumping-engines, with an inverted cylinder set over and in line with the pump-rod, thus dispensing with the beam that had been a feature in all earlier forms. But in these "Bull" engines, as they are

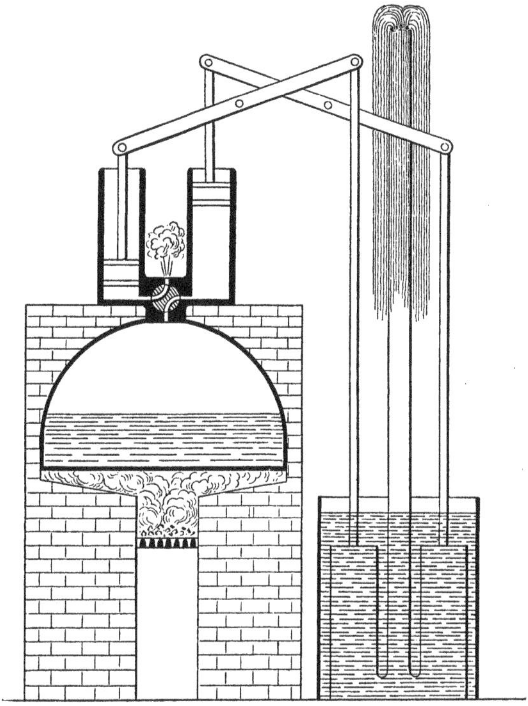

FIG. 8. Non-condensing Engine described by Leupold (1725).

called, a condenser was used, or, rather, the steam was condensed by a jet of cold water in the exhaust-pipe, and Boulton and Watt successfully opposed them as infringing the patent for condensation in a separate vessel. To Trevithick belongs the distinguished honour of being the first to use a steam-carriage on a railway; in 1804 he built a locomotive in the modern sense, to run on what had formerly been a horse-tramway in Wales; and it is noteworthy that the exhaust steam was discharged into the funnel to force the furnace draught, a device which, 25 years later, in the hands of George Stephenson, went far to make the locomotive what it is to-day. In this connexion it may be added that as early as 1769 a steam-carriage for roads had been built in France by Cugnot, who used a pair of single-acting high-pressure cylinders to turn a driving axle step by step by means of pawls and ratchet-wheels. To the initiative of Evans may be ascribed the early general use of high-pressure steam in the United States, a feature which for many years distinguished American from English practice.

17. Compound Engines. Hornblower and Woolf. Amongst the contemporaries of Watt one name deserves special mention. In 1781 Jonathan Hornblower constructed and patented what would now be called a compound engine, with two cylinders of different sizes. Steam was first admitted into the smaller cylinder, and then passed over into the larger, doing work against a piston in each. In Hornblower's engine the two cylinders were placed side by side, and both pistons acted on the same end of a beam overhead. This was an instance of the use of steam expansively, and as such was earlier than the patent, though not earlier than the invention, of expansive working by Watt. Hornblower was crushed by the Birmingham firm for infringing their patent in the use of a separate condenser and air-pump.

Soon after the expiry of Watt's master patent in 1800 the compound engine was revived by Woolf, with whose name it is often associated. Using steam of fairly high pressure, and cutting off the supply before the end of the stroke in the small cylinder, Woolf expanded the steam to six or even nine times its original volume. Mechanically the double-cylinder compound engine has this advantage over an engine in which the same amount of expansion is performed in a single cylinder, that the thrust or pull exerted by the two pistons in the compound engine varies less throughout the action than that which is exerted by the piston of the single-cylinder engine. This advantage may have been clear to Hornblower and Woolf, and to other early users of compound expansion. But another and a more important merit of the system lies in a fact of which neither they nor for many years their followers in the use of compound engines were aware—the fact that by dividing the whole range of expansion into two parts the cylinders in which these are separately performed are subject to a reduced range of fluctuation in their temperature. This, as we shall have occasion to point out more particularly in a later chapter, limits to a great extent a source of waste which is present in all steam-engines, namely, the waste which results from the heating and cooling of the metal by its alternate contact with hot and cooler steam. The system of compound expansion is now used in nearly all large engines that pretend to economy. Its introduction is the most conspicuous improvement which the steam-engine of the reciprocating type has undergone since the time of Watt; and we are now able to recognize it as a very important step in the direction set forth in his "first principle," that the cylinder should be kept as hot as the steam that enters it.

18. The Cornish Pumping-Engine. Woolf introduced the compound engine somewhat widely about 1814, as a pumping-engine in the mines of Cornwall. But it met a strong competitor there in the high-pressure single-cylinder condensing engine, which was at that time being developed, in the hands of Trevithick and others, into a machine of great efficiency, with an evident advantage over Woolf's in the simplicity of its construction. Woolf's engine fell into comparative disuse, and the single-cylinder type took a form which, under the name of the Cornish pumping-engine, was for many years famous for its great economy in fuel. In this engine the cylinder was set under one end of a beam, from the other end of which hung a heavy rod which operated a pump at the foot of the shaft. Steam was admitted above the piston for a short portion of the stroke, thereby raising the pump-rod, and was allowed to expand for the remainder. Then an equilibrium valve, connecting the spaces above and below the piston, as in fig. 6, was opened, and the pump-rod descended, doing work in the pump and raising the engine piston. The large mass which had to be started and stopped at each stroke served by its inertia to counterbalance the inequalities of steam pressure which were due to expansive working, for the pump-rods and other reciprocating parts stored up energy of motion in the early part of the stroke, when the steam pressure was greatest, and gave out energy in the latter part, when expansion had greatly lowered the pressure. The frequency of the stroke was controlled by a device called a cataract, consisting of a small plunger pump, in which the plunger, raised at each stroke by the engine, was allowed to descend more or less slowly by the escape of fluid below it through an adjustable orifice, and in its descent liberated catches which held the steam and exhaust-valves from opening. A similar device controlled the equilibrium valve. The cataract could be set to give a pause at the end of the piston's down-stroke, so that the pump-cylinder might have time to become completely filled.

The Cornish engine is interesting as the earliest form which achieved an efficiency at all comparable with that of good modern engines. For many years monthly reports were published of the "duty" of these engines, the "duty" being the number of foot-pounds of work done per bushel or (in some cases) per cwt. of coal. The performance of the engines became a matter of almost sporting interest to mining engineers, and no pains were spared to "beat the record." The average duty of engines in the Cornwall district rose from about 18 millions of foot-pounds per cwt. of coal in 1813 to 68 millions in 1844, after which less effort seems to have been made to maintain a high efficiency[1]. In individual cases much higher results were reported, as in the Fowey Consols engine, which in 1835 was stated to have a duty of 125 millions. This (to use a more modern mode of reckoning) is equivalent to the consumption of only a little more than $1\frac{3}{4}$ lb. of coal per hour per horse-power—a result surpassed by very few engines in good present-day practice. It is difficult to credit figures which, even in exceptional instances, place the Cornish engine of that period on a level with the most efficient modern engines—in which compound expansion and higher pressure combine to make a much more perfect thermo-dynamic machine; and apart from this there is room to question the accuracy of the Cornish reports. They played, however, a useful part in the process of steam-engine development by directing attention to the question of efficiency, and by demonstrating the advantage to be gained from high-pressure and expansive working, at a time when the theory of the steam-engine had not yet taken shape.

It may be added that the success of the Cornish type was no doubt largely responsible for a continued tendency on the part of many designers of engines to interpose a beam between the steam-cylinder and the pump or crank on which work was being done. For a long time the beam appears, in one form or another, as an almost inevitable part of a steam-engine. The lesson to be learnt

[1] *Min. Proc. Inst. C. E.* vol. XXIII, 1863.

from Bull's early direct-acting engine was apparently, in general, overlooked.

19. Revival of the Compound Engine. The final revival of the compound engine did not occur until about the middle of the nineteenth century, and then several agencies combined to bring it about. In 1845 M'Naught introduced a plan of improving beam engines of the original Watt type, by adding a small high-pressure cylinder with a piston acting on the beam between the centre and the fly-wheel end. Steam of higher pressure than had formerly been used, after doing work in the new cylinder, passed into the old or low-pressure cylinder, where it was further expanded. Many engines whose power was proving insufficient for the extended machinery they had to drive were "M'Naughted" in this way, and after conversion were found not only to exert more power but to show a marked economy of fuel. The compound form was selected by William Pole for the pumping-engines of Lambeth and other waterworks about 1850; in 1854 John Elder began to use it in marine engines; in 1857 E. A. Cowper added an intermediate receiver or reservoir for steam between the high and low-pressure cylinders, which made it unnecessary for the low-pressure piston to be just beginning when the other piston was just ending its stroke. To keep the steam hot in the receiver he fitted it with a steam-jacket. As the mechanical construction of engines and boilers improved and engineers found it safe as well as easy to deal with comparatively high-pressure steam, compound expansion came into more general use, its advantage becoming more conspicuous with every increase in boiler pressure. Before the end of the nineteenth century it was almost universally adopted for large stationary and marine engines. In marine practice, where economy of fuel under conditions of steady working is of special importance, the principle of compound expansion was greatly extended after 1875 by the general introduction of triple and quadruple expansion engines, in which the steam was made to expand successively in three or in four cylinders. Even in locomotives for railways, where other considerations are of greater moment, compound expansion has found employment, though its use there is somewhat rare, and does not (in 1926) tend to increase.

The growth of compound expansion has been referred to at some length, because it forms the most definite improvement which the piston and cylinder type of steam-engine has undergone since the time of Watt. For the rest, the progress of that type of engine has consisted in its adaptation to particular uses, in the invention of features of mechanical detail, in the recognition and application of thermodynamical principles, in better structural design and in improved methods of manufacture by which it has profited in common with all other machines. These have made possible the use of steam with very many times the pressure of that employed by Watt, and have allowed the mean speed of movement of the piston to be greatly increased, with consequent gains both in the amount of power obtainable from an engine of given size and in the efficiency of the action.

20. Application to Locomotives. The adaptation of the steam-engine to railways, begun by Trevithick, became a success in the hands of George Stephenson, whose engine the "Rocket," when tried along with others on a part of the Liverpool and Manchester railway at Rainhill in October 1829, not only out-distanced its competitors but settled once and for all the question whether horse traction or steam traction was to be used on railways. The principal features of the "Rocket" were an improved steam-blast for urging the combustion of coal and a boiler (suggested by Henry Booth, the secretary of the railway) in which a large heating surface was given by the use of many small tubes through which the hot gases passed. Further, the cylinders, instead of being vertical as in earlier locomotives, were set at a slope, which was afterwards altered to a position still more nearly horizontal. To these features there was added later the "link-motion," a contrivance which enabled the engine to be quickly reversed and the amount of expansion to be readily varied. In the hands of George Stephenson and his son Robert the locomotive took a form which in essentials has been maintained by the far heavier locomotives of modern practice.

21. Application to Steamboats. The first practical steamboat was the tug "Charlotte Dundas," built by William Symington, and tried in the Forth and Clyde Canal in 1802. A Watt double-acting condensing engine, placed horizontally, acted directly by a connecting-rod on the crank of a shaft at the stern, which carried a revolving paddle-wheel. The trial was successful, but steam towing was abandoned for fear of injuring the banks of the canal. Ten years later Henry Bell built the "Comet," with side paddle-wheels, which ran as a passenger steamer on the Clyde; but an earlier inventor to follow up Symington's success was the American Robert Fulton, who, after unsuccessful experiments on the Seine, fitted a steamer on the Hudson in 1807 with engines made to his designs by Boulton and Watt, and brought steam navigation for the first time to commercial success.

The American river boats soon began to use high-pressure steam, but English engineers looked askance on a practice which led to frequent explosions. They were moreover slow to realize that high pressure is a necessary condition of economical working. In 1835 it was usual for the pressure in marine boilers to be no more than 4 or 5 lb. per square inch above the pressure of the atmosphere, and for many years later pressures of 20 or 25 lb. were common. With the introduction of compound working and with the substitution of cylindrical boilers for the weak box-boiler originally used on board ship the pressure rose considerably. In 1872 Sir F. J. Bramwell, describing the typical marine practice of that time[1], gave a list of engines—all compound—in which the pressure ranged from 45 to 60 lb. The consumption of coal in these engines was generally from 2 to $2\frac{1}{2}$ lb. per hour per indicated horse-power, and the mean piston speed was about 350 feet per minute. Nine years later Mr F. C. Marshall gave a similar list[2], in which the mean pressure was 77 lb., the mean piston speed about 460 feet per minute, and the consumption of coal a trifle under 2 lb. per hour per indicated horse-power. These engines were also of the type in which steam is successively expanded in two cylinders. The triple expansion type of engine was introduced by A. C. Kirk in 1874 but did not come into general use until after 1881. It became the normal type of marine engine, and was associated with a marked advance in boiler pressure and a considerable gain in economy of fuel. Reviewing the progress of marine engineering in the decade from 1881 to 1891 Mr Blechynden[3] gave a list of triple engines with boiler pressures of about 160 lb. and piston speeds of about 500 or 600 feet per minute. These engines consumed on the average about $1\frac{1}{2}$ lb. of coal per indicated horse-power-hour[4].

[1] *Proc. Inst. Mech. Eng.* 1872. [2] *Proc. Inst. Mech. Eng.* 1881.

[3] *Proc. Inst. Mech. Eng.* 1891.

[4] On this subject see further a paper by Sir A. J. Durston on the progress of Marine Engineering, read at the International Congress of Naval Architects, 1897. *Engineering*, July 9–16, 1897.

In later practice pressures of 200 lb. and over became not unusual, with piston speeds sometimes as high as 900 or 1000 feet per minute. Many large engines employed quadruple expansion, that is to say, the steam expanded successively in four stages through four separate cylinders.

The progressive rise in steam pressure and in piston speed not only increased the efficiency of engines but greatly reduced their bulk for a given power. The rate at which work is done per square inch of piston area is equal to the mean piston speed multiplied by the mean effective pressure, and increased pressure of admission implies increased mean pressure throughout the stroke.

The piston type of marine engine became highly efficient not only for these reasons but because, with the increased size of ships, engines were built to produce large amounts of power. In big engines some of the incidental sources of loss become relatively less important. Piston engines of 10,000 horse-power were not uncommon, and in some steamships each of the twin sets developed as much as 20,000 horse-power. Triple and quadruple expansion engines of this type are still (1926) fitted in many ships, especially ships of moderate tonnage liable in service to steam at various speeds, or over routes where there may be limited facilities for repair. Early in the present century however the piston engine began to give place to the steam turbine in the largest marine installations. A notable advance was marked by the decision of the Cunard Company in 1904 to adopt the steam turbine in the "Lusitania" and "Mauritania," vessels of a then unprecedented size and speed. About 70,000 horse-power had to be generated in each ship: it was a bold experiment to design steam turbines of such power, but it was completely successful. So great a concentration of power would probably not have been practicable except for the possibilities which Parsons' invention of the turbine had by that time opened up. Even these figures are eclipsed in more recent naval practice. The turbine engines of the cruisers "Renown" and "Repulse" (1916) gave on trial 120,000 shaft horse-power, and those of the battle-cruiser "Hood" (1920) about 150,000[1].

22. The Steam Turbine. The introduction of the steam turbine as a practical engine dates from 1884 when Sir Charles Parsons took out his first patent for what became known as the

[1] *Trans. Inst. Nav. Arch.* vol. LXII, 1920, p. 8.

Parsons Compound Turbine. For some years it was made in small sizes only and the steam was discharged into the atmosphere without condensation. Under these conditions its efficiency was comparatively low. One of the most important advantages which the steam turbine possesses over the reciprocating engine is its ability to utilize the energy of low-pressure steam by expansion down to the best vacuum obtainable in a condenser, and it was only when this characteristic was turned to account that the steam turbine became a serious rival to the reciprocating engine in respect of economy of steam. In 1891 Parsons adapted his turbine for use with a condenser, and it then began to be employed on a fairly large scale as a power generator in electric supply stations. Its efficiency at that date was found, in tests by the present writer, to be comparable with that of a good reciprocating compound engine of the same capacity, but the figures then obtained were much improved on later in turbines of larger size and modified design. Before long it was recognized that the steam turbine, in addition to possessing conspicuous advantages in respect of constructive simplicity, compactness, and freedom from vibration, as well as requiring comparatively little skilled attention, was a highly economical generator of power on the largest scale, surpassing in this respect the best steam-engines of the older type. The invention of the steam turbine has consequently effected a revolution in steam-engine practice. The largest demands for power occur in central stations from which electric energy is distributed for traction or other uses, and in fast passenger vessels and war ships. In such cases the turbine has displaced the piston engine. It is to the genius of Parsons that we owe not only the leading idea of the modern steam turbine, as well as the working out of mechanical details which have been essential to its success, but also its adaptation both to dynamo driving and to the propulsion of ships.

The first application of the turbine to marine propulsion was made by Parsons in the "Turbinia" in 1897, a little experimental vessel of 100 tons which was fitted with turbines of 2100 horsepower driving three propeller shafts. The "Turbinia" attained what was then a record speed for a ship of any size. This was soon followed by further trials of turbine propulsion in the destroyers "Viper" and "Cobra," and in 1901 the first passenger vessel to be driven by steam turbines—the "King Edward"—was built on the Clyde. Her success led to the adoption of turbines first in various cross-channel packets and later in ocean-going passenger steamers of the largest size. In war ships the use of steam turbines has a special advantage in enabling the machinery to be placed at a low level beneath the protective deck, in addition to the general advantages of reduced bulk and weight, reduced vibration, reduced liability to break-down and reduced consumption of fuel and oil, which also apply to other vessels. The successful trials of the turbine driven cruiser "Amethyst" in 1904 demonstrated the superiority of steam turbines so conclusively that they were forthwith adopted in the design of all new ships for the British Navy. At first, in the marine use of the steam turbine, each propeller shaft carried its own turbine, but in 1910 Parsons showed by trial that it was practicable to introduce mechanical gearing between the turbine and propeller shafts, which made it possible to apply the turbine economically to slow as well as fast vessels, and in all cases to select the most suitable speeds for turbine and propeller separately. The introduction of gearing has also greatly increased the adaptability of the turbine for use on land, in the driving of other mechanism, such as that of rolling mills, paper mills, and textile or other factories.

In the steam turbine, of whatever form, just as in the water turbine, the force directly operative to do useful work is derived from the kinetic energy of the working fluid, either by the impulse of a jet or jets impinging on and sliding over movable blades, or by the reaction on orifices or guides from which the jets issue. The pressure of the steam, instead of being exerted on a piston, is employed in the first instance to set the fluid itself in motion. There is a conversion of pressure-energy into velocity-energy as a preliminary step towards obtaining the effective work of the machine.

When this is done in a single operation the velocity acquired by the steam is immensely greater than the velocities with which water turbines have to deal, in consequence of the much smaller density of steam as a working fluid. Early attempts to design a steam turbine fell short of practical success mainly because of the difficulty of arranging for a sufficiently high velocity in the moving parts to utilize a good proportion of the kinetic energy of the steam. There was the further difficulty of getting the energy of the steam into a suitable kinetic form without excessive waste. The problem of getting the steam to form a jet in which the particles should have a common direction, without undue dispersion, when the steam expands through an orifice from a region of high to a region of low pressure, was solved by Dr Gustaf de Laval, who in 1889 introduced a form of steam turbine in which this was accomplished and in which also the velocity of the moving blades was so high as to make it practicable to recover a fair proportion of the kinetic energy of the jet. The novel features of De Laval's turbine were the form of diverging nozzle which served to produce the jet and the mechanical devices by which an exceptionally high speed was attained in the wheel carrying the vanes or blades on which the jet impinged. His turbine, which will be described in a later chapter, consists essentially of a wheel carrying a single ring of blades on which the steam jet or jets act simply by impulse. It has met with considerable success, especially in comparatively small sizes, and its efficiency in them is fairly good, but it is not adapted for developing large amounts of power, and it has not been applied to the propulsion of ships.

Parsons, who successfully attacked the problem of designing a steam turbine at an earlier date than De Laval, proceeded in a different way. By dividing the whole range of expansion into many successive steps he prevented the steam from acquiring an inconveniently high velocity at any stage of the process. At each step in the Parsons steam turbine the steam suffers only a small drop in pressure, and acquires only a moderate velocity; its kinetic energy is therefore easily extracted by means of blades moving at a moderate speed before it passes on to the next step. Moreover in each step the drop in pressure is too small to give rise to any difficulty in the formation of the jets. To form these the steam passes through fixed guide-blades which are distributed round the whole circumference of the revolving wheel and all the revolving blades are consequently in action at once. The stream steams from end to end of the turbine through an annular space between a revolving drum and the casing which surrounds it. Parallel rings of fixed guide-blades project inwards from the casing so as almost to touch the drum, and between these rings alternate rings of moving blades, fixed to the drum, project outwards so as almost to touch the casing. At each step in the expansion the steam streams through a ring of fixed guide-blades, forming a continuous ring of jets which impinge on the adjacent moving blades and give up to them the greater part of the jets' kinetic energy. The same process is repeated many times from ring to ring as the steam progresses from the high-pressure to the low-pressure end, and the size of the annulus containing the blades increases to accommodate the increased volume of the expanded steam. The force on the moving blades is exerted partly by impulse and partly by reaction, for the steam not only impinges on them with the velocity it has acquired in the preceding guide-blades but also is accelerated relatively to the moving blades themselves in the act of passing through them. The construction, which is of great simplicity, will be described later. It lends itself remarkably well to the economical generation of power in large quantities.

The success of the Parsons turbine was followed by the development of others differing from it in certain features of design. In types of turbine which are associated with the names of Curtis and Rateau the steam acts on the moving blades wholly by impulse, but the principle of compound action is retained; that is to say, the total range of expansion is divided into a number of stages in each of which there is only a moderate drop of pressure. Some of these impulse turbines, which will be described in Chapter VIII, have come into extensive use in large sizes as alternatives to the reaction turbine, especially in power stations. So far as thermal efficiency is concerned no great difference is to be expected in the performance of the impulse and the reaction types, provided the turbine is so designed as to take full advantage of the expansion of the steam down to the best vacuum which the cooling water allows to be maintained.

One of the early uses of the steam turbine was to supplement the action of a reciprocating engine by taking steam which had already done duty under a piston, and extracting more work out of it by continued expansion through a range of pressures lower than those with which the piston and cylinder engine can properly deal. An important characteristic of steam turbines is their capacity to utilize the remaining energy of low-pressure steam, after the volume has become so great that further expansion in a cylinder would be impracticable. This led to the employment of what are called exhaust steam turbines as auxiliaries to other engines, with the result of adding materially to the output of power without increasing the expenditure of coal.

23. Development of the Theory of Heat-Engines. It is remarkable how little the infancy of the steam-engine has owed to scientific nursing. The early inventors had no theory of thermodynamics to guide them. Watt had the advantage, as he mentions in his narrative, of a knowledge of Black's doctrine of latent heat; but there was no philosophy of the relation of work to heat until long after the inventions of Watt were complete. The theory of the steam-engine as a heat-engine may be said to date from 1824, when Sadi Carnot published his *Réflexions sur la Puissance Motrice du Feu*. He there showed that heat does work only by being let down from a higher to a lower temperature, and he stated and proved the fundamental principle that the largest possible amount of work is done when this process is carried out in a strictly reversible manner. But Carnot had no idea then that any of the heat disappears in the process, and it was not until the doctrine of the conservation of energy was established in 1843 by the experiments of Joule, which determined the mechanical equivalent of heat, that the foundations of the theory were settled. Important data were furnished by Regnault's experiments on the properties of steam, the results of which were published in 1847. From 1849 onwards the science of thermodynamics was developed with extraordinary rapidity by Clausius, Rankine, and Thomson (Lord Kelvin), and was applied, especially by Rankine, to practical problems in the use of steam. The publication in 1859 of Rankine's *Manual of the Steam-Engine* formed an epoch in the philosophical treatment of the subject. While the thermodynamic theory was rigorous in itself, its direct application to steam-engine problems was limited to a greater degree than Rankine probably realized, owing to its being founded on certain simplifying assumptions which are by no means fulfilled in real engines. In the ideal engine of the theory it was assumed that the cylinder and piston might be treated as behaving to the steam like non-conducting bodies—that the transfer of heat between the steam and the metal was negligibly small. Rankine's calculations of steam-consumption, of work, and of thermodynamic efficiency involve this assumption, except in the case of steam-jacketed cylinders, where he estimates that the steam in its passage through the cylinder takes just enough heat from the jacket to prevent a small amount of condensation which would otherwise occur as the process of expansion goes on. If the transfer of heat from steam to metal could be overlooked, the steam which enters the cylinder of a reciprocating engine would remain during admission as dry as it was before it entered, and the volume of steam consumed per stroke would correspond with the volume of the cylinder up to the point of cut-off. It is here that the actual behaviour of steam in the cylinder diverges most widely from the behaviour assumed in the ideal engine. When steam enters the cylinder it finds the metal chilled by the previous exhaust, and a portion of it is at once condensed. This has the effect of increasing, often very largely, the volume of boiler steam required per stroke. As expansion goes on the water that was condensed during admission begins to be re-evaporated from the sides of the cylinder, and this action is generally continued during the escape of the steam. In later chapters the effect which this exchange of heat between the metal of the cylinder and the working fluid produces on the economy of the engine will be discussed, and an account will be given of experimental means by which we may examine the amount of steam that is initially condensed and trace its subsequent re-evaporation. The influence which the walls of the cylinder exert is in fact immense, by the alternate give and take of heat between them and the steam. The exchanges of heat are so complex that there seems little prospect of submitting them to any comprehensive theoretical treatment, and the theory with its simplifying assumptions has to be supplemented by scientific analysis of experiments made upon actual machines. Many such experiments have been made and their value has long been realized. Questions relating to the influence on thermal economy of speed, of pressure, of ratio of expansion, of jacketing, of compound expansion, or of superheating cannot be completely answered without appeal to experiment. The student must not, however, conclude that because the conditions under which an actual engine works are so complex as to make an exact theory of the action impracticable, no theory need be studied. The very complexity of conditions makes the study of theory more necessary, as a guide in judging what conditions are favourable to efficiency and what are unfavourable. Moreover the general theory of heat-engines assigns a limit of efficiency which engines of any type may approach but cannot surpass. To interpret rightly the results of experiments, or to proceed intelligently with the design of an engine, requires a knowledge of the principles of thermodynamics and of the physical properties of steam.

The calculated performance of an ideal engine not only affords a valuable criterion with which to compare the results obtained in trials of actual engines, but it serves to determine the proportions of new forms. Thanks largely to the labours of Callendar, engineers now have at their disposal very complete data regarding the properties of steam. Numerical calculations relating to the action of steam in engines, whether of the piston or the turbine type, are easy when certain simplifying conditions are assumed. In the chapters which follow calculations of this kind are explained. With steam turbines especially it is by the aid of such calculations that the process of design is in fact carried out. The turbine escapes the complication which arises in a reciprocating engine from alternate heating and cooling of the working parts. Hence in it the relation of theory to practice is more intimate, and the working can be predicted with greater certainty and precision. The progress which steam turbines have made in recent years, their rapid development in power and their adaptation to high pressures and temperatures, could not have been accomplished had not their evolution been guided by scientific theory at every step and in every detail.

REFERENCES

Dircks, *Life of the Marquis of Worcester*, 1865, containing a reprint of the *Century of Inventions* (1663). Desaguliers, *Course of Experimental Philosophy*, 1763. Robison, *System of Mechanical Philosophy*, vol. II, 1822. Stuart, *Descriptive History of the Steam-Engine*, 1825. Farey, *Treatise on the Steam-Engine*, 1827. Tredgold, *The Steam-Engine*, 1838. Muirhead, *Mechanical Inventions of James Watt*, and *Life of Watt*. Galloway, *The Steam-Engine and its Inventors*. Thurston, *History of the Growth of the Steam-Engine*. Cowper on the Steam-Engine (*Heat Lectures, Inst. C.E.*, 1884). Parsons, *The James Watt Lecture*, *Nature*, 25 Feb. 1909, also *The Rede Lecture*, Cambridge, 1911. Richardson, *The Evolution of the Parsons Steam Turbine*, 1911.

CHAPTER II

ELEMENTARY THEORY OF HEAT-ENGINES

24. Laws of Thermodynamics. The First Law. In the action of a heat-engine, heat is either taken in by the engine from a furnace or other external source or is generated by the combustion of fuel within the engine itself. A portion of the heat thus supplied is spent in doing mechanical work and so ceases to exist as heat, being converted into another form of energy; and the remainder is rejected by the engine, still in the form of heat. The relation which holds between the heat supplied, the heat converted into mechanical energy, and the heat rejected depends on two general principles which are described as the two Laws of Thermodynamics. The first law states the fact that the amount of heat which disappears in the process (as heat) is proportional to the amount of mechanical work done in the engine; in other words, it states the principle of the Conservation of Energy in relation to the doing of mechanical work by the agency of heat. This may be expressed in the following terms:—*When mechanical energy is produced from heat a definite quantity of heat goes out of existence for every unit of work done; and conversely, when heat is produced by the expenditure of mechanical energy the same definite quantity of heat comes into existence for every unit of work spent.*

To put this statement into a numerical form we must have a unit for the measurement of quantities of heat as well as a unit for the measurement of mechanical work. For engineering purposes the foot-pound is the common unit of work in British and American usage, and the metre-kilogramme (or kilogrammetre) in continental usage[1]. These convenient and familiar units are open to the objection that they have slightly different values in different places on account of differences in the intensity of gravity; but these differences are scarcely large enough to be important from a practical point of view. In cases where greater precision of statement is required a particular locality or rather a particular latitude

[1] Since 1 metre = 3·28085 ft. and 1 kilogramme = 2·20462 lb., one metre-kilogramme = 7·233 foot-pounds, when both are measured at the same place, so that gravity acts alike on the pound and the kilogramme.

has to be specified, or recourse may be had to absolute units, such as the foot-poundal, the erg, or the joule, which are independent of gravity[1].

Quantities of heat are expressed in terms of the *thermal unit*, which is the quantity of heat required to raise the temperature of unit quantity of water by one degree. The magnitude of this unit accordingly depends on what unit is used in reckoning the quantity of water, and on what scale the degree is taken by which the rise in temperature is measured. If we take one pound of water, and a degree of the Fahrenheit scale, we have what is commonly called the British Thermal Unit. It is however of great advantage to avoid using the Fahrenheit scale, and students will do well to make their calculations in terms of centigrade degrees. The Pound-Degree Centigrade (often called the pound-calory) is accordingly to be preferred: it is greater than the British Thermal Unit or Pound-Degree Fahrenheit in the proportion of 9 to 5. If we take one gramme of water, and one degree on the centigrade scale, we have a unit of heat called the gramme-calory which is in general use for scientific purposes. Engineers often use a unit one thousand times larger than this, namely the heat required to raise one kilogramme of water through one centigrade degree.

To make the definition of the thermal unit precise we have to specify at what place in the scale of temperature the change through one degree is supposed to occur, for the specific heat of water is not quite constant. As was first shown by Regnault, it takes rather more heat to raise the temperature of a pound of water one degree if the temperature is high than if it is low. Later investigations have shown that when water is warmed from the temperature of melting ice its specific heat at first decreases slightly as the temperature rises, reaches a minimum at about 35° C. and then increases continuously. In defining the thermal unit a standard temperature such as 15° C. is often taken, but a more convenient practice is to define the unit of heat as one-hundredth part of the

[1] The erg is the absolute unit of work on the centimetre-gramme-second system, namely one centimetre-dyne, the dyne being the C.G.S. unit of force, which is the force required to give one gramme a velocity of one centimetre per second in one second. The joule is ten million (10^7) ergs. At sea level in latitude 45°, where the acceleration due to gravity is 980·6 cm. per second per second, one kilogrammetre is equal to 9·806 joules. In the latitude of London it is 9·812 joules. For any latitude λ it may be calculated from the formula

$$9{\cdot}780\,(1 + 0{\cdot}0053 \sin^2 \lambda).$$

whole quantity of heat required to warm one pound of water from the melting-point to the boiling-point (0° C. to 100° C.) under a constant pressure of one atmosphere[1]. This unit, which may be called the *mean pound-calory* is used in the calculations and Tables that follow.

Our knowledge of the mechanical equivalent of heat is originally due to the experiments of Joule, which were begun in 1843 and continued for many years. Causing the potential energy of a raised weight to be spent in turning a paddle which generated heat by the agitation of the liquid in which it was immersed, and observing the increase in temperature which this brought about, Joule arrived at the figure 772 as the number of foot-pounds equivalent to one pound-degree Fahrenheit, and this was for long the commonly accepted value of the mechanical equivalent of heat. Later experiments by Joule himself gave a larger number; in 1878 an improved method of measurement, in which the mechanical stirring of water was still used, pointed to a value between 774 and 775. A comparison by Rowland[2] of the scale of the thermometer used by Joule with that of an air thermometer led to a further increase in this number, and very accurate experiments by Rowland himself, carried out by a generally similar method but dealing with larger quantities of work, gave the number 778, the standard temperature being about 60° F. and the interval of one degree being taken on the air thermometer. Subsequent determinations by Griffiths[3] and others confirmed the result that Joule's original value was too low and that a number not less than 778 should be accepted. In the experiments of Griffiths the water was heated by a measured expenditure of electrical energy, and this method was also adopted in determinations by Schuster and Gannon[4] and by Callendar and Barnes[5]. An important mechanical determination by Osborne Reynolds and W. M. Moorby[6], of the amount of work spent in raising the temperature of water from freezing- to boiling-point gives very approximately 778 foot-pounds as the mean value of the mechanical equivalent throughout that

[1] The standard atmosphere is 14·689 pounds per square inch, or 1·0327 kilogrammes per square centimetre, in the latitude of London.

[2] Rowland, *Proceedings of the American Academy*, 1879.

[3] Griffiths, *Phil. Trans.* 1893, vol. 184 A.

[4] Schuster and Gannon, *Phil. Trans.* 1895, vol. 186 A.

[5] Barnes, *Phil. Trans.* 1902, vol. 199 A.

[6] Reynolds and Moorby, *Phil. Trans.* 1898, vol. 190 A.

range: in other words, according to their measurements 180 times 778 is the number of foot-pounds of work required to raise the temperature of 1 lb. of water from 32° to 212° F. which approximately corresponds to 100×1400 foot-pounds for the same range expressed in centigrade degrees. This mode of defining and measuring the mechanical equivalent has the advantage of escaping all ambiguity arising from variations in the specific heat of water. It gives, in fact, a unit which has nearly the same value as the unit which assumes 15° C. as a standard temperature, for according to the experiments of Callendar and Barnes the mean specific heat from 0° to 100° C. is almost exactly equal to the specific heat at 15° C.

Taking the evidence as a whole we may accept 1400 as the number of foot-pounds (in the latitude of London) that are equivalent to the mean pound-calory as defined above[1]. This corresponds to using 777·8 foot-pounds as the mechanical equivalent of the mean British thermal unit, and 426·7 gramme-metres as the mechanical equivalent of the mean gramme-calory. These numbers will be used in any calculations that occur in this book. In absolute (C.G.S.) units the gramme-calory is equivalent to $4{\cdot}1868 \times 10^7$ ergs, or cm.-dynes.

Since a definite number of mechanical units of work is equivalent to one thermal unit we may, whenever it is convenient, express quantities of work in thermal units, or quantities of heat in foot-pounds or in kilogrammetres.

The mechanical equivalent of heat enters into many of the formulas of thermodynamics. It is often called Joule's Equivalent and is generally represented by the symbol J. The symbol A is used for the reciprocal of Joule's Equivalent. As the pound-calory is adopted throughout this book, the numerical value of J in any examples is 1400 and that of A is 1/1400.

25. The Second Law of Thermodynamics. *It is impossible for a self-acting machine, unaided by any external agency, to convey heat from one body to another at a higher temperature.*

[1] A summary of results of measurements of the mechanical equivalent of heat will be found in Professor Callendar's article on "Calorimetry" in the *Encyclopaedia Britannica.* See also E. H. Griffiths on the *Thermal Measurement of Energy* (Cambridge University Press, 1901), where an interesting account is given of various methods by which this physical constant has been determined.

This is the form in which the second law has been stated by Clausius[1]. Another statement of it, different in form but similar in effect, has been given by Lord Kelvin[2]. Its force may not be immediately obvious, but it will be shown below that this law sets a most important limit to the convertibility of heat into work. So far as the first law goes, there is nothing to prevent the whole heat taken in by an engine from changing into mechanical energy. In consequence of the second law, however, as we shall presently see, no heat-engine converts, or can convert, more than a small fraction of the heat supplied to it into work; a large part is necessarily rejected as heat. The ratio

$$\frac{\text{Heat converted into work}}{\text{Heat taken in by the engine}}$$

is a fraction always much less than unity. This fraction is called the *efficiency* of the engine considered as a heat-engine.

26. The Working Substance in a Heat-Engine. In every heat-engine there is a *working substance* which alternately takes in and rejects heat. In general it suffers changes of volume, and does work by overcoming resistance to these changes. The working substance may be gaseous, liquid, or solid. We can, for example, imagine a heat-engine in which the working substance is a long metallic rod, arranged to act as the pawl of a ratchet-wheel with closely pitched teeth. Let the rod be heated so that it elongates sufficiently to drive the wheel forward through the space of one tooth. Then let the rod be cooled (say by applying cold water), the ratchet-wheel being meanwhile held from returning by a separate click or detent. The rod, on cooling, will retract so as to engage itself with the next succeeding tooth, which may then be driven forward by heating the rod again, and so on. To make it evident that such an engine would do work, we have only to suppose that the ratchet-wheel carries round with it a drum by which a weight is wound up. The device forms a complete heat-engine, in which the working substance is a solid rod, which receives heat by being brought into contact with some source of heat at a comparatively high temperature, transforms a small part of this heat into work, and rejects the remainder to what we may

[1] See Clausius, *Mechanical Theory of Heat*, translated by W. R. Browne.

[2] See Lord Kelvin's (Sir W. Thomson's) *Collected Papers*, vol. I, for his early investigations in thermodynamic theory.

call a receiver of heat, which is kept at a comparatively low temperature. The greater part of the heat may be said simply to pass through the engine, from the source to the receiver, *becoming degraded as regards temperature* in doing so. It will be seen presently that this is typical of the action of all heat-engines; when they are doing work they must take in heat at a comparatively high temperature and reject heat at a comparatively low temperature. They convert some heat into work only by letting down a much larger quantity of heat from a high to a relatively low temperature. The action is, to some extent, analogous to that of a water-wheel, which does work by letting down water from a high level to a lower level, change of level in the one case being the analogue of change of temperature in the other. But there is this important difference, that whereas in the action of the water-wheel none of the water disappears, in the action of the heat-engine an amount of heat disappears which is equivalent to the work done.

27. Graphic representation of work done in the changes of volume of a fluid. In almost all actual heat-engines the working substance is a fluid. In some it is air, in some a mixture of several gases. In a steam-engine the working fluid is a mixture (in varying proportions) of water and water-vapour. When the engine is of the cylinder and piston type work is done by changes of volume only; its amount depends solely on the relation of pressure to volume during the change, and not at all on the form of the vessels in which the change takes place. Let a diagram be drawn (fig. 9) in which the relation of the intensity of pressure to the volume of any supposed working fluid is graphically exhibited by the line ABC, where AM, CN are pressures and AP, CQ are volumes, then the work done by the substance in expanding from volume AP to volume CQ is the area of the figure $MABCN$. And similarly, if the substance be compressed from volume CQ back to its original volume in such a manner that the line CDA represents the relation of pressure to volume during compression,

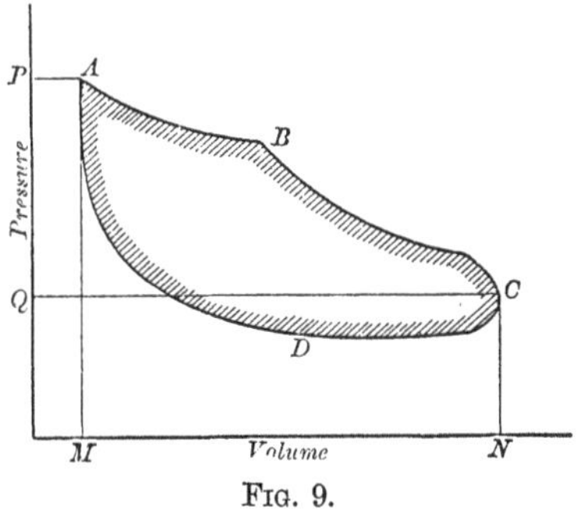

FIG. 9.

a quantity of work is done *upon* the substance which is represented by the area $NCDAM$. Taking the two operations together, we find that the substance has done a net amount of work equal to the area of the shaded figure $ABCDA$, or $\int PdV$, in the complete action which the closed figure represents. This is an example and a generalization of the method of representing work which Watt introduced by his invention of the indicator; the figure $ABCDA$ may be called the *indicator diagram* of the supposed action.

Modern forms of the indicator will be described in a later chapter. For the present it may suffice to say that the indicator draws automatically a diagram showing the relation of the pressure of the working fluid to the movement of the piston, or in other words to the volume of working fluid in the cylinder, and thus gives complete information as to the work done throughout the stroke.

28. Cycle of Operations of the working substance. In many heat-engines the working substance returns periodically to the same state of temperature, pressure, volume, and physical condition. Each time this has occurred the substance is said to have passed through a complete cycle of operations. For example, in a condensing steam-engine, water taken from the hot-well is pumped into the boiler; it then passes into the cylinder as steam, passes thence into the condenser, and thence again as water into the hot-well; it completes the cycle by returning to the same condition as at first. In other less obvious cases, as in that of the non-condensing steam-engine, a little consideration will show that the cycle is completed, not indeed by the same portion of working substance being returned to the boiler, but by an equal quantity of water being fed to it, while the steam which has been discharged into the atmosphere cools to the temperature of the feed-water. In the theory of heat-engines it is of the first importance to consider as a whole the cycle of operations performed by the working substance (as was first done by Carnot in 1824). If we stop short of the completion of a cycle matters are complicated by the fact that the substance is in a state different from its initial state, and may therefore have changed its stock of internal energy. After the cycle is completed, on the other hand, the internal energy of the substance is necessarily the same as at first, since the condition is in every respect the same. Hence in regard to the cyclic process as a whole this equation must hold good,

Heat taken in = Work done + Heat rejected.

29. Internal Energy. We have used here a phrase which requires some further explanation—the *internal energy* of a substance. No means exist by which the whole stock of energy that a substance contains can be measured. But we are concerned only with changes in that stock, changes which may arise from the substance taking in or giving out heat, or doing work, or having work done upon it. If a substance takes in heat without doing work its stock of internal energy increases by an amount equal to the heat taken in. If it does work without taking in heat, it does the work at the expense of its stock of internal energy, and the stock is diminished by an amount equal to the work done. In general, when heat is being taken in and the substance is at the same time doing work, we have

Heat taken in = Work done + Increase of Internal Energy,

which we may write

$$dQ = AdW + dE,$$

the factor A being used to convert from units of work to units of heat.

In a complete cycle there is, at the end, no change of E, and consequently

$$Q_1 - Q_2 = AW,$$

where $Q_1 - Q_2$ is the net amount of heat received in the cycle as a whole, namely the difference between the heat taken in and the heat rejected in the complete process.

30. Engine using a Perfect Gas as working substance. It is convenient to approach the theory of heat-engines by considering, in the first instance, the action of an engine in which the working substance is what is called a perfect gas. Any one of the so-called permanent gases, or a mixture of them, such as air, is very nearly perfect. The word permanent, as applied to a gas, is to be understood only as meaning that the gas is liquefied with difficulty—by the use of extremely low temperature in conjunction, generally, with high pressure. So long as gases are under conditions of pressure and temperature widely different from those which produce liquefaction, they conform very approximately to certain simple laws—laws which may be regarded as *rigorously* applicable to ideal substances called *perfect* gases. After stating these laws we shall examine the efficiency of a heat-engine using a gas in a certain manner as working substance, and then show that the results so derived have a general application to all heat-engines whatsoever. In this procedure there is no sacrifice of generality, and it provides the easiest means of establishing certain fundamental principles.

The laws which have now to be stated are very nearly though not absolutely true for air, oxygen, nitrogen, hydrogen and carbonic oxide, except when at specially high pressures or specially low temperatures. Hydrogen probably comes nearest to the ideal of a perfect gas; but no real gas is in this sense strictly perfect.

31. Laws of the Permanent Gases. Boyle's Law and Charles' Law. Two laws which are very approximately true of the permanent gases, and may be regarded as strictly true of ideal perfect gases, are the following:—

Boyle's Law:—The volume of a given mass of gas varies inversely as the pressure, provided the temperature be kept constant.

Thus, if V be the volume of a given quantity of any gas, and P the pressure, then so long as the temperature is unchanged—

$$V \text{ varies inversely as } P, \text{ or } PV = \text{constant}.$$

Charles' Law:—Under constant pressure equal volumes of different gases increase equally for the same increment of temperature.

This law is sometimes stated by saying that all gases expand alike, or have the same coefficient of expansion.

If, for example, we take a vessel containing a quantity of air and heat it from one temperature to another, taking care to arrange the experiment so that the air may expand without any change in its pressure, we shall find that a certain change of volume takes place. Let any other permanent gas then be substituted for the air in the vessel, and let the experiment be repeated by heating this other gas from the same initial to the same final temperature as before, the pressure being still kept constant. The volume will be found to have changed by sensibly the same amount as was observed in the experiment with air. And further, if the experiment be varied by using a greater or smaller interval of temperature, it will be found that different gases continue to agree in the changes of volume which they show. This is equivalent to saying that if we use a gas thermometer (in which air or any other gas is allowed to expand without change of pressure) to measure temperatures, defining equal intervals of temperature to be those which correspond to equal expansions on the part of the gas, we obtain a thermometric scale which is the same whatever gas we select. The scale of such a gas thermometer will be found to differ slightly, but only slightly, from the usual mercurial scale, which defines equal intervals of temperature to be those that correspond to equal expansions of mercury in glass.

32. Scales of Temperature. The Gas Scale. In the construction of an ordinary thermometer a fine tube of uniform bore is chosen, and a bulb is formed on it to contain the mercury or other liquid whose expansion is to be used as an indication of temperature. When it is filled the two fixed points are determined by placing the instrument (*a*) in melting ice, and (*b*) in the steam coming from water boiling under a pressure of one atmosphere. The position taken by the end of the column of liquid in the tube is marked for each of these two points. The distance between them is then divided into equal parts which are called degrees, 100 parts for the centigrade scale and 180 for the Fahrenheit scale. By this construction equal steps in temperature are defined by equal amounts of expansion on the part of the selected liquid, or rather by equal amounts of difference between the expansion of the liquid itself and that of the glass in which it is contained, for it is the difference of expansion that determines the rise of the column in the tube. This common method of measuring temperature gives results that vary for different liquids and for different sorts of glass. Each of two mercury thermometers, for example, may have the fixed points correctly marked, and be of uniform bore, and yet if they are made of different sorts of glass they may give readings that differ by as much as half a degree at the middle of the range between the fixed points, and may show still more serious discrepancies when they are applied to measure higher temperatures. This illustrates the fact that the measurement of temperature by an ordinary thermometer gives an arbitrary scale, which cannot even be relied on to be the same in different instruments.

Measurements of temperature are much less capricious if we select for the expanding substance any one of the so-called permanent gases such as air, or nitrogen, or hydrogen, taking care to keep the pressure of the gas constant while it is employed to measure temperature by its changes of volume. Such an instrument is called a constant-pressure gas thermometer. It would be inconvenient for ordinary use; but it serves to supply a scale with which the readings of an ordinary thermometer can be compared. Thus the readings of any thermometer can be corrected to bring them into agreement with the scale of a gas thermometer if that scale be adopted as the standard scale in stating temperatures.

Experiments on the expansion of various gases by heat have shown that all gases which are far from the conditions that would cause liquefaction expand very nearly alike. This is the law of Charles, as already stated. It follows that if we compare an air thermometer with a nitrogen or a hydrogen thermometer we get practically the same scale except at extremely low temperatures such as those at which the gas is approaching the liquid state. Gases expand by almost exactly the same amount between the two fixed points, and at intermediate points, or at points beyond the range, their agreement with one another is almost perfect. Hence the scale of the gas thermometer is much to be preferred to that of any mercury thermometer as a means of stating temperature. But there is another and even stronger reason for this preference. We shall see later that it is possible to imagine a scale of temperature, based on general thermodynamic principles, which does not depend on the properties of any particular substance: that scale is called the *thermodynamic scale* of temperature, and much use is made of it in thermodynamic reasoning. The scale of a gas thermometer is practically identical with the thermodynamic scale. This is true whether we use a constant-pressure gas thermometer, or what is called a constant-volume gas thermometer, in which increments of temperature are measured by the increments of pressure that are required to keep the volume of the gas constant while it is heated.

33. Reckoning of Temperature from the Absolute Zero. Experiment shows that the amount by which air or hydrogen or any other so-called permanent gas expands between the two fixed points is about 100/273 of the volume at the lower fixed point, care being taken that the pressure does not change. Hence, if we adopt the scale of the gas thermometer as our scale of temperature, and use centigrade divisions, this result may be expressed by saying that when 273 cubic inches of gas at 0° C. are heated under constant pressure to 1° C. the volume alters to 274 cubic inches. When the gas is heated to 2° C. its volume becomes 275 cubic inches, at the upper fixed point it becomes 373, and so on. Similarly, if the gas be cooled from 0° C. to − 1° C. its volume changes from the original 273 cubic inches to 272, and so on.

Putting this in a tabular form, let the volume be

	273 at 0° C.
It will become	272 at − 1° C.
	⋮ ⋮
and finally would be	0 at − 273° C.

if the same law could be held to apply down to the lowest temperatures. Any actual gas would change its physical state before so low a temperature was reached, becoming first liquid and then solid, and the volume to which it would contract would consequently be not zero but the volume of the substance in the solid state.

The above result may be concisely expressed by saying that if temperature be reckoned not from the ordinary zero but from a zero which is about 273 centigrade degrees below it (more exactly 273·1), the volume of a gas, heated under constant pressure, is proportional to the temperature reckoned from that zero. The zero in question is spoken of as the Absolute Zero of temperature. Denoting any temperature on the ordinary scale by t and the corresponding temperature reckoned from the absolute zero by T, we have

$$T = t + 273{\cdot}1 \text{ on the centigrade scale}$$

and

$$T = t + 459{\cdot}6 \text{ on the Fahrenheit scale.}$$

The absolute zero has been defined here by reference to the expansion of a gas. But it will be seen later that the thermodynamic scale of temperature starts from a zero which is absolute in the sense that no lower temperature can possibly exist, and that the zero of the thermodynamic scale coincides with the zero of the gas scale as defined above[1].

[1] The exact position of the absolute zero is uncertain to the extent of about one-tenth of a degree. Callendar places it at − 273·1° C.; that figure is generally adopted and will be used in this book.

34. Properties of a "Perfect" Gas. If real gases were ideally "perfect" they would exactly obey Charles' Law and any one of them would serve, if used as the expanding substance in a thermometer, to give a scale which would exactly agree with the thermodynamic scale of temperature.

A perfect gas would also conform exactly to Boyle's Law, making the product PV strictly constant so long as the temperature remained constant. If we define the temperature scale by reference to the expansion of the gas we should also have V varying as the temperature T (reckoned from the absolute zero) under any constant pressure. Combining these two statements we should have

$$PV = RT \quad \text{.........................(1)},$$

where R is a constant. For the present it is to be understood that the symbol T stands for temperature measured on the scale of a gas thermometer, from a zero which is 273·1 degrees (centigrade) below the melting-point of ice.

We may accordingly write, for any gas assumed to be perfect,

$$R = \frac{P_0 V_0}{273 \cdot 1},$$

where P_0 and V_0 are the pressure and volume respectively at 0° C. When the volume is reckoned per unit quantity of the gas (say per lb.) we have a definite constant value of R for each gas, depending on the units employed and on the specific density of the gas in question[1].

When a gas satisfying this equation is heated under constant pressure and consequently expands, R is a measure of the amount of work done by the gas for each degree through which the temperature rises. Let the original temperature of the gas be T_1 and its volume V_1, and let it be heated under constant pressure P till the temperature is T_2 and the volume V_2. Then we have $RT_1 = PV_1$ and $RT_2 = PV_2$, from which

$$R(T_2 - T_1) = P(V_2 - V_1),$$

[1] Since in different gases (assumed to be perfect) the density varies as the molecular weight m, and R varies inversely as the density, the product mR has one and the same value. It is called the universal gas constant. Expressed in foot-pounds per lb. it is 2779. R can be calculated for any gas by dividing that number by the molecular weight. Thus for oxygen, where $m = 32$, R so calculated is 86·8 foot-pounds per lb.; for nitrogen, where $m = 28$, it is 99·3; for hydrogen, where $m = 2 \cdot 016$, it is 1378.

which is the work done by the gas in expanding from V_1 to V_2 under the constant pressure P. Let the interval of temperature be one degree, then R is equal to the work done. Thus R is numerically expressed in units of work per unit of mass.

According to Regnault's measurements a cubic metre of dry air at 0° C. and at a pressure of one atmosphere weighs 1293 grammes, which is equivalent to 0·08072 lb. per cub. ft. Since the pressure of the standard atmosphere is 14·689 × 144 lb. per sq. ft., we should accordingly have for dry air

$$R = \frac{14{\cdot}689 \times 144}{0{\cdot}08072 \times 273{\cdot}1} = 96{\cdot}0$$

foot-pounds per lb.

35. The Internal Energy of a Gas. Joule's Law. A further important generalization about gases relates to their Internal Energy. It is known as Joule's Law, and while it is only approximately true of any real gas it is taken as applying rigorously to an ideal perfect gas:—

The Internal Energy of a given quantity of a gas depends only on the temperature.

This is an inference from the fact, established by the experiments of Joule, that *when a gas expands without doing external work, and without taking in or giving out heat* (*and therefore without changing its stock of internal energy*), *its temperature does not change.*

Joule connected a vessel containing compressed gas with another vessel which was empty, by means of a pipe with a closed stop-cock. Both vessels were immersed in a bath of water and were allowed to assume a uniform temperature. Then the stop-cock was opened, and the gas distributed itself between the two vessels, expanding without doing external work. After this the temperature of the water in the bath was found to have undergone no appreciable change. The temperature of the gas appeared unaltered, and no heat had been taken in or given out by it, and no work had been done by it.

Since the gas had neither gained nor lost heat, and had done no work, its internal energy was the same at the end as at the beginning of the experiment. The pressure and volume had changed, but the temperature had not. The conclusion follows that the internal energy of a given quantity of a gas depends only on its temperature, and not upon its pressure or volume; in other words, a change of pressure and volume not associated with a change of temperature does not alter the internal energy. Hence in any change of temperature the change of internal energy is independent of the relation of pressure to volume during the operation: it depends only on the amount by which the temperature has been changed.

The apparatus used by Joule in this experiment is shown in the figure. The vessel A was filled with air compressed to more than 20 atmospheres, and B was exhausted. The water in the bath was stirred and the temperature noted before the stop-cock C was opened. After the gas had come to rest in the two vessels the water was again stirred, and was found to show the same temperature as before, so far as tests made by a very sensitive thermometer could detect.

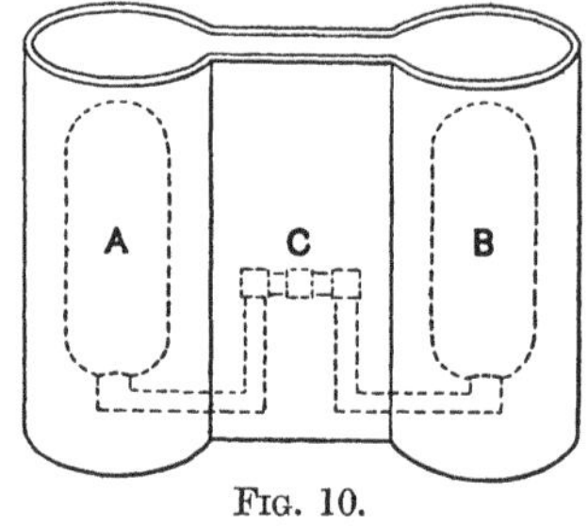

FIG. 10.

In another form of the apparatus Joule separated the bath into three portions, one portion round each of the vessels and one round the connecting pipe. When the stop-cock was opened the water surrounding A was cooled, but this was compensated by a rise of temperature in the water surrounding B and C. The gas in A became colder in the act of expanding, but heat was given up in B and C as its eddying motion settled down, and when all was still there was neither gain nor loss of heat on the whole so far as could be detected in this form of experiment.

It is now, however, known that a very slight change of temperature does in fact take place when a gas expands without doing work. In later experiments by Joule and Thomson (Lord Kelvin) a more delicate method was adopted of detecting whether there is any change of internal energy when the pressure and volume change under conditions such that external work is not done. The gas was forced to pass through a porous plug by maintaining a constant high pressure on one side of the plug and a constant low pressure on the other. Care was taken to prevent any heat being gained or lost by conduction from outside. In this operation work was done upon the gas in forcing it up to the plug, and work was done by it when it passed the plug, by its displacing gas under the lower pressure on the side beyond the plug. If no change of temperature took place, and if the gas conformed to Boyle's Law, these two quantities of work would be exactly equal, and consequently no external work would be done on the whole. For let P_1 be the pressure and V_1 the volume before passing the plug, and P_2 the pressure and V_2 the volume after passing the plug, the volumes being in both cases stated per lb. of the gas. Then the work done upon the gas (per lb.) as it approaches the plug is P_1V_1, and the work done by it as it leaves the plug is P_2V_2. If the temperature is the same on both sides these quantities are equal in a gas for which PV is constant at any one temperature. Thus a "perfect" gas, which means a gas that conforms strictly both to Boyle's Law and to Joule's, would in its passage of the plug have expanded without (on the whole) doing any work, and therefore without changing its internal energy, no heat being gained or lost. In such a gas no change of temperature should accordingly be found, as it passes the plug, and if a change of temperature is observed it is due to the fact that the real gas experimented on is not strictly "perfect."

In the experiments of Joule and Thomson[1] small changes of temperature were in fact detected and measured in air and other real gases, on passing the porous plug. This Joule-Thomson effect, as it is called, is in general a cooling.

Observations of the Joule-Thomson effect are of great value in determining exactly the properties of gases and vapours which are not perfect; and certain practical methods of liquefying gases under extreme cold depend upon the existence of this effect.

In the imaginary perfect gas, however, the Joule-Thomson effect is entirely absent. There is no change of temperature in passing the plug, and there is also no change of internal energy, for no work is done and (by assumption) no heat is taken in or given out.

It is important to notice that we assume the imaginary perfect gas to satisfy two conditions: it obeys Boyle's Law exactly and also Joule's Law exactly. These characteristics are independent of one another: it would be possible to have a gas satisfy one and not the other, but a gas is said to be perfect in the thermodynamic sense only when it satisfies both, and in that case certain other properties follow which will now be pointed out.

[1] See Lord Kelvin's *Mathematical and Physical Papers*, vol. I, p. 333.

36. Specific Heats of a Gas. The specific heat of any substance means the amount of heat required per degree to raise the temperature of unit quantity of the substance, under any assumed mode of heating. Thus when a substance is heated through a small interval of temperature dT the heat taken in (per unit quantity of the substance) is KdT, where K is the specific heat for the particular conditions and mode of hearing. In dealing with gases or other fluids two important modes of heating must be distinguished: we may heat them under conditions of constant pressure or of constant volume. We shall use the symbol K_p to represent specific heat at constant pressure, and K_v to represent specific heat at constant volume.

Consider first the operation of heating unit quantity of a perfect gas at constant volume, from temperature T_1 up to temperature T_2. The heat taken is

$$K_v (T_2 - T_1).$$

No external work is done, for the volume (by assumption) does not change, and consequently all this heat goes to increase the stock of internal energy contained in the gas. But by Joule's Law the internal energy depends only on the temperature. Therefore if we heat the same quantity of the same gas in any other manner from T_1 to T_2, the same change of internal energy must take place.

Imagine then another manner of heating, namely at constant pressure. In that case the heat taken in is

$$K_p (T_2 - T_1).$$

During this process external work is done, because the gas expands, and its amount is

$$P (V_2 - V_1),$$

where V_1 and V_2 represent the volumes at the beginning and end of the operation respectively, and P is the pressure, which by assumption is constant. Since $PV_2 = RT_2$ and $PV_1 = RT_1$, we may write the expression for the external work in the form

$$R (T_2 - T_1).$$

This is in work units: in heat units it is

$$AR (T_2 - T_1),$$

where A is the reciprocal of Joule's equivalent (§ 24).

The difference between the heat taken in and the work done, namely

$$(K_p - AR) (T_2 - T_1),$$

is simply an addition to the stock of internal energy. But as was pointed out above, the change of internal energy must be the same in both modes of heating, and therefore

$$K_v = K_p - AR \text{ ....(2).}$$

This important relation between the two specific heats in a perfect gas follows from the Laws of Boyle and of Joule.

We have here taken K_v and K_p as applying throughout a finite range of temperature from T_1 to T_2. But this range may be made infinitesimally small without affecting the argument[1], and in that case K_v and K_p become the specific heats at a definite temperature. The conclusion holds that for any condition of the gas

$$K_p - K_v = AR,$$

and this is true whether the specific heats are or are not independent of the temperature.

37. Constancy of the Specific Heats in a Perfect Gas. From the above result it follows that if either of the two specific heats is constant the other must also be constant. To be constant the specific heat has to be independent both of the pressure and of the temperature.

First as to independence of pressure: we have seen (§ 35) that the internal energy of a perfect gas depends only on the temperature and is independent of the pressure. If we heat a perfect gas through 1° at any one temperature the change of internal energy is measured (§ 36) by K_v, no matter what is the pressure. Hence K_v is independent of the pressure; and since K_p is equal to $K_v + AR$ it follows that K_p also must be independent of the pressure.

But a gas may conform to the Laws of Boyle and Joule without having K_p and K_v independent of the temperature, and if we are to treat them as constant we must make a further assumption regarding the properties of that convenient imaginary substance

[1] Suppose the heating to be through a very small interval of temperature dT. In heating at constant volume, the heat taken in is $K_v dT$, and all of it goes to increase the internal energy by an amount dE. Hence

$$K_v dT = dE.$$

In heating at constant pressure through the same interval of temperature the heat taken in (dQ) does work dW and also adds to the internal energy by the amount dE. dQ is $K_p dT$; and dW is PdV, which is equal to RdT. Hence

$$K_p dT = ARdT + dE = ARdT + K_v dT.$$

From which

$$K_p - K_v = AR.$$

a perfect gas. Regnault's experiments showed that in real gases K_p is nearly constant through a moderate range of temperature. It is now known, however, that at high temperatures, such as those which occur in the cylinders of gas-engines, the specific heat of a gas rises very considerably. This change is important in relation to the action of gas-engines: but for our present purpose it will simplify matters to think of an ideal gas in which the specific heat is constant. Accordingly, in dealing with a perfect gas, it is assumed that K_p in such a gas is strictly independent of the temperature. This is a third assumed characteristic of a perfect gas, additional to the two already described, namely, that the gas conforms to Boyle's Law and to Joule's Law. It does not in any way conflict with them: each of the three characteristics is independent of the others. With this further assumption we have, for any perfect gas, K_p constant under all conditions, and consequently K_v also constant under all conditions, since the difference between them is constant.

38. Reversible actions. The next step is to consider particular modes in which a working substance may be expanded or compressed and may take in or give out heat, and at the outset it is essential to distinguish between actions that are *reversible* and those that are irreversible.

An expansion or compression is reversible if it is carried out in such a manner that the operation can be reversed, with the result that the substance will pass back through all the stages through which it has passed during the expansion or compression and be in the same condition in all respects at each corresponding stage in both processes.

This implies that the substance must expand smoothly, without setting up any motions within itself of a kind such that their kinetic energy is frittered down into heat through internal friction. The whirls and eddies which occur as a fluid enters or expands in the cylinder of an engine are irreversible, and in ideal reversible expansion we must suppose them absent. Reversible expansion implies that there are no losses of mechanical effect from any sort of internal friction. It excludes throttling, such as occurs when a substance expands through a valve or other constricted opening into a region of lower pressure where the kinetic energy of the stream and eddies is dissipated. In such cases the motion of the stream and eddies cannot be reversed. To get the substance back to the region of higher pressure would require an expenditure of more work than was done by it during its expansion, and if we were to force it back we should find it had gained heat through the subsidence of the internal eddying motions, though no heat had come in from outside.

The kind of expansion which takes place in Joule's experiment (§ 35) is an extreme instance of irreversible expansion.

A transfer of heat to or from any substance is reversible only if the substance is at the same temperature as the body from which it is taking heat or to which it is giving heat. Suppose, for instance, that a working substance is taking in heat from a hot source and is expanding as it does so. The expansion may be reversible in itself, that is to say it may involve no internal friction, but if there is any drop of temperature between the working substance and the source the operation as a whole cannot be reversed. Any thermal contact between bodies at different temperatures involves an irreversible transfer of heat.

Neither the expansions and compressions nor the transfers of heat that occur in a real engine are ever strictly reversible, some of them indeed are far from being reversible. But the study of an ideal engine, in which all the operations are reversible, is of fundamental importance in the science of the subject, and it furnishes a basis for the critical analysis of actions in a real engine.

39. Adiabatic and Isothermal Expansion. There are two specially important kinds of reversible expansion, (1) Adiabatic, and (2) Isothermal.

Adiabatic expansion or compression means expansion or compression carried out reversibly and without allowing any heat to enter or leave the substance. A curve drawn to show the relation of pressure to volume during the process is called an adiabatic line. Adiabatic action would be realized if we had a substance expanding, or being compressed, without change of chemical state, and without any eddying motions, in a cylinder which (along with the piston) was totally impervious to heat.

From this definition it follows that the work which a substance does while it is expanding adiabatically is all done at the expense of its stock of internal energy; and the work which is spent upon a substance when it is being compressed adiabatically all goes to increase its stock of internal energy.

In actual heat-engines the action is never strictly adiabatic, for there are always some exchanges of heat between the working substance and the surface of the containing vessel. Very rapid compression or expansion may come near to being adiabatic by giving little time for any transfer of heat to occur.

After what has been said already about reversibility, it is scarcely necessary to add that expansion through a throttle-valve is not adiabatic, though it may occur without letting heat enter or leave the substance.

In the adiabatic expansion of any substance work is done at the expense of the stock of internal energy. Since no heat is taken in or given out, there must be a decrease of internal energy equivalent to the amount of the work done by the substance.

Taking the general equation (§ 29)

$$dQ = A\,dW + dE,$$

which applies to any small change of state on the part of any substance, we have $dQ = 0$ when the action is adiabatic, and hence in adiabatic expansion

$$A\,dW = -\,dE.$$

Here, as before, dW is the work done by the substance as it expands, dE is the change of internal energy, and A is the factor required to convert an expression for work into heat units.

Similarly, in adiabatic compression a substance increases its internal energy by an amount which is the thermal equivalent of the work spent in compressing it.

Isothermal expansion or compression means expansion or compression carried out reversibly (as regards internal action) and without change of temperature. A curve drawn to show the relation of pressure to volume during isothermal expansion or compression is called an isothermal line.

When a substance is expanding isothermally it takes in heat to maintain its temperature constant; it therefore must be in contact with a source of heat. When it is being compressed isothermally it gives out heat, and must be in contact with a receiver which can take heat from it.

40. Adiabatic Expansion of a Perfect Gas. Consider next the behaviour of a perfect gas during adiabatic expansion or compression. We have seen that in a small adiabatic expansion of any substance

$$dE = -A\,dW = -AP\,dV.$$

In a perfect gas $dE = K_v dT$ (§ 36). Hence in the adiabatic expansion of a perfect gas

$$AP\,dV = -K_v dT.$$

But $P = RT/V$ (§ 34). Hence

$$ART\,dV/V + K_v dT = 0,$$

or, dividing by T,

$$AR\,dV/V + K_v dT/T = 0,$$

which gives on integration

$$AR \log_\epsilon V + K_v \log_\epsilon T = \text{constant} \quad \text{............(3).}$$

Writing $K_p - K_v$ for AR (§ 36), and dividing by K_v, which is constant (§ 37),

$$(K_p/K_v - 1) \log_\epsilon V + \log_\epsilon T = \text{constant}.$$

We shall write γ for the ratio of the two specific heats, namely K_p/K_v.

Thus we have

$$\gamma \log_\epsilon V - \log_\epsilon V + \log_\epsilon T = \text{constant} \quad \text{.........(4).}$$

Further, since PV/T is constant,

$$\log_\epsilon P + \log_\epsilon V - \log_\epsilon T = \text{constant}.$$

Adding these two equations, we have

$$\log_\epsilon P + \gamma \log_\epsilon V = \text{constant} \quad \text{..............(5),}$$

which gives

$$PV^\gamma = \text{constant} \quad \text{.......................(6)}$$

as the equation of any adiabatic line in the pressure-volume diagram of a perfect gas[1].

41. Change of Temperature and Work done in the Adiabatic Expansion of a Perfect Gas. When a gas is expanding adiabatically its stock of internal energy is, as we have seen, being reduced, and hence its temperature falls, the change of internal energy being proportional to the change of temperature

[1] It is to be remembered that $\log_\epsilon$, the "hyperbolic" or "Napierian" or "natural" logarithm of any number, is 2·3026 times the common logarithm of the number.

(§ 36). Conversely, in adiabatic compression the temperature rises. The amount by which the temperature is changed (in a perfect gas) may be found by combining the equations

$$P_1V_1^\gamma = P_2V_2^\gamma \text{ and } P_1V_1/P_2V_2 = T_1/T_2.$$

Multiplying them together we have

$$\frac{T_1}{T_2} = \frac{P_1V_1P_2V_2^\gamma}{P_2V_2P_1V_1^\gamma},$$

whence $$\frac{T_1}{T_2} = \left(\frac{V_2}{V_1}\right)^{\gamma-1}, \text{ or } T_2 = T_1\left(\frac{V_1}{V_2}\right)^{\gamma-1} \qquad \text{...........(7).}$$

This result of course applies to compression as well as to expansion along an adiabatic line. It also follows directly from equation (4), which may be written

$$\log_\epsilon T + (\gamma - 1) \log_\epsilon V = \text{constant};$$

whence $$TV^{\gamma-1} = \text{constant}.$$

When any fluid expands in any manner, whether adiabatic or not, the work done in expanding from volume V_1 to volume V_2, namely the cross-hatched area in fig. 11, is

$$W = \int_{V_1}^{V_2} PdV.$$

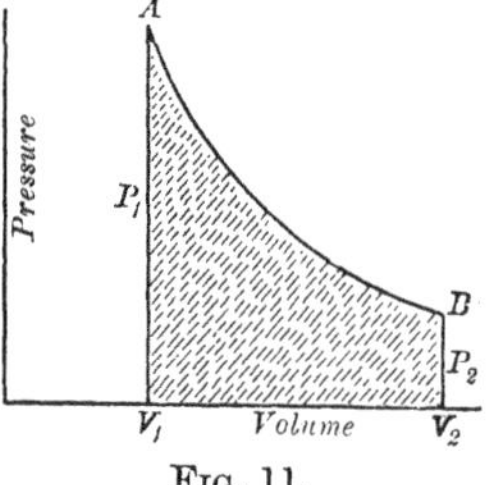

FIG. 11.

If the nature of the expansion be such that PV^n is constant, n being any index, then P at any point when the volume is V is $P_1V_1^n/V^n$, P_1 and V_1 being the pressure and volume in the initial state. In that case, for expansion from V_1 to V_2,

$$W = P_1V_1^n \int_{V_1}^{V_2} dV/V^n,$$

which gives on integration

$$W = \frac{P_1V_1 - P_2V_2}{n-1} \qquad \text{.......................(8).}$$

If we apply this result to a gas expanding adiabatically, for which $n = \gamma$, we obtain

$$W = \frac{P_1V_1 - P_2V_2}{\gamma - 1} = \frac{R(T_1 - T_2)}{\gamma - 1} \qquad \text{...........(9),}$$

since $P_1V_1 = RT_1$ and $P_2V_2 = RT_2$.

As to expansions which are not adiabatic, it follows, from the expressions given above for the external work done by an expanding gas and for the change of internal energy, that if n is less than γ the work done is greater than the loss of internal energy—that is to say, the gas is then taking in heat while it expands. On the other hand, if n is greater than γ the work done is less than the loss of internal energy; in other words, the gas is then rejecting heat by conduction to the walls of the containing vessel or in some other manner.

By way of exemplifying an adiabatic process suppose a quantity of dry air to be contained in a cylinder at a temperature of 15° C. ($T = 288$) and to be suddenly compressed to half its original volume, the process being so rapid that no appreciable part of the heat developed by compression has time to pass from the air to the cylinder walls. Here $V_1/V_2 = 2$, and taking γ for air to be 1·4 the temperature immediately after compression, before the gas has time to cool, is

$$T_2 = T_1 (V_1/V_2)^{\gamma-1} = 288 \times 2^{0\cdot4} = 380$$

or 107° C. The work spent in compressing the air, namely,

$$\frac{R (T_2 - T_1)}{\gamma - 1}, \text{ is } \frac{96\cdot0 \times 92}{0\cdot4} = 22080 \text{ foot-pounds}$$

for each pound of air in the cylinder. The internal energy of the gas becomes increased by this amount; but if the cylinder be a conductor of heat the whole of this will in time become dissipated by conduction to surrounding bodies, and the internal energy will gradually return to its original value, as the temperature of the gas sinks to 15° C.

During compression the pressure rises (following the law $PV^\gamma = \text{constant}$), and just at the end its value is greater than at the beginning in the ratio $2^{1\cdot4}$ or 2·64 to 1. If as before we then suppose the temperature to sink slowly by conduction to 15° C. while the volume does not change, the pressure will fall with the temperature until it reaches a value only twice that which it had before the air was compressed.

42. Isothermal Expansion of a Perfect Gas. In a gas which satisfies the equation $PV = RT$, PV is constant during isothermal expansion or compression, and any isothermal line on the pressure-volume diagram is a rectangular hyperbola, the pressure varying inversely as the volume. To find the work done in the isothermal expansion of such a gas we have

$$W = \int_{V_1}^{V_2} P dV$$

and $$P = \frac{P_1 V_1}{V},$$

from which $$W = P_1 V_1 \int_{V_1}^{V_2} \frac{dV}{V}.$$

Integrating, $$W = P_1 V_1 (\log_\epsilon V_2 - \log_\epsilon V_1)$$

or $$W = P_1 V_1 \log_\epsilon \frac{V_2}{V_1} \quad\text{..........................(10).}$$

Instead of $P_1 V_1$ we may write PV, since the product of P and V is constant through the process, and again, since $PV = RT$,

$$W = RT \log_\epsilon \frac{V_2}{V_1} \quad\text{....................(11).}$$

These expressions give either the work done by a gas during isothermal expansion or the work spent upon it during isothermal compression.

During isothermal expansion or compression a gas suffers no change of internal energy (by § 35, since T is constant). Hence during isothermal expansion the gas must take in an amount of heat just equal to the work it does, and during isothermal compression it must reject an amount of heat just equal to the work spent upon it. The expression for W consequently measures, not only the work done by or upon the gas, but also the heat taken in during isothermal expansion or given out during isothermal compression. In the diagram, fig. 12, the line AB is an example of a curve of isothermal expansion for a perfect gas, called for brevity an isothermal line, while AC is an adiabatic line starting from the same point A.

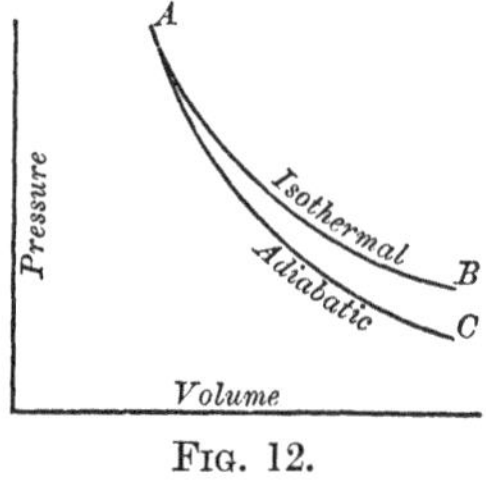

FIG. 12.

The compression of air or any other gas in a real cylinder is approximately adiabatic when the process is very quickly performed, but approximately isothermal when it is performed so slowly that the heat has time to be dissipated by conduction while the process goes on.

43. Carnot's Cycle of Operations. We shall now consider the action of an ideal engine in which the working substance, which we shall in the first place assume to be a perfect gas, is caused to pass through a cycle of changes each of which is either isothermal or adiabatic. The cycle to be described was first examined by Carnot, and is spoken of as Carnot's cycle of operations. Imagine a cylinder and piston composed of a perfectly non-conducting material, except as regards the bottom of the cylinder, which is a conductor. Imagine also a hot body or indefinitely capacious source of heat A, kept always at a temperature

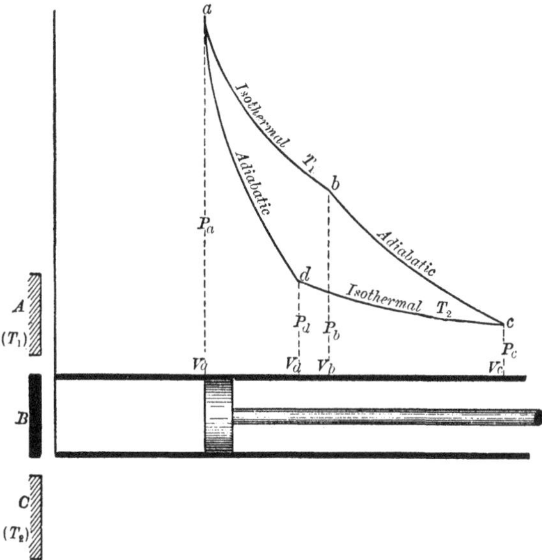

Fig. 13. Carnot's Cycle, with a gas for working substance.

T_1, also a perfectly non-conducting cover B, and a cold body or indefinitely capacious receiver of heat C, kept always at some temperature T_2 which is lower than T_1. It is supposed that A, B, or C can be applied at will to the bottom of the cylinder. Let the cylinder contain 1 lb. of a perfect gas, at temperature T_1, volume V_a, and pressure P_a to begin with. The suffixes refer to the points on the indicator diagram, fig. 13.

(1) Apply A, and allow the piston to rise slowly through any convenient distance. The gas expands isothermally at T_1, taking in heat from the hot source A and doing work. The pressure changes to P_b and the volume to V_b.

(2) Remove A and apply B. Allow the piston to go on rising. The gas expands adiabatically, doing work at the expense of its internal energy, and the temperature falls. Let this go on until the temperature is T_2. The pressure is then P_c, and the volume V_c.

(3) Remove B and apply C. Force the piston down slowly. The gas is compressed isothermally at T_2, since the smallest increase of temperature above T_2 causes heat to pass into C. Work is spent upon the gas, and heat is rejected to the cold receiver C. Let this be continued until a certain point d (fig. 13) is reached, such that the fourth operation will complete the cycle.

(4) Remove C and apply B. Continue the compression, which is now adiabatic. The pressure and temperature rise, and if the point d has been properly chosen, when the pressure is restored to its original value P_a, the temperature will also have risen to its original value T_1. [In other words, the third operation must be stopped when a point d is reached such that an adiabatic line drawn through d will pass through a.] This completes the cycle.

To find the proper place at which to stop the third operation, we have by equation (7), for the cooling during the adiabatic expansion of stage (2),

$$\frac{T_1}{T_2} = \left(\frac{V_c}{V_b}\right)^{\gamma-1},$$

and also, for the heating during the adiabatic compression of stage (4),

$$\frac{T_1}{T_2} = \left(\frac{V_d}{V_a}\right)^{\gamma-1}.$$

Hence
$$\frac{V_c}{V_b} = \frac{V_d}{V_a},$$

and therefore also
$$\frac{V_c}{V_d} = \frac{V_b}{V_a}.$$

That is to say, the ratio of isothermal compression in the third stage of the cycle is to be made equal to the ratio of isothermal expansion in the first stage, in order that an adiabatic line through d shall complete the cycle. For brevity we shall denote either of these last ratios (of isothermal expansion and compression) by r.

The following are the transfers of heat to and from the working gas, in the four successive stages of the cycle:—

(1) Heat taken in from $A = RT_1 \log_\epsilon r$ (by § 42).

(2) No heat taken in or rejected.

(3) Heat rejected to $C = RT_2 \log_\epsilon r$ (by § 42).

(4) No heat taken in or rejected.

Hence, the net amount of external work done by the gas, being the excess of the heat taken in above the heat rejected in a complete cycle, is

$$R(T_1 - T_2)\log_\epsilon r;$$

this is the area enclosed by the four curves in fig. 13.

44. Efficiency in Carnot's Cycle. The *efficiency* of the process, namely, the fraction

$$\frac{\text{Heat converted into work}}{\text{Heat taken in}},$$

is

$$\frac{R(T_1 - T_2)\log_\epsilon r}{RT_1 \log_\epsilon r} = \frac{T_1 - T_2}{T_1} \quad \text{..............(12).}$$

This is the fraction of the whole heat given to it which an engine following Carnot's cycle converts into work. The engine takes in an amount of heat, at the temperature of the source, proportional to T_1; it rejects an amount of heat, at the temperature of the receiver, proportional to T_2. It works within a range of temperature extending from T_1 to T_2, by letting down heat from T_1 to T_2 (§ 26), and in the process it converts into work a fraction of that heat, which fraction will be greater the lower the temperature T_2 at which heat is rejected is below the temperature T_1 at which heat is received.

45. Carnot's Cycle reversed. Next consider what will happen if we reverse Carnot's cycle, that is to say, if we force this imaginary engine to act so that the same indicator diagram as before is traced out, but in the direction opposite to that followed in § 43. Starting as before from the point a (fig. 13) and with the gas at T_1, we shall require the following four operations:—

(1) Apply B and allow the piston to rise. The gas expands adiabatically, the curve traced is ad, and when d is reached the temperature has fallen to T_2.

(2) Remove B and apply C. Allow the piston to go on rising. The gas expands isothermally at T_2, taking heat from C and the curve dc is traced.

(3) Remove C and apply B. Compress the gas. The process is adiabatic. The curve traced is cb, and when b is reached the temperature has risen to T_1.

(4) Remove B and apply A. Continue the compression, which is now isothermal, at T_1. Heat is now rejected to A, and the cycle is completed by the curve *ba*.

In this process the engine is not doing work; on the contrary, a quantity of work is spent upon it equal to the area of the diagram, or $R(T_1 - T_2) \log_\epsilon r$, and this work is converted into heat. Heat is taken in from C in the first operation, to the amount $RT_2 \log_\epsilon r$. Heat is rejected to A in the fourth operation, to the amount $RT_1 \log_\epsilon r$. In the first and third operations there is no transfer of heat.

The action is now in every respect the reverse of what it was before. The same work is now spent upon the engine as was formerly done by it. The same amount of heat is now given to the hot body A as was formerly taken from it. The same amount of heat is now taken from the cold body C as was formerly given to it. This will be seen by the following scheme:—

Carnot's Cycle, Direct.

Work done by the gas $= R(T_1 - T_2) \log_\epsilon r$;
Heat taken from $A = RT_1 \log_\epsilon r$;
Heat rejected to $C = RT_2 \log_\epsilon r$.

Carnot's Cycle, Reversed.

Work spent upon the gas $= R(T_1 - T_2) \log_\epsilon r$;
Heat rejected to $A = RT_1 \log_\epsilon r$;
Heat taken from $C = RT_2 \log_\epsilon r$.

The heat rejected to the cold body is now equal to the sum of the heat taken in from the hot body and the work spent on the substance. This of course follows from the principle of the conservation of energy. But what is important to observe is that the reversal of the work has been accompanied by an exact reversal of each of the transfers of heat.

46. Reversible Engine. An engine in which this is possible is called, from the thermodynamic point of view, a *reversible* engine. Every action which takes place in it is a reversible action in the sense of § 38. The result is that if it be forced to trace out its indicator diagram reversed in direction, so that the work which would be done by the engine when running direct is actually spent upon it, it will give to the source of heat the same quantity of heat as, when running direct, it would take from the source, and will take from the receiver of heat the same quantity as, when running direct, it would give to the receiver. By "the source of heat" is meant the hot body which acts as source when the engine is running direct, and by "the receiver" is meant the cold body which then acts as receiver. An engine performing Carnot's cycle of operations is one example of a reversible engine. The idea of thermodynamic reversibility in the sense here defined is of the greatest interest, for the reason that no heat-engine can be more efficient than a reversible engine when both work between the same limits of temperature; that is to say, when both engines take in heat at the same temperature and also reject heat at the same temperature. This theorem, due to Carnot, is of fundamental importance in the theory of heat-engines. It is deduced as follows from the laws of thermodynamics.

47. Carnot's Principle. To prove that no other heat-engine can be more efficient than a reversible engine when both work between the same limits of temperature, imagine two engines R and S of which R is reversible, and let them work by taking in heat from a hot body A and by rejecting heat to a cold body C. Let Q_A be the quantity of heat which the reversible engine R takes in from A for each unit of work which it does, and let Q_C be the quantity which it rejects to C.

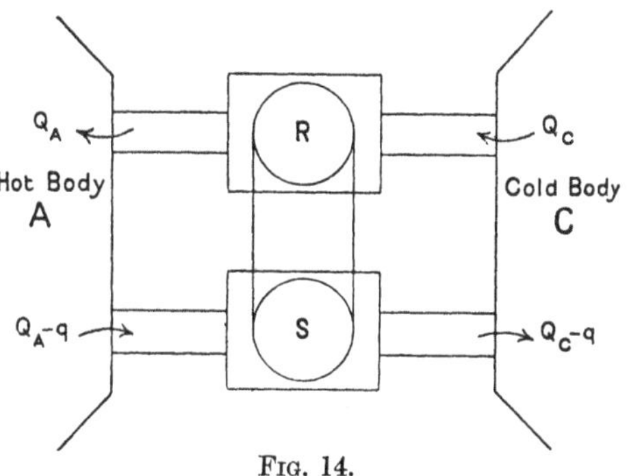

FIG. 14.

Now consider what consequences would follow if it were possible for S to be more efficient than R. It would take in less heat from A and reject correspondingly less heat to C, in doing each unit of work. Denote the heat which it would take in from A by $Q_A - q$ and the heat which it would reject to C by $Q_C - q$.

Suppose that S working direct (that is to say, converting heat into work) be set to drive R as a reversed engine, so that R converts work into heat. For every unit of work done by the engine S on the reversible engine R the quantity $Q_A - q$ would be taken from A by the engine S, and the quantity Q_A would be restored to A

by the reversed action of the engine R. This is because R being reversible restores to A when working reversed the same amount of heat as it would take from A when working direct. Hence the hot body would on the whole gain heat, by the amount q for every unit of work done by the one engine on the other. Again, S gives to C a quantity $Q_C - q$ while R takes from C a quantity Q_C, and hence the cold body C would lose an amount of heat equal to q for every unit of work done by the one engine on the other. Thus the combined action of the two engines—one working direct, as a true heat-engine, and the other reversed, as what we might call a heat-pump—would result in a transfer of heat from the cold body C to the hot body A, and this process might evidently go on without limit. Moreover the two engines taken together form a purely self-acting system, for the whole power generated in one is spent on the other and is sufficient to drive the other; if we assume that there is no mechanical friction the double machine requires no help from without. Hence the supposition that the engine S could be more efficient than the reversible engine R has led to a result inconsistent with the second law of thermodynamics, for it has led us to construct, in imagination, a self-acting machine capable of transferring heat, in any quantity, from a cold body to a hot body. The Second Law (§ 25) asserts that this is contrary to all experience, and we are therefore forced to the conclusion that no other engine S can be more efficient than a reversible engine R when both work between the same limits of temperature. In other words, when the source and receiver of heat are given, a reversible heat-engine is as efficient as any engine working between them can be.

Further, let both engines be reversible. Then the same argument shows that neither can be more efficient than the other. Hence all reversible heat-engines taking in and rejecting heat at the same two temperatures are equally efficient.

48. Reversibility the Criterion of Perfection in a Heat-engine. These results imply that reversibility, in the thermodynamic sense, is the criterion of what may be called perfection in a heat-engine. A reversible engine is perfect in the sense that it cannot be improved on as regards efficiency: no other engine, taking in and rejecting heat at the same temperatures, will convert into work a greater fraction of the heat which it takes in. Moreover, if this criterion be satisfied, it is as regards efficiency a matter of complete indifference what is the nature of the working substance, or what, in other respects, is the mode of the engine's action.

49. Efficiency of a Perfect Heat-engine. Further, since all engines that are reversible are equally efficient, provided they work between the same temperatures, an expression for the efficiency of one will apply equally to all. Now, the engine whose efficiency was found in § 44, namely, an engine having a perfect gas for working substance and performing Carnot's cycle of operations, is one example of a reversible engine. Hence the expression which was obtained for its efficiency, namely,

$$\frac{T_1 - T_2}{T_1},$$

is the efficiency of any reversible heat-engine whatsoever taking in heat at T_1 and rejecting heat at T_2. And, as no engine can be more efficient than one that is reversible, this expression measures the efficiency of an engine that is ideally *perfect* in the thermodynamic sense. We have thus arrived at the immensely important conclusion that no heat-engine can convert into work a greater fraction of the heat which it receives than is expressed by the excess of the temperature of reception above that of rejection divided by the absolute temperature of reception.

50. Summary of the argument. Briefly recapitulated, the steps of the argument by which this result has been reached are as follows. After stating the experimental laws to which gases conform, and finding that they afforded a provisional means of defining temperature upon an absolute scale, we examined the action of a heat-engine in which the working substance took in heat when at the temperature of the source and rejected heat when at the temperature of the receiver, the change of temperature from one to the other of these limits being accomplished by adiabatic expansion and adiabatic compression. Taking a special case in which the engine had for its working substance a perfect gas, we found that its efficiency was $(T_1 - T_2)/T_1$ (§ 44). We also observed that it was, in the thermodynamic sense, a reversible engine (§ 46). Then we found, by an application of the second law of thermodynamics, that no heat-engine can have a higher efficiency than a reversible engine, when taking in and giving out heat at the same two temperatures T_1 and T_2; this was shown by the fact that a contrary assumption would lead to a violation of the second law (§ 47). Hence, we concluded that all reversible heat-engines receiving and rejecting heat at the same temperatures, T_1 and T_2 respectively, are equally efficient, and hence that the efficiency

$$\frac{T_1 - T_2}{T_1},$$

already determined for one particular reversible engine, is the efficiency of any reversible engine, and is a limit of efficiency which no engine whatever can exceed.

Another way of stating the performance of a perfect engine evidently is to say that the heat taken in Q_1 is to the heat rejected Q_2 as T_1 is to T_2, or

$$\frac{Q_1}{T_1} = \frac{Q_2}{T_2} \quad \text{..........................(13).}$$

The efficiency of any heat-engine may be stated as

$$\frac{Q_1 - Q_2}{Q_1} \text{ or } 1 - \frac{Q_2}{Q_1}:$$

in the perfect engine this becomes

$$1 - \frac{T_2}{T_1}.$$

51. Availability of Heat for transformation. Conditions of maximum efficiency. It follows that the availability of heat for transformation into work depends essentially on the range of temperature through which the heat is let down from that of the hot source to that of the cold body into which heat is rejected; it is only in virtue of a difference of temperature between bodies that conversion of any part of their heat into work becomes possible. No mechanical effect could be produced from heat, however great the amount of heat present, if all bodies were at a dead level of temperature. Again, it is impossible to convert the whole of any supply of heat into work because it is impossible to have a body at the absolute zero of temperature as the sink into which heat is rejected.

If T_1 and T_2 are given as the highest and lowest temperatures of the range through which a heat-engine is to work, it is clear that the maximum of efficiency can be reached only when the engine takes in all its heat at T_1 and rejects at T_2 all that is rejected.

With respect to every portion of heat supplied the greatest ideal efficiency is

$$\frac{\text{Temperature of reception} - \text{Temperature of rejection}}{\text{Temperature of reception}}.$$

If any heat be taken in at a temperature below T_1 or any rejected at a temperature above T_2 there will be less availability for conversion into work than if all had been taken in at T_1 and none rejected above T_2. With a given pair of limiting temperatures, it is essential to maximum efficiency that no heat be taken in by the engine except at the top of the range, and no heat rejected except at the bottom of the range. Further, as we have seen in § 47, when the temperatures at which heat is received and rejected are assigned, an engine attains the greatest possible efficiency only if it be reversible.

Hence in the design and working of a heat-engine reversibility should be aimed at. It is therefore important to inquire more particularly what kinds of action are reversible in the thermodynamic sense. A little consideration will show that a transfer of heat from the source to the working substance, or from the working substance to the receiver, is reversible only when the working substance is at sensibly the same temperature as the source or the receiver, as the case may be, and an expansion is reversible only when it occurs in such a manner that the expanding fluid does not waste energy by setting itself into eddying motion. This excludes what may be termed free expansion, such as that of the gas in Joule's experiment, § 35, and it excludes also what may be called imperfectly-resisted expansion, such as would occur if the fluid were allowed to expand into a chamber in which the pressure was less than that of the fluid, or if the fluid were expanding in a cylinder under a piston which rose so fast as to cause, through the inertia of the expanding fluid, local variations of pressure throughout the cylinder. A similar condition of course applies in regard to the compression of the working fluid: neither expansion nor compression must take place in such a manner as to set up eddies within the fluid[1].

[1] It is important in this connexion to distinguish between turbulent or eddying motion, which involves dissipation of energy, and stream-line motion, such as may be produced in the nozzles of a steam turbine. Except for practical imperfection due to such causes as friction, stream-line motion is compatible with reversibility.

To make a heat-engine, working within given limits of temperature, as efficient as possible the conditions to aim at therefore are—(1) to take in no heat except at the highest temperature, and to reject no heat except at the lowest temperature; (2) to secure that the working substance shall, when receiving heat, be at the temperature of the body from which the heat comes, and that it shall, when giving up heat, be at the temperature of the body to which heat is given up; (3) to avoid free or imperfectly-resisted expansion or other causes of turbulence. If these conditions are fulfilled the engine works reversibly and is the most efficient heat-engine possible within the given range of temperatures.

The first and second of these conditions are satisfied if in the action of the engine the working substance changes its temperature from T_1 to T_2 by adiabatic expansion, and from T_2 to T_1 by adiabatic compression, thereby being enabled to take in and reject heat at the ends of the range without taking in or rejecting any by the way. This is the action in Carnot's ideal engine (§ 43).

52. Perfect Engine using Regenerator. But there is another way in which the action of a heat-engine may be made reversible. Suppose that the working substance can be caused to deposit heat in some body within the engine while passing from T_1 to T_2, in such a manner that the transfer of heat from the substance to this body is reversible (satisfying the second condition above), then when we wish the working substance to pass from T_2 to T_1 we may reverse this transfer and so recover the heat that was deposited in this body. This alternate storing and restoring of heat would serve, instead of adiabatic expansion and compression, to make the temperature of the working substance pass from T_1 to T_2 and from T_2 to T_1 respectively. The alternate storing and restoring is an action occurring wholly within the engine, and is therefore distinct from the taking in and rejecting of heat by the engine.

In 1827 the Rev. Robert Stirling designed an apparatus, called a *regenerator*, by which this process of alternate storing and restoring of heat could be actually performed. For the present purpose it will suffice to describe the regenerator as a passage (such as a group of tubes) through which the working fluid can travel in either direction, whose walls have a very large capacity for heat, so that the amount alternately given to or taken from them by the working fluid causes no more than an insensible rise or fall in their temperature. The temperature of the walls at one end of the passage is T_1, and this falls continuously down to T_2 at the other end. When the working fluid at temperature T_1 enters the hot end and passes through, it comes out at the cold end at temperature T_2, having stored in the walls of the regenerator a quantity of heat which it will pick up again when passing through in the opposite direction. During the return journey of the working fluid through the regenerator from the cold to the hot end its temperature rises from T_2 to T_1 by picking up the heat which was deposited when the working fluid passed through from the hot end to the cold. The process is strictly reversible, or rather would be so if the regenerator had an unlimited capacity for heat, if no conduction of heat took place along its walls from the hot to the cold end, and if no loss took place by conduction or radiation from its external surface. A regenerator satisfying these conditions is of course an ideal impossible to realize in practice.

53. Stirling's Regenerative Air-Engine. Using air under fairly high pressure as the working substance, and employing his regenerator, Stirling made an engine which, allowing for practical imperfections, is the earliest example of a reversible engine. The cycle of operations in Stirling's engine was substantially this:—

(1) Air (which had been heated to T_1 by passing through the regenerator) was allowed to expand isothermally through a ratio r, taking in heat from a furnace and raising a piston. Heat taken in (per lb. of air) $= RT_1 \log_\epsilon r$.

(2) The air was caused to pass through the regenerator from the hot to the cold end, depositing heat and having its temperature lowered to T_2, without change of volume. Heat stored in regenerator $= K_v (T_1 - T_2)$. The pressure of course fell in proportion to the fall in temperature.

(3) The air was then compressed isothermally to its original volume at T_2 in contact with a cooler (or receiver of heat). Heat rejected $= RT_2 \log_\epsilon r$.

(4) The air was again passed through the regenerator from the cold to the hot end, taking up heat and having its temperature raised to T_1. Heat restored by the regenerator $= K_v (T_1 - T_2)$. This completed the cycle.

The efficiency is $\dfrac{RT_1 \log_\epsilon r - RT_2 \log_\epsilon r}{RT_1 \log_\epsilon r} = \dfrac{T_1 - T_2}{T_1}$.

The indicator diagram of this action is shown in fig. 15. Stirling's engine is important, not as a present-day heat-engine (though it has been revived in small forms after a long interval of disuse), but because it is typical of the only mode, other than Carnot's plan of adiabatic expansion and compression, by which the action of a heat-engine can be made reversible.

The regenerative principle has been largely used in metallurgy and other industrial processes as a means of economizing fuel and attaining high temperatures: the Siemens steel-furnace is an example of its application on a large scale. It has also supplied a means of reaching the lowest extremes of temperature, as in Linde's method of liquefying air. Notwithstanding the immensely valuable services which the regenerator has rendered in such processes, its application to heat-engines has hitherto been very limited. Another way of using it in air-engines was tried by Ericsson, who kept the pressure instead of the volume constant while the working air was passed through the regenerator, thus getting an indicator diagram made up of two isothermal lines and two lines of constant pressure. Attempts have also been made by Siemens and by Fleeming Jenkin to apply it to steam-engines and to gas-engines. It finds application in some refrigerating machines, or reversed heat-engines, to which reference will be made in a later chapter. A regenerative method of heating the feed-water is used (Chap. VIII) in some large steam turbines. But, speaking broadly, actual heat-engines, in so far as they can be said to approach the condition of reversibility, do so, not by the use of the regenerative principle, but by more or less nearly adiabatic expansion and compression after the manner of Carnot's ideal engine.

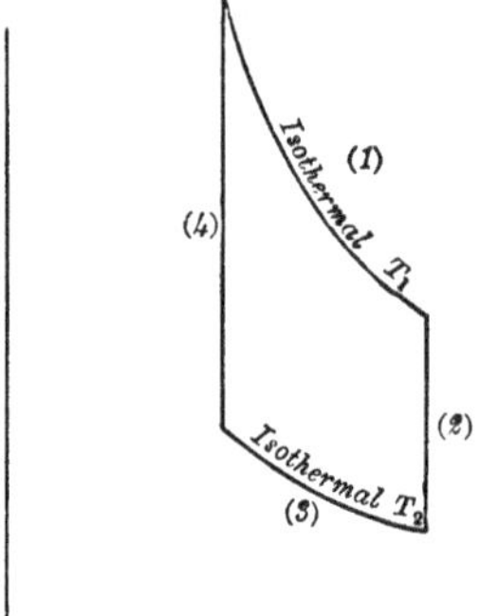

FIG. 15. Ideal Indicator diagram of Air-engine with Regenerator (Stirling).

54. Absolute Zero of Temperature. Throughout our discussion of Carnot's engine the zero from which T_1 and T_2 are measured is the zero of the gas thermometer, which was defined (§ 33) as the temperature at which the volume of the gas would vanish if the same law of expansion continued to apply. But we can now give it another meaning. Taking the expression for the efficiency of a reversible heat-engine

$$1 - T_2/T_1,$$

we see that if the cold receiver were at the temperature of the absolute zero (so that $T_2 = 0$) the efficiency would be equal to 1: in other words, all the heat supplied to the engine would be converted into work. It is clearly impossible to imagine a receiver colder than that, for it would cause the efficiency to be greater than 1 and thereby violate the First Law of Thermodynamics by making the amount of work done greater than the heat supplied. Hence the zero which we found on the gas scale is also an absolute thermodynamic zero, a temperature so low that it is inconceivable on thermodynamic grounds that there can be any lower temperature. The term "absolute zero" has consequently acquired a new meaning: without reference to the properties of any substance we see that it represents a limit below which temperature cannot go. This consideration justifies us in calling the zero of the perfect gas thermometer an absolute zero of temperature.

CHAPTER III

PROPERTIES OF STEAM AND ELEMENTARY THEORY OF THE STEAM-ENGINE

55. Formation of Steam under Constant Pressure. We have now to consider the action of heat-engines in which the working substance is water and water-vapour or steam, and as a preliminary to this it is necessary to give some account of the physical properties of steam as determined by experiment. The properties of steam are most conveniently stated by referring in the first instance to what happens when steam is formed *under constant pressure*. This is substantially the process which occurs in the boiler of a steam-engine when the engine is at work. To fix the ideas we may suppose that the vessel in which steam is to be formed is a long upright cylinder fitted with a frictionless piston which may be loaded so that it exerts a constant pressure on the fluid below. Let there be, to begin with, at the foot of the cylinder a quantity of water, which for convenience of statement we shall take as one unit of mass, namely 1 lb., at temperature 0° C.; and let the piston rest on the surface of the water with a pressure P, which includes the pressure of the atmosphere. Let heat now be applied to the bottom of the cylinder. As it enters the water it will produce the following effects in three stages:—

(1) The temperature of the water rises until a certain temperature t is reached, at which steam begins to be formed. The value of t depends on the particular pressure P which the piston exerts. Until the temperature t is reached there is nothing but water below the piston.

(2) Steam is formed, more heat being taken in. The piston, which is supposed to continue to exert the same constant pressure, rises. No further increase of temperature occurs during this stage, which continues until all the water is converted into steam. During this stage the steam which is formed is said to be *saturated*. The volume which the piston encloses at the end of this stage—the volume, namely, of unit mass of saturated steam at pressure P and consequently at temperature t—will be denoted by V_s.

(3) If after all the water has been converted into steam more heat be allowed to enter, the volume will increase and the tem-

perature will rise. The steam is then said to be *superheated*: its temperature is above the temperature of saturation.

56. Saturated and Superheated Steam. The difference between saturated and superheated steam may be expressed by saying that if water (at the temperature of the steam) be mixed with steam some of the water will be evaporated if the steam is superheated, but none if the steam is saturated. Any vapour in contact with its liquid and in thermal equilibrium is necessarily saturated. When saturated its properties differ considerably, as a rule, from those of a perfect gas, but when superheated they approach those of a perfect gas more and more closely the farther the process of superheating is carried, that is to say, the more the temperature is raised above t, the temperature of saturation corresponding to the given pressure P. Saturated steam at a given pressure can have but one temperature; superheated steam at the same pressure can have any temperature higher than that.

57. Relation of Pressure and Temperature in Saturated Steam. The temperature t is called the temperature of saturation for the given pressure. The relation of this temperature to the pressure was determined with great care by Regnault, in a series of experiments to which a great part of our knowledge of the properties of steam is due[1]. Regnault's observations extended from temperatures below the zero of the centigrade scale, where the vapour whose pressure was measured was that given off by ice, up to 220° C. The pressures found by him, expressed in millimetres of mercury, were as follows, omitting those below 0° C. as not relevant to steam-engine calculations:—

Temperature °C.	Pressure of saturated steam in mm. of Mercury
0	4·60
25	23·55
40	54·91
50	91·98
75	288·50
100	760·00
130	2030·0
160	4651·6
190	9426
220	17390

[1] *Mem. Inst. France*, 1847, vol. XXI. An account of Regnault's methods of experiment and a statement of his results expressed in British measures will be found in Dixon's *Treatise on Heat* (Dublin, 1849).

It will be seen from these figures that the pressure of saturated steam rises with the temperature at a rate which increases rapidly in the upper regions of the scale. Various empirical formulas have been devised to express the relation of pressure to temperature in saturated steam and to allow tables to be calculated in which intermediate values are shown. When a table is available, however, it is more convenient to find the pressure corresponding to a given temperature, or the temperature corresponding to a given pressure directly from it, either interpolating or drawing a portion of the curve connecting pressure with temperature when the values concerned lie between those that are stated in the table.

58. Tables of the Properties of Steam. At the end of this book a number of tables will be found showing not only the relation of the pressure to the temperature of saturation, but also various other properties of steam which are of use in engineering calculations. Tables of the properties of steam have been calculated by Professor Callendar, by methods explained in the Appendix, and have been published under the title of *The Callendar Steam Tables*. From Callendar's Tables, which give the most authoritative results now available, a selection has been made, with his permission and that of his publishers[1], for the purposes of this book.

The figures which are given for the pressure of saturated steam are not taken directly from the measurements of Regnault but are inferred from a characteristic equation due to Callendar, the validity of which is demonstrated by the general agreement of the quantities calculated from it with the best experimental results in measurements not only of the pressure but of other properties of steam. The pressures, however, which are stated in the table do agree very closely with the results of Regnault's observations quoted above. It is only at the highest pressures that an appreciable difference will be found, and even there it is not material.

In other respects the tables given here will be found to differ

[1] Edward Arnold and Co. For a fuller account of Callendar's method see the present writer's *Thermodynamics for Engineers* (Camb. Univ. Press, 1920). Since Callendar compiled his original Steam Tables in 1915 he has issued (through the same publishers) two abridged versions in Centigrade and Fahrenheit units, and also (in 1924) *The Enlarged Callendar Tables* which extend the figures for superheated steam to a pressure of 2000 lb. per sq. in. and a temperature of 1000° F.

somewhat widely from the earlier tables of such authorities as Rankine[1] or Zeuner[2], which used to be accepted as standards and copied into many text-books. When these were calculated the only available data of value were those furnished by the experiments of Regnault. But more recent researches have supplied additional data which in some particulars modify his, and it is recognized that Regnault's figures require revision and in some cases considerable amendment. The various properties of steam are linked together in such a manner that relations between them can be expressed in the form of a number of thermodynamic equations. When tested in this way the figures given in the old tables are found to be mutually inconsistent. The formulas of Callendar give a set of values which are consistent amongst themselves and are also in good agreement with the most trustworthy experimental results. Further researches may in time lead to a still closer adjustment of the figures to the results of observation, but the values in Callendar's Tables may be accepted as very much more correct than those that were formerly available[3].

59. Relation of Pressure and Volume in Saturated Steam. Among the quantities shown in the tables is the volume V_s in cubic feet per lb. at various temperatures and at various pressures. The volume of a given quantity of saturated steam at any assigned pressure is a quantity difficult to measure by direct experiment, and the volumes which are given in steam tables are generally inferred from the results of experiments on other properties of steam which can be more easily measured. Successful measurements of volume have however been carried out in the laboratory of the Technical High School at Munich[4] and the results are in general agreement with the figures stated in the table. Reference must again be made to the Appendix as to the manner in which the tabulated figures are obtained.

[1] Rankine, *A Manual of the Steam Engine and other Prime Movers.*

[2] Zeuner, *Technische Thermodynamik*, vol. II.

[3] Tables giving the properties of steam in metric units, calculated by R. Mollier on the basis of Callendar's methods, were included in former editions of this work.

[4] O. Knoblauch, R. Linde and H. Klebe, *Mitteilungen über Forschungsarbeiten herausgegeben vom Verein deutscher Ingenieure*, Heft 21, 1905. For references to other experimental data see Callendar's book on *The Properties of Steam*, 1920.

The relation of P to V_s in saturated steam may be approximately expressed by the formula

$$PV_s^{\frac{16}{15}} = \text{constant}.$$

When the pressure is stated in kilogrammes per square centimetre and the volume in cubic metres per kilogramme, this may be written

$$PV_s^{\frac{16}{15}} = 1{\cdot}786.$$

With P in lb. per square inch and V_s in cubic feet per lb. it becomes

$$PV_s^{\frac{16}{15}} = 490.$$

The student will find it useful to draw curves, with the data of the table, showing the relation between the pressure and the temperature of saturated steam and also the relation of pressure to volume, especially within the range usual in steam-engine practice. He will observe that $\frac{dP}{dt}$, the rate of change of pressure with respect to change of temperature, increases rapidly as the temperature rises, and hence that in the upper part of the range a very small elevation of temperature in a boiler is necessarily associated with a large increment of pressure. The familiar case of water boiling in a kettle or other open vessel is only a special case of the formation of steam under constant pressure. There the constant pressure is that of the atmosphere, which is 14·7 lb. per square inch or thereabouts (as indicated by the barometer) and consequently the temperature at which the water boils is about 100° C.[1]

The pressure shown by a pressure-gauge on a boiler is the excess of pressure in the boiler above the pressure of the atmosphere. Consequently the true or "absolute" pressure in the boiler is to be found by adding, to the reading of a correct gauge, the pressure which corresponds to the height of the barometer at the time; this is generally about 14·7 lb. per square inch or 1·033 kilogrammes per square centimetre.

Water in the open *evaporates* slowly at any temperature lower than that at which it boils. Though the pressure of the vapour so

[1] Water in the open boils at 100° C. when the atmospheric pressure has its standard value, which corresponds to a barometer reading (corrected to 0° C.) of 760 mm. at sea level in latitude 45° or 759·6 mm. in London. This is equivalent to 14·689 lb. per square inch in London.

formed is lower than that of the atmosphere—and may be very much lower—the vapour is able to escape from the surface by diffusion: the atmosphere is not displaced and the pressure on the surface of the water is still that of the air. As the temperature of water in the open is raised this slow evaporation from the surface becomes more rapid, but it is only when the temperature reaches the value which corresponds (for saturated steam) to the given atmospheric pressure that the water boils: the vapour is then formed in bubbles at the pressure of the atmosphere, and it escapes not by diffusion but by displacing the superincumbent air.

60. Supply of Heat in the formation of Steam under Constant Pressure. Heat of the Liquid and Latent Heat. We have next to consider what quantity of heat is taken in, in the imaginary experiment of § 55, when water is first heated to the boiling-point and then converted into steam under a constant pressure P.

In stating quantities of heat we shall, as was explained in § 24, use for thermal unit one-hundredth part of the quantity of heat that is required to warm 1 lb. of water from 0° C. to 100° C., under a pressure of one atmosphere. In the *first stage* the substance is wholly in the condition of water which is being heated from the initial temperature to T_2, the temperature at which the second stage begins. During this first stage the heat taken in (per lb. of the water) is approximately equal to one thermal unit for each degree by which the temperature of the water rises. It would be exactly equal to that if the specific heat of water were constant and equal to unity, but this is not the case. At about 30° C. the specific heat of water is less than unity; it passes a minimum value thereabouts of 0·9967, and then increases, becoming appreciably greater than unity at such temperatures as are found in steam boilers. Thus for instance to heat 1 lb. of water from 0° C. to 80° C. requires 79·9 thermal units instead of 80. On the other hand, to heat it from 0° C. to 200° C., under a pressure sufficient to prevent steam from forming, requires nearly 203·2 thermal units instead of 200. These figures will indicate how far it is legitimate to estimate the heat taken in during the first stage as one unit per degree. More accurate values of the heat of the liquid, that is to say the heat taken in during the first stage, can be found by means of the Steam Tables (see § 63).

During this first stage, while the substance is still liquid, nearly all the heat that is taken in goes to increase the stock of internal energy. There is scarcely any external work done, for the volume is only slightly increased. Thus in heating water from 0° C. to 200° C. (under a pressure of 225·24 lb. per sq. inch) the volume of the water changes from 0·0160 cub. ft. per lb. to 0·0185. The external work done during this heating is therefore 225·24 × 144 × 0·0025 or 81 foot-pounds. This is equivalent to barely 0·06 thermal unit, and is negligible in comparison with the 203·2 units of heat that are taken in.

In the *second stage*, the water changes into saturated steam without change of temperature. The heat that is taken in during this stage constitutes what is called the Latent Heat of steam, at the given pressure. We shall denote it by L. Values of the latent heat for various pressures are given in the tables. For steam formed under a pressure of one atmosphere (saturation temperature 100° C.) the latent heat is 539·3: with lower pressures of formation it is greater, and with higher pressures it is less. At the end of the second stage the substance contains no liquid; it is spoken of as dry saturated steam: at any earlier point, when the substance consists partly of saturated steam and partly of water, it may be spoken of as wet steam.

The latent heat of a vapour may be defined as the amount of heat which is taken in by unit mass of the liquid while it all changes into saturated vapour under constant pressure, the liquid having been previously heated up to the temperature at which the vapour is formed.

A considerable part of the heat taken in during this process is spent in doing external work, since the substance expands against the constant pressure P. It is only the remainder of the so-called latent heat L that can be said to remain in the fluid and to constitute an addition to its stock of internal energy. The amount spent in doing external work during the second stage is

$$AP\,(V_s - V_w),$$

where V_s is the volume of the saturated vapour and V_w is the volume of the liquid at the same temperature and pressure, A being the factor for converting units of work into thermal units, namely 1/1400 when the unit of work is the foot-pound. The excess of L above this quantity measures the amount by which the internal energy increases during the second stage.

Thus for instance when water at 200° C. and a pressure of 225·24 lb. per sq. inch is converted into steam, the volume changes from 0·0185 cub. ft. to 2·0738 cub. ft.; 467·41 thermal units are taken in, of which 47·61 units are spent in doing external work and 419·8 units go to increase the stock of internal energy.

61. Change of Internal Energy, and External Work done. In the two stages together the whole amount of external work done is to be found by taking the whole increase of volume and multiplying it by the pressure. Since the water was assumed to be originally at 0° C. its initial volume was 0·0160. In converting water from 0° C. to saturated steam at 200° C. under constant pressure the external work done is found thus to be equivalent to 47·67 thermal units: this is 0·06 unit more than the external work of the second stage, for it includes the small amount already referred to as having been done during the first stage. The whole increase of internal energy, in the process of passing from the condition of water at 0° C. to that of saturated steam at any temperature, is equal to the whole amount of heat taken in, less the equivalent of the external work done. This in fact is only a particular example of the general principle stated in § 29, that when a substance expands in any manner, taking in heat and doing work, the heat taken in is equal to the work done *plus* the increase of internal energy. In the case here considered the action is going on under constant pressure, but the statement applies to any change of state whatever.

No matter what changes a substance may undergo, its internal energy will return to the same value when the substance returns to the same condition in all respects. In other words the internal energy is a function of the actual state of the substance and is independent of the way in which that state has been reached. Thus the internal energy of 1 lb. of saturated steam at a particular pressure is a definite quantity which is the same whether the steam has been formed by boiling under constant pressure or in any other manner. Steam formed in a closed vessel of constant volume, for example, would have the same internal energy as steam at the same pressure but formed under conditions of constant pressure, though the amount of heat taken in during its formation would be different, for no external work is done in the process of formation in a closed vessel of constant volume. In that case the heat taken in would be equal to the change of internal energy.

We have no means of measuring the total stock of internal energy in a substance, and can deal only with changes in the stock. But by taking some arbitrary starting-point as a zero from which the internal energy E is reckoned we can give E numerical values for any other states of the substance. These values really express differences from the internal energy in the zero state. The usual convention is to write $E = 0$ when the substance is in the liquid condition at a temperature of 0° C., and at a pressure equal to the vapour-pressure corresponding to that temperature. We may call this, for brevity, the zero state of the substance.

Following this convention we take $E = 0$ for water at 0° C. The value of E for saturated water-vapour at 0° C. will then be 564·21 thermal units. That this agrees with other figures given in the tables will be seen by considering the conversion of water at 0° C. to steam at 0° C. under constant pressure. The only heat taken in is L, which is 594·27 units, and of this the external work $AP(V_s - V_w)$ represents 30·06 units: the difference measures E.

Values of E for saturated steam at various temperatures are given in the tables. It will be seen that they increase slowly with the temperature.

62. The "Total Heat" of a fluid. We come now to another function of the state of any substance, a function which is of very great use in thermodynamic calculations. It is called the *Total Heat* and is represented[1] by the letter I.

The total heat I is defined for any state of the substance by the equation

$$I = E + APV.$$

That is to say I is equal to the sum of the internal energy and the external work which would be done if the substance could be imagined to start from no volume at all and to expand to its actual volume, under a constant pressure equal to its actual pressure. Since the pressure, volume, and internal energy are all functions of the actual state, I is also a function of the actual state: its value is independent of how the state has been reached. In steam, for example, the amount of heat that is taken in during formation

[1] Callendar in his Tables uses H to represent this function. In view of the fact that Rankine and other writers have used H in another sense a different symbol is to be preferred, and I is generally used.

depends on how the steam is formed, but the "total heat" I depends only on the final condition. The total heat can be calculated for any condition of a substance, whether in the state of liquid or of saturated or superheated vapour. It is measured in thermal units per lb. Values of the total heat of saturated steam I_s and also of water under saturation pressure I_w at various temperatures are given in the tables. They show the total heat of steam increasing progressively with the temperature, rather more rapidly than does the internal energy.

It follows from the definition of I that in the zero state of any substance, at which E is reckoned to be zero, I is not equal to zero but to a small positive quantity depending on the volume of the liquid and its pressure at that state. Since E is then zero I is equal to AP_0V_0, where P_0 is the pressure at the zero state, namely the vapour-pressure at 0° C., and V_0 is the volume of the liquid at 0° C. and pressure P_0. For water this quantity AP_0V_0 is quite negligible, amounting as it does to 0·000146 thermal unit. Hence in the Steam Tables, in the column for I_w, the total heat of water at 0° C. is given as 0.

An important property of the function I, for steam or any other substance, is that when the substance is heated under constant pressure the change of I is equal to the amount of heat taken in. To prove this, let Q be the amount of heat taken in while the substance expands under constant pressure P from a state in which the volume is V_1 and the internal energy is E_1 to another state in which the volume is V_2 and the internal energy is E_2. Then the amount of external work done is $P(V_2 - V_1)$ and, by the conservation of energy,

$$Q = E_2 - E_1 + AP(V_2 - V_1),$$

which may be written

$$Q = E_2 + APV_2 - (E_1 + APV_1)$$

or

$$Q = I_2 - I_1,$$

where I_1 is the total heat in the first state and I_2 is the total heat in the second state.

63. Application to steam formed under constant pressure from water at 0° C. The above proposition applies to every stage of the imaginary experiment of § 55. In that experiment we assumed that to begin with there was under the piston

1 lb. of water at 0° C. and at the pressure P at which steam was to be formed. By definition of the total heat,

$$I = E + APV.$$

E at the beginning may be taken as zero[1]. Hence the value of I for the water at 0° C. may be taken as APV_0, where V_0 is the volume of 1 lb. of water at 0° C. and P is the pressure at which steam is to be formed. At the end of the first stage

$$I_w = Q_1 + APV_0,$$

where I_w represents the value of I for water at the temperature at which steam is about to form[2], and Q_1 is the heat taken in during the first stage. When values of I_w are known this allows Q_1 to be more accurately calculated than by the rough method of allowing one unit of heat for each degree of rise in temperature. Values of I_w are included in the Steam Tables.

During the second stage an amount of heat equal to L is taken in at constant pressure, and the total heat changes from I_w to I_s, where I_s is the total heat of saturated steam. Hence

$$\begin{aligned} I_s &= L + I_w \\ &= L + Q_1 + APV_0. \end{aligned}$$

64. Heat of Formation. The quantity $Q_1 + L$ is the whole amount of heat taken in while 1 lb. of saturated steam is formed under constant pressure, starting from the condition of water at 0° C. This is called the *heat of formation* of steam under constant pressure. It is equal to $I_s - APV_0$.

Thus the total heat of steam may be regarded as equal to the whole heat of formation under constant pressure, *plus* a small quantity which is the thermal equivalent of the work that would be done in lifting the piston far enough to admit the original volume of the water. The quantity APV_0 forms a very small part of the whole: it is only 0·37 thermal unit when the temperature of formation is 200° C., corresponding to a pressure of 225 lb. per sq. inch, and it is much less at lower pressures.

A tabular scheme follows which will serve to show how the total

[1] The convention of § 61 makes $E = 0$ for water at 0° C. and pressure P_0. Here we have water at 0° C. and pressure P, which is higher than P_0: but the higher pressure does not cause the internal energy of water at 0° C. to differ appreciably from zero.

[2] In Callendar's Tables this quantity I_w is written h.

heat of saturated steam is related to the heat of formation under constant pressure. But the student should accustom himself to think of the total heat without reference to any process of formation, as a property which the steam possesses in its actual state—a property which is just as simply a function of the state as is the temperature, or the pressure, or the volume, or the internal energy[1].

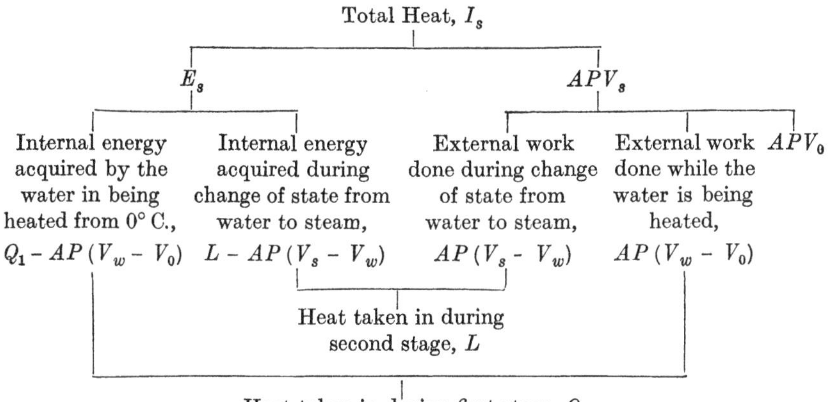

The scheme accordingly shows, as before,

$$I_s = Q_1 + L + APV_0.$$

E_s and V_s are, of course, the internal energy and the volume of steam in the saturated states.

It will help towards an understanding of these quantities, and of the use of the tables, to consider a numerical example. Suppose that a boiler is supplying saturated steam at a pressure of 250 lb. per sq. inch. By Table B the temperature of the steam is 205·1° C., its total heat I_s is 672·07 calories and its latent heat L is 463·00 calories. Hence I_w, which is equal to $I_s - L$, is 209·07 calories. This agrees with the figure for I_w which might be found from Table A* by interpolating between 200° and 210° C.

If the feed-water were supplied at 0° C. the heat of formation,

[1] The function here called the total heat I, namely $E + APV$, was introduced by Willard Gibbs (*Trans. of the Connecticut Academy*, vol. III; *Collected Scientific Papers*, vol. I, p. 92), and was first called the "Total Heat" by Callendar (*Phil. Mag.* 1903, vol. V, p. 50). Its great importance in technical thermodynamics was emphasized by Mollier, who employed it in charts for exhibiting the properties of steam and other substances. The use of such charts is described in Chapter V.

or quantity of heat required to convert each pound into saturated steam, would be

$$I_s - APV_0 = 672{\cdot}07 - \frac{250 \times 144 \times 0{\cdot}016}{1400}$$

$$= 672{\cdot}07 - 0{\cdot}41 = 671{\cdot}66 \text{ calories.}$$

Of this, the quantity $I_w - APV_0$ or 208·66 would be taken in during the first stage, namely, while the water is being warmed from 0° C. to the temperature at which steam is formed (205·1° C.) and the quantity L or 463·0 during the second stage.

In general, however, the feed-water is supplied at some higher temperature than 0° C. The amount of heat which has then to be taken in is readily found from the fact that it is equal to the change in the total heat I which the substance undergoes in passing, at constant pressure, from the state of water at the temperature of supply to the state of steam. The values of I_w in Table A* give the total heat of water at the pressure of saturation corresponding in each case to the stated temperature, a pressure which we may call P_w to distinguish it from the (much higher) boiler pressure P to which the feed-water is raised in the feed-pump. This raising of pressure increases the value of I from I_{wf} to $I_{wf} + A(P - P_w) V_{wf}$, where I_{wf} is the tabulated value of I_w at the temperature of the feed, and V_{wf} is volume of 1 lb. of water at that temperature. Hence the heat that has to be supplied (per lb.) in the boiler is

$$I_s - \{I_{wf} + A(P - P_w) V_{wf}\}.$$

The term $A(P - P_w) V_{wf}$ is so small that it is often omitted in making this calculation, and the heat of formation is reckoned simply as $I_s - I_{wf}$.

Suppose, for example that the feed-water is supplied at 30° C., the boiler pressure being 250 lb. as before. Then P_w is 0·34 lb. per sq. inch and the volume of the feed-water, per lb., is 0·0161 cub. ft., which makes $A(P - P_w) V_{wf} = 0{\cdot}41$. By Table A* I_{wf} is 29·91. Hence the number of heat units taken up in the boiler by each lb. of water that is converted into steam is

$$672{\cdot}07 - (29{\cdot}91 + 0{\cdot}41) = 641{\cdot}75.$$

65. Wet steam. In calculations which relate to the action of steam in engines we are often concerned, not with *dry* saturated steam, but with *wet* steam, or steam which either carries in suspension, or is otherwise mixed with, a greater or less proportion of water. In the cylinder of a steam-engine, for example, the working substance is a wet mixture, and the proportion of water to steam in it varies as the stroke proceeds. When any such mixture is in a state of thermal equilibrium the steam and water have the same temperature, and the steam is saturated. The *dryness* of wet steam is measured by the fraction q of dry steam in each unit of mass of the mixed substance. When the dryness is known it is easy to determine the other physical constants. Thus, reckoning in every case per lb. of the mixture, we have

Latent heat of wet steam $= qL$;

Total heat of wet steam $= I_w + qL = qI_s + (1 - q) I_w$;

Volume of wet steam $= qV_s + (1 - q) V_w$

$= qV$ very nearly, unless the steam is so wet as to consist mainly of water.

66. Superheated steam. When steam is superheated (as in the third stage of the imaginary experiment of § 55) by continuing the heating process under constant pressure after the saturated condition has been reached, the total heat I becomes increased above the value I_s by an amount equal to the heat that is taken in during the process of superheating. The volume is also increased, nearly in proportion to the rise in temperature.

Tables and formulas are given in the Appendix which allow the volume and the total heat of superheated steam to be ascertained, when the pressure and the temperature are known, throughout the range that applies in practice. Sometimes, however, it may be convenient to calculate the total heat in a different manner, by finding first the total heat of steam in the saturated state at the same pressure, and then adding a quantity which represents the heat taken in during the process of superheating. This requires a knowledge of the specific heat of the vapour, which in general varies as the temperature rises from the point of saturation. What is needed therefore is to know the average specific heat over the range through which superheating takes place. The table below (adapted from Mollier) gives values of this average specific heat for various ranges of superheating at various pressures, or in other words starting from various saturation temperatures. In each case the pressure is assumed to be constant while the steam is being superheated, a condition which is in fact generally fulfilled in steam-engine practice. The values given in the table are accordingly the mean specific heats at constant pressure, in each case for a range of temperature extending from the temperature of saturation t up to the temperature t' to which superheating is carried.

Table of Mean Specific Heat of Steam in superheating at constant pressure from the saturation temperature t to temperature t'.

Temperature of superheat t'	Temperature of Saturation °C.				
°C.	80°	120°	160°	180°	200°
100°	·49				
150°	·49	·51			
200°	·49	·51	·54	·57	
250°	·48	·50	·53	·56	·59
300°	·48	·50	·52	·54	·57
350°	·48	·49	·51	·53	·56
400°	·48	·49	·51	·52	·55
450°	·48	·49	·51	·52	·54

Taking κ as the mean specific heat from this table we may readily find the total heat of superheated steam by the equation

$$I' = I_s + \kappa (t' - t),$$

where I' and t' refer to the superheated condition and I_s and t to the saturated condition at the same pressure. As κ changes only slowly it is easy by inspection of the table to find a value suitable for any assigned conditions as to initial pressure and degree of superheat.

In practice, superheating is usually done at constant pressure: the steam on leaving the boiler passes through a group of tubes in which the superheat is given to it while its pressure remains equal (or closely equal) to that in the boiler. Superheating has now become a very usual feature in steam-engine practice, for reasons which will be apparent later when the actual behaviour of steam in the engine is discussed in detail. It is rarely carried further than 400° C. and not often so far.

The *heat of formation* of superheated steam when formed in this way is equal, by the principle explained in § 62, to the difference between the total heat in the superheated state (I') and that of the feed-water at the same pressure. It is therefore equal to

$$I' - \{I_{wf} + A\,(P - P_w)\,V_{wf}\},$$

where the quantity in brackets has the meaning explained in § 63.

It is generally best to find I' directly, in any condition of superheating, by using a table such as Table D (Appendix) which gives the total heat for various temperatures and pressures. In dealing with super-heated steam it is important to realize that when the temperature and pressure are known the material is fully defined, and all the other properties, such as volume, total heat, energy, and so forth, can be calculated from the equations in the Appendix without consideration of the particular starting-point from which superheating began.

67. Isothermal lines for wet steam. The expansion of volume which occurs during the conversion of water into steam under constant pressure—the second stage of the process described in § 53—is isothermal. From what has been already said it is obvious that a mixture of steam and water can be expanded or compressed isothermally only under constant pressure, and that evaporation (in the one case) or condensation (in the other) must accompany the process. Isothermal lines for a working substance which consists of a liquid and its vapour are straight lines of uniform pressure.

68. Adiabatic lines for wet steam. The form of adiabatic lines for a working substance consisting of a vapour or of a mixture of a liquid and its vapour depends not only on the particular fluid, but also on the proportion of liquid to vapour in the mixture. When steam initially in the dry saturated state is allowed to expand adiabatically it becomes wet, and if initially wet (unless very wet) it becomes wetter. To keep steam dry while it expands, doing work, some heat must be supplied during the process of expansion. If the expansion is adiabatic, so that no heat reaches the expanding fluid, a part of the steam is condensed, forming either minute particles of water suspended throughout the mass or a dew upon the surface of the containing vessel. The temperature and pressure fall; and, as that part of the substance which remains uncondensed is saturated, the relation of pressure to temperature throughout the expansion is that which holds for saturated steam. The following approximate formula, a proof of which will be given in the next chapter, shows the extent to which condensation takes place during adiabatic expansion, and so allows the relation of pressure to volume to be determined.

Before expansion, let the initial dryness of the steam be q_1 and its absolute temperature T_1. Then, if it expand adiabatically until its absolute temperature falls to any value T, its dryness after expansion is (nearly)

$$q = \frac{T}{L}\left(\frac{q_1 L_1}{T_1} + \log_\epsilon \frac{T_1}{T}\right).$$

L_1 and L are the latent heats (in thermal units) of unit mass of steam before and after expansion respectively. When the steam is dry to begin with, $q_1 = 1$.

This formula is valid only if the steam does not become supersaturated during its expansion, in other words, it assumes that there is, at every stage, thermal equilibrium between the part which is condensed and the part which is still vapour, so that the part which is vapour remains saturated throughout the process. It does not directly give the relation of pressure to volume, but it allows the dryness at any stage of the process to be calculated, and from that (together with the fact that the part which remains in the condition of vapour is saturated) it is easy to find the volume which the mixture will fill when its pressure has changed to any assigned value. An example may help to make this clear. Suppose for instance that originally dry saturated steam at an absolute pressure of 182·1 lb. per sq. inch is made to expand adiabatically. Its initial temperature is 190° C. We wish to find the relation of pressure to volume at any stage in the expansion. Take for example the stage which is reached when the temperature has fallen to 120° C. We then have

$$T = 120 + 273{\cdot}1 = 393{\cdot}1.$$

By Table A

$$L = 526{\cdot}85.$$

Also $q_1 = 1, \quad T_1 = 190 + 273{\cdot}1 = 463{\cdot}1,$

$$L_1 = 475{\cdot}82.$$

Hence $$q = \frac{393{\cdot}1}{526{\cdot}85}\left(\frac{1 \times 475{\cdot}82}{463{\cdot}1} + 2{\cdot}303 \log_{10} \frac{463{\cdot}1}{393{\cdot}1}\right)$$

$$= 0{\cdot}89.$$

This means that by the time the temperature has fallen to 120° C. eleven per cent. of the originally dry steam has become condensed into water. The pressure is then 28·8 lb. per sq. inch by the table. The volume occupied by that part of the substance which is still in the state of steam is qV per lb. of the mixture, where V is the volume of 1 lb. of dry steam at that pressure. Taking the value of V given in the table, namely, 14·27 cub. ft., qV is 12·68 cub. ft. To obtain the whole volume of 1 lb. of the working substance we ought in strictness to add to this the volume occupied by that part which has been converted into water, namely, by the fraction of a pound which is represented by $1 - q$. But this is only 0·11 lb., and its volume is $0{\cdot}11 \times 0{\cdot}017$ or 0·0019 cub. ft.[1]—a quantity which is negligible in comparison with the volume occupied by the still uncondensed steam. We conclude that 12·68 cub. ft. is the volume of the mixture (per lb.) when its pressure has fallen to 28·8 lb. per sq. inch by adiabatic expansion; in other words, these numbers determine one point on the adiabatic line which begins with dry saturated steam at a pressure of 182·1 lb. per sq. inch.

In the same way we may go on to find as many points on an adiabatic line as we please, by taking a series of pressures, each lower than the initial pressure, first finding q for each, and then from q calculating the volume of the mixture, which in ordinary cases is practically equal to qV, V being the volume which unit mass of saturated steam would occupy at the same pressure and temperature.

The steam may be wet to begin with, and if q_1 have a value much less than unity it will be found on working out examples that q may turn out greater than q_1. This means that in very wet steam adiabatic expansion may reduce the amount of water as the net result of two opposing actions: as the temperature falls during expansion part of the steam initially present becomes condensed; on the other hand, part of the water initially present becomes evaporated because its initial temperature is higher than the temperature which the mixture takes as it expands. With very wet steam the result may be on the whole to make the mixture become drier. An extreme case occurs when all the substance is in the state of water to begin with. Then if adiabatic expansion

[1] The volume of 1 lb. of water, which is 0·016 cub. ft. when cold, rises to 0·017 at about 120° C. and to 0·018 at about 180° C.

be allowed to take place steam is formed, and the above formula may be applied, by writing $q_1 = 0$, to find how much of the water will be evaporated when the pressure, or the temperature, has fallen to any assigned value.

It will be shown in Chapter v how, by using the idea of *Entropy*, which is a property that does not change during adiabatic expansion, we obtain an exact method of determining the wetness at any stage of the process when steam is expanded or compressed adiabatically, and also how the same results may be arrived at by graphic methods.

These graphic methods have the additional advantage that they can be applied with great convenience to cases where the steam instead of being saturated to begin with is initially superheated to any extent. Such cases frequently occur in steam-engine practice. When superheated steam expands adiabatically its superheat disappears as expansion proceeds, and after the steam has passed through a stage when it is just saturated any further expansion causes a part of it to condense.

In all cases of adiabatic expansion the fluid is doing work at the expense of its stock of internal energy. After drawing the pressure-volume curve for steam expanding adiabatically the student should calculate the internal energy at the beginning and also at the end of the expansion, taking account of the proportion of water to steam in the mixture, and see that the difference, representing the loss of internal energy which has occurred in the process, agrees with the work done, which is the area under the curve.

Adiabatic curves for steam, whether initially dry or wet, may be calculated in the way that has just been explained, and may then be represented with a fair degree of approximation by empirical equations of the form

$$Pv^n = \text{constant},$$

where v is the volume of the wet mixture, by choosing such values for the index n as will give curves which nearly agree with the actual adiabatic curves. An expression of this kind is sometimes useful for application to cases where the data are the initial pressure and the ratio of expansion r, and it is required to find the pressure after expansion. To find P when the substance has expanded to r times its initial volume,

$$P = \frac{P_1}{r^n}.$$

The index n has a value which depends on q_1, the initial degree of dryness of the steam. According to the calculations of Zeuner[1] suitable values of n are given by the formula

$$n = 1{\cdot}035 + 0{\cdot}1q_1,$$

so that for

$q_1 =$ 1	0·95	0·9	0·85	0·8	0·75	0·7
$n =$ 1·135	1·130	1·125	1·120	1·115	1·110	1·105.

It must not be supposed that the expansion of steam in an actual engine is adiabatic, for there is a transfer of heat between the working fluid and the metal of the cylinder and piston, which will be discussed in a later chapter. If it were practicable to make the surfaces of the cylinder and piston non-conducting the idea of adiabatic expansion could be realized. When expansion goes on too rapidly to allow any considerable transfer of heat to take place it may be nearly adiabatic.

69. Carnot's Cycle with steam for working substance. We are now in a position to study the action of a heat-engine employing water and steam (or any other liquid and its vapour) as the working substance. To simplify the first consideration of the subject as far as possible, let it be supposed that we have, as before, a long cylinder composed of non-conducting material except at the base, and fitted with a non-conducting piston; also a source of heat A at some temperature T_1; a receiver of heat, or as we may now call it, a condenser, C, at some lower temperature T_2; and also a non-conducting cover B (as in § 43). To fix the ideas, suppose that there is unit quantity of water in the cylinder to begin with, at the temperature T_1. Then Carnot's cycle of operations can be performed as follows.

(1) Apply A, and allow the piston to rise against the constant pressure P_1 which corresponds to the temperature T_1. The water will take in heat and be converted into steam, expanding isothermally at the temperature T_1. This part of the operation is shown by the line ab in fig. 16.

(2) Remove A and apply B. Allow the expansion to continue adiabatically (bc), with falling pressure, until the temperature falls

[1] Zeuner's *Technical Thermodynamics*, vol. II. See, however, the author's *Thermodynamics for Engineers*, Art. 78, where it is shown that no constant index fits the curve accurately.

to T_2. The pressure will then be P_2, namely, the pressure which corresponds in the Steam Table to T_2 which is the temperature of the cold body C.

(3) Remove B, apply C, and compress. Steam is condensed by rejecting heat to C. The action is isothermal, and the pressure remains P_2. Let this be continued until a certain point d is reached, after which adiabatic compression will complete the cycle.

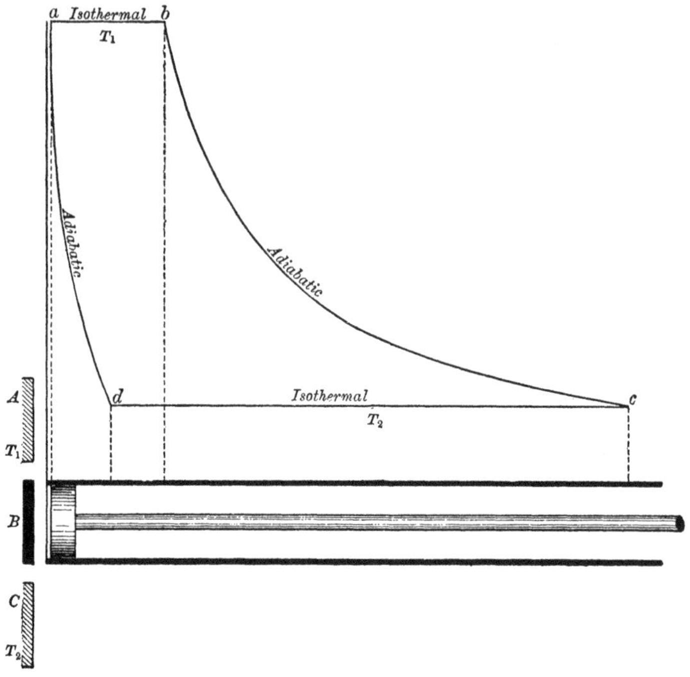

FIG. 16. Carnot's Cycle with water and steam for working substance.

(4) Remove C and apply B. Continue the compression, which is now adiabatic. If the point d has been rightly chosen, this will complete the cycle by restoring the working fluid to the state of water at temperature T_1.

The indicator diagram for the cycle is drawn in fig. 16, the lines bc and da having been calculated by the help of the equations in §§ 67 and 68, for a particular example, in which $p_1 = 90$ lb. per sq. inch ($T_1 = 433$), and the expansion is continued down to the pressure of the atmosphere, 14·7 lb. per sq. inch ($T_2 = 373$).

Since the process is reversible, and since heat is taken in only at T_1 and rejected only at T_2, the efficiency (by § 49) is

$$\frac{T_1 - T_2}{T_1}.$$

The heat taken in per unit mass of the fluid is L_1, and therefore the work done is

$$\frac{L_1 (T_1 - T_2)}{T_1},$$

a result which may be used to check the calculation of the lines in the diagram by comparing it with the area which they enclose. It will be seen that the whole operation is strictly reversible in the thermodynamic sense.

Instead of supposing the working substance to consist wholly of water at a and wholly of steam at b, the operation ab might be taken to represent the partial evaporation of what was originally a mixture of steam and water. The heat taken in would then be $(q_b - q_a) L$, and as the cycle would still be reversible the area of the diagram would be

$$\frac{L (q_b - q_a)(T_1 - T_2)}{T_1}.$$

70. Efficiency of a perfect Steam-engine. Limits of temperature. If the action here described could be realized in practice, we should have a thermodynamically perfect steam-engine using saturated steam. Like any other perfect heat-engine an ideal engine of this kind has an efficiency which depends upon the temperatures between which it works, and upon nothing else. The fraction of the heat supplied to it which such an engine would convert into work would depend simply on the two temperatures, and therefore on the pressures, at which the steam was produced and condensed respectively.

It is interesting therefore to consider what are the limits of temperature between which steam-engines may be made to work. The temperature of condensation is limited by the consideration that there must be an abundant supply of some substance to absorb the rejected heat; water is actually used for this purpose, so that T_2 has for its lower limit the temperature of the available water-supply.

To the higher temperature T_1 and pressure P_1 a practical limit is set by the mechanical difficulties, with regard to strength and to lubrication, which attend the use of high-pressure steam. In special instances pressures up to 1200 lb. per sq. inch have been used, but with engines and boilers of the ordinary construction the pressure very rarely exceeds 500 lb. per sq. inch and is commonly under 300.

This means that the upper limit of temperature, at which the steam takes in heat, is usually between 200° and 250° C. A steam-engine, therefore, under the most favourable conditions, comes very far short of taking full advantage of the high temperature at which heat is produced in the combustion of coal. From the thermodynamic point of view the worst thing about a steam-engine is the irreversible drop of temperature between the furnace and the boiler. The combustion of the fuel supplies heat at a high temperature: but a great part of the convertibility of that heat into work is at once sacrificed by the fall in temperature which is allowed to take place before the conversion into work begins.

If the temperature of condensation be taken as 15° C., as a lower limit, the efficiency of a perfect steam-engine, using saturated steam and following the Carnot cycle, would depend on the value of P_1, the absolute pressure of production of the steam, as follows:

Perfect steam-engine, with condensation at 15° C.,

P_1 in lb. per sq. inch being	50	100	200	300	400	500
Highest ideal efficiency =	·300	·341	·384	·409	·427	·441

But it must not be supposed that these values of the efficiency are actually attained, or are even attainable. Many causes conspire to prevent steam-engines from being thermodynamically perfect, and some of the causes of imperfection cannot be removed. These numbers will serve, however, as one standard of comparison in judging of the performance of actual engines, and as setting forth the advantage of high-pressure steam from the thermodynamic point of view. We shall see later (§ 75) that there is another standard with which the performance of a real steam-engine may more appropriately be compared.

71. Efficiency of an engine using steam non-expansively. As a contrast to the ideally perfect steam-engine of § 69 we may next consider a cyclic action such as occurred in the early engines of Newcomen or Leupold, when steam was used non-expansively—or rather, such an action as would have occurred in engines of this type had the cylinder been a perfect non-conductor of heat. In that case the volume of steam formed is equal to the volume swept through by the piston. We may represent the action of such an engine thus:

(1) Apply the hot body A and evaporate the water as before at P_1. Heat taken in, per unit mass of the working fluid, $= L_1$.

(2) Remove A and apply the cold body C. This at once condenses a part of the steam, and reduces the pressure to P_2.

(3) Compress at P_2, in contact with C, till condensation is complete, and water at T_2 is left.

(4) Remove B and apply A. This heats the water again to T_1 and completes the cycle. Heat taken in $= I_{w_1} - I_{w_2}$.

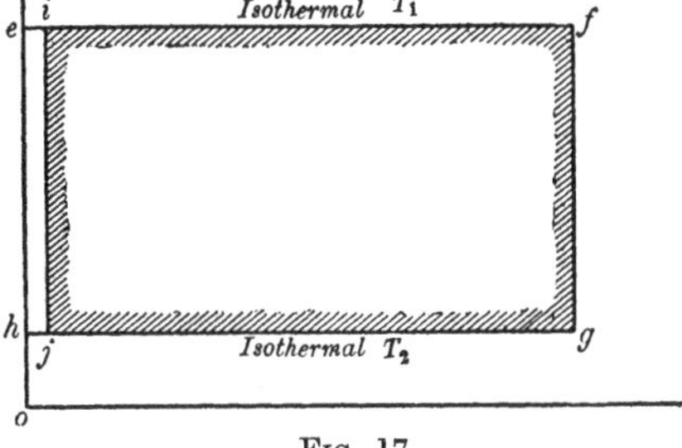

FIG. 17.

The indicator diagram for this series of operations is shown in fig. 17, where $oe = P_1$ and $oh = P_2$.

Here the action is not reversible. To calculate the efficiency,

$$\frac{\text{Heat converted into work}}{\text{Heat taken in}} = \frac{A\,(P_1 - P_2)\,(V_1 - V_w)}{I_{s_1} - I_{w_2}}.$$

The values of this will be found to vary from 0·068 to 0·074 for the same range of pressure (50 lb. to 500 lb. per sq. inch) as was taken in the last paragraph, the temperature of condensation being 15° C. as before. Contrast these numbers with the much higher efficiencies given there for a perfect steam-engine, following Carnot's cycle.

The efficiency of the actual Newcomen engine was much lower even than this calculation indicates, because in every stroke of the piston a large part of the steam entering the cylinder was at once condensed upon the sides, and the volume of steam which had to be supplied from the boiler was therefore much greater than the volume swept through by the piston.

72. Engine with separate organs. In the ideal engine whose action is represented in fig. 16 the functions of boiler, cylinder, and condenser are combined in a single vessel; but after what has been said in Chapter II it is scarcely necessary to remark that, provided the working substance passes through the same cycle of operations, it is indifferent whether these are performed in several vessels or in one. To approach a little more closely the conditions which hold in practice, we may think of the engine which performs the cycle of § 71 as consisting of a boiler A (fig. 18) kept at T_1, a non-conducting cylinder and piston B, a surface condenser C kept at T_2, and a feed-pump D which restores the condensed water to the boiler. Then for every unit mass of steam supplied and used non-expansively as in § 71, we have

$$\text{Work done on the piston} = (P_1 - P_2) V_1;$$

but the amount of work which has to be expended in driving the feed-pump is $(P_1 - P_2) V_w$. Deducting this, the net amount of work done per unit mass of steam is the same as before, and the heat taken in is also the same. An indicator diagram taken from the cylinder would give the area *efgh* (fig. 17), where

$$oe = P_1, \; ef = V_1, \; oh = P_2;$$

an indicator diagram taken from the pump would give the negative area *hjie*, where *ei* is the volume of the feed-water. The difference between these two areas, namely, the area *ifgh* which is shaded in the figure, is the diagram of the complete cycle gone through by

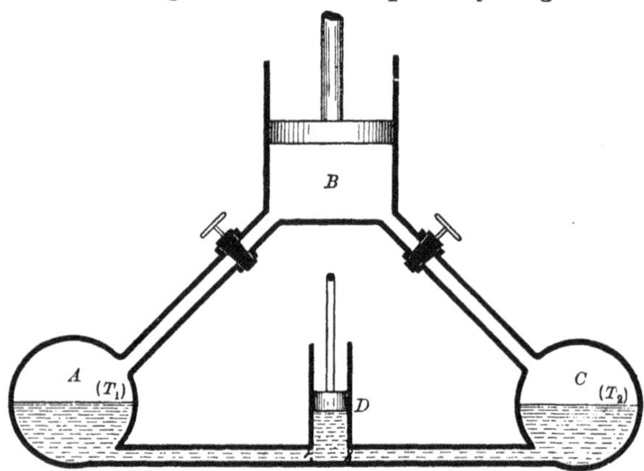

FIG. 18. Organs of a steam-engine.

each unit of the working substance. In experimental measurements of the work done in steam-engines, only the action which occurs within the cylinder is shown on the indicator diagram. From this the work spent on the feed-pump is to be subtracted if we wish to make a determination of the thermodynamic efficiency of the engine as a whole, by comparing the net amount of work done with the heat taken in. The heat taken in is determined by the method explained in § 64.

73. How nearly may the process in a steam-engine be reversible? We have now to inquire how nearly, with the engine of fig. 18, that is to say, with an engine in which the boiler and condenser are separated from the cylinder, we can approach the reversible cycle of § 69. The first stage of that cycle corresponds to the *admission* of steam from the boiler into the cylinder, for during admission of steam to the cylinder a corresponding quantity of steam is being formed in the boiler. Then the point known as the point of *cut-off* is reached, at which admission ceases, and the steam already in the cylinder is allowed to expand, exerting a diminishing pressure on the piston. This is the second stage, or the stage of *expansion*. The process of expansion may be carried on until the pressure falls to that of the condenser, in which case the expansion is said to be complete. At the end of the expansion *release* takes place, that is to say, communication is opened with the condenser. Then the return stroke begins, and a period termed the *exhaust* occurs, during which steam passes out of the cylinder into the condenser, where it is condensed at pressure P_2, which is felt as a *back pressure* opposing the return of the piston. So far, all has been essentially reversible, and identical with the corresponding parts of Carnot's cycle.

But we cannot complete the cycle as Carnot's cycle was completed. The existence of a separate condenser makes the fourth stage, that of adiabatic compression, impracticable, and the best we can do is to continue the exhaust until condensation is complete, and then return the condensed water to the boiler by means of the feed-pump.

It is true that we may, and in actual practice do, stop the exhaust before the return stroke is complete, and compress that portion of the steam which remains below the piston. It is generally only a small part of the working substance that is so compressed. This compression does not materially affect the thermodynamic efficiency; it is done partly for mechanical reasons, and partly to avoid loss of power through clearance (see Chap. VII). By clearance is meant the small volume which, in a real cylinder, is left below the piston at the end of the return stroke. In the present instance it is supposed that there is no clearance, in which case any such compression is out of the question, for there is no volume which the compressed steam could occupy. The indicator diagram given by a cylinder in which steam goes through the action described above is drawn to scale in fig. 19 for a particular example, in

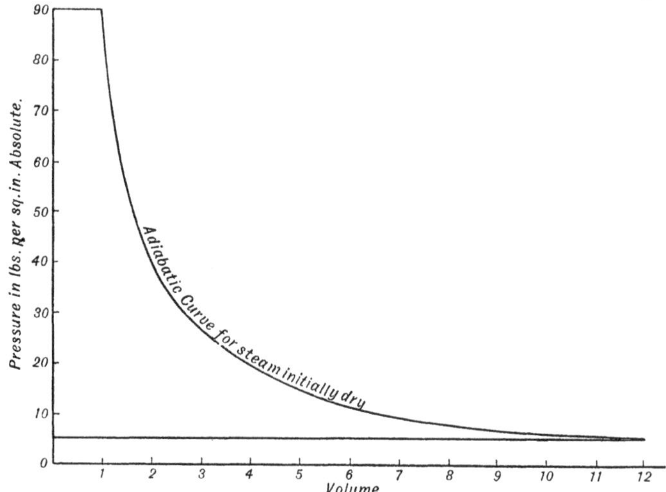

FIG. 19. Ideal Indicator Diagram for steam used expansively.

which it is supposed that dry saturated steam is admitted to the cylinder at an absolute pressure of 90 lb. per sq. inch, and is then expanded adiabatically to twelve times its original volume. This brings it down to a pressure of 5·4 lb. per sq. inch, at which pressure it is discharged to the condenser. As we have assumed the cylinder to be non-conducting, and the steam to be initially dry, the expansion curve may be found by the formula $PV^{1\cdot135} =$ constant (§ 68). The advantage of expansion is obvious, that part of the diagram which lies under the curve being so much clear gain, as compared with the case dealt with in § 71.

We might proceed to calculate the performance as follows:

Work done during admission $= P_1V_1$;

,, ,, expansion to volume $rV_1 = \dfrac{P_1V_1 - P_2V_2}{n-1}$

(by equation 8 of § 41), $= \dfrac{P_1V_1 - P_2V_2}{0\cdot135}$;

Work spent during return stroke $= P_2V_2$;

,, ,, on the feed-pump $= (P_1 - P_2)\,V_w$;

and so deduce the net amount of work done.

It is however much more instructive to consider the action from another point of view, namely with reference to the changes of energy and of total heat. In what follows there is no restriction as to the steam being initially in the dry saturated state: it may be dry, wet, or superheated; but the expansion is assumed to be adiabatic, and to continue until the pressure of condensation is reached.

Let I_1 represent the total heat of the working substance in the condition before expansion. The numerical value of I_1 will depend not only on the pressure of admission but on whether the steam is initially dry, or wet, or superheated: in any case it is readily calculated when the initial conditions are known. I_1 is made up of the internal energy E_1 which the substance possesses at B (fig. 20) and the heat equivalent of the work done during admission, namely AP_1V_1,

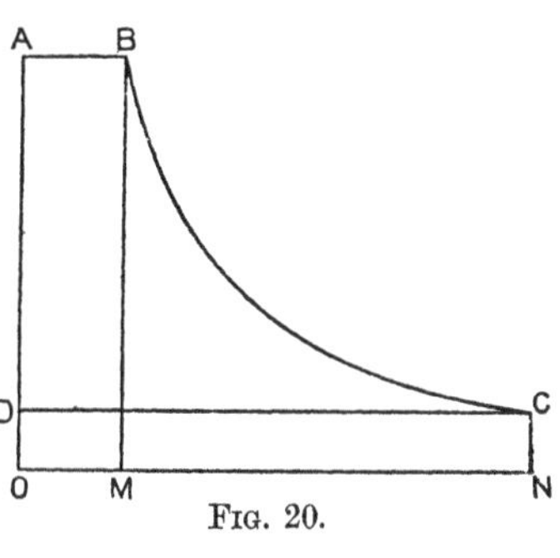

FIG. 20.

$$I_1 = E_1 + AP_1V_1 = E_1 + A \text{ (area } ABMO).$$

From B to C no heat is taken in, since the expansion is adiabatic, but work is done equal to the area under the expansion curve, and therefore the internal energy falls to a value E_2 at C such that

$$E_1 - E_2 = A \text{ (area } BCNM).$$

At C the total heat of the mixed substance is

$$I_2 = E_2 + AP_2V_2 = E_2 + A \text{ (area } DCNO).$$

Hence, subtracting,

$$\begin{aligned} I_1 - I_2 &= E_1 - E_2 + A\,(P_1V_1 - P_2V_2) \\ &= A \text{ (area } BCNM + \text{area } ABMO - \text{area } DCNO) \\ &= A \text{ (area } ABCD). \end{aligned}$$

That is to say, when there is complete adiabatic expansion down to the pressure of the condenser, the area of the indicator diagram for the whole operation in the cylinder, which measures the work done on the piston, when converted into heat units is equal to $I_1 - I_2$, where I_1 is the total heat of the working substance before expansion and I_2 is the total heat of the working substance after expansion, all the quantities being reckoned per unit of mass of the working substance[1].

74. Ideal performance measured by the Heat-Drop. We have, then, the very important conclusion that the amount of work ideally obtainable per unit quantity of steam in an ideal engine, using adiabatic expansion from the initial pressure at which the steam is supplied to the final pressure at which it is condensed, is (in heat units)

$$I_1 - I_2.$$

This quantity is often called the *heat-drop*. The conclusion holds whatever be the state of the steam on supply, whether wet or saturated or superheated. All that is assumed is that there is a constant pressure of supply, a constant pressure of condensation, that there are no transfers of heat between steam and metal, and that the expansion is complete from the pressure of supply to the pressure of condensation. Both I_1 and I_2 have to be reckoned with reference to the state of the substance, in the initial and final stages respectively. As a rule at the end of expansion we have a wet mixture for which to reckon I_2 (see § 68).

By the help of graphic and other processes which are explained in Chapter v, the problem of finding I_2 after any given amount of adiabatic expansion is rendered exceedingly simple, and it becomes easy to work out the theoretical performance $I_1 - I_2$ per unit mass of steam under any assigned conditions as to initial pressure, initial superheat, and final pressure[2]. In this way we can obtain

[1] This result may be obtained more shortly as follows:

The general equation, applicable to any process,

Heat taken in = gain of internal energy + work done,

may be written

$$dQ = dE + APdV,$$

where dQ represents a small gain of heat on the part of any substance, dE the corresponding gain of internal energy, and dV the change of volume.

From this

$$dQ = d(E + APV) - AVdP$$
$$= dI - AVdP.$$

But in an adiabatic process $dQ = 0$: hence in such a process,

$$dI = AVdP,$$

and, integrating,

$$I_1 - I_2 = A \int_{P_2}^{P_1} VdP,$$

which is the equivalent in heat units of the work done, namely the area $ABCD$.

[2] *Heat-Drop Tables* based on Callendar's figures have been published (E. Arnold, 1917) for a large range of initial and final states.

figures for comparison with the actual performances of real engines. The performance of any real engine, in which the cylinder takes heat from the steam during admission and in which the expansion is not adiabatic, is necessarily less than this theoretical limit $I_1 - I_2$. But in engines of a particular class the work actually done per lb. bears a fairly constant ratio to the heat-drop, and consequently calculation of the heat-drop is of great assistance to the designer in estimating probable performance.

Another reason why the performance of a real engine is less than the heat-drop is that in it the expansion is, in general, not complete. Instead of the steam being caused to expand until its temperature falls to that of the condenser it is allowed to escape by the opening of the exhaust-valve when its temperature (and pressure) is still somewhat above that of the condenser. The pressure accordingly suffers a sudden drop, and an irreversible transfer of heat takes place. Incomplete expansion is illustrated by fig. 21, where the steam is supposed to escape after expanding to five times its initial volume. It results simply in a loss of the work which is represented by the toe of the diagram, that is to say, by the difference of areas between this and fig. 19.

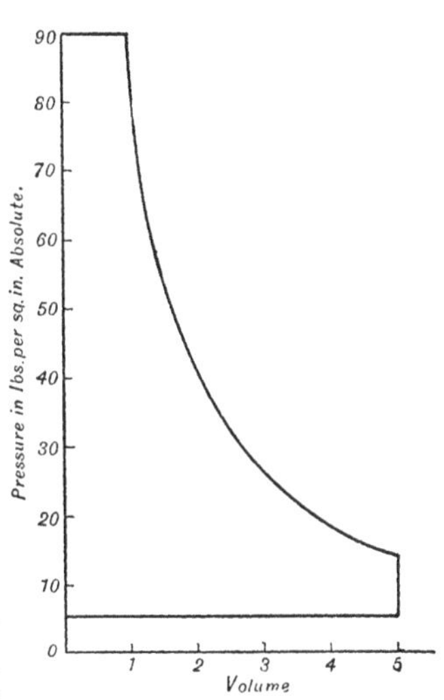

FIG. 21. Incomplete expansion.

75. Rankine Cycle. This name is given to the ideal cyclical process of § 73, in which the steam is passed into the cylinder at the initial pressure P_1 at which it is formed, then expands adiabatically to the final pressure P_2, at which it is condensed, and finally is returned to the boiler as water, at the temperature at which it has been condensed. The net amount of work done by the substance in going through this cycle is the work done by it in the cylinder, which is the area $ABCD$ of fig. 20, less the work spent upon it in the feed-pump, which is $(P_1 - P_2)\, V_w$. In other words, the net amount of work done by the steam is, in heat units,

$$I_1 - I_2 - A\,(P_1 - P_2)\, V_w.$$

If we divide this by the heat taken in, we have the efficiency of the ideal Rankine cycle. It will be shown later (§ 94) how to calculate these quantities.

The feed-pump term $A (P_1 - P_2) V_w$ is relatively so small that no material error is introduced by ignoring it, and treating the net amount of work done in the Rankine cycle as equivalent to $I_1 - I_2$ for purposes of practical calculation.

The efficiency of the Rankine cycle is always less than the ideal highest efficiency of a perfect engine working between the same limits of temperature. This is because of the absence of the compression which formed the fourth stage in Carnot's cycle, and had the effect of bringing the temperature up to the top of the range before the substance began to take in heat. Without compression some of the heat is taken in at temperatures below the highest temperature T_1, and any heat taken in at a lower temperature cannot contribute so much work as if it had been taken in at T_1.

The most useful criterion that can be applied to tests of real engines is to find the ratio between the actual amount of indicated work that is done per lb. of steam in passing through the engine and the amount of work that is ideally obtainable with complete adiabatic expansion, namely $I_1 - I_2$. This is called the *Efficiency Ratio*. In the comparison we leave out of account the work spent on the substance in the feed-pump, both in the real and the ideal engine.

Only in very favourable circumstances does the Efficiency Ratio reach so high a figure as 0·7. In Chapter VII we shall consider why it is that the actual performance of engines falls so much short of the ideal limit.

76. Changes of state in a liquid and its vapour: Critical Point. It will assist the student to understand the properties of steam or any other vapour if he will consider what happens when the pressure is gradually reduced under conditions such that the liquid or vapour remains at a constant temperature during the process. Imagine for instance a cylinder to contain a quantity of the liquid under pressure applied by a loaded piston, and let the cylinder stand on a body at a definite constant temperature, which will supply enough heat to it to maintain the temperature unchanged when the pressure of the piston is gradually relaxed and the volume consequently increases. Starting from a condition of very high pressure, say at A_1 (fig. 22), when the contents of the cylinder are wholly liquid, let the load on the piston be slowly reduced so that the pressure gradually falls. The contents at first remain liquid, until the pressure falls to a certain value depending on the temperature, at which vapour begins to form. Thus we have in the pressure-volume diagram a line A_1B_1 to represent what happens while the pressure is falling during this first stage and the contents are still liquid. The volume of the liquid increases, but only very slightly, in consequence of the pressure being relaxed, and hence A_1B_1 in the diagram is nearly but not quite vertical. At B_1 steam begins to form, and continues forming until all the liquid becomes vapour. This is represented by B_1C_1, a stage during which there is no change of pressure. At C_1 we have saturated vapour. Then, if the fall of pressure continues, a line C_1D_1 is traced, the progressive fall of pressure being associated with a progressive increase of volume. The temperature, by assumption, is kept constant throughout. At D_1, or at any point beyond C_1, the vapour has become superheated, because its pressure is now lower than the pressure corresponding to saturation, and hence its temperature is higher than the temperature corresponding to saturation at the actual pressure. The line $ABCD$ is an *isothermal* for the substance in the successive states of liquid (A to B), liquid and vapour mixed (B to C), saturated vapour (at C), superheated vapour (C to D). Now take a much higher temperature. We get a similar isothermal $A_2B_2C_2D_2$; at a higher temperature still another isothermal $A_3B_3C_3D_3$, and so on. The higher the temperature the nearer do B and C approach each other, and if the temperature is made high enough the line BC vanishes. A curve (shown by the broken line) drawn through $B_1B_2B_3$, etc. is continuous with one passing through $C_1C_2C_3$, and

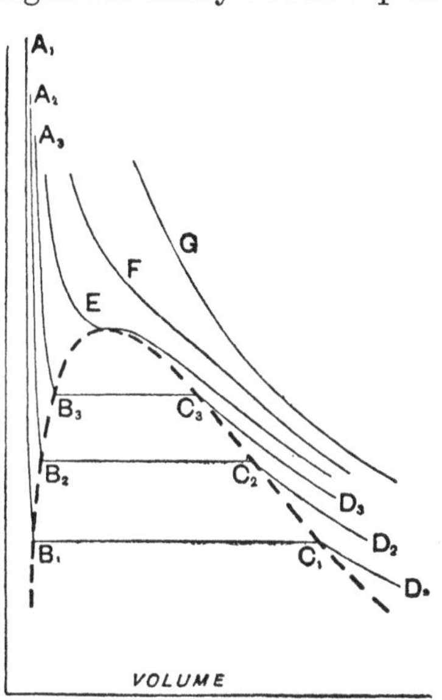

FIG. 22. Isothermal Lines.

it is only within the region of which this curve is the upper boundary that any change from liquid to vapour takes place. The branch $B_1B_2B_3$, which shows the volume of the liquid, meets the branch $C_1C_2C_3$, which shows the volume of the vapour, in a rounded top. The summit of the curve is called the *Critical Point*. The temperature for an isothermal line E that would just touch the top of this curve is called the *Critical Temperature*. We might define the critical temperature in another way by saying that if the temperature of a vapour is above the critical temperature no pressure, however great, will cause it to liquefy.

Starting from D and increasing the pressure, the temperature being kept constant, we may trace any of the isothermals backwards. The initial state is then that of a gas (a superheated vapour): as the pressure increases C is reached when it is saturated and condensation begins: at B condensation is complete, and from B upwards towards A we are compressing liquid. At any point between C and B the substance exists in two states; part is liquid and part is vapour. But if the isothermal is one that lies altogether outside of the boundary curve, shown by the broken line, the substance does not suffer any sharp change of state as the pressure rises. It follows a course such as is indicated by the lines F or G, and at no stage in the process is it other than homogeneous.

The critical temperature for steam is about 374° C., and the *critical pressure*, or pressure at the critical point, is about 3158 lb. per sq. inch. In the action of a steam-engine we are concerned only with lower pressures. But with carbonic acid, whose critical temperature is only about 31° C., the behaviour above the critical point is of great practical importance in connexion with refrigerating machines which employ carbonic acid as working substance.

Gases such as air, oxygen and so forth, are vapours which under ordinary conditions are very highly superheated. Their critical temperatures are so low that it is only by extreme refrigeration that they can be brought into a condition which makes liquefaction possible. The critical temperature of hydrogen is − 241° C. or 32° absolute. Even helium, the most refractory of the gases, has been liquefied, but only by cooling it to a temperature within about 5° of the absolute zero.

77. Steam at very high pressures. Of the properties of steam when the saturation pressure exceeds 500 lb. per sq. inch but little is precisely known. Exact measurements in the higher region present great difficulties. Callendar in his *Properties of Steam* (1920) has reviewed the results of various observers, and has given reasons for taking 374° C. as the critical temperature in place of 365°, which was formerly accepted. As the critical conditions are approached the latent heat L rapidly diminishes and becomes zero at the critical point. This consideration led Thiesen to devise an empirical formula for L at any high temperature t:—

$$L = C\,(t_c - t)^n; \text{ or } \log L = \log C + n \log (t_c - t),$$

where t_c is the critical temperature and n and C are constants. Callendar adopts a formula of this type for pressures higher than the range covered by his steam tables, making

$$\log L = 1{\cdot}9638 + 0{\cdot}3151 \log (374 - t).$$

With this, and with the help of other empirical formulas, he has calculated tentative values of the properties of saturated steam up to the critical point, from which the following table is compiled. The constants in the various formulas are adjusted to make the figures agree with his other tables at 200° C.

Saturated Steam at High Pressures

Temp. Cent. t	Pressure lb. per sq. in. p_s	Volume cub. ft. per lb.		Latent Heat L	Total Heat	
		Steam V_s	Water V_w		Steam I_s	Water I_w
200	225·2	2·074	0·0186	467·4	671·0	203·5
250	575	0·820	0·020	420	680	260
260	680	0·692	0·021	409	681	272
270	799	0·587	0·021	397	681	284
280	932	0·500	0·022	385	681	296
290	1082	0·427	0·022	372	681	309
300	1249	0·366	0·023	357	680	323
310	1435	0·314	0·024	341	678	337
320	1642	0·269	0·025	323	675	352
330	1870	0·230	0·026	303	670	367
340	2120	0·196	0·027	280	663	383
350	2394	0·165	0·029	250	652	401
360	2693	0·135	0·031	211	632	421
370	3020	0·098	0·035	142	589	447
374	3158	0·052	0·052	0	494	494

Reference has already been made to Callendar's *Enlarged Steam Tables* (1924) in which the properties of steam with various amounts of superheat are stated for a series of pressures ranging up to 2000 lb. per sq. inch. When steam at these high pressures is considerably superheated, as it generally is in engineering uses, the values of its total heat and other properties can be estimated with more probability of accuracy than when it is at or near the temperature of saturation. But all such numbers, though in the meantime of service to engineers, should be regarded as liable to revision when fuller experimental data become available.

CHAPTER IV

FURTHER POINTS IN THE THEORY OF HEAT-ENGINES

78. Rankine's statement of the Second Law. Rankine, to whom with Clausius and Lord Kelvin is due the development of the theory of heat-engines from the point at which it was left by the *Réflexions* of Carnot and the experiments of Joule, has, in his *Manual of the Steam-Engine and other Prime Movers*, stated the second law of thermodynamics in a form which is neither easy to understand, nor obvious, as an experimental result, when understood. His statement runs:—

"If the absolute temperature of any uniformly hot substance be divided into any number of equal parts, the effects of those parts in causing work to be performed are equal."

To make this intelligible we may suppose that any quantity Q of heat from a source at temperature T_1 is taken by the first of a series of perfect heat-engines, and that this engine rejects heat at a temperature T_2 which is less than T_1 by a certain interval ΔT. Let the heat so rejected by the first engine form the heat-supply of a second perfect engine working from T_2 to T_3 through an equal interval ΔT; let the heat which it in turn rejects form the heat-supply of a third perfect engine working again through an equal interval from T_3 to T_4; and so on. The efficiencies of the several engines are (by § 49)

$$\frac{\Delta T}{T_1}, \quad \frac{\Delta T}{T_2}, \quad \frac{\Delta T}{T_3}, \quad \text{etc.}$$

The amounts of heat supplied to them are

$$Q, \quad Q\frac{T_2}{T_1}, \quad Q\frac{T_3}{T_1}, \quad \text{etc.}$$

Hence the amount of work done by each engine is the same, namely,

$$Q\frac{\Delta T}{T_1}.$$

Thus Rankine's statement is to be understood as meaning that each of the equal intervals into which any range of temperature may be divided is equally effective in allowing work to be produced from heat when heat is made to pass, doing work in the most efficient possible way, through all the intervals from the top to the bottom of the range.

79. Absolute Temperature: Lord Kelvin's scale. In the preceding chapters we have been using the imaginary perfect gas thermometer as the means of framing a scale of temperatures. In other words, our scale has been such that equal intervals of temperature are defined as those which correspond to equal amounts of expansion of a perfect gas under constant pressure. We have defined temperature by taking T as proportional to V when P is kept constant. And seeing that air, or hydrogen, behaves as a nearly perfect gas, this scale is practically realized by the air thermometer, or by the hydrogen thermometer.

Starting from this definition of temperature we have found by an application of Carnot's principle that a reversible engine working between a hot source A and cold receiver of heat C takes in from the source and gives out to the receiver quantities of heat Q_A and Q_C which are proportional to the absolute temperatures of the source and receiver respectively, as defined by reference to the perfect gas thermometer.

Hence we might have defined temperature in a very different way and still have arrived at just the same scale. We might have said, let the temperatures of A and C be specified by two numbers which shall be proportional to the heat taken in and given out respectively by a reversible heat-engine when working with A for source and C for receiver of heat. This method of defining absolute temperature was proposed by Lord Kelvin. It gives a scale which is truly absolute in the sense of being independent of the properties of any gas or other substance, real or imaginary. The scale so obtained coincides with the scale of the perfect gas thermometer.

Lord Kelvin's method of devising a scale of absolute temperatures may also be put in a somewhat different fashion, thus:— Starting with any arbitrary temperature let a series of intervals be taken such that equal amounts of work will be done by every one of a series of reversible engines, each working with one of these intervals for its range and each handing on to the engine below it the heat which it rejects, so that the heat rejected by the first forms the supply of the second, and so on. Then call these intervals equal. This is only another way of putting the definition of absolute temperature which has just been quoted: it is suggested by what was said in the previous paragraph about Rankine's statement of the Second Law.

To make this aspect of the matter more intelligible, think of a quantity of heat supplied at some high temperature and used to drive a chain of perfect (reversible) heat-engines. As the heat goes down from engine to engine in the chain part of it is converted into work at each step and the remainder passes on to form the heat-supply of the next engine. We have to think of the steps as being such that the amount of heat converted into work is the same for each step. Thus if we have n engines in the chain and if the whole quantity of heat supplied to the first engine is Q, then the steps are such that each engine converts the quantity Q/n of heat into work. When n steps are completed there is no heat left: all is converted into work. This means that the absolute zero of temperature has been reached: we may in fact define the absolute zero as the temperature which is reached in this manner. We have reached it by coming down through a finite number of steps of temperature, and each step represents a finite fall in temperature. We define the absolute or thermodynamic scale by saying that these steps are to be taken as equal to one another. From this it will be seen that the conception of an absolute zero and of an absolute thermodynamic scale with uniform intervals does not depend on any notion about perfect gases or about the properties of any particular substance. We reach the absolute zero when, on going down through the chain of perfect engines, we come to a point at which the last fraction of the heat has been converted into work. That fixes the absolute zero. And we call the steps by which we have come equal steps of temperature. That fixes the scale. Moreover the steps can be so taken, by choosing a suitable number of them, that the scale so obtained will agree at two fixed points with the ordinary thermometric scale, and will contain between those fixed points the same number of steps as the ordinary scale contains degrees. Thus suppose the initial temperature, at the top of the chain, is that of the boiling-point of water, and that we have 373 engines in the chain. Then we find that it takes 100 steps to come down to the temperature of melting ice, and 273 more steps to complete the conversion of the remaining heat into work. This means that the uniform step of temperature on the thermodynamic scale is equal to the average of the intervals called degrees on the centigrade thermometer, when that average is taken between the freezing-point and the boiling-point (0° and 100°), and that the absolute zero is at a point 373 of such steps below the boiling-point, or 273 below the freezing-point.

The scale of the actual air thermometer would be in perfect agreement with Lord Kelvin's absolute scale if the laws stated in Chapter II were rigorously true of air, namely, Boyle's law, according to which the product PV is constant at any one temperature, and Joule's law, according to which there is no change of temperature when a gas expands without doing external work and without receiving or rejecting heat. The experiments by which Joule established his law have been already described. Reference has also been made to the subsequent experiments of a more searching kind, devised by Lord Kelvin, and carried out by him in conjunction with Joule, in which air was forced slowly through a porous plug to see whether its temperature became changed, which have shown that air does not conform with perfect exactness to Joule's law[1]; but the deviations are so slight that for all practical purposes the scale of the air thermometer may be taken as agreeing with the absolute scale. The agreement is still closer if hydrogen, which is more nearly "perfect," be used in the thermometer instead of air[2].

Actual air or gas thermometers may be made for use in two ways: In one the pressure is kept constant and the volume is allowed to expand or contract as the temperature varies; in the other the volume is kept constant by adapting the pressure to the temperature which is being measured, and the temperature is then taken to be proportional to the pressure. This latter is the more practicable form: it is called the constant-volume gas thermometer. The air or other gas must be perfectly dry: if there is any water-

[1] See Lord Kelvin's *Collected Papers*, vol. I, p. 333.

[2] See Callendar, "On the Thermodynamical Correction of the Gas Thermometer," *Phil. Mag.* Jan. 1903. Tables are there given showing the correction to be applied, for air, hydrogen, and other gases, to convert the scale reading of the gas thermometer to the absolute thermodynamic scale. The correction is exceedingly small: with the constant-volume hydrogen thermometer it is not more than 0·001 of a degree at any points throughout the range from − 10° C. to 200° C., and even at 1000° C. it is barely one-tenth of a degree. With the constant-volume air thermometer it is about six times as great, but even with air it is only at temperatures below − 100° C. or above 300° C. that the correction is so much as one-tenth of a degree. The corrections are some what greater when the gas thermometer is of the constant pressure type.

vapour in it the volume in the one case or the pressure in the other may be far from proportional to the temperature.

80. Clapeyron's equation. This name is given to an important relation between the Latent Heat of steam or any other vapour, the change of volume which it undergoes in being vaporized, and the rate at which its saturation pressure varies with the temperature. To establish it we may revert to the ideally perfect steam-engine of § 69, in which Carnot's cycle is followed with water and steam for working substance. We saw that this gave an indicator diagram (fig. 16) with two lines of uniform pressure (isothermals) connected by two adiabatic curves. The heat taken in was L per unit mass of working substance, and since the engine was reversible its efficiency was

$$\frac{T_1 - T_2}{T_1},$$

from which it followed that the work done, or the area of the diagram, was

$$\frac{L(T_1 - T_2)}{T_1}.$$

This is in thermal units: to reduce it to units of work we multiply by J. Now suppose that the engine works between two temperatures which differ by only a very small amount. We may call the temperatures T and $T - \delta T$, δT being the small interval through which the engine works. The above expression for the work done becomes

$$\frac{JL\delta T}{T}.$$

The indicator diagram is now a long narrow strip (fig. 23). Its length ab is $V_s - V_w$, V_s being the volume of unit mass of steam and V_w the volume of unit mass of water. Its height is δP, where δP is the difference between the pressure in ab and that in cd. In other words, since the steam is saturated in cd as well as in ab, δP is the difference in the pressure of saturated steam due to the difference in temperature δT. When δP is made very small, the area of the diagram becomes more and more

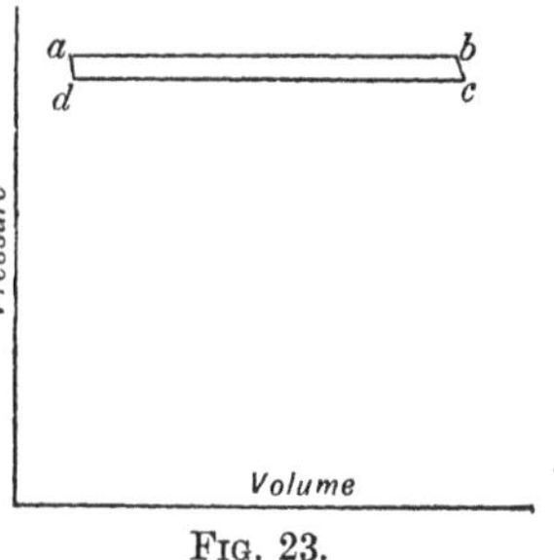

FIG. 23.

nearly equal to the product of the length by the height, namely, $\delta P\,(V_s - V_w)$. This is equal to the work done, whence

$$\delta P\,(V_s - V_w) = \frac{JL\delta T}{T} \quad\text{..................... (1).}$$

This equation is only approximate when the interval δT (or δP) is a small finite interval. In the limit, when the interval is made indefinitely small, it becomes exact and may then be written

$$V_s - V_w = \frac{JL}{T}\,\frac{dT}{dP} \quad\text{...................... (2),}$$

$\frac{dT}{dP}$ being the rate at which the temperature of saturated steam alters relatively to the pressure when the temperature is T.

Thus we have the Clapeyron equation

$$V_s = V_w + \frac{JL}{T}\,\frac{dT}{dP}$$

as a relation between the volume, the latent heat, and $\frac{dP}{dT}$. It may be applied to find the volume of saturated vapour when we know the latent heat and also know the relation of pressure to temperature so well as to be able to find $\frac{dP}{dT}$ by drawing a tangent to the pressure-temperature curve or by differentiating an expression connecting P and T. Or it may be applied to find the latent heat when the data are the volume and $\frac{dP}{dT}$. The values of pressure, volume, and latent heat, given in steam tables in relation to temperature, must, if the tables are accurate, be such as will satisfy this equation. Take, for example, steam at 100° C. From the tables it may be found, by taking differences or by plotting the pressure-temperature curve, that a rise in temperature of one degree at that point corresponds to a rise in saturation pressure of 0·525 lb. per sq. inch, or 0·525 × 144 lb. per sq. foot. The latent heat is given as 539·3. Hence, since J is 1400, we should have

$$V_s - V_w = \frac{1400 \times 539{\cdot}3}{373{\cdot}1} \cdot \frac{1}{0{\cdot}525 \times 144} = 26{\cdot}77 \text{ cub. ft.}$$

The volume given in the table for steam is 26·789, and that of water at the same temperature is 0·0167, making $V_s - V_w = 26{\cdot}772$. The agreement is as good as can be wished; a considerably greater difference would be within the limits of error which apply in the process of finding $\frac{dP}{dT}$ from the pressure-temperature curve[1].

81. Extension of the above result to other changes of physical state. In equation (2), above, the left-hand side is positive, since the volume of steam is greater than that of water. The right-hand side must also be positive, and hence it is that $\frac{dT}{dP}$ is positive, or in other words, that increasing the pressure under which steam is formed raises the boiling-point. The equation might evidently be applied in the reverse way to that indicated above (for finding V); in other words, if the amount by which the volume increases when water changes into steam were given we might employ that to calculate $\frac{dT}{dP}$, the rate at which the boiling-point is raised by increase of pressure.

Further, the reasoning by which this equation was arrived at was perfectly general and was in no way restricted to the case of steam. The engine whose indicator diagram is sketched in fig. 23 might have anything for working substance, the isothermal line of the first operation, during which heat is taken in, representing in the most general way the change of volume which occurs while any working substance changes its physical state. In the example already dealt with the change is from liquid to vapour. But we might begin with a solid substance previously raised to the tem perature T at which it begins to melt and let the first stage in the cycle consist in the expansion of the substance while it passes from the solid to the liquid state, the substance doing external work by overcoming a constant pressure as it expands. All the steps in the argument remain unaffected, and hence the equation may be written thus with reference to any transformation of state under constant pressure on the part of any substance,

$$U - U' = \frac{J\lambda}{T}\frac{dT}{dP} \quad \text{.......................(3)},$$

where U' is the volume of unit mass of the substance in the original

[1] In old steam tables such as those in Rankine's book the volumes given were calculated by means of the Clapeyron equation, using the data of Regnault's experiments for L and $\frac{dP}{dT}$, but a different procedure is followed in the modern tables (see Appendix).

state, U is the volume after the transformation has taken place, λ is the heat absorbed while the transformation is going on (the latent heat of fusion or of evaporation as the case may be), and $\frac{dT}{dP}$ is the rate at which the temperature of the transformation (say the melting-point or the boiling-point) is affected by altering the pressure under which the change of state occurs.

If a solid substance expands on melting, U is greater than U', and consequently $\frac{dT}{dP}$ must be positive: in other words, the melting-point will in that case be raised by applying pressure.

On the other hand if the substance contracts on melting, $U - U'$ is negative and T must then *decrease* relatively to P, that is to say, the melting-point is then lowered by applying pressure. This is the case with ice. From the known amount by which ice contracts when it melts James Thomson (in 1849) first applied this method of reasoning to show that the melting-point of ice must be lowered to a definite extent when the ice is melted under any assigned pressure, and the result was afterwards verified by an experiment of his brother, Lord Kelvin. The amount by which the melting-point of ice is lowered is about 0·0074° C. for each atmosphere of pressure[1].

82. Drying and superheating of steam by throttling. No change of the total heat in a throttling process. When dry steam is forced through a throttle-valve or constricted orifice without receiving or rejecting heat it becomes superheated; and if wet to begin with it becomes drier. This is because the total heat of saturated steam is less at low pressure than at high, and, as we shall see presently, the total heat of the fluid as a whole undergoes no change in the process of throttling. Suppose for instance that steam is flowing through a small pipe or orifice from a chamber where the pressure is P_1 to another where there is a lower pressure

[1] See Lord Kelvin's *Collected Papers*, vol. I, p. 156 and p. 165. The numerical result stated in the text is obtained as follows:—A pound of water changes its volume in freezing from 0·016 to 0·0174 cub. ft., and gives out 80 (centigrade) units of heat. Hence

$$\frac{dT}{dP} = \frac{0{\cdot}0014 \times 273}{80 \times 1400} = 0{\cdot}00000341,$$

and for an additional pressure of one atmosphere or 2160 lb. per sq. ft., the melting-point is lowered by 2160 × 0·00000341 or 0·0074° C.

P_2. Such an action happens in steam-engines in the movement of steam through contracted pipes or passages, such as a partially closed stop-valve or a reducing valve between the boiler and the engine: the steam becomes reduced in pressure and is said to be throttled or "wire-drawn." Eddies are formed in rushing through the constricted opening and the energy expended in forming them is frittered down into heat as the eddies subside. To calculate the effect of throttling, assume that a steady condition exists before and also after the throttling and that the chambers on both sides of the constricted opening have a relatively large cross-section, so that the stream of steam has no kinetic energy worth taking account of either before it passes the opening or after it has passed and the eddies have subsided.

To fix the ideas, imagine the substance to be transferred from one side of the opening to the other by piston A (fig. 24) coming

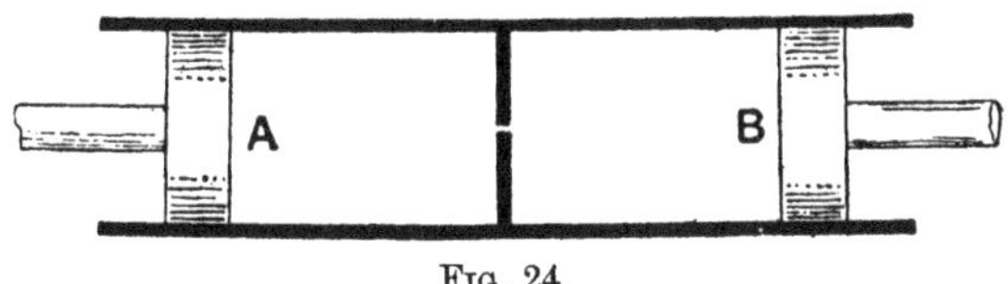

FIG. 24.

up towards the opening, and piston B receding. On the side A there is pressure P_1, on the side B a lower pressure P_2. Let V_1 and V_2 represent the volumes of unit mass of the fluid before and after transfer. Before passing through the opening the fluid has a stock of internal energy (per unit of mass) which we shall call E_1, and as it approaches the orifice the piston A does work upon it equal to P_1V_1. On the other side the fluid does work against the piston B equal to P_2V_2. Hence E_2, the stock of energy which it has after passing the opening, is given by the equation

$$E_2 = E_1 + AP_1V_1 - AP_2V_2$$

or $$E_2 + AP_2V_2 = E_1 + AP_1V_1,$$

which may be written $$I_2 = I_1,$$

I_1 and I_2 being the total heats of the fluid, per unit mass, before and after passing respectively. In other words the total heat I does not change in the process. This is a highly important characteristic of I, that it does not change in the kind of expansion which occurs when the fluid passes a throttle-valve or along a pipe offering frictional resistance. The imaginary pistons were introduced only to make the argument clear; the result is of general application to all cases of throttling, or frictional drop of pressure, and to any fluid.

Applying this now to the case of wet steam, let q_1 and q_2 be the dryness before and after throttling. By § 65 we have

$$I_1 = I_{w_1} + q_1 L_1 \text{ and } I_2 = I_{w_2} + q_2 L_2.$$

Hence the dryness after throttling is found from the equation

$$q_2 = \frac{I_{w_1} - I_{w_2} + q_1 L_1}{L_2}.$$

To take a numerical example, suppose that the steam in its first state has a pressure of 130 lb. per sq. inch (temperature 175° C.), and contains 6 per cent. of moisture, and suppose it to fall by throttling to atmospheric pressure. Then q_1 is 0·94 and by the tables $L_1 = 487{\cdot}8$, $L_2 = 539{\cdot}3$, $I_{w_1} = 177{\cdot}0$ and $I_{w_2} = 100{\cdot}0$. With these data q_2 becomes 0·993: in other words the steam after throttling contains only 0·7 per cent. of moisture. If the initial dryness had been 0·948 a similar calculation shows that with the same drop in pressure the steam would become just dry.

When the steam is dry to begin with or when the drop of pressure is more than sufficient to remove its initial moisture the result of the throttling is to superheat it. Though the temperature falls, it falls less than corresponds to the fall in pressure and consequently there is superheat. In that case the total heat after throttling is $I_{s_2} + \kappa\,(t' - t_2)$, where t' is the actual temperature after throttling, t_2 is the temperature of saturation corresponding to the pressure in the second state, I_{s_2} the total heat of saturated steam at that pressure, and κ is the mean specific heat in superheating over the range from t_2 to t'. Equating this to the total heat before throttling, we can calculate the amount of superheating which will result, provided κ is known.

Conversely, if throttling experiments are made and the temperature t' is observed, the observation can be used as a means of determining κ. We have the equation

$$\kappa = \frac{\text{Total heat before throttling} - I_{s_2}}{t' - t_2}.$$

This method of measuring κ has been employed by various investigators[1], but it presents considerable difficulties, due mainly to uncertainty as to the exact condition of dryness of the steam. It is difficult to secure that the steam will be initially dry, and even when this difficulty is overcome it appears that errors may arise through the throttled steam holding particles of moisture in suspension although the vapour containing them is itself superheated, a condition of complete thermal equilibrium not having been reached. We have assumed throughout that the steam neither gains nor loses heat by conduction or radiation, a condition very hard to secure in any actual experiment.

Dry steam escaping at high pressure from a boiler into the atmosphere is superheated near the orifice, when its kinetic energy has been frittered down to heat, but further off it becomes wet by condensation through loss of heat to the air. Before the kinetic energy is frittered down the jet may be wet, because work has been done in setting it in motion. We shall return to this point later in dealing with the theory of jets in relation to the steam turbine.

83. Engine receiving heat at various temperatures. In Carnot's cycle it was assumed that the working substance took all its heat in at the higher limit of temperature T_1. Important cases arise in which heat is taken in partly at one and partly at other temperatures in a single cycle of operations. With regard to every such quantity of heat the result still applies that the greatest fraction that can be converted into work under ideally favourable conditions is represented by the difference between its temperatures of reception and rejection, divided by the absolute temperature of reception.

Thus if Q_1 represents that part of the whole heat which is taken in at T_1, and Q_2 represents what is taken in at some other temperature T_2, Q_3 at T_3 and so on, and if T_0 be the temperature at which the engine rejects heat, the whole work done, if the processes within the engine are reversible, is

$$W = \frac{Q_1 (T_1 - T_0)}{T_1} + \frac{Q_2 (T_2 - T_0)}{T_2} + \frac{Q_3 (T_3 - T_0)}{T_3} + \dots \text{etc.} \dots (4).$$

[1] Ewing and Dunkerley, *Brit. Assoc. Rep.* 1897; J. H. Grindley, *Phil. Trans. Roy. Soc.* vol. 194 A; A. Griessmann, *Mitteilung aus dem Maschinenlaboratorium der Tech. Hochschule in Dresden*, 1903; A. H. Peake, *Proc. Roy. Soc.* vol. 76 A, 1905. Observations of the fall of temperature which steam undergoes when it is throttled have also been made by Callendar and have been used by him in determining the constants of his characteristic equation for water-vapour, on which the formulas and tables in the Appendix are based.

It is perhaps worth while to point out the analogy here to the supposititious case of a water-wheel working by gravity and receiving water into its buckets at different heights above the level at which water is discharged from them. Let M_1, M_2 and so on be the quantities of water received at heights l_1, l_2 etc. above any datum level, and let l_0 be the height above the same datum level at which the water leaves the wheel. If the wheel is perfectly efficient (and here again the test of perfect efficiency is reversibility) the work done is

$$M_1 (l_1 - l_0) + M_2 (l_2 - l_0) + M_3 (l_3 - l_0) + \dots \text{etc.}$$

Comparing the two cases we see that the quantity $\frac{Q_1}{T_1}$ is the analogue in the heat-engine of M_1 in the water-wheel and so on. The amount of work which can be got out of a given quantity of heat by letting it down to an assigned level of temperature is not simply proportional to the product of the quantity of heat by the fall of temperature, but to the product of $\frac{Q}{T_1}$ by the fall of temperature. On the strength of this analogy Zeuner has called the quantity $\frac{Q}{T}$ the "heat weight" of a quantity of heat Q obtainable at a temperature T.

Another way of expressing the matter has a wider application. Let the engine as before take in quantities of heat represented by Q_1, Q_2, Q_3 etc. at T_1, T_2, T_3 and let Q_0 represent the heat rejected at T_0. The heat rejected is negative if we regard heat taken in as positive. Then by the principle that in a reversible cycle the heat rejected is to the heat taken in as the absolute temperature of rejection is to the absolute temperature of reception, we have

$$\frac{Q_0}{T_0} = \frac{Q_1}{T_1} + \frac{Q_2}{T_2} + \frac{Q_3}{T_3} + \dots,$$

from which

$$\Sigma \frac{Q}{T} = 0 \quad \dots\dots\dots\dots (5),$$

when the summation is effected all round the reversible cycle. It is clear that this result may be at once extended to cases where heat is given out at various temperatures as well as taken in at various temperatures, Q being taken positive or negative according as heat is being received or rejected.

In cases where changes of temperature are going on continuously while heat is being taken in or given out, we cannot divide the reception or rejection of heat into a limited number of steps, as has been done above. But the equation may be adapted to this most general case by writing it, for any reversible cycle,

$$\int \frac{dQ}{T} = 0 \quad \text{..............................(6)},$$

integration being performed round the whole cycle. In a cycle which is not reversible this integral for the cycle as a whole is not zero but a negative quantity, because the amount of heat rejected is relatively larger than when the cycle is reversible.

84. Application to a steam-engine working without compression, but with complete adiabatic expansion. In § 73 we considered the action of an ideal steam-engine following the Rankine cycle, in which the steam formed at T_1 was expanded adiabatically and fully, that is to say, down to the pressure corresponding to the temperature of the condenser T_2 and was there condensed, the condensed water being then restored to the boiler by a feed-pump and there heated again to T_1 to complete the cycle. This cycle is specially important in the discussion of steam-engines because it forms the ideally best performance of an engine which returns the condensed water in its cold state directly from the condenser to the boiler. It is what such an engine might achieve provided the expansion were complete, so that there should be no sudden drop of pressure at release, and provided the cylinder and piston were perfect non-conductors. The efficiency in this cycle falls short of the ideal Carnot limit $(T_1 - T_2)/T_1$ because in the fourth stage of the cycle the working substance has its temperature raised from T_2 to T_1, not by adiabatic compression, as in Carnot's cycle (§ 69), nor by any equivalent of that[1], but by being brought into contact with the contents of the boiler, which are kept at T_1. Consequently heat enters it in this stage by a non-reversible process: in all other respects however the cycle is reversible.

But we may regard this as a strictly reversible cycle if we think

[1] It will be shown in the chapter on steam turbines that there is a possible regenerative method of heating the feed-water which converts the Rankine cycle into what is virtually a Carnot cycle. Such regenerative feed-heating is (in theory) reversible, and is equivalent in this respect to adiabatic compression.

of the feed-water as taking up its heat by infinitesimal instalments at a series of temperatures ranging from T_2 up to T_1 from a series of imaginary sources each of which has the same temperature as the water when the water is brought into contact with it. One may realize this notion by thinking of the feed-pipe as passing through a heated channel the temperature of which is T_1 close to the boiler and tapers down to T_2 close to the condenser. Thus the feed-water would have its temperature raised gradually and would nowhere be brought into contact with a source at a temperature different from the temperature which it had itself then reached. With such an arrangement as this it is clear that the engine becomes a strictly reversible engine, receiving portions of its heat, however, at various temperatures. But the action of the engine is in no way altered by this imaginary arrangement of the feed-pipe, nor is the total supply of heat in any way altered. The notion of gradual heating in the feed-pipe has been introduced merely to show that the cycle is a reversible cycle if we take account of the fact that heat is received not all at the top of the range of temperature, but partly at lower temperatures. Every part of the heat which the substance receives is used in the most efficient possible way, *after it has been taken in*, so that the expression

$$\frac{T - T_2}{T}$$

measures the efficiency of the transformation into work of each portion of the heat, T being the absolute temperature at which the working substance happens to be when it takes in that portion of the heat. The only non-reversible feature in the action of this engine is the flow of heat from a source at T_1 into the feed-water while the temperature of the feed-water is less than T_1; and we get rid of this partial non-reversibility by taking as the temperature of reception of each portion of the heat that temperature which the working substance has when the portion in question was taken in. It will be evident that these remarks are of general application, and that when this understanding is accepted, with regard to the temperatures of both reception and rejection of heat, the process in any heat-engine is to be taken as reversible provided the expansions and compressions which occur in it are themselves reversible in the sense which has been explained in § 51. With a source of heat at a given temperature the heat can be turned to account most efficiently only when all the heat is taken in while the working substance is at that temperature, and it is only then that the greatest value of the efficiency, namely, $\frac{T_1 - T_2}{T_1}$, can be reached. But the engine may take in part of its supply of heat at temperatures below T_1 and still act reversibly in the conversion of the heat so received into work, in which case the efficiency of the whole action will be less than $\frac{T_1 - T_2}{T_1}$ though the general formula $\frac{T - T_2}{T}$ is still applicable in respect of every separate portion of the heat, when proper values are assigned to T.

The ideal steam-engine which we are now considering is a case in point. It takes in the greater part of its heat at T_1, but some is taken in at temperatures ranging between T_2 and T_1. So far as actions occurring within the engine are concerned it is reversible. The amount of heat it converts into work is therefore to be found by calculating

$$\Sigma \frac{\delta Q\,(T - T_2)}{T},$$

where δQ represents any part of the heat taken in and T the temperature at which it is taken in. The whole heat taken in, per lb. of working substance, is, first, the amount of heat which is required to heat the water from T_2 to T_1, namely the heat which is taken in while the temperature is varying, and, second, the latent heat L, which is taken in at the temperature T_1. During the heating of the water through any small interval dT the heat taken in may be expressed as dh or σdT, where σ is the specific heat of water. Hence the whole amount of work done per unit mass of working steam (expressed in thermal units) is

$$W = \int_{T_2}^{T_1} \frac{(T_1 - T_2)\,\sigma dT}{T} + \frac{L_1 (T_1 - T_2)}{T_1}$$

$$= \int_{T_2}^{T_1} \sigma dT - T_2 \int_{T_2}^{T_1} \frac{\sigma dT}{T} + \frac{L_1 (T_1 - T_2)}{T_1} \quad \text{.........(7).}$$

No serious error is introduced in this calculation if we treat the specific heat of water as constant and equal to unity throughout the range usual in steam-engine practice. Writing $\sigma = 1$ we have, on that approximate basis,

$$W = T_1 - T_2 - T_2 \log_\epsilon \frac{T_1}{T_2} + \frac{L_1 (T_1 - T_2)}{T_1},$$

or

$$W = (T_1 - T_2)\left(1 + \frac{L_1}{T_1}\right) - T_2 \log_\epsilon \frac{T_1}{T_2} \quad \text{...........(8).}$$

This is the greatest amount of work which can be done, per unit mass of steam, under ideally favourable conditions by an engine which takes in saturated steam from a boiler at temperature T_1 and restores condensed water to the boiler at temperature T_2. The result accordingly supplies a standard with which the performance of actual steam-engines may be compared.

This is only another way of working out the ideal performance in the Rankine cycle (§ 75), for the particular case in which steam is supplied in the saturated state. It is the way adopted by a Committee of the Institution of Civil Engineers[1] in their report recommending the Rankine cycle as a standard for comparison with the performance of real engines. To compare the results of tests with this standard is more logical and useful than to compare them with the standard of the Carnot cycle, because of the absence of compression in the practical steam-engine cycle. If the action in the real steam-engine were strictly adiabatic, and if no losses occurred through leakage, throttling, or unresisted expansion into clearance spaces, the ideal efficiency of the Rankine cycle would be actually attained.

The efficiency of a steam-engine working in the Rankine cycle may be found by dividing the above expression for W by the heat taken in per unit mass of steam, which, for an engine using saturated steam, may be written

$$L_1 + T_1 - T_2,$$

if, as before, we take the specific heat of water to be equal to unity as a first approximation.

The efficiency of the Rankine cycle for saturated steam may accordingly be expressed in this form

$$\frac{(T_1 - T_2)\left(1 + \frac{L_1}{T_1}\right) - T_2 \log_\epsilon \frac{T_1}{T_2}}{L_1 + T_1 - T_2}.$$

This method of treating the Rankine cycle is only approximate because it ignores variations in the specific heat of water. It is however instructive, as a supplement to the exact treatment based on the idea of heat-drop, which was indicated in § 75 and will be more fully developed in Chapter V.

It is easy to extend this way of treating the Rankine cycle

[1] *Report of Committee on the Thermal Efficiency of Steam-Engines*, 1898. (*Min. Proc. Inst. C. E.* vol. CXXXIV.)

to an ideal engine in which the steam is superheated before admission. Let κ be the specific heat for the range through which superheating is carried: values of κ are given in § 66. For the present purpose we shall treat κ as sensibly constant throughout the range. Then the extra heat taken in is $\kappa\,(T_1' - T_1)$, where T_1' is the temperature of superheat and T_1 the saturation temperature at the pressure of supply. In the expression for W there is now the additional term

$$\int_{T_1}^{T_1'} \frac{(T - T_2)\,\kappa dT}{T},$$

and hence

$$W = (T_1 - T_2)\left(1 + \frac{L_1}{T_1}\right) + \kappa\,(T_1' - T_1) - T_2\left(\log_\epsilon \frac{T_1}{T_2} + \kappa \log_\epsilon \frac{T_1'}{T_1}\right).$$

The efficiency is found by dividing this expression by

$$T_1 - T_2 + L_1 + \kappa\,(T_1' - T_1).$$

As a numerical example of this method of working out the efficiency in a Rankine cycle, take an engine receiving saturated steam at 180° C. (pressure 146 lb. per sq. inch), and condensing it at 40° C., and calculate the ideal performance when there is complete adiabatic expansion from the top to the bottom of this range. Here T_1 is 453, T_2 is 313, L_1 is 483·9, and hence the expression for W (equation 8) gives 173·8 thermal units as the equivalent of the work ideally obtainable per unit mass of steam. The supply of heat, reckoned as $L_1 + T_1 - T_2$, is 623·9. The efficiency is therefore 0·279. Compare this with the number 0·309 which represents the value of $\frac{T_1 - T_2}{T_1}$, namely, the efficiency of a reversible (Carnot) cycle completed by adiabatic compression as in the engine of § 69. The absence of adiabatic compression has in this case reduced the efficiency by about ten per cent. This comparison shows what is lost by the partial misapplication of heat which results from letting the feed-water come into the boiler cold, to be heated by contact with the hot water already there, so that the portion of the heat-supply received at that stage is taken in at temperatures lower than the top of the range.

The expression for W serves to show how much work it is ideally possible to get from steam when the temperature of the boiler and the temperature of the condenser are assigned. But it is further useful as a standard with which we may compare the action of the steam-cylinder, taken by itself, without reference to what has happened before the steam reaches the cylinder, or to what happens in the exhaust-pipe and in the condenser. For that purpose T_1 is to be taken as the temperature of the steam on reaching the cylinder and T_2 the temperature of the steam on leaving the cylinder, although these temperatures may differ from those of the boiler and condenser respectively, and in fact have a narrower range. Then the formula serves to show the limiting amount of work obtainable in working through that range. The calculation furnishes a useful check on the results of engine trials, the work actually done being necessarily less than the quantity so calculated.

We shall see in Chapter v that the numerical value of W is better found by taking advantage of the fact that it is equal to the heat-drop $I_1 - I_2$, where I_1 is the total heat of the steam as supplied, whether saturated, wet, or superheated, and I_2 is the total heat of the fluid after expansion, the fluid being then a mixture of steam and water. When a suitable table or chart is available for finding the heat-drop this is an easier way of finding W. It is also more accurate, for the method given above does not take account of variations in specific heat.

85. Extension to steam not initially dry. The result arrived at in the last paragraph may be readily extended to ideal engines in which the steam is not dry when the adiabatic expansion begins. Let q_1 be the dryness at this stage: then the heat taken in during evaporation is $q_1 L_1$ per unit mass of working substance, but the heat taken in during the heating of the water up to T_1 remains what it was before. The expression for the work done (assuming complete adiabatic expansion as before) is therefore found by substituting $q_1 L_1$ for L_1 in equation (7) or (8), giving,

$$W = T_1 - T_2 - T_2 \log_\epsilon \frac{T_1}{T_2} + \frac{q_1 L_1 (T_1 - T_2)}{T_1} \quad \text{......(9).}$$

86. Derivation of the adiabatic equation from this result. This result may be applied to prove the equation for the adiabatic expansion of steam which was stated, without proof, in § 68. The whole heat taken in, per unit of mass, in raising the water from any temperature T_2 to T_1 and in evaporating the fraction q_1 of it at the temperature T_1 is (if we take the specific heat of water as unity)

$$T_1 - T_2 + q_1 L_1.$$

By expanding this mixture adiabatically to the temperature T_2 and then condensing it, we get an amount of work equal (by the equation which has just been given) to

$$T_1 - T_2 + q_1 L_1 - \frac{q_1 L_1 T_2}{T_1} - T_2 \log_\epsilon \frac{T_1}{T_2}.$$

Hence, subtracting the work done from the heat supplied we find that the heat rejected is

$$\frac{q_1 L_1 T_2}{T_1} + T_2 \log_\epsilon \frac{T_1}{T_2}.$$

But the only rejection of heat in the cycle takes place during the condensation at T_2 after adiabatic expansion, and the amount of heat so rejected is

$$q_2 L_2,$$

where q_2 is the dryness after adiabatic expansion to the temperature T_2.

Hence $$q_2 L_2 = \frac{q_1 L_1 T_2}{T_1} + T_2 \log_\epsilon \frac{T_1}{T_2},$$

or $$\frac{q_2 L_2}{T_2} = \frac{q_1 L_1}{T_1} + \log_\epsilon \frac{T_1}{T_2}.$$

Now T_2 may be any temperature lower than T_1, for the adiabatic expansion might be stopped at any point along the curve and the cycle completed by condensing the mixture at the temperature it had then reached. Hence this equation serves to show in a perfectly general way the change of dryness which takes place during adiabatic expansion, and, dropping the second suffix, we may write it

$$\frac{qL}{T} = \frac{q_1 L_1}{T_1} + \log_\epsilon \frac{T_1}{T} \quad\text{.....................(10)},$$

which is the same as the equation in § 68. It is to be noticed that in deriving this expression the specific heat of water has been treated as constant. The result is therefore (to a very small extent) inexact, especially at high temperatures.

87. Heat-engines employing more than one working substance. So far as general thermodynamic principles are concerned the choice of working substance for a heat-engine is indifferent. With a reversible cycle the same efficiency is given by one substance as by another, and with the Rankine cycle this is nearly though not quite true. But the consideration that the pressure should be neither excessively high nor excessively low often determines whether one or another working substance is to be preferred. Vaporizable liquids have the advantage over air or any other permanent gas that heat can be more readily communicated to and extracted from them; but any such liquid has a comparatively limited range of temperature within which it is practicable for it to work. The efficiency of the steam-engine is, as we have already seen, largely conditioned by the fact that even when steam is highly superheated most of the heat is necessarily taken in at a comparatively low temperature; hence full advantage is not taken of the high-temperature heat which is generated in the combustion of fuel in boiler furnaces. From this point of view a less easily vaporizable liquid would be better, for it would allow heat to be taken in at a higher temperature without making the pressure excessive.

Going to the other end of the range, it will be seen by reference to the table of pressure and temperature for steam that a steam-engine of the cylinder and piston type is not well fitted to take full benefit of the low temperature which may be reached when there is condensing water at hand. A more volatile liquid would do this better, because its vapour could be expanded to the bottom of the range of temperature without making the pressure fall inconveniently low. With steam complete expansion in a piston and cylinder engine would be useless, because in the last stages of the expansion the pressure would barely suffice to move the piston against its own frictional resistance; and therefore the *indicated* work which would be saved by completing the expansion would contribute nothing to the output of the engine. Apart from this fundamental difficulty, complete expansion would be impracticable owing to the excessive bulk required in the cylinder. One of the great merits of the steam turbine is that it enables the last stages of expansion to be effective for the performance of useful work, for in the turbine the difficulty as to friction does not arise, and it is practicable to deal with the large volume which the steam assumes when its pressure approaches the pressure in the condenser. In the reciprocating steam-engine considerations both of friction and bulk make it impracticable to work the steam down nearly so far, and a considerable part of the theoretically possible output $I_1 - I_2$ is lost because of the high temperature at which steam is discharged.

For this reason it has been proposed to use what is called a "binary" heat-engine, that is, an engine with two working fluids, one to work through the upper part of the range, and another—a more volatile fluid—to work through the lower part. The less volatile fluid, after being evaporated in a boiler and after doing work in its cylinder, is condensed by passing through tubes in a vessel containing the more volatile fluid, to which it rejects heat. The more volatile fluid is thereby evaporated and does work in another cylinder, after which it is passed into a surface-condenser supplied with cold circulating water. A binary engine using water as the less volatile fluid and ether as the more volatile fluid was introduced by Du Tremblay about 1850[1]; the type has more than once been revived on a small scale[2], but has never come into serious use.

A modern application of the idea of binary working on very different lines is found in the "Still" engine[3], which is primarily an internal combustion engine using oil fuel, the auxiliary fluid being steam which is formed by utilizing the heat of the oil-engine exhaust and also the heat that is given up to the cooling jacket of the oil-engine. In an ordinary internal combustion engine the working cylinder is kept cool enough to allow of lubrication by having a jacket through which water circulates; and the gases that result from the combustion are discharged to the atmosphere, after expansion, at a temperature which is so comparatively high that they carry off much heat. In the Still engine the heat that would go to waste in these two ways is turned to account for generating steam, which is then caused to act on one side of the main piston, while on the other side and in the same cylinder the internal combustion working takes place. There is a small boiler for the steam formed by the waste heat; from its water-drum

[1] See *Min. Proc. Inst. C. E.* vol. XVIII, p. 233. Also Rankine's *Steam-Engine*, p. 444.

[2] See *Min. Proc. Inst. C. E.* vol. CXII, 1893, pp. 481, 482.

[3] See Ackland, *Trans. Roy. Soc. of Arts*, May 1919; Denny, *Trans. Inst. of Nav. Arch.* July 1920; Second Report of the Oil-Engine Trials Committee, *Proc. Inst. Mech. Eng.*, March 1925.

water circulates through a system of tubes heated by the exhaust gases and through the cylinder jackets. In normal running it is only the waste heat so collected that goes to form steam, but to provide for special conditions, such as starting, the boiler has a fire-box in which some oil fuel may be burnt. The steam is used as in an ordinary condensing engine, and in normal conditions contributes about ten or fifteen per cent. of the total indicated power.

In Chapter VIII reference will be made to an experimental binary-fluid turbine employing mercury vapour and steam. There the idea is to widen the working range of temperature by raising the upper limit to the temperature at which mercury boils under a moderate pressure, the lower limit being as usual the temperature at which steam is condensed.

CHAPTER V

ENTROPY

88. Entropy. The Entropy of a substance is a function of its state which is most conveniently defined by reference to the heat taken in or given out when the state changes in a reversible manner. In any such change the quantity of heat received or rejected, divided by the absolute temperature of the substance at the time, measures the change of entropy. Thus if a substance which is expanding under a piston takes in δQ of heat at any absolute temperature T, its entropy increases by the amount $\frac{\delta Q}{T}$. The entropy of a substance in any definite state is a definite quantity, to be reckoned per unit of mass. Just as in dealing with total heat, we take an arbitrary starting-point at which the entropy of the substance is reckoned as zero. Thus in reckoning the entropy of water or of steam in any given state, we take the condition of water at 0° C. as the starting-point, and calculate the entropy by considering what heat is taken in, and at what temperatures, in passing from that to the state in which the substance is. Each element δQ of the heat taken in is to be divided by T, which is the absolute temperature of the substance when that element is taken in, and the sum

$$\Sigma \frac{\delta Q}{T},$$

reckoned from the starting-point till the given state is reached then measures the entropy. When any quantity of heat δQ is given out at temperature T the entropy is reduced by the amount $\frac{\delta Q}{T}$. The entropy of water, per unit of mass, at any temperature T_1, is $\int_{T_0}^{T_1} \frac{\sigma dT}{T}$, where σ is the specific heat of water and T_0 is 273·1°. In passing from the condition of water at any temperature T to steam at the same temperature the substance takes in a quantity of heat equal to L: consequently the entropy increases by $\frac{L}{T}$.

We shall denote entropy by ϕ. Using ϕ_s for the entropy of saturated steam at any temperature T and ϕ_w for the entropy of water at the same temperature we have

$$\phi_s = \phi_w + \frac{L}{T}.$$

Values of ϕ_s and ϕ_w are given in the tables.

It follows from this definition of entropy that when any substance expands or is compressed in an adiabatic manner its entropy does not change. An adiabatic line in the pressure-volume diagram is consequently a line of constant entropy, or as it is sometimes called, an *isentropic* line.

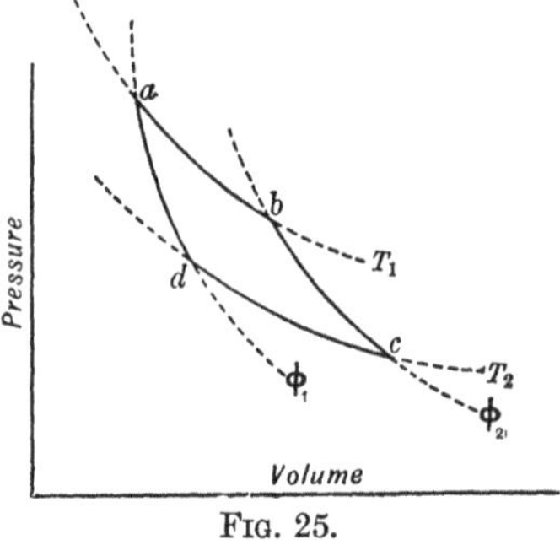

FIG. 25.

Consider now a cycle consisting of two isothermal and two adiabatic operations, fig. 25. In passing from a to b by the isothermal line T_1 the substance gains entropy $\frac{Q_1}{T_1}$, where Q_1 is the heat taken in during this operation. Along the adiabatic line from b to c there is no change of entropy. In the isothermal line cd the entropy is reduced by $\frac{Q_2}{T_2}$, and from d to a there is again no change of entropy. Since this is a reversible cycle (§ 50) $\frac{Q_1}{T_1} = \frac{Q_2}{T_2}$, and hence the entropy changes by the same amount whether we pass from one adiabatic line ad to another adiabatic line bc by one isothermal path ab or by any other isothermal path dc. And moreover the change of entropy between one adiabatic and another will be the same whether the cross-path be isothermal or not, for a curve expressing any relation between P and V may be regarded as made up of a succession of minute isothermal and adiabatic elements, and the change of entropy along such a curve is the sum of the changes which occur during the isothermal elements of the process, and is still equal to $\frac{Q}{T}$ for any single isothermal path between the same pair of adiabatic lines.

We see, then, that not only is there no change of entropy during an adiabatic process, but there is a perfectly definite change of entropy when a given substance passes from one adiabatic line to another, by whatever path. Just as isothermal lines can be distinguished by numbers T_1, T_2, etc. denoting the particular temperature for which each is drawn, so adiabatic lines can be distinguished by numbers ϕ_1, ϕ_2, etc. denoting the particular value of the entropy on each. The conception of entropy as that characteristic of a substance which does not change during adiabatic expansion or compression is of the greatest service in problems relating to heat-engines. We shall see presently some of the uses to which this notion may be put.

When a substance expands in an *irreversible* manner, as by passing through a throttle-valve from a region of high pressure to a region of lower pressure, it gains entropy. Work is then done by the substance on itself, namely in giving energy of motion to each particle as it passes through the valve, and this energy of motion is frittered down into heat as the motion subsides through internal friction. The effect is therefore equivalent to the communication of a quantity of heat, though no heat reaches the substance from outside. We may regard expansion through a throttle-valve as consisting of two stages. The first stage is adiabatic expansion in which the substance does work in setting itself in motion: the second stage is the loss of this motion and the acquiring of an equivalent amount of heat. There is accordingly a gain of entropy.

It was shown in § 83 that when a substance goes through any cycle consisting entirely of reversible operations

$$\int \frac{dQ}{T} = 0$$

for the cycle as a whole. It follows that in passing from any state (a) to any other state (b) by taking in or giving out heat the change of entropy $\phi_b - \phi_a$ which is $\int_a^b \frac{dQ}{T}$ must have the same value whatever path is followed in the operation, provided it be reversible. For if we choose any two such paths we may go from (a) to (b) by one and return by the other, thus completing a reversible cycle for which the integral vanishes. Hence the numerical value of the entropy as defined in § 88 for any state of a substance is the same whatever reversible process be followed in passing to the state in question from the starting-point at which the entropy is taken as zero.

In applying the definition of § 88 to reckon numerical values of ϕ it is generally convenient to think of the substance as passing to the state considered from the zero state by being heated under constant pressure, this being one sort of reversible operation, and to find the value of $\int \frac{dQ}{T}$ in the process. But the same result would be arrived at if we were to assume any other process of heating, provided it included no irreversible step.

89. Sum of the entropies in a system. It is instructive, in this connexion, to inquire how the sum of the entropies of all parts of the system is affected when we include not only the working substance but also the source of heat and the sink or receiver to which heat is rejected. Consider a cyclic action in which the working substance takes in a quantity of heat Q_1 from a source at T_1 and rejects a quantity Q_2 to a sink at T_2. When the cycle is completed the source has lost entropy to the amount $\frac{Q_1}{T_1}$: the working substance has returned to the initial state, and therefore has neither gained nor lost entropy: the sink has gained entropy to the amount $\frac{Q_2}{T_2}$. If the cycle is a reversible one $\frac{Q_1}{T_1} = \frac{Q_2}{T_2}$, and therefore the system taken as a whole consisting of source, substance and sink has suffered no change in the sum of the entropies of its parts. But if the cycle is not reversible the action is less efficient, Q_2 bears a larger proportion to Q_1 and $\frac{Q_2}{T_2}$ is greater than $\frac{Q_1}{T_1}$. Hence in an irreversible action the sum of the entropies of the system as a whole becomes increased. This conclusion is a perfectly general one: it may be expressed in general terms by saying that when a substance undergoes any change the sum of the entropies of the substance and of such other substances as take part in the action remains unaltered if the action is reversible, but becomes increased if the action is not reversible. No real action is strictly reversible, and hence any real action occurring within a system of bodies has the effect of increasing the sum of the entropies of the bodies which make up the system. This is a statement, in terms of entropy, of the principle that there is in all actual transformations of energy a universal tendency towards what Lord Kelvin called the dissipation of energy. The sum of the entropies in any system, considered as a whole, tends towards a maximum, which would be reached if all the energy of the system were to take the form of uniformly diffused heat, and if this state were reached no further transformations would be possible. Any action within the system, by increasing what may be called the aggregate entropy, which is obtained by summing the entropies of the various parts, brings the system a step nearer to this state, and to that extent diminishes the availability of the energy in the system for further transformations.

As an extreme case of thermodynamic waste take the direct conduction of a quantity of heat Q from a hot part of the system, at T_1, to a colder part at T_2. The hot part loses entropy by the amount $\frac{Q}{T_1}$: the cold part gains entropy by the amount $\frac{Q}{T_2}$, and as the latter is greater there is an increase of the aggregate entropy in the system as a whole.

90. Entropy of steam: derivation of the adiabatic equation. Reckoning from water at any initial temperature T_0 the entropy of steam (taken wet, for greater generality)

$$\phi = \int_{T_0}^{T_1} \frac{\sigma dT}{T} + \frac{q_1 L_1}{T_1} \quad \text{.......................(1)}.$$

The first term represents the entropy which is acquired during the heating of the water from T_0 to T_1, which is the temperature of evaporation, and the second term represents what is acquired during evaporation, q_1 being the dryness of the steam. Treating the specific heat of water as unity we can write dT for σdT; then integrating,

$$\phi = \log_\epsilon T_1 - \log_\epsilon T_0 + \frac{q_1 L_1}{T_1} \quad \text{..............(2)}.$$

Now in adiabatic expansion we have

$$\phi = \text{constant},$$

and hence if the steam be expanded adiabatically to any temperature T

$$\log_\epsilon T - \log_\epsilon T_0 + \frac{qL}{T} = \log_\epsilon T_1 - \log_\epsilon T_0 + \frac{q_1 L_1}{T_1},$$

from which
$$\frac{qL}{T} = \frac{q_1 L_1}{T_1} + \log_\epsilon \frac{T_1}{T},$$

which is the adiabatic equation of § 68, already derived by another and longer method in § 86. As was pointed out there, it is only approximate, because the specific heat σ is not constant.

Since σ is appreciably greater than unity at high temperatures the entropy of water ϕ_{w_1}, which is $\int_{T_0}^{T_1} \frac{\sigma dT}{T}$, becomes somewhat greater than $\log_\epsilon T_1 - \log_\epsilon T_0$.

An exact expression is, however, easily obtained as follows, for use with the tabulated values of the entropy. In the initial state let the entropy be ϕ; if the steam is wet to begin with this will be

$$\phi = \phi_{w_1} + \frac{q_1 L_1}{T_1} \qquad \text{......................(3).}$$

Let the steam expand adiabatically to a lower temperature T_2, at which the latent heat is L_2 and the entropy of water is ϕ_{w_2}. Call the dryness then q_2. The entropy in this second state may be written

$$\phi_{w_2} + \frac{q_2 L_2}{T_2}$$

Since there has been no change in the entropy of the mixture as a whole this is still equal to ϕ, and hence

$$q_2 = \frac{T_2}{L_2}(\phi - \phi_{w_2}) \qquad \text{......................(4).}$$

This determines the dryness after expansion, and from it the volume per lb. of mixture is readily found, being $q_2 V_{s_2} + (1 - q_2) V_{w_2}$ which is practically equal in ordinary cases to $q_2 V_{s_2}$. In the special case where the steam is dry saturated to begin with, $\phi = \phi_{s_1}$.

Apply this method to the same example as in § 68, where dry saturated steam at 190° expands to 120°. Here $\phi = \phi_{s_1} = 1{\cdot}5613$, $T_2 = 393{\cdot}1$, $L_2 = 526{\cdot}85$, $\phi_{w_2} = 0{\cdot}3646$, making

$$q_2 = \frac{393{\cdot}1}{526{\cdot}85}(1{\cdot}5613 - 0{\cdot}3646) = 0{\cdot}893.$$

It will be seen that when tables giving the entropy are used the solution becomes shorter as well as more exact.

The expression $q_2 = \frac{T_2}{L_2}(\phi - \phi_{w_2})$ is also applicable to steam that was initially superheated, provided that expansion has been carried far enough to make it become wet. With superheated steam the initial value of ϕ, which remains constant during the expansion, may be found from Table E when the temperature, as well as the pressure, is assigned.

91. Entropy-temperature diagrams. The familiar way to represent graphically those changes which a working substance undergoes in the action of a heat-engine is to draw the indicator diagram, which shows pressure in relation to volume. Another way is to draw a diagram showing the relation of the temperature of the substance to its entropy. Diagrams of this kind form an interesting and often useful alternative to the ordinary indicator diagram[1]. Let $\delta\phi$ be the small change in entropy which a substance undergoes when it takes in any small quantity δQ of heat at any temperature T. By the definition of entropy, in any reversible action,

$$\delta\phi = \frac{\delta Q}{T},$$

whence $$T\delta\phi = \delta Q$$

and $$\int T d\phi = \int dQ \quad \text{.............................(5)},$$

the integration being performed between any assigned limits. Now if a curve be drawn with T and ϕ for ordinates, $\int T d\phi$ is the area under the curve. This by the above equation is equal to $\int dQ$, in other words, the area under any portion of the entropy-temperature curve is equal to the whole quantity of heat taken in while the substance passes through the states which that portion of the curve represents. Let ab, fig. 26, be any portion of the curve of ϕ and T. The area of the cross-hatched strip whose breadth is $\delta\phi$ and height T, is $T\delta\phi$, which is equal to δQ, the heat taken in during the small change $\delta\phi$. The whole area *mabn* or $\int T d\phi$ between the limits a and b is the whole heat taken in while the substance changes from the state represented by a to the state represented by b. Similarly, in changing from state b to state a by the line

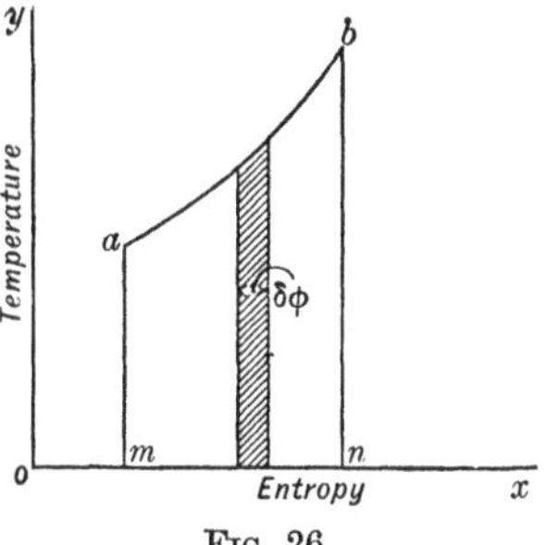

FIG. 26.
Entropy-temperature curve.

[1] Entropy-temperature diagrams were described along with other graphic methods in thermodynamics by Professor J. Willard Gibbs (*Trans. of the Connecticut Acad. of Sciences*, vol. II, 1873, p. 309). Their application to steam-engine problems is mainly due to Mr J. Macfarlane Gray (see *Proc. Inst. Mech. Eng.* 1889, p. 399). Professor Boulvin (*Cours de Mécanique appliquée: Théorie des Machines thermiques*) ascribes their earliest use to M. Th. Belpaire (*Bulletin de l'Académie royale de Belgique*, 1872, V, 34).

ba the substance rejects an amount of heat which is measured by the area *nbam*. The base line *ox* corresponds to the absolute zero of temperature.

When an entropy-temperature curve is drawn for a complete cycle of changes it forms a closed figure, since the substance returns to its initial state. To find the area of the figure we have to integrate throughout the complete cycle, when

$$\int T d\phi = Q_1 - Q_2,$$

Q_1 being the heat taken in and Q_2 the heat rejected[1]. But the difference between these is the heat converted into work, hence

$$\int T d\phi = W \quad \text{.........................(6)},$$

when the integration extends round a reversible cycle and W is expressed in thermal units. Thus entropy-temperature diagrams have the important property that the enclosed area measures the thermal equivalent of the work done in a complete cycle, provided every step in the cycle be reversible.

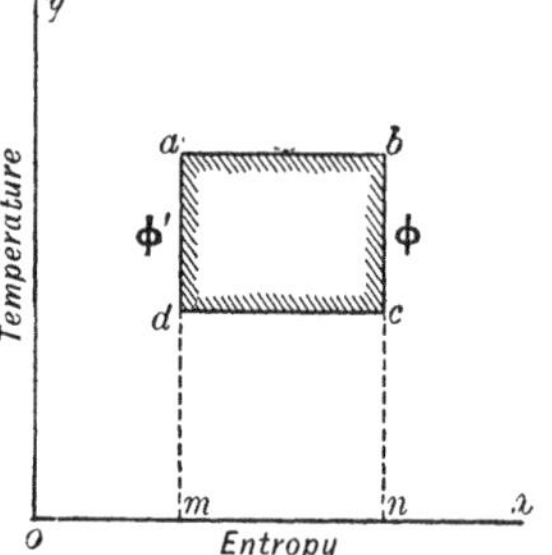

Fig. 27. Carnot's Cycle on the entropy-temperature diagram.

Isothermal lines on an entropy-temperature diagram are straight lines parallel to *ox* whatever be the working substance: adiabatic lines are straight lines parallel to *oy*, being lines of constant entropy. Hence Carnot's cycle, whether with air or steam or any other substance, would be represented by a rectangle *abcd*, fig. 27, in which the heat received

$$Q_1 = \text{area } mabn = T_1 (\phi - \phi'),$$

heat rejected

$$Q_2 = \text{area } ncdm = T_2 (\phi - \phi'),$$

and work done

$$W = \text{area } abcd = (T_1 - T_2)(\phi - \phi'),$$

ϕ being the entropy in the adiabatic process of expansion and ϕ' the entropy in the adiabatic process of compression. The efficiency is

$$\frac{\text{area } abcd}{\text{area } mabn} = \frac{T_1 - T_2}{T_1}.$$

[1] We are dealing here with cycles in which there is no irreversible operation such as throttling.

92. Entropy-temperature diagram for steam; application to ideal steam-engine working without compression but with complete expansion. A more interesting example of the use of entropy-temperature diagrams is given by the engine of § 73 using the Rankine cycle of operations. In that cycle, after complete adiabatic expansion from T_1 to T_2, the steam is condensed isothermally at T_2, and is then returned as water to the boiler. In drawing the diagram for the cycle we shall begin at the point where the water, at T_2, is about to be heated. At that stage the state of the working substance is represented on the diagram (fig. 28) by the point a, which may be drawn by taking as coordinates ϕ_{w_2} and T_2. The line ab represents the heating of the feed-water up to the temperature at which steam is formed. To draw it we may take values of ϕ_w for various temperatures between T_2 and T_1, the temperature of the boiler[1]. The coordinates of b are ϕ_{w_1} and T_1. At b steam begins to be formed, and bc is the change of entropy which the substance undergoes in passing from water to steam at the constant temperature T_1. bc is therefore equal to $\frac{L_1}{T_1}$, assuming the evaporation to be complete. If the evaporation were incomplete, bc would be equal to $\frac{q_1 L_1}{T_1}$. The adiabatic process of complete expansion from the condition of saturated steam at T_1 down to the temperature T_2 is represented by cd, and da is the process of condensation which completes the cycle.

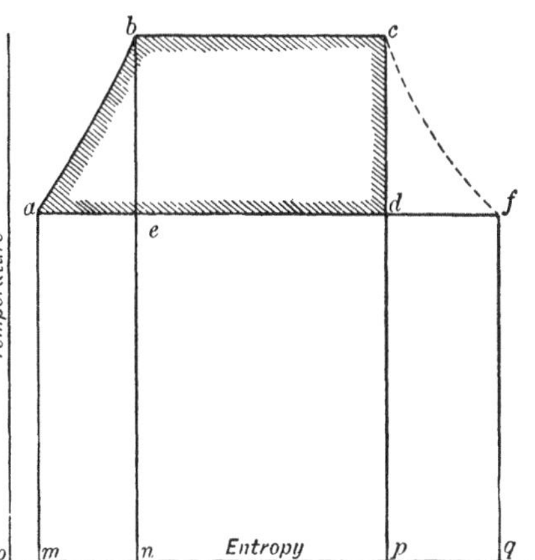

FIG. 28.
Entropy-temperature diagram for steam.

The heat taken in during the heating of the feed-water to the

[1] There is, in strictness, a distinction between the entropy of water at saturation pressure and that of water at the same temperature but at the pressure of the boiler. The difference, however, is too small to be represented on the diagram.

boiler temperature is the area *mabn*. The heat taken in during evaporation is *nbcp*. The work done is the enclosed area *abcda*. The heat rejected is *pdam*. Of the heat taken in during the process of evaporation, namely, the area *bp*, the part measured by the area *bd* is converted into work: it represents the fraction $\frac{be}{bn}$ or $\frac{T_1 - T_2}{T_1}$ of the whole, as we should expect. Of the heat taken in during the process of warming the feed-water a smaller fraction is converted into work, namely, the fraction $\frac{abe}{mabn}$. This is because the heat is less advantageously supplied during this operation, the temperature being then less than T_1. An engine going through Carnot's cycle would have the diagram *ebcd*. The engine of the Rankine cycle does more work (by the area *abe*), but to do this it has to take in a more than proportionally larger amount of heat and is therefore less efficient. The heat which it takes in is the area *mabcp*, against the area *nbcp* for the Carnot cycle. It will be seen that the diagram exhibits in a very simple way results which we have already arrived at by other routes.

Further, let a curve *cf* be drawn such that the distance from any point in *ab* to it, measured horizontally (that is, parallel to *op*), is equal to the value of $\frac{L}{T}$ corresponding to that point. Thus let *af* be equal to $\frac{L_2}{T_2}$, *f* being the point where *ad* produced meets this curve. This curve may be called the *saturation* curve, for it shows the entropy of saturated steam at various temperatures. Then the dryness of the steam after the process of adiabatic expansion represented by *cd* is given by the fraction $\frac{ad}{af}$. This follows from the fact that if the steam were perfectly dry at T_2 the heat given out during its condensation would be equal to the area *qfam*, whereas the heat actually given out is equal to *pdam*. In other words, the former area is L_2 and the latter

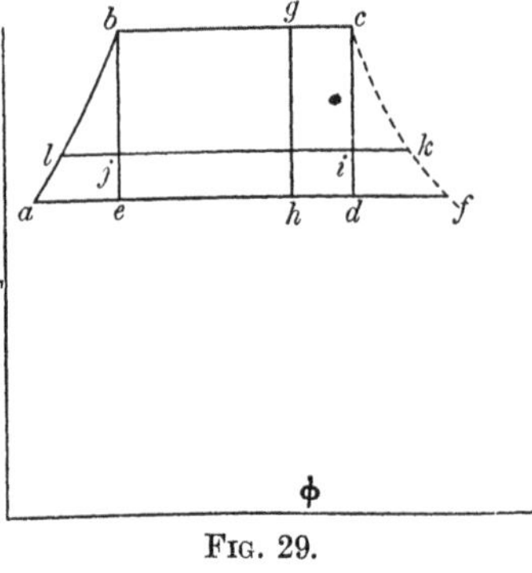

FIG. 29.

is q_2L_2, q_2 being the dryness when the stage d is reached, whence $q_2 = \frac{ad}{af}$. In the same way a straight line drawn horizontally from any point i in cd (fig. 29) to meet the curves cf and ab is divided by cd into segments il and ik. These are proportional to the quantities of steam and water respectively which make up the working substance when the expansion has advanced as far as the point i. In other words, at i the dryness $q = \frac{li}{lk}$. The temperature-entropy diagram thus affords a convenient method of finding q graphically at any stage in adiabatic expansion.

Further, suppose the steam has not been dry when adiabatic expansion begins. This state of things is represented on the diagram by making the horizontal line from b terminate at a point g such that $bg = \frac{q_1L_1}{T_1}$: in other words, $\frac{bg}{bc} = q_1$. The line gh now represents the process of adiabatic expansion and the construction just described is still applicable to find q at any stage. Thus at h, $q = \frac{ah}{af}$, and $\frac{hf}{af}$ is the proportion then present as water.

Again, reverting to the Carnot cycle of fig. 16, § 69, we can use the entropy-temperature diagram to determine the point at which condensation at T_2 must be stopped in that cycle in order that adiabatic compression may bring the substance to the state of water at T_1. The process of compression required for this is eb (fig. 28 or 29), and hence compression must begin when the proportion of steam still uncondensed is $\frac{ae}{af}$. Similarly the fraction $\frac{lj}{lk}$ measures the dryness at any stage j of this adiabatic compression.

93. Application of the entropy-temperature diagram to superheated steam. The entropy of steam superheated to any temperature T' is greater than the entropy in the saturated state at the same pressure by the amount

$$\int_{T_1}^{T'} \frac{\kappa dT}{T},$$

where κ is the specific heat of the steam during superheating, that is to say, the amount of heat required to raise unit mass of the steam 1° C. when its temperature exceeds T_1 the temperature of saturation, the process of superheating being conducted at constant pressure. It has been pointed out in § 65 that κ is not strictly constant, but its variations, for any given constant pressure, are not very great, and an approximate reckoning of the entropy may be made by treating κ as a constant with the value given in the table in § 65 for the mean specific heat over the range through which superheating occurs. Using this approximate method we should have, for the entropy after superheating at constant pressure from T_1 the temperature of saturation to T',

$$\phi = \phi_s + \kappa (\log_\epsilon T' - \log_\epsilon T_1),$$

where ϕ_s is the entropy of dry saturated steam at the same pressure.

It is however more convenient and generally more accurate to take the entropy of superheated steam, when the temperature and pressure are known, from the Callendar Tables, reading it directly or by interpolation (see Table E of the Appendix). Having found the entropy in the superheated state, we can extend the entropy-temperature diagram in the manner shown in fig. 30, where starting from c, which represents dry saturated steam at a particular pressure, cr is drawn to show the increased amount of entropy produced by superheating as calculated for a series of temperatures above the temperature of saturation.

We may now apply the entropy-temperature diagram as a means of representing the ideal action of an engine supplied with superheated steam. Starting as before with feed-water at T_2 whose temperature and entropy are represented by a, ab is the process of heating the feed-water up to the temperature at which steam is formed, bc is the process of evaporation, cr is the subsequent superheating, during which an extra amount of heat is taken in represented by the area under cr, namely $pcru$.

After superheating to any extent let the cycle be completed by the processes rs and sa, namely, by adiabatic expansion to temperature T_2 and condensation at that temperature. The diagram shows that, in consequence of superheating, the work done by the substance is increased by the area $dcrs$, while the heat taken in is increased by $pcru$. The efficiency is slightly increased, since this additional heat is received at temperatures somewhat higher than those at which the other portions of the heat were received. But unless superheating be carried very far the extra supply of heat is too small a part of the whole to make any large difference in the efficiency of the ideal engine we are dealing with here. In the case sketched in fig. 30 the steam is supposed to be superheated as much as 110 degrees centigrade above the boiler temperature, but the diagram shows this makes little improvement in the ideal efficiency. In real engines superheating does make a marked difference, but its influence is indirect, being mainly due to the fact that it tends to prevent the steam from being condensed by contact with the metal of the cylinder and piston. This effect of superheating will be considered in the next chapter. Nothing of the kind takes place in the ideal case now dealt with, because here we postulate adiabatic expansion, or, in other words, a perfectly non-conducting cylinder and piston.

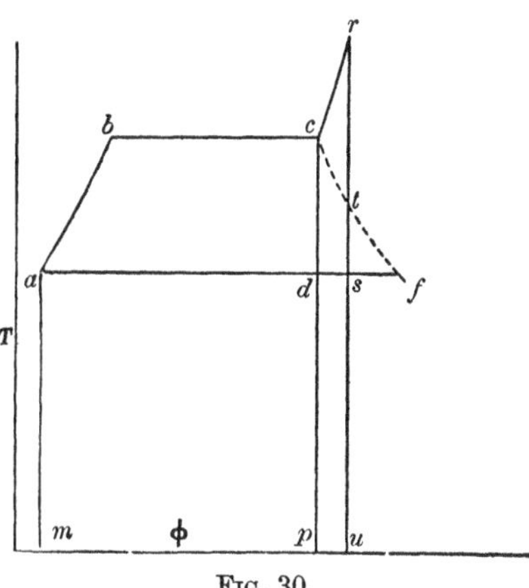

FIG. 30.

It would evidently be fallacious to suppose that when superheating is applied to the steam of the ideal engine the increased range of temperature implies anything like a corresponding gain of efficiency, for the chief part of the heat is still taken in at the temperature of saturation, and its value for conversion into work depends on the temperature at which it is taken in, not upon the temperature to which the working substance is subsequently raised.

The efficiency of the cycle may in fact be expressed thus, in terms of the *average* temperature at which heat is taken in, namely, a temperature T_m which represents the average height of the figure *mabcru*. The area of that figure, which measures the heat taken in, is equal to $T_m . mu$. The heat rejected is $T_2 . mu$. The heat converted into work is the difference between these quantities, namely, $(T_m - T_2)\, mu$. Hence the efficiency is

$$\frac{T_m - T_2}{T_m}, \text{ or } 1 - \frac{T_2}{T_m},$$

an expression directly comparable with that which was found for a perfect engine working on the Carnot cycle (§ 50).

In the diagram, fig. 30, the adiabatic line *rs* shows by its intersection of the curve *cf* at *t* the stage in the expansion at which the steam will cease to be superheated. At the point t it is dry and saturated: as the expansion proceeds it becomes wet, and at the end of expansion the condensed part is $\frac{sf}{af}$ of the whole. The extent to which superheating has to be carried if the steam is just to be dry, and no more, at the end of expansion, is readily found by drawing a vertical line through f to meet the continuation of the curve cr.

94. Work done in the Rankine cycle. Calculation of the heat-drop and the efficiency. Though the work done in the Rankine cycle—or rather the heat converted into work—is exhibited by the area of the enclosed figure in the entropy-temperature diagram whose boundary represents the cyclic process, its amount is best determined, as in § 75, by finding the heat-drop $I_1 - I_2$, which is the thermal equivalent of the work done by the working substance in the cylinder, and subtracting from that the thermal equivalent of the work spent upon the substance in the feed-pump, namely the small term $A\,(P_1 - P_2)\,V_{w_2}$. Thus the net amount of work done in the cycle as a whole is, in heat units,

$$I_1 - I_2 - A\,(P_1 - P_2)\,V_{w_2}.$$

Here I_1 is the total heat of the steam on admission to the cylinder, whether wet, dry, or superheated, and I_2 is its total heat after expanding to the lower limit of temperature, namely the temperature of condensation, at which it is supposed to be returned through the feed-pump to the boiler. V_{w_2} is the volume of water at that temperature. To interpret this expression numerically we must of course know the initial conditions, which determine I_1, and also be able to calculate I_2 for the wet mixture after expansion. Referring to fig. 30, if the steam before expansion was saturated, then I_1 is the total heat of saturated steam in the state c, and I_2 is the total heat of the wet mixture in the state d. But if the steam was superheated to the state r before expansion, I_1 is the total heat in that state and I_2 is the total heat of the wet mixture in the state s. In either case the entropy does not change during expansion, and this gives a convenient means of calculating I_2. Let ϕ_1 represent the entropy in the initial state (to be found from the tables), and let ϕ_{s_2} represent the entropy of saturated steam at the final pressure, and ϕ_{w_2} the entropy of water at the final pressure.

After expansion the mixture still has the entropy ϕ_1. To bring it to the state of dry saturated steam at the same pressure, namely the state f, would require the addition of a quantity of heat equal to $T(\phi_{s_2} - \phi_1)$. Its total heat would then be I_{s_2}. Hence

$$I_2 = I_{s_2} - T_2(\phi_{s_2} - \phi_1) \quad\quad (7).$$

Or, alternatively, we may think of the heat which would be removed from the wet mixture if it were completely condensed, namely $T_2(\phi_1 - \phi_{w_2})$. Its total heat would then be I_{w_2}. Hence

$$I_2 = I_{w_2} + T_2(\phi_1 - \phi_{w_2}) \quad\quad (8).$$

Either of these expressions will serve for calculating I_2.

To take a numerical example, let steam be supplied at a pressure of 180 lb. per sq. inch, superheated to 250° C., and let it expand adiabatically to a pressure of 1 lb. per sq. inch. With these data the tables show that I_1 is 702·95, ϕ_1 is 1·6188, T_2 is 311·8, I_{s_2} is 612·46 and ϕ_{s_2} is 1·9724. Hence I_{s_2} is 502·2 and the heat-drop $I_1 - I_2$ is 200·7. If the steam had not been superheated we should have had $I_1 = 668·5$ and $\phi_1 = 1·5620$. I_2 would in that case have been 484·5, making the heat-drop 184·0.

The feed-pump term $A(P_1 - P_2) V_{w_2}$ is

$$\frac{(180 - 1)\, 144 \times 0·0161}{1400} = 0·30,$$

or less than one six-hundredth of the work done in the cylinder.

To find the efficiency of the Rankine cycle we have to compare the work done with the heat taken in. The heat taken in is found by the method of §§ 64–66 to be

$$I_1 - I_{w_2} - A(P_1 - P_2) V_{w_2}.$$

Hence the efficiency is

$$\frac{I_1 - I_2 - A(P_1 - P_2) V_{w_2}}{I_1 - I_{w_2} - A(P_1 - P_2) V_{w_2}},$$

which applies whether the steam is initially saturated, wet, or superheated.

The importance of this calculation lies in the fact that the efficiency of an ideal steam-engine working on the Rankine cycle is the highest possible efficiency that is compatible with the condition that the steam is to be completely condensed at the lower limit of temperature, the cycle being completed by pumping the condensed water back to the boiler.

When the efficiency of the Rankine cycle is worked out for various values of the initial and final pressures the advantage will be seen of raising the pressure of admission and lowering the pressure of exhaust. This is illustrated by the following tables, both of which relate to steam saturated on admission to the engine. In the first table the initial pressure is varied and the final pressure is kept constant at 1 lb. per sq. inch. In the second table the initial pressure is kept constant at 180 lb. per sq. inch and the final pressure is varied.

Rankine Cycle for Saturated Steam. Effect of varying the Initial Pressure.

Initial pressure (pounds per square inch absolute)	Heat-drop, to a final pressure of 1 lb. per square inch (lb. calories)	Work done per lb. of steam allowing for work spent in feed-pump (lb. calories)	Heat supplied per lb. of steam (lb. calories)	Efficiency of the Rankine cycle
300	202·0	201·5	634·8	0·317
280	199·6	199·1	634·2	0·314
260	197·0	196·6	633·4	0·310
240	194·2	193·8	632·6	0·306
220	191·1	190·8	631·7	0·302
200	187·7	187·4	630·7	0·297
180	184·0	183·7	629·6	0·292
160	179·8	179·5	628·3	0·286
140	175·0	174·8	626·8	0·279
120	169·5	169·3	625·1	0·271
100	162·9	162·8	623·0	0·261

These figures show that when the admission pressure is fairly high it takes a large increase of pressure to effect much improvement in the efficiency. The twenty-pound rise, for example, from 280 to 300 augments the efficiency by only one per cent.

On the other hand it is of great advantage to have what engineers call a "high vacuum"—that is to say to make the pressure of condensation as low as possible. If a high vacuum can be maintained and effectively utilized we obtain from the steam the work which it is capable of doing under conditions of low pressure but of very large volume in the last stages of the expansion. The second table illustrates the gain in heat-drop and in efficiency that results, in the Rankine cycle, from reducing the lower limit of pressure.

Rankine Cycle for Saturated Steam. Effect of varying the Final Pressure.

Final Pressure (pounds per square inch, absolute)	Heat-drop from an initial pressure of 180 lb. per square inch	Work done per lb. of steam allowing for work spent in feed-pump	Heat supplied per lb. of steam	Efficiency of the Rankine cycle
	(lb. calories)	(lb. calories)	(lb. calories)	
4	144·5	144·2	601·1	0·240
3	153·1	152·8	607·5	0·251
2	164·9	164·6	616·1	0·267
1·5	173·0	172·7	621·8	0·278
1	184·0	183·7	629·6	0·292
0·5	201·8	201·5	642·0	0·314

The last figure corresponds to a vacuum of nearly 29 inches of mercury with the barometer at 30 inches.

To secure in practice the benefit of a high vacuum the steam must continue to do useful work in expanding down to the pressure at which condensation is to take place. In engines of the cylinder and piston type this is impracticable for two reasons: the volume of the steam becomes excessive, and the mechanical friction of the piston against the cylinder becomes relatively so great as to absorb all the work done in the final stages. But with the steam turbine these considerations do not apply; there is then nothing to prevent the steam from continuing to do useful work as it expands right down to the pressure of the condenser, and special pains are accordingly taken to maintain a good vacuum in the condenser of a steam turbine. It is largely for this reason that steam turbines are able to achieve in practice a greater efficiency than even the best engines of the piston type.

95. Values of the entropy of water and steam: entropy-temperature chart. In applying this useful graphic method to the investigation of particular cases in the expansion of steam it is convenient to have an entropy-temperature chart for water and steam drawn on section-paper throughout the range of pressures which are found in practice: the construction for particular cases is then readily made by adding horizontal straight lines to correspond with the formation and condensation of the steam, while any adiabatic process is represented by a vertical line. The student will find it instructive to draw for himself, to a large scale, a chart of this kind, using the values given in the table for ϕ_w and ϕ_s, and also showing the entropy of superheated steam at various pressures and temperatures so as sufficiently to map out the region of superheating in the chart.

A skeleton of such a chart is shown in miniature in fig. 31.

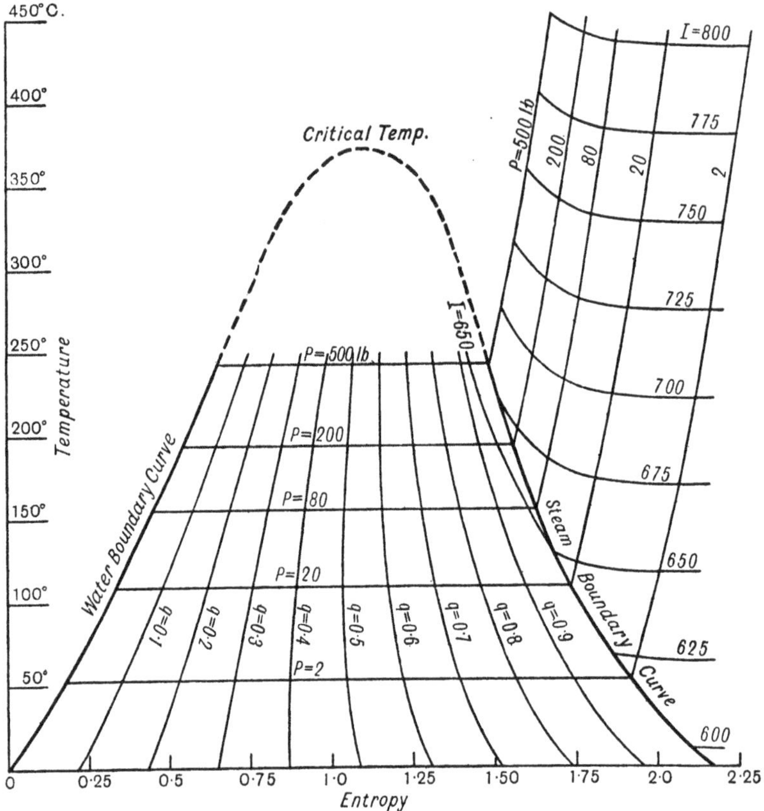

FIG. 31. Entropy-temperature chart for water and steam.

There the curve on the left marked "water boundary curve" shows the relation of entropy to temperature before steam begins to form: the curve on the right marked "steam boundary curve" shows the same relation when all the water is converted into steam. Between these two curves is what may be called the wet steam region. The horizontal distance between the two curves at any point, or $\phi_s - \phi_w$, represents the gain of entropy which occurs while the water is changing into steam (namely L/T).

The steam boundary curve separates the saturated and superheated states. To the right of it lies the region of superheating which may most conveniently be mapped out by drawing a system of *lines of constant pressure* as shown. In the wet region the constant pressure lines are horizontal straight lines (of uniform temperature). To the left of this the constant pressure lines are practically coincident with the water boundary curve. Each constant pressure line may be regarded as running up almost along that curve, it then continues horizontally to the steam boundary curve, and then rises in the manner shown in the region of superheating. Such a line shows completely the changes which the entropy undergoes as the water is successively heated, evaporated and superheated, all at constant pressure. When a sufficient number of constant pressure lines are drawn it is easy to mark at once on the chart a point corresponding to any assigned state of pressure and superheat. *Lines of constant volume* may be drawn without difficulty (see § 96 below), but they are of less general importance and are omitted from the figure to avoid confusion. *Lines of constant total heat* (I) may also be added to the chart: examples of them are shown ranging from $I = 800$ downwards. The line for $I = 650$ lies partly in the wet steam region, taking a sudden bend where it crosses the boundary curve. The wet region as well as the region of superheat may be completely mapped out by a system of such lines. In the wet steam region it is very useful to draw a system of *lines of constant dryness*, namely lines showing the proportion of dry steam in the mixture, expressed as a fraction of the whole. These are drawn in the figure for values of $q = 0{\cdot}9$, $0{\cdot}8$, $0{\cdot}7$, etc., down to $q = 0{\cdot}1$.

With the aid of such a chart it is easy to trace the changes which steam undergoes in adiabatic expansion. Taking for example steam whose pressure is 200 lb. per sq. inch, initially superheated to 400° C., we mark the corresponding point and draw from it a vertical line towards the base to represent the adiabatic process. This cuts the boundary curve at about $t = 102°$ C., showing that the steam has become just saturated when its pressure has fallen to 16 lb. per sq. inch. The line then enters the wet steam region. If the expansion be continued till the temperature falls to 50° C., we find that the adiabatic line cuts the line of constant dryness $q = 0{\cdot}9$, showing that the proportion of water is 10 per cent. when the pressure has fallen to 1·8 lb.

Having thus found the dryness after expansion we might proceed to deduce the total heat in the wet mixture and so to determine the heat-drop. But numerical results for the ideal process are better found by calculation, in the manner already explained, than by measurement from the diagram. The chief merit of the entropy-temperature diagram is not that it provides a means of solving problems graphically, but that it helps to make the thermodynamic action of an ideal heat-engine intelligible by showing it as a whole, exhibiting the nature of each step and the effect of any changes in the initial or final conditions.

In fig. 31 a conjectural curve has been added (shown by a broken line) to connect the water and steam boundary curves in the region of high pressures where, at present, there are not enough data for a precise determination of the entropy. It is a smooth curve forming a continuation of each boundary curve and drawn so as to be tangent to the line $t = 374$, which is the critical temperature of water, that being the temperature at which the distinction between ϕ_w and ϕ_s disappears. The line as drawn uses tentative values of ϕ_s and ϕ_w calculated by Callendar (*Properties of Steam*, p. 196) on certain assumptions which, however, await experimental verification. The water and steam boundary lines are two parts of one continuous curve whose summit is at the critical point[1].

Not the least merit of the entropy-temperature diagram as a means of representing graphically the cycle of operations in a heat-engine is that it shows the heat taken in and the heat rejected, as well as the work done, and so allows estimates of efficiency to be made by inspection of the diagram itself. The advantage, for instance, which results from raising the initial pressure of the steam

[1] The following provisional values of ϕ_s and ϕ_w are quoted to supplement the data for high-pressure steam in § 77:

Temp. Cent. t	Entropy		Temp. Cent. t	Entropy	
	Steam ϕ_s	Water ϕ_w		Steam ϕ_s	Water ϕ_w
250	1·471	0·668	320	1·373	0·828
260	1·458	0·690	330	1·355	0·853
270	1·444	0·712	340	1·334	0·878
280	1·431	0·735	350	1·308	0·906
290	1·417	0·757	360	1·270	0·936
300	1·404	0·780	370	1·196	0·975
310	1·389	0·804	374	1·047	1·047

is readily shown in a diagram such as fig. 28 by drawing horizontal lines at temperatures corresponding to the initial pressures which are to be compared, and vertical lines through the points where they meet the steam boundary curve, the vertical lines being continued to meet the base, which is the absolute zero of temperature. Comparison of the enclosed areas then shows that while the heat taken in is but slightly increased with higher boiler pressure there is a more considerable gain of work, a result which is of course to be expected from the fact that the general temperature of reception of the heat is raised.

When a vertical line is drawn to represent the adiabatic expansion of a mixture of steam and water, it is clear from the diagram that if the mixture consists mostly of water to begin with it will become drier as it expands, instead of wetter as is the case when the initial proportion of water is less. In the region of ordinary working pressures the "water" and "steam" curves of fig. 31 have nearly equal inclinations to the vertical line which represents an adiabatic process. Hence if such a line be drawn starting from a point midway between the two curves it will continue to lie nearly midway between them: in other words, if there is about 50 per cent. of water present at the beginning of adiabatic expansion, nearly the same percentage will be found as the expansion goes on. When the steam is much wetter than this to begin with, adiabatic expansion makes it drier.

It will be shown later that the entropy-temperature diagram is also of service in exhibiting the changes of dryness which occur in real steam-engines, where the action is by no means adiabatic.

96. Entropy-temperature diagram for steam used non-expansively, or with incomplete expansion. By way of contrast with the cases treated in §§ 92 and 93, we may draw the entropy-temperature diagram for a steam-engine working without expansion. The four steps of the cycle have been stated in § 71, and the volume-pressure diagram is drawn there (fig. 17). In the entropy-

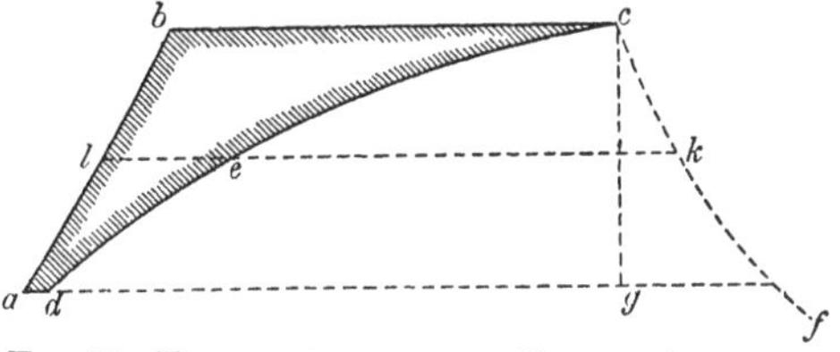

FIG. 32. Entropy-temperature diagram for steam used non-expansively.

temperature diagram (fig. 32) we have the four corresponding lines *ab*, *bc*, *cd*, *da*. *ab* is the heating of the water from T_2 to T_1. *bc* is the conversion of the water into steam, *cd* the partial condensation which takes place when the cold body is applied, the piston meanwhile remaining at the end of its forward stroke, and *da* is the remainder of the condensation, which occurs while the piston is pressed in, the cold body being still applied. *cd* is a *line of constant volume*, for throughout the change which it represents the substance as a whole suffers no change of volume, since it remains in the cylinder and there is no movement of the piston. To find points in *cd*, draw the saturation curve *cf* as in former examples, and at any temperature T intermediate between T_1 and T_2 draw the line *lk*. We have to divide *lk* in a point *e* such that $\frac{le}{lk}$ shall represent q, the dryness of the steam at the time its temperature has fallen to T. The dryness q is determined by the consideration that the volume of the substance as a whole is sensibly equal to qV, where V is the volume of unit mass of saturated steam at T. This remains equal to V_1, which is the volume originally occupied by unit mass before the process of condensation began, since there is no change of volume in the process. Hence $q = \frac{V_1}{V}$, and e is found by making

$$le = \frac{V_1}{V} lk.$$

The work which is lost through the absence of adiabatic expansion is, in heat units, equal to the area *cgd*.

The much more usual case in which the steam is incompletely expanded admits of similar treatment. Let adiabatic expansion be carried on until, at the end of the stroke, the temperature has fallen to the level indicated by c' in the entropy-temperature diagram, fig. 33. This process is represented by the line cc'. Then let the steam be suddenly cooled by applying the cold body. The constant-volume curve $c'd$ shows this cooling; after which the return

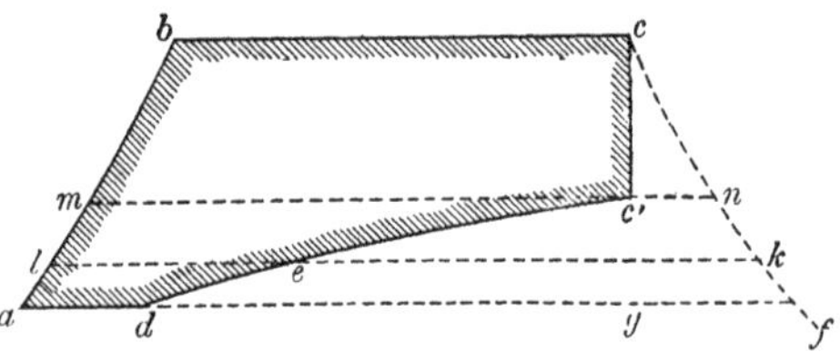

FIG. 33. Entropy-temperature diagram showing incomplete expansion of steam.

stroke takes place, which is shown by *da*. To draw the curve *c'd* take a line *lk* at any lower temperature T and take e in it, such that

$$le = \frac{q' V_s'}{V_s} lk,$$

where V_s' is the volume of unit mass of saturated steam at the temperature corresponding to c', and q' is the dryness at c', which is equal to $\frac{mc'}{mn}$, and V_s is the volume of unit mass of saturated steam at temperature T.

In dealing with the process of sudden condensation represented by the line *cd* in fig. 32 and *c'd* in fig. 33 we have supposed, to simplify the statement, that the steam is retained in the cylinder and the cold body is applied to it. But it makes no difference if the steam be allowed to escape into a separate vessel, to be condensed there. Just the same amount of work is done, for the pressure on the piston is the same in that case as in the other. Hence the area of the entropy-temperature diagram is unaffected, and since that is true whatever be the value of T the form of the curve *cd* or *c'd* is unchanged[1].

97. Derivation of Clapeyron's equation. As another example of the use of the entropy-temperature diagram we may

[1] The constant-volume curve *cd* or *c'd* in the entropy-temperature diagram may be drawn as follows by an application of equation (3) of § 81. Let U represent the volume of the mixture of steam and water at any stage in the process of condensation, the temperature then being T. Let λ represent the heat which would be given out if the condensation of the mixture were completed at the temperature T. Then by that equation

$$U - V_w = \frac{J\lambda}{T} \frac{dT}{dP},$$

V_w being the volume when the substance is all water. Hence

$$\frac{\lambda}{T} = \frac{U - V_w}{J} \frac{dP}{dT}.$$

But $\frac{\lambda}{T}$ is the length *le*, if the line *le* be drawn at the level T, and U is the volume of the cylinder, which is constant. We therefore have a relation which allows *le* at any level of temperature to be readily determined when the values of $\frac{dP}{dT}$ for saturated steam are known. These may be found by measurement of the slope of the pressure-temperature curve, or approximately from tables of the properties of saturated steam by dividing small differences of pressure by corresponding differences of temperature.

apply it in obtaining the Clapeyron equation of § 80. Take as before the action of an engine following Carnot's cycle and working through a very small interval of temperature δT. The indicator diagram, sketched in fig. 23, has the area $\delta P\,(V_s - V_w)$. The entropy-temperature diagram is a long narrow strip whose length is $\frac{L}{T}$ and height δT. Its area is $\frac{L\delta T}{T}$, equivalent in work units to $\frac{JL\delta T}{T}$. Hence

$$\delta P\,(V_s - V_w) = \frac{JL\delta T}{T},$$

and

$$V_s - V_w = \frac{JL}{T}\frac{dT}{dP},$$

which is Clapeyron's equation.

98. Adiabatic equation for superheated steam. In the adiabatic expansion of steam that is initially superheated two stages have to be considered. The first stage extends down to the temperature of saturation. That temperature may be found graphically as in § 93, or by inspection from the tables, namely by observing at what temperature ϕ_s is equal to the initial entropy ϕ_1. During the first stage the steam behaves as a nearly perfect gas, in which, according to Callendar, the following equations apply:

$$P\,(V - b) = RT, \qquad P\,(V - b)^{1\cdot3} = \text{const.},$$

$$\frac{P}{T^{\frac{13}{3}}} = \text{const.}, \quad T\,(V - b)^{0\cdot3} = \text{const.}$$

Here R is 154·17 when V is expressed in cub. feet per lb. and P in pounds per sq. foot. b is 0·016, a small constant which may be omitted from the formulas unless V is small. Intermediate points may also be found from Table E, noting at what temperature the entropy is equal to ϕ_1 when the pressure has fallen to any assigned figure.

In the second stage the saturation line has been crossed. We are then dealing with a wet mixture, and the method of § 90 applies.

99. Mollier's diagrams. In 1904 Dr R. Mollier[1] introduced two novel methods of representing the properties of steam by charts

[1] R. Mollier, "Neue Diagramme zur technischen Wärmelehre," *Zeitschrift des Vereins deutscher Ingenieure*, 1904, p. 271. See also his *Neue Tabellen und Diagramme für Wasserdampf*, Berlin (Julius Springer), 1906.

which furnish very valuable alternatives to the chart of entropy and temperature. Both are interesting and useful, and one of them in particular may be said to offer advantages in point of convenience which entitle it to the first place among these devices. It puts into the hands of engineers an extremely simple method of determining the state of steam which has expanded adiabatically from any initial condition, superheated or not, and of finding the

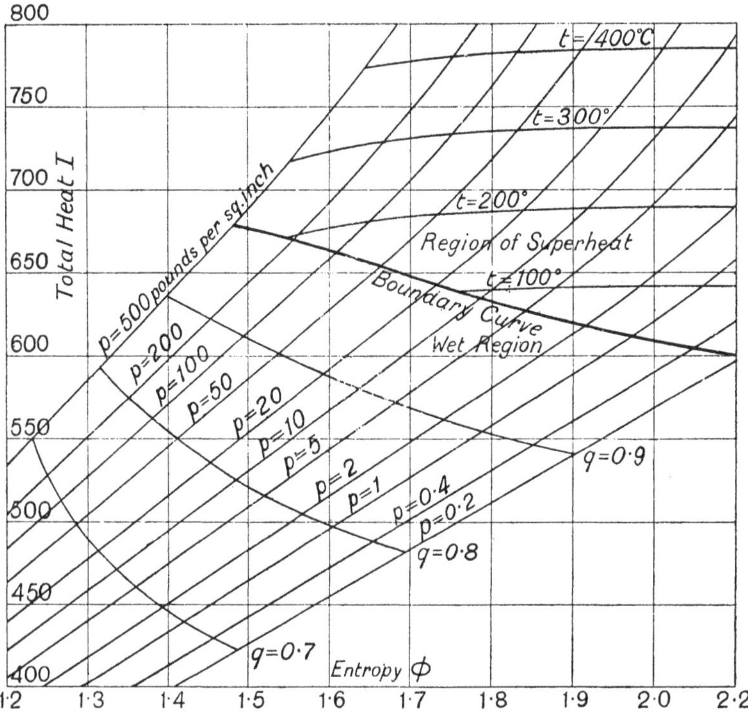

FIG. 34. Mollier's chart of entropy and total heat.

greatest theoretical output attainable from steam when the initial condition and the lower limit of temperature are assigned. For it allows the heat-drop to be found graphically by direct measurement of the length of a straight line which represents the process of adiabatic expansion.

In this chart, which is shown in a skeleton form in fig. 34, Mollier takes as coordinates the entropy and the total heat.

For practical purposes, in steam engineering, only the useful region near the steam boundary curve need be drawn, namely from

$\phi = 1{\cdot}4$ to $2{\cdot}2$ and from $I = 400$ to 800. The vertical lines being lines of constant entropy are adiabatics. Lines of constant pressure are drawn extending throughout the wet steam region, where they are straight, over the boundary curve into the region of superheat where they become curved, but with no abrupt change of direction in crossing the boundary. In the wet region there is also a system of lines of constant dryness, illustrated here for $q = 0{\cdot}9$, $0{\cdot}8$ and $0{\cdot}7$. In the region of superheat there is a system of lines of constant temperature drawn for intervals of 100° C. in fig. 34. On a large-scale chart with many such lines, a point may be readily found to correspond with any given initial condition of pressure and superheat. Once this point is marked, a vertical line drawn down through it shows an adiabatic process of expansion and allows the consequent changes of condition to be traced, giving the temperature for each pressure so long as there is superheat, and also the wetness for each pressure in the final stages after the boundary curve has been crossed. Further, the length of this vertical line is the "heat-drop"; it is a direct measure of the work done in the cylinder in an adiabatic process, being $I_1 - I_2$. This fact alone gives the diagram much value as an aid in calculations of theoretical performance. Again, since in a throttling process I is constant, the diagram exhibits such processes by horizontal straight lines. It will be seen from these that in any throttling process the entropy is increased, the steam becomes drier if it is wet, and superheated if it is already dry, although its temperature falls.

In Mollier's other chart the coordinates are the pressure and the total heat. This construction misses the feature possessed by charts in which the entropy is one of the coordinates, of exhibiting adiabatic processes by straight lines, but it has certain advantages, and forms in any case an interesting adjunct to the ϕ-I chart. Any of these charts may be regarded as a graphic way of stating the properties of steam, superheated as well as saturated. The P-I chart appeals to engineers as being a graphic table with a pressure base, and is particularly convenient not only in showing the total heat and the heat of superheating for given pressure and temperature, but also in tracing changes of volume. It has the property that isotherms or lines of constant temperature, and also lines of constant volume are straight, and consequently systems of such lines are very readily drawn.

A skeleton of the P-I chart is shown in fig. 35 for the region that is useful in practice, from $I = 600$ to 800. The space below the boundary curve is the wet region, and is mapped out as in the other charts by lines of constant dryness. Throughout this space, as well as in the region of superheat above, lines of constant volume are drawn. These are straight lines in the region above the boundary curve, and they remain sensibly straight in the region of wetness but suffer an abrupt change of direction on

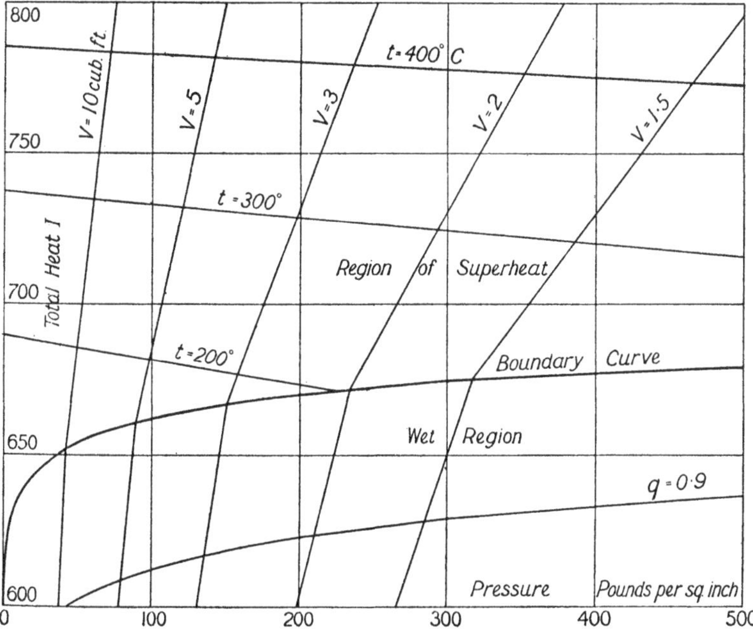

FIG. 35. Mollier's chart of pressure and total heat.

crossing the boundary curve. A few representative lines of constant volume are sketched in the figure. The lines of constant temperature are also straight and slope down to the right in the region of superheat; in the wet region they of course coincide with lines of constant pressure.

With regard to the practical use of Mollier charts on a large scale, of the kinds represented in miniature and in skeleton form in figs. 34 and 35, it should be said that the publication of full numerical tables of the properties of steam, and also tables of heat-drop under various conditions, has now made it easy to solve the problems that are met with in steam practice without resorting to graphic methods. But charts of a similar kind, especially those which show I in relation to ϕ for other substances than steam, namely ammonia, carbonic acid, and other working fluids used in the mechanical production of cold, for which the available tables are less complete, are of great service in the design of refrigerating machinery. For the purpose of drawing diagrams which will exhibit the thermodynamic cycle used in such machines the chart of entropy and total heat has special advantages.

100. Entropy-temperature diagrams in engines using a regenerator. An engine such as Stirling's, which substitutes the use of a regenerator for the adiabatic expansion and compression in Carnot's cycle, has an entropy-temperature diagram of the type shown in fig. 36. The isothermal operation of taking in heat at T_1 is represented by ab; bc is the cooling of the substance from T_1 to T_2 in its passage through the regenerator, where it deposits heat: cd is the isothermal rejection of heat at T_2; and da is the restoration of heat by the regenerator while the substance passes through it in the opposite direction, by which the temperature is raised from T_2 to T_1. Assuming the action of the regenerator to be reversible and therefore ideally perfect, bc and ad are precisely similar curves whatever be their form. The area of the figure is then equal to the area of the rectangle which would represent the ordinary Carnot cycle (fig. 27). The equal areas $pbcq$ and $ndam$ measure the heat stored and restored by the regenerator.

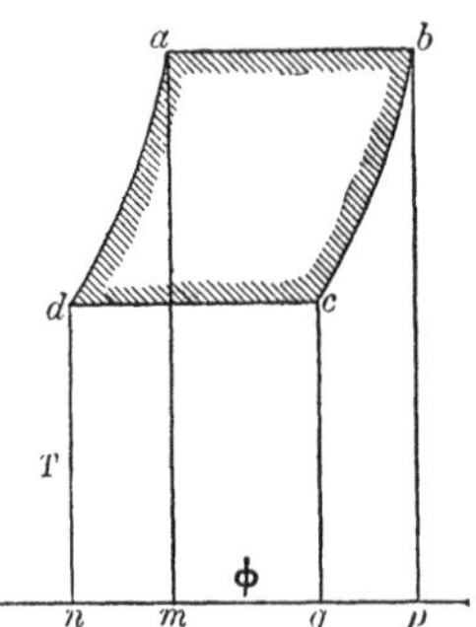

FIG. 36. Entropy-temperature diagram of perfect engine using a regenerator.

When the working substance is air and the regenerative changes take place either under constant volume, as in Stirling's engine, or under constant pressure, as in Ericsson's, the specific heat K being treated as constant, ad and bc are logarithmic curves with the equation

$$\phi = \int \frac{KdT}{T},$$

K being K_v in one case and K_p in the other.

101. Joule's air-engine. A type of air-engine was proposed by Joule which, for several reasons, possesses much theoretical interest. Imagine a chamber C (fig. 37) full of air (temperature T_2), which is kept cold by circulating water or otherwise; another chamber A heated by a furnace and full of hot air in a state of compression (temperature T_1); a compressing cylinder M by which air may be pumped from C into A, and a working cylinder N in which air from A may be allowed to expand before passing back into the cold chamber C. We shall suppose the chambers A and C to be large, in comparison with the volume of air that passes in each stroke, so that the pressure in each of them may be taken as sensibly constant. The pump M takes in air from C, compresses it adiabatically until its pressure becomes equal to the pressure in A, and then, the valve v being opened, delivers it into A. The indicator diagram for this action on the part of the pump is the diagram $fdae$ in fig. 38. While this is going on, the same quantity of hot air from A is admitted to the cylinder N, the valve u is then closed, and the air is allowed to expand adiabatically in N until its pressure falls to the pressure in the cold chamber C. During the back stroke of N this air is discharged into C. The operation of N is shown by the indicator diagram $ebcf$ in fig. 38. The area $fdae$ measures the work spent in driving the pump; the area $ebcf$ is the work done by the air in the working cylinder N. The difference, namely, the area $abcd$, is the net amount of work obtained by carrying the given quantity of air through a complete cycle. Heat is taken in when the air has its temperature raised

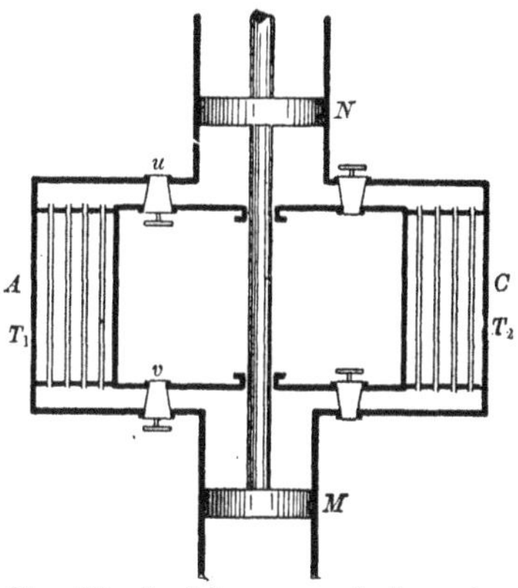

FIG. 37. Joule's proposed air-engine.

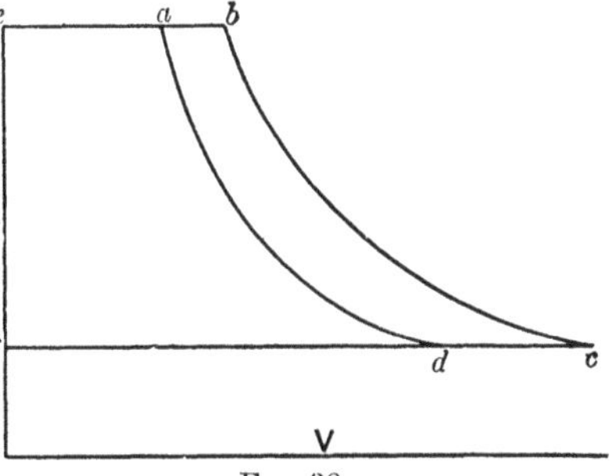

FIG. 38.
Indicator diagram for Joule's air-engine.

on entering the hot chamber A. Since this happens at a pressure which is sensibly constant,

$$Q_A = K_p (T_b - T_a),$$

where T_b is T_1, the temperature of A, and T_a is the temperature reached by adiabatic compression in the pump. Similarly, the heat rejected

$$Q_C = K_p (T_c - T_d),$$

where $T_d = T_2$, the temperature of C, and T_c is the temperature reached by adiabatic expansion in N. Since the expansion and compression both take place between the same terminal pressures, the ratio of expansion and compression is the same. Calling it r, we have

$$\frac{T_a}{T_d} = \frac{T_b}{T_c} = r^{\gamma - 1}$$

(§ 41), and hence also

$$\frac{T_b}{T_a} = \frac{T_c}{T_d}, \text{ and } \frac{T_b - T_a}{T_a} = \frac{T_c - T_d}{T_d}.$$

Hence

$$\frac{Q_A}{Q_C} = \frac{T_a}{T_d} = \frac{T_b}{T_c},$$

and the efficiency

$$\frac{Q_A - Q_C}{Q_A} = \frac{T_a - T_d}{T_a} = \frac{T_b - T_c}{T_b}.$$

This is less than the efficiency of a perfect engine working between the same limits of temperature $\left(\frac{T_1 - T_2}{T_1}\right)$ because the heat is not taken in and rejected at the extreme temperatures.

The atmosphere may take the place of the chamber C: that is to say, instead of having a cold chamber, with circulating water to absorb the rejected heat, the engine may draw a fresh supply at each stroke from the atmosphere and discharge into the atmosphere the air which has been expanded adiabatically in N.

The entropy-temperature diagram for this cycle is drawn in fig. 39, where the letters refer to the same stages as in fig. 38. After adiabatic compression da, the air is heated in the hot chamber A and the curve ab for this process has the equation

$$\phi = \int_{T_a}^{T} \frac{K_p dT}{T} = K_p (\log_\epsilon T - \log_\epsilon T_a).$$

Then adiabatic expansion gives the line bc, and cd is another logarithmic curve for the rejection of heat to C by cooling under constant pressure. The ratio $\frac{T_a}{T_b}$ which is represented by $\frac{ea}{eb}$ in fig. 38 and by $\frac{ma}{nb}$ in fig. 39, shows the proportion which the volume of the pump M must bear to the volume of the working cylinder N. The need of a large pump would be a serious drawback in practice, for it would not only make the engine bulky but would cause a relatively large part of the net indicated work to be expended in overcoming friction within the engine itself.

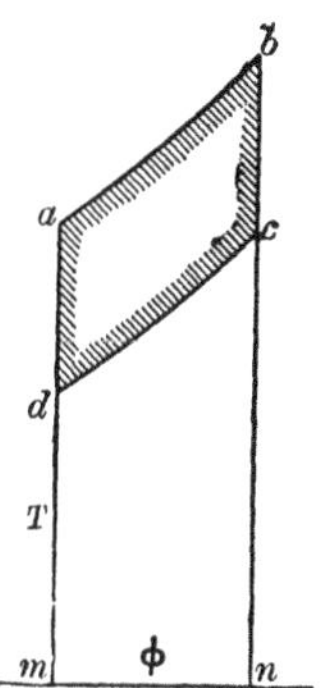

FIG. 39. Entropy-temperature diagram for Joule's air-engine.

In the original conception of this engine by Joule it was intended that the heat should reach the working air through the walls of the hot chamber, from an external source. But instead of this we may have combustion of fuel going on within the hot chamber itself, the combustion being kept up by the supply of fresh air which comes in through the compressing pump, and, of course, by supplying fuel either in a solid form from time to time through a hopper, or in a gaseous or liquid form. In other words, the engine may take the form of an *internal combustion* engine. Internal combustion engines, essentially of the Joule type, employing solid fuel have been used on a small scale, but by far the most important development of this type is the explosive gas or oil-engine. Its cycle is substantially Joule's, considerably modified, however, by features which will be noticed in a later chapter.

This, however, is not the only reason why Joule's cycle is interesting. It has found application in the reversed form as a practical process for cooling air. Refrigerating machines in which air is the working substance were at one time extensively used to keep the temperature of rooms on board ship below the freezing-point, to allow frozen meat to be carried over seas, and such machines work, as we shall see in the next chapter, by reversing the cycle suggested by Joule.

102. The heat account in a real process. The importance, in thermodynamic theory, of considering a complete cycle was pointed out in § 28 and has been illustrated by many examples.

In the action of a steam-engine the complete cycle involves the boiler and condenser as well as the engine itself. But it is often useful to deal with the engine separately, regarding it as an apparatus which begins at the steam-pipe and ends at the exhaust-pipe, an apparatus through which steam flows from a region of high pressure to a region of low pressure, entering in one state and being discharged in another state. It is instructive to draw up a heat account or balance sheet for the passage of steam through an engine or any other thermodynamic apparatus, setting down the energy that comes in at one end, the energy that passes off at the other end, the heat that is converted into work during the passage, and the heat that is lost by conduction or radiation to the outside. In considering the performance of ideal engines we simplified matters by supposing there was no heat so lost, but in any real process there are heat losses to be taken into account as well as irreversible actions within the engine, both of which affect the output of mechanical work.

Whether the apparatus considered be an engine cylinder, or the series of cylinders of a compound engine, or a turbine, or a throttling device, we may in all cases compare the state of the fluid at entry and at exit, as for example in the admission-pipe of an engine and the exhaust-pipe. Imagine a steady flow of the working fluid through the apparatus. At entry let its pressure be P_1, its volume (per lb.) V_1, and its internal energy E_1. At exit let its pressure be P_2, its volume V_2, and its internal energy E_2. To make the comparison complete we may write K_1 for the kinetic energy (also per lb.) of the stream as it enters, and K_2 for its kinetic energy as it leaves. In passing through the apparatus the fluid will, in general, do external work, and also lose by conduction some heat to external space. Let W represent the output of work, and let Q_l represent the heat lost by conduction to external space, both of these quantities (like the others) being reckoned, in thermal units, per pound of the fluid that passes through.

Each pound that enters the apparatus represents a supply of energy equal to $K_1 + E_1 + AP_1V_1$, for E_1 is the internal energy it carries, and P_1V_1 is the work done by the fluid behind in pushing it in. But $E_1 + AP_1V_1$ is equal to I_1, the total heat per pound of the fluid in its actual state at entry. Similarly, each pound that leaves the apparatus represents a rejection of energy to the extent $K_2 + E_2 + AP_2V_2$, for E_2 is the internal energy which the fluid carries out, and P_2V_2 is the work spent upon it by the fluid behind in pushing it out. $E_2 + AP_2V_2$ is equal to I_2, the total heat per pound of the fluid in its actual state at exit. Hence, by the conservation of energy for the apparatus as a whole,

$$K_1 + I_1 = K_2 + I_2 + W + Q_l.$$

The terms on the left of this equation represent the energy that enters the apparatus; the terms on the right show how it is disposed of in the issuing stream, in output of useful work, and in leakage of heat.

The terms K_1 and K_2 are usually very small, except when the apparatus is one for forming a steam jet, in which case K_2 is the useful term: this will be considered in a later chapter. When the change of kinetic energy in the stream is practically negligible, as it is between the admission-pipe and exhaust-pipe of an engine, we have

$$I_1 = I_2 + Q_l.$$

And when, in addition, the apparatus does not allow any appreciable amount of heat to escape to the outside ($Q_l = 0$), we have

$$I_1 - I_2 = W.$$

This means that when there is a steady flow of a working substance through any thermodynamic apparatus, the output of work is measured by the *actual heat-drop, whether the internal action is or is not reversible,* provided there is no loss of heat to the outside by conduction or radiation, and no change of kinetic energy.

The actual heat-drop must not be confused with the adiabatic heat-drop, which is the difference between I_1 and that value which the total heat would reach if there were adiabatic expansion to the exit pressure P_2. The actual heat-drop $I_1 - I_2$ is identical with the adiabatic heat-drop only when there is no loss of heat to the outside and when, in addition, the internal action is wholly reversible. In that case the working substance passes through the apparatus without undergoing any change of entropy.

Any irreversible feature in the internal action will increase I_2 above the value which would be reached by adiabatic expansion, and will consequently diminish the output of work. It will also make the entropy greater at exit than at entrance.

In the extreme case of a throttling process there is no output of work and therefore $I_2 = I_1$, provided there is no loss of heat to the outside. Any loss of heat to the outside in a throttling process will make I_2 correspondingly less, for we then have $I_2 = I_1 - Q_l$.

The losses of thermodynamic effect in a real engine, which make W less than the ideal limit, namely the value corresponding to the adiabatic heat-drop, arise partly from loss of heat to the outside and partly from two kinds of irreversible internal action. One of these two kinds is mechanical; the other is thermal. In the mechanical kind, the action involves fluid friction within the working substance. It is of the same nature as that which occurs in throttling: there is irreversible passage of the working substance from one part of the engine to another where the pressure is lower, as for instance the passage of steam through somewhat constricted openings into the cylinder, or its passage, on release after incomplete expansion, into the exhaust-pipe, with a sudden drop of pressure: or again, there is the same kind of irreversibility in the frictional losses that attend the formation of steam jets in a turbine or the friction of the jets on the turbine blades. These are all instances of mechanical irreversibility. In the second kind of irreversible action there is exchange of heat between the working substance and the internal surface of the engine walls. The hot steam, on admission to a cylinder which has just been vacated by a less hot mixture of steam and water, finds the surfaces colder than itself. A part of it is accordingly condensed on them, which re-evaporates after the pressure has fallen through expansion. This alternate condensation and re-evaporation involves a considerable deposit and recovery of heat in a manner that is not reversible, for it takes place by contact between fluid and metal at different temperatures. The action, which will be considered more fully in a later chapter, may occur without loss of heat to the outside: it would occur, for instance, in an engine with a conducting cylinder covered externally with a "lagging" of non-conducting material. Its effect, like that of throttling or fluid friction generally, is to reduce the output of work below the limit that is attainable only in a reversible process, and it does this by making the actual heat-drop $I_1 - I_2$ less than the adiabatic heat-drop.

The equation $W = I_1 - I_2$ takes account of the effect of thermal exchanges within the apparatus, as well as of any throttling or frictional effects in the action of the working substance. But it does not take account of heat lost to the outside, and for that the term Q_l has to be deducted, making

$$W = I_1 - I_2 - Q_l.$$

The full statement of the heat account in a real process may be expressed as follows: When there is a steady flow of a working substance through any thermodynamic apparatus the output of work is measured by the actual heat-drop from entrance to exit, less any heat that escapes by conduction to the outside, and less any gain of kinetic energy of the issuing stream over the entering stream; or, in symbols

$$W = I_1 - I_2 - Q_l - (K_2 - K_1),$$

all these quantities being expressed in thermal units, and reckoned per unit quantity of the working substance.

This equation also applies to the reversed heat-engines, or heat-pumps, which will be considered in the next chapter, but in them the quantity W is negative: there as an input instead of an output of work. Q_l is also generally negative, for as a rule the apparatus is colder than its surroundings and the leakage of heat is inwards.

CHAPTER VI

REVERSAL OF THE HEAT-ENGINE: MECHANICAL REFRIGERATION[1]

103. Reversal of the cycle in heat-engines. Refrigerating machines or heat-pumps. By a refrigerating machine or heat-pump is meant a machine which will carry heat from a cold to a hotter body. This, as the second law of thermodynamics asserts, cannot be done by a self-acting process, but it can be done by the expenditure of mechanical work. Any heat-engine will serve as a heat-pump if it be forced to trace its indicator diagram backwards, so that the area of the diagram represents work spent on, instead of done by, the working substance. Heat is then taken in from the cold body and heat is rejected to the hot body.

Take for instance the Carnot cycle, using air as working substance (fig. 40), and let the cycle be performed in the order *dcba*, so that the area of the diagram is negative, and represents work spent upon the machine. In stage *dc*, which is isothermal expansion in contact with the cold body C, the gas takes in a quantity of heat from C equal to $RT_2 \log_\epsilon r$ (§ 43), and in stage *ba* it gives out to the hotter body A a quantity of heat equal to $RT_1 \log_\epsilon r$. There is no transfer of heat in stages *cb* and *ad*. Thus C, the cold body, is constantly being drawn upon for heat and can therefore be maintained at a temperature lower than its surroundings. Suppose that such a machine were to be applied to the making of ice, then C might consist of a coil of pipe immersed in brine. The brine could in this way be kept by the action of the machine at a temperature below 0° C., and be used, in its turn, to extract heat by conduction from the water which is to be frozen. The "cooler" A, which is the relatively hot body, is kept at as

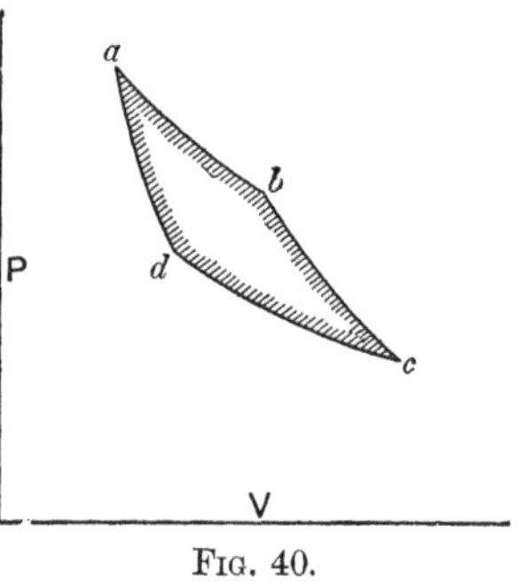

FIG. 40.

[1] On the subject of refrigerating processes generally reference should be made to the author's book on the *Mechanical Production of Cold* (Camb. Univ. Press, second edition 1921), and his *Thermodynamics for Engineers*, Chap. IV.

low a temperature as possible by means of circulating water, which absorbs the heat rejected to A by the working air. This is substantially the process which is used in actual ice-making machines, except that the cycle of operations is not a reversed Carnot cycle, but more nearly a reversed Rankine cycle, and the working substance is a vaporizable liquid instead of air.

A machine using air as working substance and following Carnot's cycle would be exceedingly bulky. Its size would be considerably reduced if a regenerator, as in Stirling's engine, were resorted to in place of the two adiabatic stages of the Carnot cycle. Refrigerating machines of this kind, using air as working substance, with a regenerator, were introduced by A. C. Kirk, and were at one time considerably used[1]. The working air was completely enclosed, which allowed it to be in a compressed state throughout, so that even its lowest pressure was much above that of the atmosphere. This made a greater mass of air pass through the cycle in each revolution of the machine, and hence increased the performance of a machine of given size.

Kirk's type of refrigerating machine has not survived, and those machines which now use air as working substance follow the reversed Joule cycle as described below in § 107.

104. Vapour compression refrigerating machines. In most modern refrigerating machines, however, the working substance, instead of being air, consists of a liquid and its vapour, and the action proceeds by alternate evaporation under a low pressure and condensation under a relatively high pressure. A liquid must be chosen which evaporates at the lower extreme of temperature under a pressure which is not so low as to make the bulk of the engine excessive. Sulphuric ether was one of the earliest liquids to be used in this way, but ether machines were inconveniently bulky and could not be used to produce intense cold, for the pressure of that vapour is only about 1·3 lb. per sq. inch at − 15° C., and to make it evaporate at a temperature considerably below zero would require the cylinder to be excessively large in proportion to the performance. This would not only make the machine clumsy and costly but would involve much waste of power in mechanical friction. The tendency of the air outside to leak

[1] See Kirk, "On the Mechanical Production of Cold," *Min. Proc. Inst. C. E.* vol. XXXVII, 1874.

into the machine is another practical objection to the use of so low a pressure. The fluids now most commonly used are ammonia and carbonic acid. With ammonia it is easy to reach as low a limit of temperature as is required in any of the usual industrial applications of cold: the pressures are fairly but not excessively high, and the apparatus is compact. With carbonic acid the apparatus is even more compact, but the pressures are much higher and the thermodynamic efficiency of the operation is not quite so good. Carbonic acid however is frequently preferred, especially on board ship, on account of its being more harmless should any of the working substance escape by leakage into the atmosphere of the room containing the machine. Other fluids with lower vapour-pressures are occasionally used in small plants, especially sulphurous acid and ethyl chloride.

Machines of this type are usually arranged to act as follows, in a cycle which is almost exactly the reverse of the Rankine cycle. The organs, which are shown diagrammatically in fig. 41, are

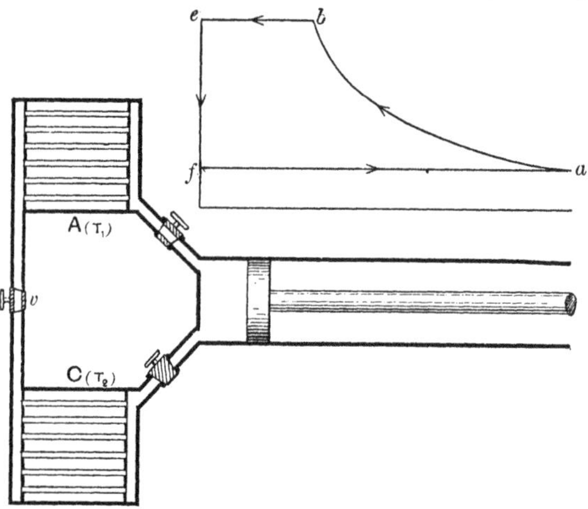

FIG. 41. Refrigerating machine using the vapour of a liquid.

(1) a cold body C which serves as an evaporator for the volatile working fluid and allows heat to pass into the working fluid from the water or other substance that is to be made cold, (2) a compressing cylinder, (3) a condenser A such as a coil of pipe surrounded by circulating water, in which the working fluid is condensed under pressure, and (4) a throttle-valve v called the expansion valve. The steps of the cycle are shown by the indicator diagram in the same

figure; *fa* is the forward stroke, during which the cylinder is taking in vapour from *C* at a uniform pressure corresponding to the lower limit of temperature T_2. Compression of the vapour occurs during *ab*, which is the first part of the back stroke, and during which the valves leading to both chambers are shut. This continues till the pressure in the cylinder becomes equal to the pressure in *A*.

Next, the communication with *A* opens and the back stroke is completed under a uniform pressure which corresponds to the temperature in *A*, the working substance passing into *A* and being condensed there.

To complete the cycle, the same quantity of the substance is allowed to pass through the valve *v* directly from *A* to *C*.

This last step in the process is not reversible, but it is a simpler way of completing the cycle than to complete it reversibly by letting the fluid do work in an expansion cylinder in passing from *A* to *C*, and the amount of work which would be saved if that were done is inconsiderable.

The operation of such machines may be represented by an entropy-temperature diagram like that of fig. 42, which is sketched for ammonia as the working substance, or fig. 43 which is sketched for carbonic acid. If the evaporation were complete, a horizontal line extending to *h* in either figure would represent the process of evaporation, during which heat is being taken in from the body to be cooled. More generally however evaporation is incomplete; what is taken into the cylinder and compressed is a mixture of vapour with some of the unevaporated liquid. This reduces the superheating which compression would otherwise cause, and may even prevent superheating entirely, provided enough liquid be present in the mixture. Thus, in fig. 42 or 43, the point *a* represents the condition as regards wetness of the mixture which is taken into the cylinder, and the adiabatic process of compression, which is represented by the line *ab*, brings the vapour to a moderately superheated state when its pressure is raised to the upper limit, namely to the pressure of the condenser *A*. By regulating the flow through the expansion valve *v* the compression may be made more or less "dry" or "wet," the point *a* being brought nearer to or further from *h*.

The next process consists of cooling and condensation at the pressure of the condenser: it is made up of three stages, *bc*, *cd*, and *de* (figs. 42 and 43). In the stage *bc* the superheated vapour is cooled to the temperature at which condensation begins; in the stage *cd* it is condensed; and in the stage *de* it is cooled to the lowest available temperature before it passes the expansion valve.

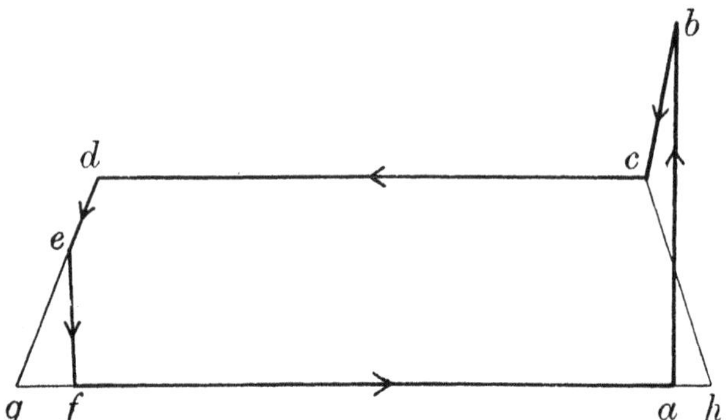

FIG. 42. The Vapour-compression Cycle, using Ammonia.

The lines *bc*, *cd* and *de* form parts of one line of constant pressure. In fig. 42 *de* is practically indistinguishable from the boundary line, but in fig. 43 the distinction is apparent, for we are there dealing with a liquid that is highly compressible because it is not far from the critical state. The line *ef* represents the process of passing through the expansion valve, in which the pressure falls to that of the evaporator. This is a throttling process, for which in any fluid the total heat I is constant (§ 82); *ef* is therefore a line of constant total heat. Its direction changes at the point where it crosses the boundary curve (fig. 43). As a result of passing the

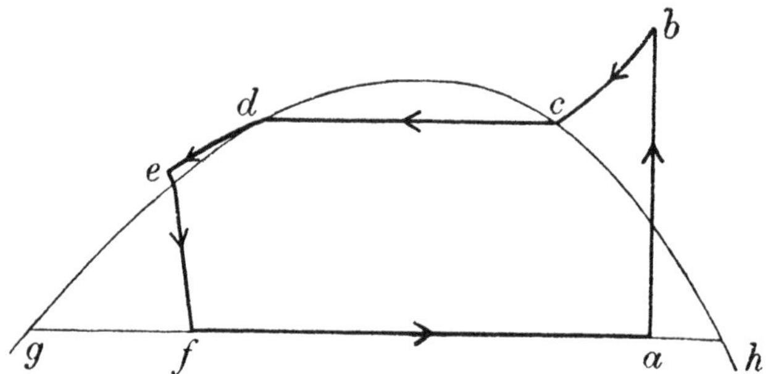

FIG. 43. The Vapour-compression Cycle, using Carbonic Acid.

expansion valve the working substance comes into the condition shown by the point *f*. It is then a wet mixture, for part of it is converted into vapour by the mere act of passing the expansion valve, namely the fraction *gf/gh*. Lastly we have the process of effective evaporation when the substance is usefully extracting heat from the brine or other cold body by evaporating in the refrigerator. This process is represented by the line *fa*, and during it the dryness increases to the fraction *ga/gh*.

The refrigerating effect, that is to say the amount of heat taken in from the cold body, is represented by the area under the line *fa* measured down to the absolute zero of temperature (in the manner which was illustrated in fig. 28). The amount of heat rejected during cooling and condensation and subsequent cooling of the condensed liquid before it passed the expansion valve is the area under the lines *bc*, *cd* and *de*. The thermal equivalent of the work spent in carrying the substance through the complete cycle—namely the work spent on it in the compressor—is the difference between those two quantities. It should be noted that the work spent is not measured by the area *abcdefa* enclosed by lines which represent the complete cycle, because the cycle includes an irreversible step *ef*. In consequence of that the work spent is greater than the enclosed area by the area under the line *ef*.

Comparing this with the reversed Carnot cycle, which would represent an ideally efficient refrigerating process, we see that the net refrigerating effect is less, and also that the work spent is greater. For both these reasons the efficiency is less. The difference represents what is sacrificed by using a throttle-valve instead of an expansion cylinder between the condenser and the refrigerator.

As a further example we may take a compression process (fig. 44)

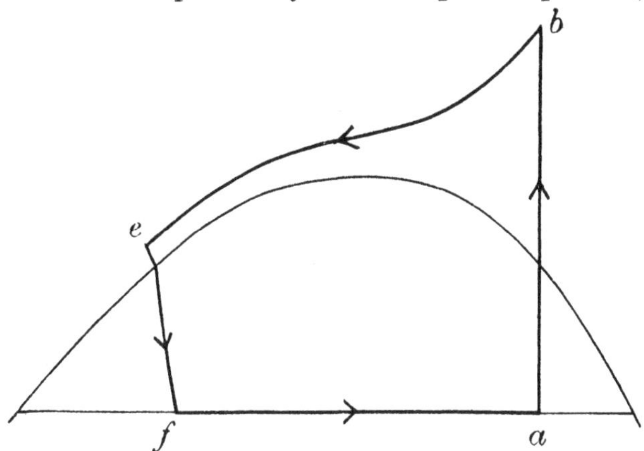

FIG. 44. Cycle for Carbonic Acid, with compression above the critical pressure.

with carbonic acid for working substance, in which the temperature of the cooling water is so high that the pressure during cooling is above the critical pressure. The line *be* is accordingly a continuous curve lying entirely outside of the boundary curve. The working substance passes from the state of a superheated vapour at *b* to its state at *e* without any stage, corresponding to *cd* in fig. 43, in which it is a mixture of liquid and vapour. As before, the refrigerating effect is measured by the area under *fa*: the heat rejected to the cooling water is measured by the area under *be*: the difference between these two quantities measures the work spent, and is greater than the area of the closed figure *abcfa* by the area under the irreversible step *ef*.

105. Coefficient of performance of refrigerating machines. The ratio

$$\frac{\text{Heat extracted from the cold body}}{\text{Thermal equivalent of work expended}}$$

may be taken as a coefficient of performance in estimating the merit of a refrigerating machine from the thermodynamic point of view. When the limits of temperature T_1 and T_2 are assigned it is easy to show by a slight variation of the argument used in § 47 that no refrigerating machine can have a higher coefficient of performance than one which is reversible in Carnot's sense. For let a refrigerating machine S be driven by another R which is reversible and is used as a heat-engine in driving S. Then if S had a higher coefficient of performance than R it would take from the cold body more heat than R (working reversed) rejects to the cold body, and hence the double machine, though purely self-acting, would go on extracting heat from the cold body in violation of the Second Law. Reversibility, then, is the test of perfection in a refrigerating machine just as it is in a heat-engine.

When a reversible refrigerating machine takes in all its heat, namely Q_C, at T_2 and rejects all, namely Q_A, at T_1, $\frac{Q_C}{T_2} = \frac{Q_A}{T_1}$ and the coefficient of performance,

$$\frac{Q_C}{W} = \frac{Q_C}{Q_A - Q_C} = \frac{T_2}{T_1 - T_2}.$$

Hence the smaller the range of temperature the higher is the limit of efficiency in the refrigerating process. This has a very important practical bearing. To cool a large mass of any substance through a few degrees will require much less expenditure of energy than to cool one-tenth of the mass through ten times as many degrees, though the amount of heat extracted is the same in both cases. If we wish to cool a large quantity, say of water or of air, it is better to do it by the direct action of a refrigerating engine working through the desired range of temperature, than to cool a portion through a wider range and then let that mix with the rest. This is only another instance of a wide general principle, of which we have had examples before, that any mixture or contact of substances at different temperatures is thermodynamically wasteful because the interchange of heat between them is irreversible. An ice-making machine, for example, should have for its lower limit a temperature only so much below the freezing-point as will allow heat to be conducted with sufficient rapidity to the working fluid from the water that is to be frozen.

106. Refrigerating effect and work of compression expressed in terms of the total heats. While it is instructive to state the refrigerating effect, the work of compression, and the heat rejected, in terms of areas on the entropy-temperature diagram as in § 104, it is much more useful, for purposes of practical calculation, to express these as follows in terms of the total heat of the substance at the various stages of the operation.

The refrigerating effect, that is to say the amount of heat taken in from the cold body, is $I_a - I_f$, where I_a is the total heat at a and I_f is the total heat at f. This is because the (reversible) operation fa is effected at constant pressure (§ 62). For the same reason the amount of heat rejected to the condenser and cooler is $I_b - I_e$, where those quantities designate the total heat at b and at e respectively. Further, in the process ef of passing the expansion valve there is no change of total heat, by the principle proved in § 82. Consequently, $I_f = I_e$. We may therefore state the amount of heat rejected as $I_b - I_f$.

Again, the work spent in the compressor is (in thermal units) $I_b - I_a$. It is the thermal equivalent of the area of the indicator diagram in fig. 41, namely $A\int_a^b VdP$, which is equal to $I_b - I_a$ by the general principle proved in § 73. We are dealing here with the increase of total heat in adiabatic compression instead of its decrease in adiabatic expansion.

That these results are in agreement with one another is seen by considering the heat-account of the cycle as a whole:

$$\text{Work spent} = \text{Heat rejected} - \text{Heat taken in.}$$

$$I_b - I_a = (I_b - I_f) - (I_a - I_f).$$

The coefficient of performance, which is the ratio of heat taken in from the cold body to the work spent in compression, is

$$\frac{I_a - I_f}{I_b - I_a}.$$

From these results it will be seen that calculations of performance, as regards refrigerating effect, heat rejected, and work expended, become very easy when we can find the total heat of the liquid just before the expansion valve and that of the vapour before and after compression. This is readily done if data are available for drawing a Mollier chart of entropy and total heat for the working substance. Fairly complete data are available for ammonia, carbonic acid, and sulphurous acid. $I\phi$ charts for these substances will be found in a Report of the Refrigeration Research Committee of the Institution of Mechanical Engineers[1].

In drawing these charts a geometrical device is resorted to for the purpose of making them at once open and compact, with the effect that measurements may be made with sufficient accuracy on a chart of reasonable size. This device, which Mollier originally adopted in drawing his $I\phi$ chart for carbonic acid, is to use oblique coordinates. The lines of constant I are horizontal: the lines of constant ϕ instead of being perpendicular to them are inclined at a small angle. The result is that when the chart is drawn the curves on it are sheared over, as compared with the

[1] *Min. Proc. Inst. Mech. Eng.* Oct. 1914. The charts given there are drawn by Professor C. F. Jenkin. The chart for carbonic acid embodies results of experiments by Messrs Jenkin and Pye on the thermal properties of that substance (*Phil. Trans. Roy. Soc.* vols. A 499, p. 67, and A 534, p. 353, which involve some correction of an earlier chart published by Dr Mollier. The data for ammonia are those given by Messrs Goodenough and Mosher (*Bulletin* No. 66 of the University of Illinois, 1913). Complete tables of the thermodynamic properties of ammonia have been calculated with somewhat different numerical results by Messrs Keyes and Brownlee (New York, John Wiley and Sons, 1916). In each of these publications a Mollier $I\phi$ chart is included. See also *Scientific Papers of the U.S. Bureau of Standards*, Nos. 313, 314, 315 (1917) and 369 (1920); and *Properties of Refrigerants*, by H. D. Edwards (New York, 1924).

positions they would take with rectangular axes and there is a gain in the clearness and precision with which one may measure

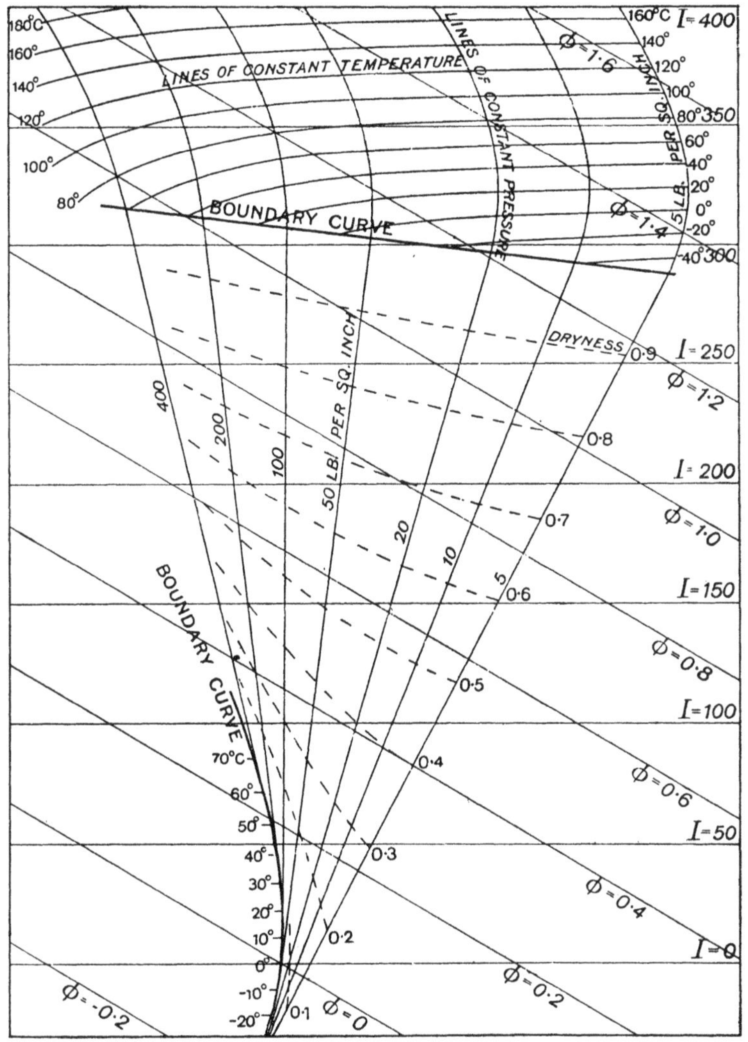

FIG. 45. $I\phi$ chart for Ammonia.

from them the changes of total heat which occur in the successive stages of the ideal vapour-compression process.

An $I\phi$ chart on a small scale, with oblique coordinates, is shown for ammonia in fig. 45, and for carbonic in fig. 46. In each case the region useful in refrigeration is included: in fig. 46 that region extends above as well as below the critical point. On such charts it is easy to draw the ideal diagram for assigned temperatures of evaporation, condensation, and subsequent cooling, and for any

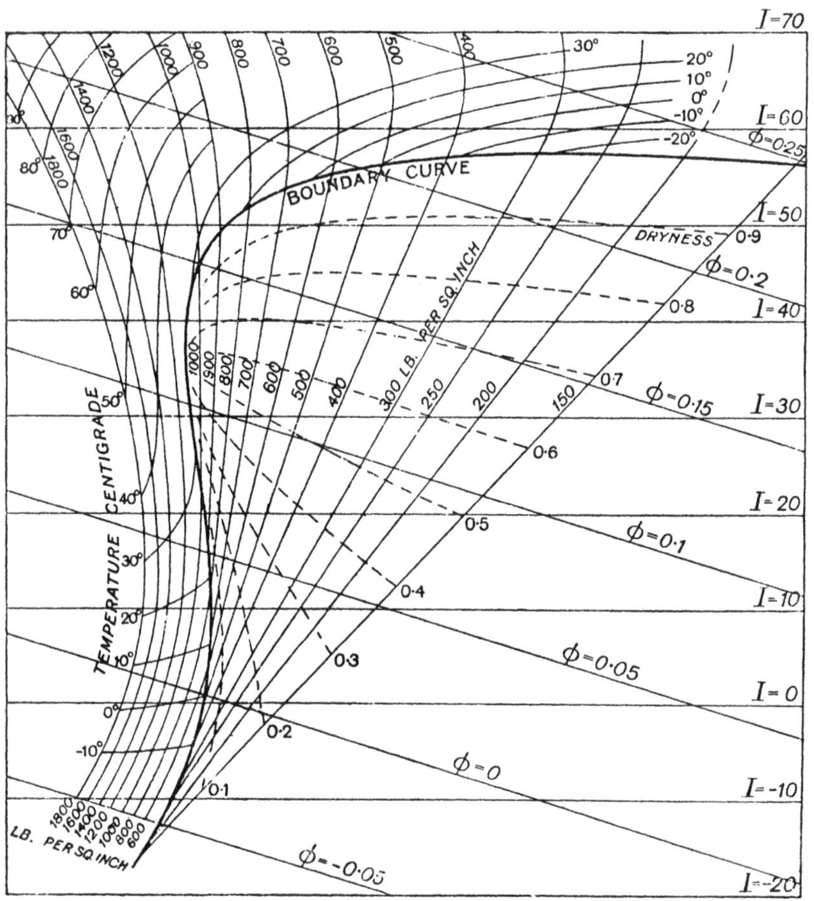

FIG. 46. $I\phi$ chart for Carbonic Acid.

assigned wetness at the beginning of compression, and so determine graphically those values of the total heat which determine the amount of the refrigerating effect and the coefficient of performance.

Fig. 47 illustrates how the ideal action of a vapour-compression refrigerating machine is represented on the $I\phi$ chart. In this example the working substance is carbonic acid, the lower limit of temperature at which evaporation occurs is supposed to be − 10° C., the temperature of condensation is 25° C., and the condensed liquid is cooled to 15° C. before it streams through the expansion valve. The cycle begins at the point *a* which represents the state of the substance when it is about to enter the compressor. This point is on the constant-pressure line corresponding to the process of evaporation in the evaporator, and its distance from the two boundary curves corresponds to the proportion of vapour to liquid in the mixture. If the compression is to be completely "dry," the starting-point will be at a_1: more generally the substance is slightly wet when compression begins. The straight line *ab*, drawn parallel to the lines of constant entropy on the chart, is the process of adiabatic compression. The position of *b* is determined by the intersection of this line with a line of constant pressure corresponding to the known upper limit of pressure at which condensation is to occur. The temperature reached in the process of compression is seen by the position of *b* among the lines of constant temperature. When compression begins at a point such as *a* it involves some superheating. But if the mixture is so wet to begin with that the adiabatic line through *a* does not cross the boundary curve during compression before the upper limit of pressure is reached there is no superheating, and the process is then spoken of as "wet" compression. This would require the compression to have begun at a_c instead of *a*. By beginning at *a* it carries the substance into the region of superheat before compression is completed at *b*. Next we have the constant-pressure process of cooling and con-

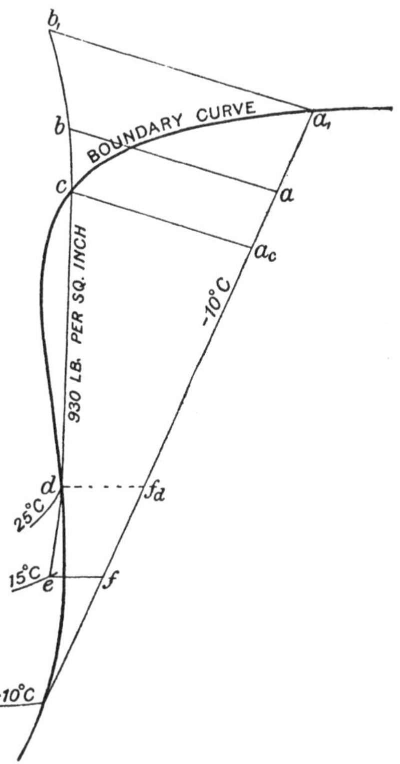

FIG. 47. Refrigeration cycle traced on the $I\phi$ chart for Carbonic Acid.

densation and further cooling, represented in its three stages by the lines *bc*, *cd*, and *de*, the position of *e* being fixed by the temperature to which the liquid is known to be cooled before it reaches the expansion valve. Then a horizontal straight line through *e* (a line of constant total heat) represents the process of streaming through the expansion valve, and determines a point *f*, on the evaporation line, which exhibits the condition in which the substance enters the evaporator. The process of evaporation *fa*, which is the effective refrigerating process, completes the cycle. The values of I_a, I_b, I_e, and I_f (which is the same as I_e) are read directly by measurement from the chart, and from them the work spent in compressing the substance, which is $I_b - I_a$, and the refrigerating effect, which is $I_a - I_f$, are determined. The position of the starting-point *a*, between a_c and a_1, which determines how far the compression will be wet or dry, does not greatly affect the thermodynamic efficiency of the process. Between the two extremes there is a certain degree of initial dryness which gives a slightly higher coefficient of performance than is obtained either by starting as at a_1 with dry vapour or as at a_c with a mixture so wet that compression does no more than vaporize the liquid it contains. The position of *a* for maximum efficiency may be found thus: the refrigerating effect for any position of the compression starting-point *a* is proportional on some scale to the length *fa*. The work spent in compressing the fluid is proportional, on another scale, to the length *ab*. Hence the position of the compression line *ab* which will give the highest coefficient of performance is that which gives the smallest ratio of *ab* to *fa*. This position is found by drawing a tangent from *f* to the curve of constant pressure dcb_1. The compression line *ab* is then drawn through the point of contact *b*, and this fixes *a* as the starting-point for maximum efficiency in the ideal cycle with adiabatic compression.

It does not, of course, follow that the same degree of initial wetness would give the maximum coefficient in a real compressor, for the performance of a real machine is complicated by transfers of heat between the working substance and the metal.

107. Reversed Joule Engine: the Bell-Coleman Refrigerating Machine. This machine was briefly mentioned in § 101 as one which has been employed to maintain a cold atmosphere in the frozen-meat chambers of ocean steamships. It acts by drawing in a small portion of the air of the chamber, compressing that and extracting as far as possible by means of a cooler the heat developed by compression, then expanding the air until its pressure falls to that of the chamber. The temperature of the expanded air is then lower than the temperature of the chamber in consequence of the removal of heat which took place while it was compressed. The air thus chilled by expansion is returned to the chamber, and in this way the temperature of the chamber is kept down notwithstanding the heat which reaches it by conduction from outside. The chamber has a thick lining of a heat-insulating substance in order to reduce as far as may be the work which has to be spent on refrigeration.

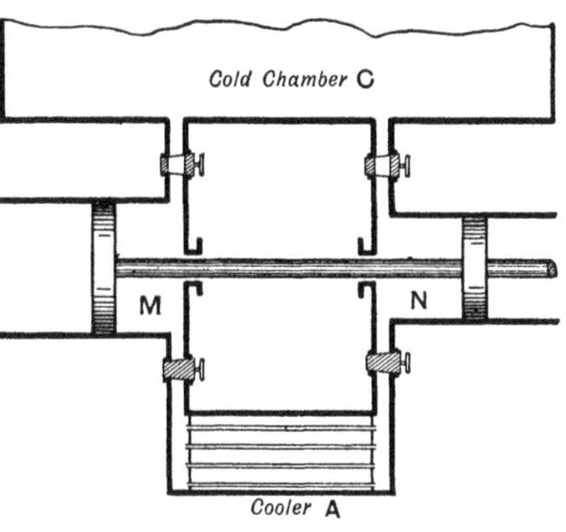

Fig. 48. Organs of the Bell-Coleman refrigerating machine.

The sketch, fig. 48, shows the organs diagrammatically. C is part of the cold chamber, which is at or about atmospheric pressure, and A is the cooler, a set of pipes with circulating water. Compression takes place in M and expansion in N. M takes in air from C at temperature T_2 during its out-stroke, and compresses that during part of its in-stroke till the pressure becomes equal to the pressure in A. These two operations are represented by the lines *fc* and *cb* in the indicator diagram, fig. 49. The compression *cb* has the effect of raising the temperature of the air above that of A. Consequently when the pump delivers the compressed air into A, by completing its return stroke (*bc*), which is the next operation, the temperature of the air falls and a quantity of heat is rejected to A, namely

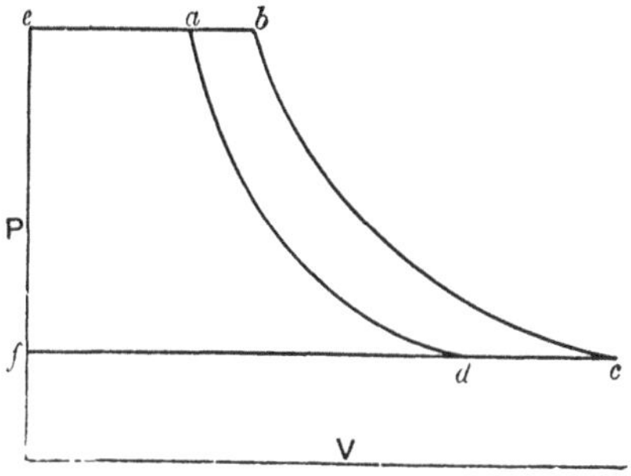

Fig. 49.

$$K_p (T_b - T_a),$$

where T_b is the temperature reached by compressing, and T_a is the temperature of A. While this is going on, the cylinder N takes an equal quantity of air from A and expands it to the pressure of C: these operations are shown by the lines *ea* and *ad* in the indicator diagram. At the end of this expansion the temperature T_d is lower than that of the cold chamber. Finally the chilled air is discharged into C during the return stroke of N, which is shown by the line *df* in the indicator diagram. The net amount of work expended is *badc*, *fcbe* being the indicator diagram of work spent upon the pump M and *eadf* being the diagram of work recovered in the expansion cylinder N. The net amount of heat taken from the cold chamber is $K_p\ (T_c - T_d)$. Assuming the processes *cb* and *ad* to be adiabatic, the ratio of expansion in N is equal to the ratio of compression in M, and hence $\frac{T_a}{T_d} = \frac{T_b}{T_c}$, as we have already seen in treating of the Joule cycle (§ 101) of which this is simply a reversal. Also $\frac{Q_A}{Q_C} = \frac{T_a}{T_d} = \frac{T_b}{T_c}$, and the coefficient of performance $\frac{Q_C}{Q_A - Q_C} = \frac{T_d}{T_a - T_d}$ or $\frac{T_c}{T_b - T_c}$.

This coefficient of performance is low because of the very large range of temperature through which the working air is carried.

Considered as a means of pumping up heat from T_c the temperature of the cold chamber to T_a the temperature of the circulating water to which heat is discharged, the air machine has two serious thermodynamic defects. There is an irreversible transfer of heat when the working air, after being heated by compression to T_b, comes into thermal contact with the circulating water at T_a; and there is another irreversible transfer when the working air, chilled by expansion to T_d, mixes with the less cold atmosphere of the chamber at T_c. An ideally efficient refrigerating machine, namely a reversed Carnot engine, working between T_a and T_c as upper and lower limits, would have (§ 45) a coefficient of performance equal to

$$\frac{T_c}{T_a - T_c}.$$

The coefficient found above for the reversed Joule cycle is substantially less because T_b is higher than T_a.

In the practical working of such machines the presence of moisture in the air has to be reckoned with. The air coming from the cold chamber is more or less saturated: during expansion it becomes supersaturated and the water from it would be deposited as snow in the expansion cylinder, and might interfere with the action of the mechanism, if preventive devices were not introduced. One such device is to divide the whole expansion into two stages by making it compound. In the first stage the expansion is carried only far enough to cool the air to a temperature just above the freezing-point. In that way nearly all the moisture is deposited in the form of water, and is easily drained away before the final stage, which would freeze it, begins. Another device is to condense out most of the moisture before expansion, by passing the compressed air through pipes which bring its temperature down to near the freezing-point before it enters the expansion cylinder. These "drying pipes" are kept cold by air from the cold chamber: that air is consequently warmed by them, but the loss is made good by the lower temperature which the working air reaches in expansion, as a consequence of the precooling it has undergone in the drying pipes. This arrangement of "drying pipes" has been employed by Messrs Haslam in their construction of the Coleman apparatus, but other makers of such machines have been content with a mechanical separation of the air from any water deposited in the cooling which precedes admission to the expansion cylinder. Provided the air entering that cylinder is merely saturated and does not carry with it water in a state of mechanical suspension, the deposit of snow is not so great as to be seriously troublesome[1].

The actual coefficient of performance of a machine of this class is much less than that of a machine using for working substance a vaporizable liquid such as ammonia or carbonic acid. This is partly due to the relatively great waste of power, through friction, in air machines, and partly due to the practical necessity of using, in them, a much wider range of temperature than the range through which refrigeration is to be carried on. To keep the dimensions of the machine within reasonable bounds, the air is cooled by expansion to a temperature much lower than that of the cold chamber, and is heated by compression to a temperature considerably higher than that of the cooling water. When the working substance is a liquid which is being alternately vaporized and condensed, the heat is much more easily got into and out of it. The efficiency of a vapour

[1] For particulars of the construction and performance of these machines see Coleman, *Min. Proc. Inst. C. E.* vol. LXVIII, 1882, p. 146; Lightfoot, *Proc. Inst. Mech. Eng.* 1881, p. 105, and 1886, p. 201.

machine can be made to approach more closely to the ideal of a perfect refrigerator, and the coefficient of performance of a machine using ammonia is found in practice to be about five times that of an air machine. In consequence of their higher efficiency and smaller bulk vapour-compression machines have almost entirely displaced air machines in cold storage both on land and on board ship.

108. The reversed heat-engine as a warming machine. It was pointed out by Lord Kelvin in 1852 that the reversed heat-engine cycle might serve not only as a means of cooling but as a means of warming[1]. Let it be required for instance to raise and keep the temperature of a room above the temperature of the surrounding air. A machine of the Bell-Coleman type may take in air from the atmosphere, expand it so as to lower the temperature somewhat, and allow the temperature to rise again by conduction from external air. Then let it compress the air so as to restore it to atmospheric pressure. The temperature of the air will be thereby raised above the temperature of its surroundings, and it may then be discharged into the room which is to be warmed. The effect is, that by expending some mechanical work a quantity of heat is transferred from the cold atmosphere to the warmer room—a quantity which may be far greater than the thermal equivalent of the work spent in driving the machine.

The importance of the suggestion lies in the fact that the necessary power may be obtained, by means of a heat-engine, with a smaller supply of heat than would be required to effect the warming directly, provided the range of temperature of the warming be much less than the range through which the heat-engine works in generating the required power. Burning fuel to warm a room by a few degrees is a wasteful way to utilize heat, even if all the heat of combustion be conceived to pass into the air of the room. The high-temperature heat produced in the combustion of coal or gas could warm a much larger volume of air to the same extent if it were applied to drive an efficient heat-engine, which in its turn drove a reversed heat-engine or warming machine to pump up heat through a short range of temperature from the diffused store of heat which is contained in the atmosphere or in the ocean.

[1] *Proc. of the Phil. Soc. of Glasgow*, vol. III, p. 269, or *Collected Papers*, vol. I, p. 515.

This is because a heat-engine can be arranged to take advantage of the high temperature at which heat is produced in the burning of fuel, whereas any direct communication of this high-temperature heat to a comparatively cool body, such as the air of a room—a typically irreversible operation—is thermodynamically bad. The thermodynamic advantage of high-temperature heat is wasted if we allow it directly to enter a comparatively cold substance. The suggestion that some of the coal which is used for heating rooms might be saved by applying heat in this indirect manner cannot be said to have more than a theoretical interest.

CHAPTER VII

ACTUAL BEHAVIOUR OF STEAM IN THE CYLINDER

109. Comparison of actual and ideal indicator diagrams. We have now to consider in what respects the action of steam in a real engine differs from the ideal action discussed in §§ 73–75, and more fully in Chapter v, where an engine was imagined in which the Rankine cycle of operations was followed, and the work done in the cylinder was measured by the heat-drop. As was mentioned in § 75 the ratio of the work actually done in the cylinder to the heat-drop is called the efficiency ratio. It is a ratio always much less than unity: even in favourable cases only about 70 per cent. of the ideal output is obtained. This is true both of reciprocating engines and of turbines. In considering reasons it will be convenient first to deal with engines of the piston and cylinder class.

In the first place, the expansion in real engines is not (except in rare cases) complete: the steam at release has a pressure which is higher than the pressure in the condenser if the engine is a condensing engine, or higher than the pressure of the atmosphere if the engine is non-condensing. Reasons for this have been already indicated: complete expansion would increase unduly the bulk and weight of the engine; the work done by the steam in the last stages would add nothing to the net mechanical output for it would be used up in overcoming the friction of the piston; further, complete expansion would aggravate certain evils to be described later which arise from the cooling of the cylinder during expansion and exhaust. For these reasons it is practically desirable to cut off the toe of the ideal diagram sketched in fig. 19 (§ 73). The effect which incompleteness in the expansion produces by itself on the efficiency of the ideal process has already been considered in reference to the indicator diagram, fig. 21 (§ 74), and to the entropy-temperature diagram, fig. 33 (§ 96).

Other features of difference are most conveniently noticed by comparing stage by stage the ideal diagram of fig. 21 with a diagram taken from a real engine. In the action to which figs. 19 and 21 refer it was assumed—(1) that the steam was supplied in the dry saturated state, and had during admission the full (uniform) pressure of the boiler P_1; (2) that there was no transfer of heat to or from the steam except in the boiler and in the condenser; (3) that after more or less complete expansion all the steam was discharged by the return stroke of the piston, during which the back pressure was the (uniform) pressure in the condenser P_2; (4) that the whole volume of the cylinder was swept through by the piston. It remains to be seen how far these assumptions are untrue in practice, and how the efficiency is affected in consequence.

The actual conditions of working differ from these in the following main respects, some of which are illustrated by the practical indicator diagram of fig. 50, which is taken from an actual engine.

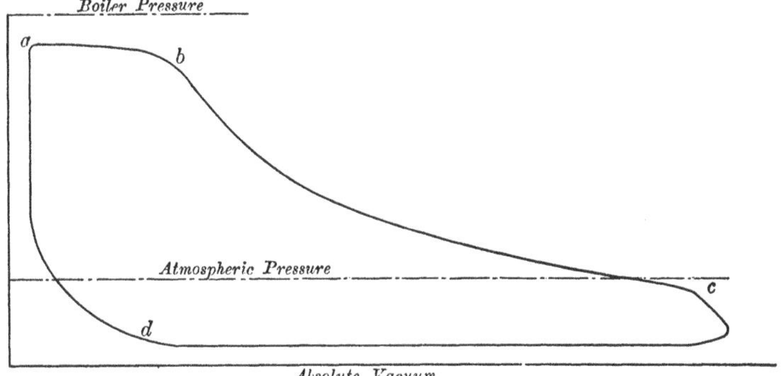

FIG. 50. Typical Indicator Diagram from a condensing steam-engine.

110. Wire-drawing during admission and exhaust. Owing to the resistance of the ports and passages, and to the inertia of the steam, the pressure within the cylinder is less than P_1 during admission and greater than P_2 during exhaust.

Moreover P_1 and P_2 are themselves not absolutely uniform, and P_2 is greater than the pressure of steam at the temperature of the condenser, on account of the presence of some air in the condenser. The presence of air is accounted for partly by its entering the boiler dissolved in the feed-water, and partly by its leaking into the cylinder and other parts of the engine at times when the pressure within is less than the pressure of the atmosphere.

During admission the pressure of steam in the cylinder is less than the boiler pressure by an amount which often increases a little as the piston advances, on account of the increased velocity of the piston's motion and the consequently increased demand for steam. When the ports and passages offer much resistance the steam is expressively said to be "throttled" or "wire-drawn." The steam is dried by the process to a small extent, as was shown in § 82, and if initially dry it becomes superheated. In an indicator diagram wire-drawing causes the line of admission to lie below a line drawn at the boiler pressure, and generally to slope a little downwards. In fairly good practical instances the mean absolute pressure during admission is about nine-tenths of the pressure in the boiler. With a long steam-pipe or a badly designed valve the fall of pressure may be greater, and the effect is aggravated when the steam is allowed to become wet by leaving the steam-pipe bare or insufficiently covered, instead of having the pipe properly "lagged" with some material which is a poor conductor of heat. Even under the best conditions unless the steam is supplied in a superheated state some of it is condensed on its way to the engine by loss of heat from the pipe. There may also be some additional water present in the steam through what is called "priming" on the part of the boiler, that is to say the delivery of steam in which particles of water are mechanically suspended. Whatever water is present, from either cause, may be more or less completely removed by the use of what is called a "separator," but usually the steam is to some extent wet when it enters the cylinder, notwithstanding the slight tendency which wire-drawing has to dry it. The separator is a vessel through which the steam passes on its way to the engine and in which the suspended particles of moisture settle, the accumulated water being drained off from time to time. In many cases the steam is made to take such a course through the separator that centrifugal action assists in causing the particles of water to be thrown off.

Again, during the exhaust the actual back-pressure exceeds the pressure in the condenser by an amount that depends on the freedom with which the steam makes its exit from the cylinder. In condensing engines with a good vacuum the back-pressure is often as much as 3 lb. per sq. inch and even more, and in non-condensing engines it is 16 to 18 lb. in place of the 14·7 lb. or so which is the pressure of the atmosphere. The excess of back-pressure may be greatly increased by the presence of water in the cylinder. The effects of wire-drawing do not stop here. The valves open and close more or less slowly; the points of cut-off and release are therefore not absolutely sharp, and the diagram has rounded corners at b and c in place of the sharp angles which mark those events in fig. 21. For this reason release is allowed in practice to begin a little before the end of the forward stroke, hence the toe of the diagram takes a form like that shown in fig. 50. The sharpness of the cut-off, and to a less extent the sharpness of the release, depends greatly on the kind of valves and valve-gear used; valves operated by a trip gear, for instance, such as will be described in a later chapter, stop the admission of steam more suddenly than the ordinary slide-valve does and therefore produce a diagram in which the events of the stroke are more sharply defined.

111. Clearance. When the piston is at either end of its stroke there is a small space left between it and the cylinder cover. This space, together with the volume of the passage or passages leading thence to the steam and exhaust valves, is called the *clearance*. It constitutes a volume through which the piston does not sweep, but which is nevertheless filled with steam when admission occurs, and the steam in the clearance forms a part of the whole steam which expands after the supply from the boiler is cut off. If AC be the volume swept through by the piston up to release, OA the volume of the clearance, and AB the volume swept through during admission, the apparent ratio of expansion is $\frac{AC}{AB}$, but the real ratio is $\frac{OA + AC}{OA + AB}$.

Clearance must obviously be taken account of in any calculation of curves of expansion. It is conveniently allowed for in indicator diagrams by shifting the line of no volume back through a distance corresponding to the clearance in the manner illustrated in fig. 51. In actual engines the volume of the clearance OA is usually from $\frac{1}{10}$ to $\frac{1}{50}$ of the volume of the cylinder. Its size depends largely on the kind of valve that is used. As a rule small engines have relatively more clearance than large ones.

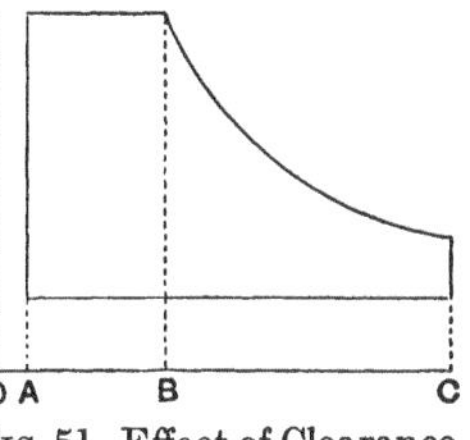

FIG. 51. Effect of Clearance.

112. Compression. Clearance affects the thermodynamic efficiency of the engine chiefly by altering the amount of steam that is consumed per stroke, and its influence depends materially on the extent to which the *compression* of part of the steam during the return stroke, referred to in § 73, is carried on. If there were no compression: if, in other words, the exhaust-pipe leading to the condenser or to the atmosphere were left open throughout the whole of the back stroke, at the end of that stroke the clearance space would have nothing more in it than steam at a pressure equal to the back-pressure, and consequently at the next admission enough steam would have to be drawn from the boiler to bring up the pressure in the clearance as well as to fill the volume which is swept through by the piston up to the point of cut-off. With compression this cause of waste is more or less completely avoided. During the back stroke the process of exhaust is discontinued before the end as at *d* in fig. 50, and the steam remaining in the cylinder is compressed. The cushion of steam thus shut in finally occupies the volume of the clearance; and by a proper selection of the point at which compression begins the pressure of this cushion may be made to rise just up to the pressure at which steam is admitted when the valve opens. This may be called complete compression, and when it occurs the existence of clearance has no direct effect on the consumption of steam nor on the efficiency; for there is then simply a permanent cushion which is alternately expanded and compressed without net gain or loss of work, in addition to the working steam proper, which on admission fills the volume AB (fig. 51), and which enters and leaves the cylinder in each stroke. But if compression be incomplete or absent there is, on the opening of the admission-valve, an inrush of steam to fill up the clearance space. This increases the consumption to an extent which is only partly counterbalanced by the increased area of the diagram, and the result is that the efficiency is reduced. The action is, in fact, a case of unresisted expansion (§ 52), and consequently tends, so far as its direct effects go, to make the engine less than ever reversible. Incidentally, compression has the mechanical advantage that it obviates the shock which the admission of steam would otherwise produce, and increases the smoothness of running by giving the piston work to do while its velocity is being rapidly reduced—an action which receives the name of "cushioning."

The opening of the steam-valve for admission being a somewhat gradual process, it generally begins before the back stroke is quite complete, in order that the valve may be widely enough open to let the steam in freely when the piston begins to move forwards.

The valve is then said to have *lead*, and the effect is to produce what is called *pre-admission*. Pre-admission tends to increase the mechanical effect of cushioning which has just been referred to.

113. Cushion steam and cylinder feed. In dealing with the influence of clearance, whether the compression be complete or incomplete or even altogether wanting, it is convenient to think of the working substance in the cylinder as made up of two parts, namely, (1) the part that has been shut up in the clearance from the previous stroke, and (2) the part that is freshly supplied from the boiler. For brevity we shall refer to these in what follows as (1) the cushion steam, and (2) the cylinder feed. During expansion the whole quantity of working substance in the cylinder is the sum of these two; during compression the cushion steam only is present. If the steam which leaves the engine is condensed and the condensed water weighed, its quantity forms a measure of the cylinder feed, from which the amount of steam passing through the cylinder per stroke may be deduced. But to this amount the cushion steam must be added when it is desired to know the whole quantity of steam (or rather of working substance) present in the cylinder.

114. Influence of the cylinder walls. Condensation and re-evaporation in the cylinder. The missing quantity. Generally by far the most important element of difference between the action of a real engine and that of our hypothetical engine is that which was alluded to at the end of Chapter I, the difference, namely, which proceeds from the fact that the cylinder and piston are not non-conductors. As the steam fluctuates in temperature in the phases of admission, expansion and exhaust there is a complex give-and-take of heat between it and the metal it touches, and the effects of this, though not very conspicuous on the apparent form of the indicator diagram, have an enormous influence in reducing the efficiency by increasing the consumption of steam. Attention was drawn to this action by D. K. Clark as early as 1855[1], and the results of his experiments on locomotives were confirmed and extended in 1860 by Isherwood's trials of the engines of the United States steamer "Michigan[2]." Rankine in his

[1] *Railway Machinery*, or art. STEAM-ENGINE, *Ency. Brit.* 8th edition. See also *Min. Proc. Inst. C. E.* vol. LXXII, p. 275.

[2] See Isherwood's *Experimental Researches in Steam Engineering*, Philadelphia, 1863, which describes a great number of experiments, undertaken at a time when engineers in general were but little alive to their value.

classical work on the steam-engine notices the subject only very briefly, and takes no account of the action of the cylinder walls in his calculations. Its importance became established beyond dispute, notably, among early experiments, by those of Messrs Loring and Emery on the engines of certain revenue steamers of the United States[1], and by a protracted series of investigations carried out by Hallauer and other Alsatian engineers under the direction of Hirn[2], whose name should be specially associated with the rational analysis of engine tests, and who was one of the first to recognize the losses that result from condensation of steam on the surface of the cylinder. The evidence afforded by these experiments has been confirmed by trials made on all kinds of engines and under every variety of working. The following is, in general terms, what experiments with actual engines show to take place.

When the amount of steam that has passed through the engine is measured, by weighing either the feed-water or the condensed steam discharged from the condenser, it is found to be greatly in excess of the quantity of dry steam that would suffice to fill the cylinder volume up to the point of cut-off, at the pressure which the steam then has, even when the effects of clearance are fully allowed for. The difference between the two is called the *missing quantity*. It is to be accounted for in two ways:—

(1) Some steam leaks directly across from the steam side to the exhaust side and so escapes being measured by the displacement of the piston.

(2) At the point of cut-off the steam in the cylinder is by no means dry: consequently more working substance passes through the engine at each stroke than corresponds to the volume of dry steam that would fill the admission space.

When steam is supplied in the saturated state the missing quantity is rarely less than 20 per cent. of the steam supplied by the boiler, often as much as 30 per cent., and is sometimes as much as 50 per cent. Even 69 per cent. has been recorded in trials of a small engine[3].

[1] An abstract of Messrs Loring and Emery's reports is given in *Engineering*, vols. XIX and XXI, and in Maw's *Recent Practice in Marine Engineering.*

[2] *Bull. Soc. Industr. de Mulhouse*, from 1877.

[3] See papers by Col. English (*Proc. Inst. Mech. Eng.* Sept. 1887, Oct. 1889, May 1892), which describe experiments on this subject. In several cases examined by him the amount is over 60 per cent.

Of the two factors which make it up the second is in general the more important, namely the effect of initial condensation in making the steam wet at the point of cut-off. The excessive amount of the missing quantity in some engines, especially where slide-valves are used, has been shown by the experiments of Callendar and Nicolson[1] to be partly due to direct leakage, but in most cases it is the give-and-take of heat between steam and metal to which the chief effect is to be ascribed.

When steam is admitted at the beginning of the stroke, it finds the metallic surfaces of the port, the cylinder and the piston chilled by having been exposed to low-pressure steam during the exhaust of the previous stroke. A portion of it is therefore at once condensed, and, as the piston advances, more and more of the chilled cylinder surface is exposed and more and more of the hot steam is condensed. At the end of the admission, when communication with the boiler is cut off, the cylinder consequently contains a film of water spread over the exposed surface, in addition to saturated steam. The boiler has therefore been drawn upon for a supply of steam greater by perhaps 20 or 25 per cent. than that which corresponds to the volume of the admission space.

Then, as expansion begins, more cold metal is uncovered, and some of the remaining steam is condensed upon it. There is in addition a further condensation which takes place in consequence of the work the steam is doing during expansion—a condensation which would be found even if the walls were perfect non-conductors and the process were strictly adiabatic. So far as these two actions are concerned, the mixture is getting wetter as it begins to expand. But the pressure of the steam now falls, and the layer of water which has been previously deposited begins to be re-evaporated as soon as the temperature of the expanding steam falls below that of the liquid layer. Hence, on the whole, the amount of water present increases during the earliest part of the expansion, but a stage is soon reached when the condensation which occurs on the newly exposed metal or throughout the steam as a whole in consequence of expansion is balanced by re-evaporation of older portions of the layer. The percentage of water present is then a maximum; and from this point onwards the mixture of steam and

[1] H. L. Callendar and J. T. Nicolson, "On the law of condensation of steam deduced from measurements of temperature-cycles of the walls and steam in the cylinder of a steam-engine," *Min. Proc. Inst. C. E.* vol. CXXXI, 1897.

water in the cylinder becomes more and more dried by re-evaporation of the layer.

If the amount of initial condensation has been small this re-evaporation may be complete before release occurs. Very usually, however, there is still an undried layer at the end of the forward stroke, and the process of re-evaporation continues during the return stroke, while exhaust is taking place. In extreme cases, if the amount of initial condensation has been very great, the cylinder walls may fail to become quite dry even during the exhaust, and a residue of the layer of condensed water may either be carried over as water into the condenser, or, if the exhaust-valves are not arranged so that it can be discharged, this unevaporated residue may gather in the clearance space, and in very bad cases may even require the drain-cocks to be left open to allow of its escape. When any water is retained in this way it may be conjectured that the initial condensation will be increased, for the hot steam then meets not only comparatively cold metal but comparatively cold water when it enters the cylinder. The latter tends to cause much condensation, partly because of its high specific heat, and partly because it is brought into intimate mixture with the entering steam.

Apart, however, from this extreme case, whatever water is re-evaporated during expansion and exhaust takes heat from the metal of the cylinder, and so brings it into a state that makes condensation inevitable when steam is next admitted from the boiler. It is in fact the condensation of the layer and its re-evaporation, whether during expansion or during exhaust, that is the means of exchange of heat between the metal of the cylinder and the working substance. Mere contact with low-pressure steam during the later stages of expansion and during the exhaust stroke would cool the metal but little, for communication of heat between dry metal and any gaseous substance is slow even when the difference of temperature between them is large. The cooling of the cylinder walls which actually occurs is due mainly to the re-evaporation of the condensed water. Thus if an engine were set in action, after being heated beforehand to the boiler temperature, the cylinder would be only slightly cooled during the first exhaust stroke, and little condensation would occur during the next admission. But the metal would be more cooled in the subsequent expansion and exhaust, since it would part with heat in re-evaporating this water. In the third admission more still would be condensed, and so on, until a permanent *régime* would be established in which condensation and re-evaporation were exactly balanced. The same permanent *régime* is reached when the engine starts cold.

However early the re-evaporation of the condensed film is completed it results in some chilling of the cylinder walls, leaving them to be re-heated by condensation of fresh steam in the next stroke. The evils of initial condensation are greater the later this re-evaporation is completed. If the steam in the condensed layer is all evaporated before the release but little further cooling of the metal will occur during the exhaust stroke: if water remains to be evaporated during exhaust the whole action of the walls is intensified. It is only in exceptionally favourable cases that the water condensed during admission is completely evaporated before release.

115. The balance of heat in the action of the cylinder walls. When steam is alternately condensed and evaporated by contact with the cylinder walls in the action of a steam-engine more heat is given to the metal by each pound that is condensed than is taken from the metal by each pound that is evaporated. This is because the condensation takes place at a higher temperature and pressure than the re-evaporation. The heat given up to the metal for each unit quantity of steam condensed upon it by contact at any temperature t_1 is the latent heat L_1 together with the heat which the layer of condensed water gives up in falling from t_1 to the temperature t_2 at which re-evaporation occurs, namely $I_{w_1} - I_{w_2}$. The heat taken from the metal, by re-evaporation of the layer, is L_2. But $L_1 + I_{w_1} - I_{w_2}$ is greater than L_2, and consequently the process could not go on unless the metal were losing heat in some other way, for so far as this alternate condensation and re-evaporation is concerned the result would be an accumulation of heat in the metal which would increase without limit.

This accumulation does in fact go on for a short time when the engine starts cold and while the metal is being warmed up. But when a permanent *régime* has become established it is clear that in each revolution of the engine the total gains and losses of heat on the part of the metal must just balance one another. This implies that the metal is losing more heat than is accounted for by the re-evaporation of the initially condensed layer. In an unjacketed engine, and especially if the cylinder is not well lagged, there is a loss by conduction to the outside, which is an important item in the general balance of losses and gains. But by using a steam-jacket the metal may be prevented from losing heat to the outside, and may even be made to take up heat. Under these conditions it is on the fact that more water is re-evaporated than was condensed by contact that the balance depends, and it is this that makes alternate condensation and re-evaporation possible.

This excess of water to be re-evaporated may arise simply as a consequence of the work done during expansion. We have seen that in a strictly adiabatic process steam becomes partially condensed as it expands doing work. Consequently, in an actual engine there is a greater quantity of water available for evaporation in the later stages of the cycle than was condensed by contact with the walls in the early stages. But the excess may also arise from the presence of water in the steam-supply. Any water that enters the cylinder as water will be available for evaporation in the later stages of the cycle and may therefore be a factor in establishing the balance of give-and-take of heat to which initial condensation is due.

On the other hand, it is important to notice that if the cylinder valves are so situated as to facilitate drainage of water from the cylinder there will be a tendency to check the whole action by removing the excess of water before re-evaporation has taken place.

A comparatively small excess of water to be re-evaporated is competent to make a large amount of alternate condensation and re-evaporation possible. To illustrate this take a numerical example. Suppose we have steam admitted at a temperature of 160° C. (say 90 lb. per sq. inch absolute), and expanded to 70° C. (4·5 lb. per sq. inch). To simplify the problem we shall ignore the action at intermediate temperatures and think of the condensation on the walls as taking place at 160° C. and the re-evaporation as taking place at 70° C. On that basis L_1 is 499·3 and $I_{w_1} - I_{w_2}$ is 91·4, so that for each unit of steam condensed by contact the metal gains 590·7 units of heat. It loses heat equal to L_2, or 556·7 units, by the re-evaporation of the same quantity of water. Hence to preserve the balance 34 additional units of heat must be taken from the metal for every unit quantity of water that is initially condensed by contact. The proportion of 34 to 557 is about one-seventeenth, and hence under these conditions any extra quantity of water that is available to be evaporated at the lower temperature, in addition to the quantity initially condensed by contact, is competent to account for the alternate condensation and re-evaporation of about seventeen times its own weight of the working steam.

The work done during expansion, between the limits of temperature chosen in this example, would itself produce 14 per cent. of wetness if the steam were dry at cut-off, and 8 per cent. if the steam at cut-off contained 20 per cent. of water: it would therefore be much more than sufficient to supply the amount of extra water for re-evaporation which is necessary to account for a large amount of initial condensation. It is however open to question how far the water that is due to the work done during expansion takes a form which permits of re-evaporation in the later stages of the cycle. In so far as it is deposited on the walls it is of course available, but whatever part of it is mechanically suspended as a mist throughout the cylinder would not contribute to the action.

The investigations of Callendar and Nicolson led them to conclude that the amount of cylinder condensation is in general limited by the time rate at which condensation can occur on the metallic surfaces with which the steam comes in contact, having regard to the difference of temperature between the metal and the steam. By examining the fluctuations of temperature in the metal, and also in the steam, during the revolution of the engine they deduced a limit to the amount of condensation that should occur in the cycle. When there is a sufficient excess of water available for re-evaporation this limit may be expected to be reached in the action of the engine, and the amount of initial condensation occurring per stroke will then not be much affected by small changes in the conditions of operation.

But the case is different if the amount of the action is restricted to something short of that limit by being dependent on the heat balance. In that case a small change in the conditions may produce a relatively large change in the amount of the initial condensation. The balance, as Callendar and Nicolson remark, is extremely delicate and is very easily turned. A little priming in the steam-supply may increase the whole effect substantially: on the other hand, a little removal of water by drainage, or a little addition of heat by superheating or by jacketing may greatly reduce it. It is to considerations of this kind that much of the advantage actually found in superheating or in the use of the steam jacket is to be ascribed.

116. Leakage affecting the missing quantity. In Callendar and Nicolson's experiments the missing quantity was much larger than they could attribute to cylinder condensation, in view of the limit which their observations of the temperature of the cylinder walls imposed on the amount of that condensation. The discrepancy was explained by the discovery that part of the missing quantity was caused by leakage past the valve, direct from the steam chest to the exhaust. A slide-valve of the ordinary type was used. When tested in a stationary position it was steam-tight, but experiments devised to examine the leakage when it moved showed that while in action it allowed a large quantity of steam to pass to the exhaust without entering the cylinder at all. The leakage appears to take place not as steam but as water, a film of which is formed on the metal surface over which the valve slides in consequence of the surface having been chilled by previous contact with exhaust steam. Then the film is driven through under the valve face by the difference of pressure between the steam and exhaust sides. These experiments suggest that in some cases when the missing quantity is very large the greater part of it may be due to valve leakage: they also show the practical advantage of having separate valves for admission and exhaust, an arrangement which is in fact adopted in engines of the most efficient class. Apart from questions of leakage, the use of separate valves has the advantage of reducing condensation by not exposing the surface of the same steam-port alternately to hot and colder steam.

117. Graphic representation, on the indicator diagram, of the water present during expansion. In testing engines the amount of steam is measured which passes through the cylinder per stroke—that is, the quantity which we have called the "cylinder feed." The whole quantity of steam and water present during expansion is the cylinder feed *plus* the cushion steam. To estimate the amount of the cushion steam we take, on the indicator diagram, a point after compression has begun, when the exhaust-valve has become completely closed, and note the pressure and the volume there, remembering that the true volume is the sum of the uncompleted portion of the stroke and the clearance. From this pressure and volume the quantity of the cushion steam is readily calculated, if we may assume that the steam is simply saturated and that no water is present when compression begins. As a rule, this assumption is probably correct: occasionally the cushion steam may be wet, which would make its amount greater, but in most cases the supposition that the steam is dry when compression begins may be accepted as involving at least no serious error. The total quantity of steam which is or should be in the cylinder during expansion is next found by adding the amount of this cushion steam to the cylinder feed. A "saturation curve" can then be drawn on the indicator diagram to show the volume which this total quantity would fill if it were dry and saturated at each pressure reached during the expansion. An example is shown in the indicator diagram of fig. 52, where *SS* is the saturation curve.

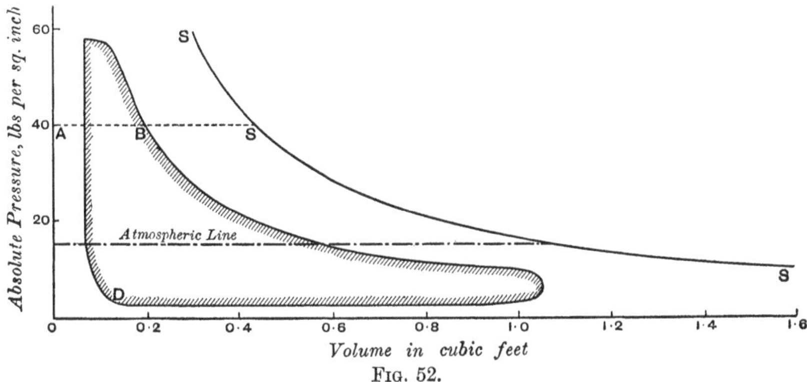

FIG. 52.

In drawing this line the axis of no volume is to be taken to the left of the diagram which the indicator traces, by a distance which represents the volume of the clearance. Then if a horizontal line *ABS* be drawn to intersect the expansion curve at any point *B*, *AB* is the actual volume which the expanding mixture filled at this pressure, *AS* is the volume it would have filled if dry and saturated; *BS* is the volume of the missing quantity, which is due to wetness and to valve leakage. If it is all due to wetness the proportion of water in the mixture is sensibly $\frac{BS}{AS}$, and the dryness q is $\frac{AB}{AS}$. Thus the proportion of water present at any stage of the expansion is determined and is shown in the diagram.

Fig. 52 relates to a trial of a small engine of the marine type supplied with saturated steam. The amount of cylinder feed per single stroke was 0·0404 lb. The pressure at the point D was found to be 4 lb. per sq. inch, and the volume there was 0·12 cub. ft. Since the volume of 1 lb. at that pressure is 90 cub. ft., it follows that the amount of cushion steam was 0·0013 lb. This gives a total of 0·0417 lb., for which the curve SS is drawn. By measuring values of $\frac{BS}{AS}$ at points along the curve it is found that the proportion of water in the mixture was 52 per cent. at cut-off, then increased to about 55 per cent. during the early stages of expansion, then became less, and finally sank to 37 per cent. just before release. These figures assume the missing quantity to be all due to wetness, but valve leakage may account for part.

Again, knowing the wetness of the mixture at the point of cut-off we may draw an adiabatic line through that point using the equation $Pv^n = $ constant with a suitable value of n (see § 68). This curve will in general be found to lie a trifle above the actual expansion curve at first, but to cross it early and lie distinctly below it towards the end of expansion. This is because the metal continues for some time after cut-off to take heat from the working fluid, but later gives up heat to it through the re-evaporation of the condensed film.

By comparing the adiabatic with the actual expansion curve it is possible to examine the give-and-take of heat between the metal and the working fluid. But this is more conveniently done after the entropy-temperature curve has been drawn, as will be presently described.

When tests of compound engines are in question it is useful to modify the construction shown in fig. 52 by separating the cylinder feed from the cushion steam, and drawing the diagram for the former. This allows a combined diagram for the several cylinders to be drawn, along with a single saturation curve. The reason is that the amount of cylinder feed is the same for both or all the cylinders, whereas the amount of cushion steam may be very different. An example of this construction will be given later in dealing with compound engine trials (§ 189).

118. Use of the entropy-temperature diagram in exhibiting the behaviour of steam during expansion and the exchanges of heat between it and the cylinder walls. In the entropy-temperature diagram, fig. 53, let ab be drawn at the temperature which corresponds to the pressure at the point of cut-off, and let it be divided at c so that $\frac{ac}{cb}$ represents the proportion of dry steam to water in the total quantity of working fluid present in the cylinder. Similarly, at any lower temperatures reached during expansion let lines $a'b'$, $a''b''$ be divided at points c', c'' in the proportion of steam to water then present, making

$$\frac{a'c'}{a'b'} = \frac{AB}{AS}$$

at the corresponding pressure in the indicator diagram (fig. 52). In this way the curve $cc'c''$ is determined, which represents the real process of expansion, and this is readily compared with the ideal adiabatic process represented by the straight vertical line cg.

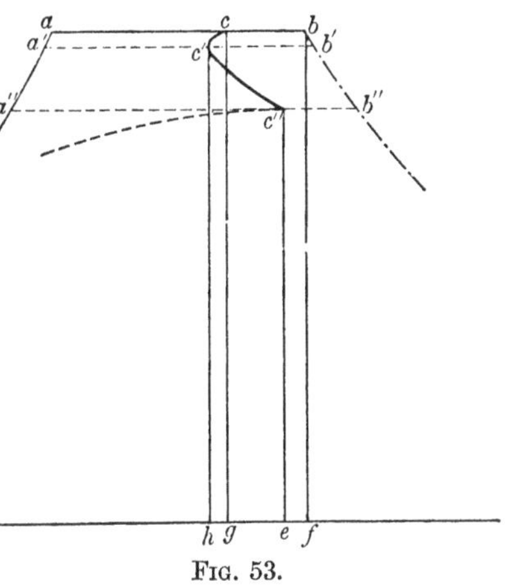

FIG. 53.

Taking c'' as the point of release, the diagram may be continued by drawing a constant-volume curve as described in § 96. In the first stages of expansion, namely from c to c' in the sketch, the proportion of water in the cylinder is increasing, and the heat abstracted by the cylinder walls from the steam is the area $cghc'$. From this point onwards the steam becomes drier, and takes up heat from the metal, the whole amount recovered up to the point of release being the area $c'c''eh$. It will be seen that a diagram of this type is particularly well fitted to allow the transfer of heat between metal and fluid to be traced throughout all stages of the expansion, the heat given up or recovered in any part of the process being equal to the area under the corresponding portion of the expansion curve $cc'c''$. When this curve slopes down to the left heat is passing from the steam to the metal; when it slopes down to the right the exchange is the other way. The heat abstracted from the steam during compression and admission is nearly equal to the area $fbcg$—nearly, but not exactly, because all the condensation in these stages does not occur at the pressure of cut-off. During compression condensation is going on at lower pressures because the temperature of the cushion steam—necessarily rising with the pressure—is being raised above the temperature to which the walls have been chilled during exhaust.

119. Thermodynamic loss due to initial condensation. From a thermodynamic point of view all initial condensation of the steam is bad, for, however early the film of water be re-evaporated, this can take place only after its temperature has cooled below that of the boiler. The process consequently involves a misapplication of heat, since the substance, after parting with high-temperature heat, takes it up again at a temperature lower than the top of its range. This causes a loss of efficiency, and the loss is greater the later in the stroke re-evaporation occurs. The heat that is drawn from the cylinder by re-evaporation of the condensed film becomes less and less effective for doing work as the end of the expansion is approached, and finally, whatever evaporation continues during the back stroke is an unmitigated source of waste. The heat it takes from the cylinder does no work[1]; its only effect, indeed, is to increase the back-pressure by augmenting the volume of steam to be expelled. A small amount of initial condensation reduces the efficiency of the engine but little; a large amount causes a much more than proportionally larger loss.

120. Action of a steam-jacket. The action of the cylinder walls is increased by any loss of heat which the engine may suffer by radiation and conduction from its external surface. We have already seen how any such loss tends to disturb the balance of heat in a cyclical process of condensation and re-evaporation, and consequently to promote condensation. More steam is initially condensed in a cylinder which is losing heat externally. The loss of efficiency due to the action of the cylinder walls will therefore be greater in an unprotected cylinder than in one which is well lagged or covered with non-conducting material. On the other hand, if the engine have a steam-jacket the deleterious action of the walls is reduced. The working substance is then on the whole gaining instead of losing heat by conduction during its passage through the cylinder. The jacket maintains a higher mean temperature on the inner surface of the cylinder, reduces condensation,

[1] Unless, of course, the cylinder in question is one of a compound series, and the steam that leaves it passes on to another cylinder to undergo further expansion there.

and accelerates the process of re-evaporation, tending to make it occur while the temperature and pressure of the steam are still comparatively high. After the process of re-evaporation is complete the jacket cannot superheat the steam in the cylinder to any material extent, for conduction and radiation between dry steam and the metal of the cylinder are incompetent to cause any considerable exchange of heat. The earlier, therefore, that re-evaporation is complete the less is the metal chilled, and the less is the subsequent condensation. But after re-evaporation is completed the steam in the jacket continues to give heat to the metal during the remainder of the cycle, and so warms it to a temperature more nearly equal to that of the boiler steam before the next admission takes place.

Thus a steam-jacket, though in itself a thermodynamically imperfect contrivance, inasmuch as it supplies heat to the working substance at temperatures lower than the top of the range, acts beneficially by counteracting, to some extent, the more serious misapplication of heat which occurs through the alternate cooling and heating of the cylinder walls. The heat which a jacket communicates to the working steam often increases the power of the engine to an extent far greater than corresponds to the extra supply of heat which the jacket itself requires. A jacket has the obvious drawback that it increases waste by external radiation, since it both enlarges the area of radiating surface and raises its temperature; notwithstanding this, however, many experiments have shown that the influence of a steam-jacket on the efficiency is good, especially in slow running engines and in engines where there is a large ratio of expansion in a single cylinder. This is to be ascribed to the fact that it reduces, though it does not entirely remove, the evils of initial condensation. To quote once more Watt's words, the jacket does good by helping to keep the cylinder as hot as the steam that enters it. To be effective, however, jackets must be well drained and kept full of "live" steam, so that they do not become traps for condensed water or for air. The action is kept up by condensation of steam in the jacket itself. When the jacket is acting effectively the amount of steam which is condensed in it generally ranges from about 7 to 12 per cent. of the whole steam-supply. The most economical treatment of the jacket-water is to allow it to drain directly back into the boiler. In some cases the activity of the jacket has been secured by letting all the steam-supply pass through the jacket on its way to the cylinder, an arrangement which makes particular care necessary to prevent the water which is formed in the jacket from passing into the cylinder. The parts of the cylinder where the application of a steam-jacket is most beneficial are the ends, for it is on the end surfaces rather than on the cylindrical surface that initial condensation mainly takes place.

We shall refer presently to experiments which show the influence of steam-jackets on the efficiency of engines of various types. Meanwhile it may be said that in no trials has it appeared that a jacket has done harm: in other words, the saving of steam in the cylinder feed brought about by the use of a jacket is always greater than the amount of steam which the jacket itself uses, and in many instances the net saving is as much as 10 or 20 per cent. The best results are found in cases where, if the jacket were absent, the conditions are such as would give rise to much initial condensation. In engines which make a great number of strokes per minute the influence of the jacket is necessarily small[1].

The advantage of the jacket may be increased by making its temperature higher than that of the steam during admission to the cylinder. Re-evaporation of the condensed layer is further hastened, and after it is over the jacket gives up but little heat. Bryan Donkin obtained good results in experiments where the cylinder of a small engine was kept hot by gas flames[2].

121. Influence of speed, size, and ratio of expansion. It is interesting to notice, if only in general terms, the effects which the particular conditions of working in different engines may be expected to produce on the loss that occurs through the action of the cylinder walls. Initial condensation will be increased by anything that augments the range of temperature through which the inner surface of the cylinder fluctuates in each stroke, or that exposes a larger surface of metal to the action of a given quantity of steam, or that prolongs the contacts in which heat is exchanged. The influence of time is specially important; for the whole action depends on the rate at which heat is taken up and given up by the substance of the metal. The changes of

[1] See the Reports of the Inst. of Mechanical Engineers' Research Committee on the Value of the Steam-jacket. *Proc. Inst. Mech. Eng.* 1889, 1892, 1895.

[2] *Min. Proc. Inst. C. E.* 1889, vol. XCVIII.

temperature which the metal undergoes are in every case mainly superficial; the alternate heating and cooling of the inner surface initiates waves of high and low temperature in the iron whose effects are sensible only to a small depth; and the faster the alternate states succeed each other the more superficial are the effects[1]. In an engine making an indefinitely large number of strokes per minute the cylinder sides would behave like non-conductors and the action of the working substance would be adiabatic.

We may conclude, then, that in general an engine running at a high speed will have a higher thermodynamic efficiency than the same engine running at a low speed, all the other conditions of working being the same in both cases.

Again, as regards range of temperature, the influence of the cylinder walls will be greater (other things being equal) with high than with low-pressure steam, and in condensing than in non-condensing engines.

In large engines the action of the walls will be less than in small engines, since the proportion of wall surface to cylinder volume is less. This conclusion agrees with the well-known fact that small engines do not readily achieve the economy that is reached in many larger forms.

Cylinder condensation is increased when the ratio of expansion is increased, all the other circumstances of working being left unaltered. The quantity of water formed by adiabatic condensation is then greater, and it is on this that the whole action mainly depends. Further, the metal is then brought into rather more prolonged contact with low-temperature steam. The volume of admission is reduced to a greater extent than the surface that is exposed to the entering steam, since that surface includes two constant quantities, the surface of the cylinder cover and of the piston. For these and perhaps other reasons we may expect that with an early cut-off the initial condensation will be relatively large, and this conclusion is amply borne out by experiment. An important result is that increase of expansion does not, beyond

[1] The temperature of the cylinder walls formed the subject of an interesting experimental study by Bryan Donkin, who examined the general gradient of temperature across the walls, both with and without steam in the jacket. See his papers, *Min. Proc. Inst. C. E.* 1890 and 1891, also *Proc. Inst. Mech. Eng.* 1895.

a certain limit, involve increase of thermodynamic efficiency; when that limit is passed the augmentation of waste through the action of the cylinder walls more than balances the increased economy to which, on general principles, expansion should give rise, and the result is a net loss. With a given engine, boiler pressure, and speed, a certain ratio of expansion will give maximum efficiency. But the conditions on which this maximum depends are too complex to admit of theoretical solution; the best ratio is a matter rather for experiment.

These remarks apply only to reciprocating engines. In steam turbines there is no alternate give-and-take of heat between the working substance and the metal, and there is a clear gain of efficiency in having the steam continue to do work while it expands down to the lowest pressure attainable in the condenser.

122. Results of experiments with various ratios of expansion. The effect of increased expansion in augmenting the action of the sides and so reducing the efficiency, when carried beyond a certain moderate grade, was clearly shown by the American and Alsatian experiments alluded to above (p. 192). The following figures, relating to a single-cylinder Corliss engine, are reduced from one of Hallauer's papers[1]:—

Single-cylinder Corliss engine; effect of varying the expansion.

Ratio of expansion	Percentage of water present		Consumption of steam in lb. per hour per I.H.P.
	At end of admission	At end of expansion	
7·3	24·2	17·8	17·8
9·4	30·8	18·6	17·6
15·1	37·5	20·8	17·7

Here, in consequence of the amount of initial condensation increasing with increased expansion, a maximum of efficiency lies between the extreme grades of expansion to which the test extends, but the efficiency varies exceedingly little even through this wide range. In the American experiments the best results were obtained with even more moderate ratios of expansion. The compound engines of the United States revenue steamer "Bache," when

[1] *Bull. Soc. Industr. de Mulhouse,* May 26, 1880.

tested with steam in the jacket of the large cylinder, with the boiler pressure nearly uniform at 80 lb. by gauge, or 95 lb. per sq. inch absolute, and the speed not greatly varied, gave results which are shown in the table. Here the efficiency is very little

U. S. Revenue steamer "Bache"; effect of varying the expansion.

Ratio of total expansion	Consumption of steam in lb. per hour per I.H.P.
4·2	21·2
5·7	20·0
7·0	20·3
9·2	20·7
16·8	25·1

affected by a large variation in the position of the cut-off, but when the ratio of expansion becomes excessive a distinct loss is incurred.

Again—to take an instance relating to a very different type of machine—trials made by Willans with one of his high-speed compound non-condensing single-acting engines, using steam with an absolute initial pressure of 130 lb., gave these results:—

Willans' engine (non-condensing); effect of varying the expansion, the initial pressure and speed being constant[1].

Ratio of total expansion	Percentage of water present at end of admission in high-pressure cylinder	Consumption of steam in lb. per hour per I.H.P.
4	8·9	20·7
4·4	10·2	20·5
4·8	11·7	20·35
5·2	14·2	20·26
5·6	14·3	20·0
6	18·4	20·3
8	25·0	23·1

The initial condensation is comparatively small here, mainly because of the exceptional speed (404 revolutions per minute), and for the same reason the economy in steam consumption is

[1] Willans on Non-Condensing Steam-Engine Trials, *Min. Proc. Inst. C. E.* March, 1888.

remarkably high for a small non-condensing engine. In another series of trials in which a compound engine of this type was worked with a condenser[1], and with steam at about 170 lbs. (absolute), Willans found a slight increase in the steam consumption from 14·26 to 14·72 lb. per hour per I.H.P. when the ratio of expansion was increased from $15\frac{1}{2}$ to 20; at the same time the percentage of water present at cut-off in the high-pressure cylinder increased from 31 to 37. All these results agree in showing that the ratio of expansion may be varied through a large range with but little influence on the efficiency, because the gain that comes of making the expansion more complete is counterbalanced by the bad effects of increased initial condensation. The ratio of expansion which gives a maximum of efficiency is never sharply defined, and its value depends much on the initial steam pressure and the particular features of the engine under trial.

123. Advantage of high speed. The advantage of high speed in making the action of an engine more nearly adiabatic has been demonstrated by experiment. Among the trials described by Willans in the papers cited are the following two sets made with one of his compound non-condensing engines, in the first set with an absolute admission pressure of 90 lb. per square inch and 3·2 as the ratio of expansion; in the second set with 130 lb. pressure and 4·8 as the ratio. In the three trials of each set the only condition varied was the speed.

Willans' non-condensing engine trials: influence of speed.

	I. Trials with steam of 90 lb. pressure			II. Trials with steam of 130 lb. pressure		
Speed: revolutions per minute	401	211	122	405	216	131
Percentage of water present at cut-off in the high-pressure cylinder	5·0	12·6	20·2	11·7	19·1	29·7
Consumption of steam in lb. per hour per I.H.P.	24·2	25·3	27·0	20·3	21·3	23·7

The increase of steam consumption as the speed is reduced is

[1] Willans on Steam-Engine Trials, *Min. Proc. Inst. C. E.* April, 1893.

considerable, and still more marked is the greater initial condensation. The same features are apparent in the trials quoted below, which relate to a condensing engine with an absolute admission pressure of 90 lb. and a very moderate ratio of expansion (4·8).

Willans' condensing engine trials: influence of speed.

Speed: revolutions per minute	401	301	198	116
Percentage of water present at cut-off in the high-pressure cylinder	8·9	12·2	17·9	20·9
Consumption of steam in lb. per hour per I.H.P.	17·3	17·6	18·9	20·0

124. Experiments on the value of the steam-jacket. Evidence of the advantage of a steam-jacket was given in Reports by a Committee appointed by the Institution of Mechanical Engineers to inquire into the subject. Individual tests varied widely, but it appeared that the saving usually secured by jackets in condensing engines was something like 12 or 15 per cent. In non-condensing engines it was less. The following results of special

Influence of steam-jacket.

Engine	Total steam in lb. per hour per I.H.P.		Percentage less with jackets	Proportion of jacket feed to total consumption per cent.
	Without jackets	With jackets		
Two-cylinder compound[1]	18·2	16·6	9	7
Two-cylinder compound[1]	24·7	20·0	19	6
Triple compound[1]	17·2	15·4	10	11
Triple compound[2]	16·4	13·6	17	—
{ Two-cylinder compound[3]	21·1	19·5	7	12
{ Same engine run non-compound, the large cylinder only being used	32·1	26·7	17	7
Small single-cylinder[4] engine	39	29	25	7

[1] *Proceedings Inst. Mech. Eng.* 1889, 1892, and 1895.

[2] Prof. O. Reynolds' tests. For particulars see *Min. Proc. Inst. C. E.* vol. XCIX, 1889.

[3] Prof. Unwin's test: *Proc. Inst. Mech. Eng.* 1892, p. 460.

[4] Mr B. Donkin's tests: *Proc. Inst. Mech. Eng.* 1892, p. 464.

trials with condensing engines are stated in the Reports. In several of these cases, notably in the last, it is remarkable how large a net saving of steam was secured by a comparatively small consumption in the jackets. In other trials of the same small engine, using an earlier cut-off, Donkin found that 8 or 9 per cent. used in the jackets was capable of saving as much as 40 per cent. of the whole steam. In this instance there was excessive initial condensation when the jackets were out of use.

In compound engines the jackets are most effective when both or all of them are filled with steam at the boiler pressure. In Osborne Reynolds' triple engine trials, it was found that steam of the full boiler pressure (200 lb. per sq. inch) in all the jackets reduced the initial condensation in the second cylinder to about one-third or one-fourth of the amount that occurred without jackets, made the steam practically dry before the end of expansion in the second cylinder, and almost entirely prevented condensation in the third cylinder. Without steam in the jackets the second and third cylinder had been very wet, the proportion of water in them being about 40 per cent. of the whole. Indicator diagrams relating to these trials will be found in Chapter x, figs. 124 and 125.

125. Superheating. The foregoing results all relate to engines supplied with saturated steam. Superheating the steam before its admission much reduces and may even entirely prevent initial condensation. It also tends to keep the steam drier during its whole action by reducing the condensation which results from expansion. It consequently lessens the losses that are due to exchange of heat between the working substance and the cylinder walls. The improvement in efficiency that is actually brought about by superheating greatly exceeds the gain that might be anticipated on purely thermodynamic grounds from the fact that when steam is superheated a part of the whole supply of heat is taken in at a temperature higher than that of the boiler. The main advantage is indirect: the use of superheated vapour makes the action in the cylinder more nearly adiabatic. That superheating has a marked advantage was first experimentally demonstrated by Hirn, who found that the consumption of steam was reduced from 19·4 to 16·2 lb. per horse-power-hour in a condensing engine by superheating the steam some 45° C. About the year 1860 superheating was frequently used in marine practice, but it was abandoned, mainly on account of difficulties in regard to lubrication. The importance of taking means to avoid or rather to reduce initial condensation was less generally understood in those days, and with the mineral oils that are now used as lubricants the old objections to superheating have little force. Towards the end of the nineteenth century the practice of superheating was revived and it is now common in engines of all types. The development of the steam turbine contributed to this revival, for it was found that in the turbine no less than in the reciprocating engine superheating caused a marked improvement in efficiency, mainly (in the turbine) through its effect in reducing internal friction by preventing condensation. Even in locomotives, where the ratio of expansion is only moderate since the steam is discharged at a pressure above that of the atmosphere, superheating has notably reduced the consumption of coal and its use has become general. All large steam plants in power stations, where economy of coal is a chief concern, employ superheating as a matter of course. The amount of superheat usually ranges from 80° to 150° C., but in some cases this figure is exceeded. Mr R. J. Walker, referring to marine steam turbines, gives results from which it appears that with a superheat of 85° the saving of steam is 13 or 14 per cent.[1] The advantage is substantial, after allowing for the 7 per cent. or so of additional heat that is required to form each pound of the superheated supply. Experiments made in 1892 by the Alsatian Association of Steam Users on a large number of engines furnished with superheaters showed that superheating effected a saving of coal to the extent of about 20 per cent. when the superheater was simply placed in the boiler flue, so that it enabled what would otherwise be waste heat to be utilized, and about 12 per cent., on the average, when the superheater was separately fired. One of the trials, dealing with a triple-expansion Sulzer engine of 300 horse-power, records a consumption of 14·6 lb. of steam per I.H.P.-hour without superheating, and 11·6 lb. when the steam was superheated 55° C.

In a trial by Prof. Schröter a triple expansion factory engine indicating 1000 H.P. and supplied with steam at 100 lb. per sq. inch was tested with saturated steam and with steam superheated by

[1] *Trans. Inst. Nav. Arch.* vol. LVI, p. 139, 1914.

52° C. Even this moderate superheat brought the consumption of steam down from 13·2 lb. to 12·0 lb. per I.H.P.-hour[1]. His analysis of the test shows that the amount of the superheating was insufficient to prevent the steam from becoming somewhat wet during admission. At the point of cut-off in the first cylinder the steam was nearly, but not quite dry, and as expansion went on in this cylinder its wetness increased. The advantage, such as it is, of moderate superheating lies in reducing the losses which proceed from exchanges of heat between the steam and the cylinder walls. Although the steam retains its superheat until it reaches the engine, it at once falls to the temperature of saturation when it meets the cylinder walls unless the amount of initial superheat is large. To keep it dry during admission would require an amount of superheating probably never less than 60° C. and often much more, the amount that is necessary depending on the ratio of expansion in the cylinder and on the rate at which heat is lost from the external surfaces. To keep the steam dry during expansion a much higher degree of superheat would be needed. Superheating, in any moderate degree, may be regarded as a device for reducing the action of the cylinder walls by bringing the expansion curve of the indicator diagram nearer to the saturation curve (fig. 52) than it would otherwise come. In many cases it barely makes the expansion curve come out so far as to reach the saturation curve even at cut-off, and as expansion proceeds the interval between the two increases. The ideal diagram sketched in fig. 30 is widely departed from in real engines using moderately superheated steam, for the action of the cylinder walls in general keeps the whole process of expansion to the left of the saturation line cf. The working substance, after being taken up the line cr before it reaches the engine, immediately comes down that line when it is admitted to the cylinder and before it begins to expand, giving up to the walls the heat (represented by the area under cr) which it has received in being superheated, and often going on to give up a part of its latent heat by becoming slightly wet. In cases where the superheat is high the steam comes only part of the way back from r towards c before it begins to expand.

[1] *Zeitschrift des Vereins deutscher Ingenieure*, vol. XL, 1896. In reducing to British measure the results of tests stated in the metric system the number of kilogrammes per metric H.P.-hour has to be multiplied by 2·235 to bring it to pounds per British H.P.-hour, the metric H.P. being 0·9863 of a British H.P.

Towards the end of the nineteenth century the employment of much higher degrees of superheat than had been usual was successfully initiated by W. Schmidt, who introduced simple forms of tubular superheaters in which the steam could have its temperature raised, by taking heat from the furnace gases or by separate firing, to 400° C. or thereabouts. He demonstrated that with slight modifications in some details of design it was practicable to utilize steam supplied at this temperature in an ordinary piston and cylinder engine, and that a remarkable improvement in thermodynamic efficiency might thereby be obtained. He also showed that it was advantageous to re-superheat the steam at an intermediate stage in the process of expansion.

In some of Schmidt's engines there is a re-heater in the intermediate receiver between the high and low-pressure cylinders, consisting of a number of pipes through which part of the supply of highly superheated steam is made to pass before it reaches the high-pressure valve-chest. The effect is to transfer a portion of the heat of the highly superheated steam to the steam in the receiver, which has already performed the first stage of its expansion, and consequently to reduce the degree of superheat on admission to the high-pressure cylinder.

In a two-cylinder compound condensing Schmidt engine tested by the present writer the boiler pressure was 140 lb. per sq. inch by gauge and the steam was initially superheated to 396° C. Before entering the first cylinder, however, the steam lost 104° of its superheat, by being used in the reheating device, so that its temperature in the high-pressure valve-chest was 292° C. The engine developed 184 indicated horse-power. Under these conditions 10·4 lb. of steam were used per hour per I.H.P. The same engine using saturated steam of the same pressure consumed 17·2 lb. per hour per I.H.P. In the superheated trial the steam remained superheated throughout the whole course of expansion in the high-pressure cylinder, but was somewhat wet from the point of cut-off onwards in the low-pressure cylinder. In the trial with saturated steam the missing quantity at cut-off in the high-pressure cylinder was fully 30 per cent.

In another of the writer's tests, a two-cylinder compound Schmidt engine developing 300 horse-power took steam of 142 lb. pressure by gauge, which was initially superheated to 425° C. but reduced to 317° C. after passing the re-heater. The consumption was

9·0 lb. of steam per I.H.P.-hour and 14·9 lb. per electrical unit (kilowatt-hour) generated by the dynamo which the engine was employed to drive. Taking the mean specific heat in the superheating process to be 0·52 (see § 66) the extra heat required for superheating was 126 units. As the feed-water was supplied at 24° C. the heat taken in by the steam was consequently 643 + 126 or 769 units, compared with 643 for saturated steam. Thus the actual consumption of 9·0 lb. per horse-power-hour is equivalent in point of heat to $\frac{9{\cdot}0 \times 769}{643}$ or 10·8 lb. of saturated steam per horse-power-hour at the same pressure and with the same temperature of feed, a figure which is much better than the best results obtained in trials of engines of comparable size using saturated steam. In trials of Schmidt engines by Prof. Lewicki and others like results have been observed, the consumption of steam being in some instances rather less than 9 lb. per I.H.P.-hour.

Taking the above result it is interesting to compare the actual performance with the ideal performance of an engine in which there is complete adiabatic expansion and no losses. That, as we have seen in § 74, is equal to the heat-drop $I_1 - I_2$. In the example just cited I_1 is 793 and I_2 is 558, when the ideal expansion is reckoned as being carried down to the pressure in the condenser, which was 1 lb. per sq. inch. Hence in the ideal process each lb. of steam would do work equivalent to 235 thermal units, reckoning from the boiler pressure and the full initial condition of superheat. Since one horse-power-hour is equivalent to 1414 thermal units (lb.-deg. Cent.) the indicated work actually done per lb. of steam was 157 units. The observed performance therefore represents 67 per cent. of the ideal: that is to say the efficiency ratio was 0·67.

A slightly more favourable comparative estimate is arrived at if we reckon the ideal performance with reference to the condition of the steam as it reached the engine. Its pressure had then fallen to 154 lb. per sq. inch and its temperature to 405°, through losses in the steam-pipe. This was before it entered the reheating device. Reckoning I_1 with reference to this state it is 786 and I_2 is 555, by the Mollier diagram: the ideal performance is therefore 231. The actual performance was accordingly 68 per cent. of this ideal, which forms a fairer criterion than the other in judging of the losses for which the engine itself is responsible, the difference between the two being due to losses in a rather long steam-pipe.

126. Advantage of compound expansion. Another highly important means of preventing cylinder condensation from becoming excessive is the use of compound expansion. If the vessels were perfect non-conductors of heat it would be, from the thermodynamic point of view, a matter of indifference whether expansion was completed in a single vessel or divided between two or more, provided the passage of steam from one to the other was performed without introducing unresisted expansion. In practice, the transfer of steam from one cylinder to another during its expansion cannot be accomplished without some more or less wasteful drop in pressure. But the loss which this entails is more than counterbalanced by the gain that results from the reduced influence of the cylinder walls. Compound working acts beneficially by diminishing the range through which the temperature of any part of the cylinder metal varies. For this reason the amount of steam initially condensed in the high-pressure cylinder of a compound engine is less than if admission were to take place at once into the low-pressure cylinder and the whole expansion were to be performed there. Further, the steam which is re-evaporated from the first cylinder during its exhaust does work in the second, and it is only the re-evaporation that occurs during the exhaust from the last cylinder that is absolutely wasteful. The same remark applies to the effects of valve leakage. When a compound engine is tested first with compound expansion and then with the same grade of expansion in the large cylinder alone it is found that more steam is required per horse-power-hour in the second case. Prof. Unwin's tests referred to in the table of § 124 furnish an instance: an engine taking 21·1 lb. of steam per horse-power-hour when working compound required 32·1 lb. when the large cylinder only was used, no jackets being then in action. With steam in the jackets the difference was rather less, for the jacket checked that excessive cylinder condensation which reduced the efficiency in the non-compound trials.

The general subject of compound expansion will be considered more particularly in a later chapter: at present we are concerned with the influence of compounding on efficiency. When high-pressure steam is used in a non-compound engine the waste due to initial condensation is apt to become excessive because of the great range of temperature through which the metallic surfaces fluctuate in every stroke. The advantage of compounding becomes greater the higher is the boiler pressure. So long as the initial pressure does not much exceed 100 lb. per sq. inch it suffices to reduce the range of temperature into two parts by employing two-cylinder compound engines; with pressures of 150 lb. or so there is economy in triple expansion, and quadruple expansion has advantages when pressures of 200 lb. or more are employed.

127. Advantage of "Uniflow" working. The periodic give-and-take of heat between the working substance and the surfaces of metal with which it comes in contact, which is the chief source of loss in a piston and cylinder engine, can be reduced by having separate valves for admission and exhaust, so that the same port-surfaces and valve-surfaces are not brought into contact with hot and comparatively cold steam. This idea is carried further in what is called the "uniflow" engine, by an ingenious arrangement which secures that the ends of the cylinder, where the steam is admitted, shall always be kept hot, the exhaust taking place at the middle of the cylinder through ports in the circumference which are uncovered by the motion of the piston (see § 301). Thus the piston itself acts as exhaust-valve: its width in the direction of the travel is made nearly equal to the length of the stroke, and it uncovers the exhaust ports just as the stroke approaches completion. On its other side there has been compression into a small clearance, and the pressure and temperature of the clearance steam are raised by compression to the admission level. The admission valves are in the cylinder ends, which are jacketed by steam chests. Thus the ends are kept hot, and there is a steady gradient of temperature along the barrel from each of the ends to the centre, resembling the gradient of temperature in the casing of a steam turbine. By limiting condensation losses the uniflow device allows a large ratio of expansion to be used with advantage, and secures a high efficiency, especially with moderate superheat. In normal working the admission valve is closed at about 10 per cent. of the stroke, but an earlier cut-off down to 5 per cent. or less is still highly efficient. In a published test[1] of a uniflow engine working at 684 horse-power, with steam at an absolute pressure of 190 pounds, and temperature 293° C.—corresponding to a superheat of 101° C., the consumption was 10·2 lb. per hour per I.H.P. The hot-well temperature was 39·5° C. Thus each pound of steam took up in its formation 686 units (centigrade) of heat and developed work equivalent to 138·6 units,

[1] F. B. Perry, *Proc. Inst. Mech. Eng.* 1920, p. 763.

giving a thermal efficiency of 20·2 per cent. Other tests of uniflow engines confirm this remarkably good result.

It should be noted that the uniflow design allows high-pressure steam to be expanded in a single cylinder without the drawbacks ordinarily entailed in non-compound working.

Although losses due to initial condensation are greatly reduced in the uniflow engine, they are not absent, and hence there is much advantage in using superheat. This is illustrated by the figures quoted below, which relate to a series of trials of a uniflow engine supplied with steam at a gauge pressure of 170 lb. per square inch, with various degrees of superheat. For a cut-off at 5 per cent. of the stroke in each instance the consumption of steam was found to be as follows:

Condition of supply		Steam in lb. per hour per I.H.P.
Saturated		12·9
Superheated	13° C.	12·4
	42° C.	11·6
	98° C.	10·4
	155° C.	9·6

In another uniflow engine, working without a condenser, the consumption was reduced from 21·6 to 15·4 lb. per. I.H.P.-hour by superheating the steam 110° C.

128. Summary of Sources of Loss. The principal reasons have now been named which make the actual results of engine performance differ from the results which would be obtained if the action conformed in every respect to the ideal Rankine cycle. The sources of loss may be summarized as follows:—

(1) Wire-drawing in admission and exhaust.

(2) Incomplete expansion before release.

(3) Incomplete compression of the cushion steam, through which the clearance becomes a cause of waste.

(4) The action of the metallic surfaces of the cylinder, steam-ports and piston, causing condensation during compression and admission, with re-evaporation during expansion and exhaust.

(5) Radiation and aerial convection of heat from steam-pipe, valve-chest and all hot parts of the engine.

(6) Escape of the working fluid by leakage to the atmosphere, and leakage of air into the condenser.

(7) Leakage of steam to the exhaust, so that it escapes acting on the piston; and leakage past the piston.

(8) In compound engines, additional wire-drawing or unresisted expansion in the transfer of steam from one cylinder to another.

If in drawing a comparison between the real engine and the ideal we take as the lower limit of temperature that of the discharge from the condenser, we have a further item, namely, the loss that comes from the pressure in the condenser being higher than the pressure corresponding to this lower limit.

Some representative examples will be given later of results obtained in trials of steam-engines. In the most favourable cases about 20 per cent. of the heat supplied to the steam is converted into work, by engines of the piston and cylinder type. Of the various sources of loss which have been enumerated the second and the fourth are in general the most important in such engines. In steam turbines the chief causes of loss are different, but, except with large powers, the aggregate amount of loss is not very much less. Turbines escape the alternations of temperature that make for waste in a reciprocating engine, and they are able to utilize the expansion much more completely: on the other hand, they suffer from internal leakage of steam over the tips of the blades and from losses due to fluid friction and to the velocity of the steam at exit.

Taking by way of comparison a steam turbine built for the power station of the Wembley Exhibition in 1924 to generate 1500 kilowatts, the guaranteed consumption of steam supplied at a (gauge) pressure of 200 lb. and superheated to 315° C. is reported to have been 13·1 lb. per kilowatt-hour at full load, with a vacuum of 28 inches[1]. These figures make the heat-drop 218 centigrade units. Since one kilowatt-hour is equivalent to 1896 heat units each lb. of steam under the conditions of the guarantee generates electrical energy equivalent to 1896/13·1 or 144·7 heat units, which represents a little over 66 per cent. of the heat-drop. If we assume the feed-water to be supplied at say 30° C. the heat required to form each lb. of steam would be 707 units, and 20·5 per cent. of this would be converted into electrical energy. When allowance is made for loss in the conversion of mechanical into electrical power this figure corresponds to an output of mechanical work equivalent to about 22 per cent. of the heat of formation. It will be seen later that large turbines, using steam of higher pressure and with more superheat, may convert into mechanical work from 25 to 30 per cent. of the heat supplied to the steam.

[1] *Engineering*, March 28, 1924, p. 389.

CHAPTER VIII

STEAM TURBINES

129. Theory of the Steam Jet. The essential feature which distinguishes steam turbines from steam-engines of other types is that in the turbine the action of the steam is kinetic, depending on its inertia. In other engines it acts as an elastic fluid pressing statically against an expanding envelope. In the turbine it acts as a mass that is set in motion in consequence of its own elasticity, forming a jet or group of jets, and the force which is utilized is produced by the impulse of the jets on moving vanes or by their reaction on moving orifices from which the jets issue. The heat energy present in the steam is employed to set the steam in motion, and the moving steam, in virtue of its momentum, drives the mechanism by impulse or reaction or both. In treating of the theory we have first to consider the manner in which a jet is formed by the discharge of steam or any other gas under pressure. To simplify matters we shall assume that no heat is taken in from or given out to external bodies during the operation.

Consider the flow of steam, or any other gas, through a nozzle or channel of any form, from a region where the pressure is relatively high to one where it is lower. The steam acquires velocity in the process and also increases in volume. Let A and B (fig. 54) be imaginary partitions, across which it flows, taken at right angles to the direction of the stream-lines, A being in the region of higher pressure. Let P_a be the pressure at A, v_a the velocity there, and V_a the volume which unit quantity of the gas has as it passes the imaginary partition at A. Similarly let P_b, v_b and V_b be the pressure, velocity, and volume of unit quantity at B. Let E_a and E_b be the internal energy of the gas at A and B respectively. In flowing from A to B the velocity changes from v_a to v_b and there is consequently a gain of kinetic energy amounting, per unit of mass, to $\frac{v_b^2 - v_a^2}{2g}$.

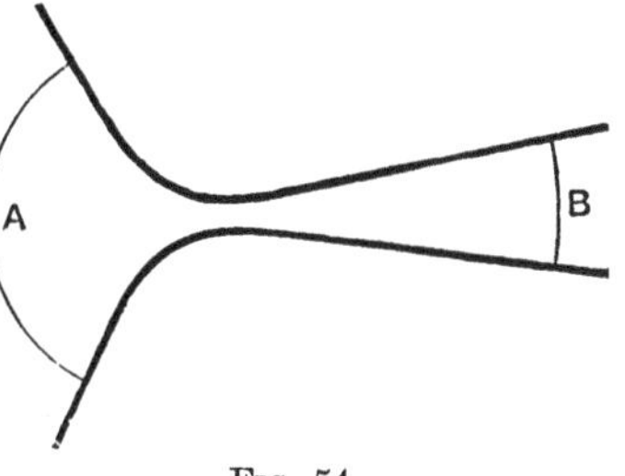

FIG. 54.

Each unit quantity of gas that enters the space between A and B has work done upon it by the gas behind amounting to P_aV_a. In passing out of this space at B it does work on the gas in front amounting to P_bV_b. In flowing from A to B it loses internal energy amounting to $E_a - E_b$. Hence by the principle of the conservation of energy, since by assumption no heat is taken in or given out,

$$\frac{v_b^2 - v_a^2}{2g} = E_a - E_b + P_aV_a - P_bV_b \quad \ldots\ldots\ldots\ldots(1).$$

But $E_a + P_aV_a$ is I_a, the total heat at A, and $E_b + P_bV_b$ is I_b, the total heat at B, and the equation may consequently be written

$$\frac{v_b^2 - v_a^2}{2g} = I_a - I_b \quad \ldots\ldots\ldots\ldots\ldots\ldots\ldots\ldots(2).$$

The gain in kinetic energy is therefore equal to the loss of total heat, or what is commonly called the heat-drop. We are treating E and I as if they were expressed in work units: when expressed in heat units they have to be multiplied by the mechanical equivalent J.

The equation applies as between any two places in the flow, and taking the process as a whole, from the initial condition in which the velocity is v_1 and total heat I_1 to the final condition in which the velocity is v_2 and total heat I_2, we have

$$\frac{v_2^2 - v_1^2}{2g} = I_1 - I_2 \quad \ldots\ldots\ldots\ldots\ldots\ldots\ldots\ldots(3).$$

In many practical cases the initial velocity is zero or negligibly small, and then

$$\frac{v^2}{2g} = I_1 - I_2 \quad \ldots\ldots\ldots\ldots\ldots\ldots\ldots\ldots\ldots(4),$$

where v is the velocity acquired in consequence of the heat-drop.

Using thermal units for the heat-drop we accordingly have

$$v = \sqrt{2gJ\,(I_1 - I_2)} \quad \ldots\ldots\ldots\ldots\ldots\ldots\ldots(4\,a),$$

which gives, in feet per second, $v = 300{\cdot}2\sqrt{I_1 - I_2}$ when the heat is expressed in pound-calories, or $v = 223{\cdot}8\sqrt{I_1 - I_2}$ when it is expressed in British Thermal Units.

So far there has been no assumption as to absence of losses through friction or eddy currents. If we assume, as an ideal case, that there is not only no conduction of heat to or from or within the fluid while the jet is being formed, but also no dissipation of energy through friction or eddies, the heat-drop in the equation

$$\frac{v_2^2 - v_1^2}{2g} = I_1 - I_2$$

is the "adiabatic heat-drop," namely, that which occurs in expansion with constant entropy, and the value of $I_1 - I_2$ is readily found from the Mollier diagram (fig. 34), provided the fluid may be treated as remaining in thermal equilibrium throughout the process. The initial condition as to pressure and superheat (if any) determines I_1, and the final pressure then allows I_2 to be found by drawing a vertical line (ϕ = const.) through the point representing the initial condition, until the final pressure is reached.

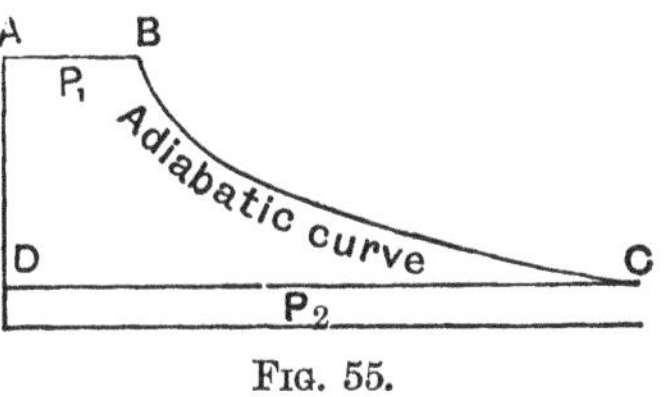

FIG. 55.

We have already seen (§ 73) that this heat-drop is equal to the area $ABCD$ of the ideal indicator diagram (fig. 55) for complete adiabatic expansion from the initial to the final state, or $\int_{P_2}^{P_1} VdP$.

Hence $$\frac{v_2^2 - v_1^2}{2g} = \text{area } ABCD = \int_{P_2}^{P_1} VdP \quad \text{............(5)}.$$

This result might also be obtained from the consideration that under the assumed conditions the steam or other gas is doing all the work of which it is ideally capable, as it expands from the first to the second state, in giving kinetic energy to the jet, and the gain of kinetic energy is therefore equal to the area of the ideal diagram.

Assume that we may, with sufficient accuracy, express the expansion in the ideal indicator diagram by a formula of the type PV^m = constant. Then the area of the diagram, namely

$$\int_{P_2}^{P_1} VdP = \frac{m}{m-1}(P_1V_1 - P_2V_2)$$

$$= \frac{m}{m-1}\left[1 - \left(\frac{P_2}{P_1}\right)^{\frac{m-1}{m}}\right]P_1V_1.$$

Hence if the expanding fluid starts from rest, at pressure P_1, to form a jet we have

$$\frac{v^2}{2g} = \frac{m}{m-1}\left[1 - \left(\frac{P}{P_1}\right)^{\frac{m-1}{m}}\right]P_1V_1 \quad \text{..................(6)},$$

or $$v = \sqrt{\frac{2gm}{m-1}\left[1 - \left(\frac{P}{P_1}\right)^{\frac{m-1}{m}}\right]P_1V_1} \quad \text{.........(6 } a\text{)},$$

as an equation from which to find the velocity v when the pressure has fallen to any lower pressure P, under the assumed conditions of flow without friction or eddies and with no conduction of heat. Eq. (6) is a particular case of Eq. (4); it is there assumed that the expansion is isentropic and that the relation of pressure to volume in isentropic expansion admits of being expressed by the formula $PV^m =$ constant.

130. Form of the jet. As expansion proceeds the volume and velocity of the steam both increase. It is easy in frictionless adiabatic flow to calculate both, and in that way to determine the proper form to give to the nozzle or channel, to make provision for the increased volume, having regard to the increased velocity. At any stage the area of cross-section of the channel required for the discharge is equal to the volume divided by the velocity. It is convenient to reckon this per lb. of steam in the discharge, and afterwards multiply by the number of pounds.

Let M represent the discharge, namely the mass which passes through the nozzle per second, X the area of cross-section of the stream at any part of the nozzle, v the velocity there, and V the volume of the fluid there (per lb.); then

$$M = \frac{vX}{V} \text{ and } \frac{X}{M} = \frac{V}{v}.$$

On making the calculation for a gaseous fluid which starts from rest and discharges into a region of much lower pressure, it will be found that in the earliest stages the gain of velocity is relatively great, but as expansion proceeds the increase of volume outstrips the increase of velocity. The result is that the ratio of volume to velocity at first diminishes, passes a minimum value, and then increases; and hence the channel to be provided for the discharge, after passing a minimum of cross-section, should expand in the later stages. The proper form for the nozzle to allow the heat-drop corresponding to a large drop of pressure to be utilized as fully as possible in giving kinetic energy to the stream is therefore one in which the area of section at first contracts to a narrow neck or "throat" and afterwards becomes enlarged to an extent that is determined by the available fall of pressure.

131. De Laval's divergent nozzle. It is on this principle that De Laval's divergent nozzle (fig. 56) is designed. The throat or smallest section is approached through a more or less rounded entrance which allows the stream-lines to converge, and from the throat outwards to the discharge end the nozzle expands in any gradual manner, generally in fact as a simple cone, until an area of section is reached which will correspond to the proper area of discharge for the final volume and velocity, the values of which depend upon the available fall of pressure.

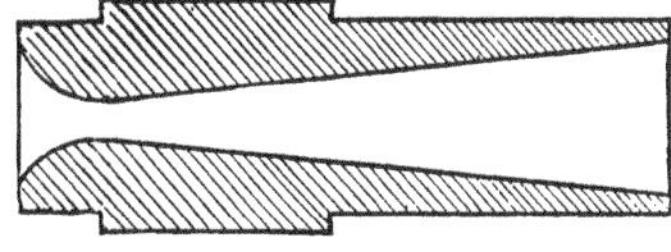

FIG. 56.

The divergent taper from the throat onwards is made sufficiently gradual to preserve stream-line motion as completely as is practicable and thus avoid the formation of eddies which would dissipate the kinetic energy of the stream. A very short rounded entrance to the throat is sufficient to guard against eddies in the convergent portion of the stream, but in the divergent portion a much more gradual change of section is required. The nozzle shown in the figure was designed for an initial pressure of 250 lb. per sq. inch and a back-pressure of about $1\frac{1}{2}$ lb. per sq. inch. By the back-pressure is meant the pressure in the space into which the fluid is discharged.

In the design of such a nozzle the purpose is (1) to cause a given amount to be discharged, and (2) to give the stream as high a final velocity as possible by utilizing completely the energy of the fluid in expanding down to the back-pressure. The data for the design are the initial pressure, the back-pressure, and the intended amount of the discharge. It will be shown as we proceed that the area of section at the throat depends only on the initial pressure and the intended discharge; and that the enlargement from the throat to the final section depends further on the back-pressure against which the stream is to escape.

At any place in the nozzle the discharge per unit area of cross-section is

$$\frac{M}{X}=\frac{v}{V}.$$

At the throat, where the cross-section is least, this is a maximum.

Consider now the ideal case of isentropic expansion in a nozzle when the fluid is one for which PV^m is constant during such expansion. Eq. (6) is then applicable. The velocity at any point, the pressure there having fallen to P, is

$$v = \sqrt{\frac{2gm}{m-1}\left[1 - \left(\frac{P}{P_1}\right)^{\frac{m-1}{m}}\right] P_1 V_1},$$

and the volume is $$V = V_1 \left(\frac{P_1}{P}\right)^{\frac{1}{m}}.$$

Hence for the discharge per unit area of section at the place where the pressure is P, we have

$$\frac{M}{X} = \frac{1}{V_1}\left(\frac{P}{P_1}\right)^{\frac{1}{m}} \sqrt{\frac{2gm}{m-1}\left[1 - \left(\frac{P}{P_1}\right)^{\frac{m-1}{m}}\right] P_1 V_1} \quad \text{......(7)}.$$

This may be applied to calculate the proper section X for a given discharge M when the pressure has fallen from the initial pressure P_1 to any assigned lower pressure P. For the purpose of designing a nozzle there are only two places where this calculation need be made, namely at the throat, and at the end where the fluid escapes against the assigned back-pressure. When the throat section X_t and the final section X_f have been calculated a suitable form for the nozzle is readily drawn; any smooth curve will serve for the convergent entrance, and any conical taper may be selected for the divergent extension from the throat to the end, provided it is neither so abrupt as to interfere with stream-line flow, nor so gradual as to make the nozzle unduly long and thereby introduce unnecessary friction.

To calculate the final section X_f which will allow the energy of the fluid to be fully utilized by expansion down to the assigned back-pressure, that pressure is to be taken for the value of P in Eq. (7). To calculate the section at the throat the pressure there has first to be found. The pressure at the throat is determined by the consideration that the discharge per unit of section (M/X) is there a maximum. If the expression for M/X in Eq. (7) is differentiated with respect to P/P_1 and the differential written equal to zero, the resulting value of P/P_1 will be that for which M/X is a maximum; in other words it will be the value of P_t/P_1, where P_t is the pressure at the throat.

Eq. (7) may be written

$$\frac{M}{X} = \sqrt{\frac{2gm}{m-1} \cdot \frac{P_1}{V_1}} \sqrt{\left(\frac{P}{P_1}\right)^{\frac{2}{m}} - \left(\frac{P}{P_1}\right)^{\frac{m+1}{m}}} \quad \text{......(7 a)}.$$

The condition for a maximum is found by differentiating the quantity under the second root:—

$$\frac{2}{m}\left(\frac{P_t}{P_1}\right)^{\frac{2-m}{m}} - \left(\frac{m+1}{m}\right)\left(\frac{P_t}{P_1}\right)^{\frac{1}{m}} = 0,$$

$$2\left(\frac{P_t}{P_1}\right)^{\frac{1-m}{m}} - (m+1) = 0,$$

from which $$\frac{P_t}{P_1} = \left(\frac{m+1}{2}\right)^{\frac{m}{1-m}} = \left(\frac{2}{m+1}\right)^{\frac{m}{m-1}} \quad \text{............(8).}$$

Further, by substituting this in Eq. (6 *a*), we have for the velocity at the throat

$$v_t = \sqrt{\frac{2gmP_1V_1}{m+1}} \quad \text{........................(9).}$$

The volume (per lb.) of the fluid at the throat is

$$V_t = V_1\left(\frac{P_1}{P_t}\right)^{\frac{1}{m}} = V_1\left(\frac{m+1}{2}\right)^{\frac{1}{m-1}} \quad \text{...........(10).}$$

By combining these an equation is obtained for the discharge per unit of cross-section at the throat,

$$\frac{M}{X_t} = \frac{v_t}{V_t} = \left(\frac{2}{m+1}\right)^{\frac{1}{m-1}}\sqrt{\frac{2gmP_1}{(m+1)\,V_1}} \quad \text{.........(11).}$$

From this equation the cross-section at the throat is found which will give an assigned discharge when the initial pressure is known. The ratio of the cross-section at any place, where the pressure is P, to the cross-section at the throat, is readily found from Eq. (7 *a*):—

$$\frac{X}{X_t} = \frac{\sqrt{\left(\frac{P_t}{P_1}\right)^{\frac{2}{m}} - \left(\frac{P_t}{P_1}\right)^{\frac{m+1}{m}}}}{\sqrt{\left(\frac{P}{P_1}\right)^{\frac{2}{m}} - \left(\frac{P}{P_1}\right)^{\frac{m+1}{m}}}} \quad \text{..............(12).}$$

This expression is convenient in determining the proper amount of enlargement of the nozzle from the throat to the end when the back-pressure is assigned.

132. Limit to the discharge through an orifice. It follows from these equations that the discharge through a given orifice under a given initial pressure P_1 depends only on the cross-section at the narrowest part of the orifice, and is independent of the back-pressure, provided the back-pressure is not greater than P_t as calculated by Eq. (8). By continuing the expansion in a divergent nozzle after the throat is passed, the amount of the discharge is not increased, but the fluid acquires a greater velocity before it leaves the nozzle, because the range of pressure which is effective for producing velocity is increased. To put it in another way we may say that the heat-drop down to the pressure at the throat determines the amount of the discharge, and the remainder of the heat-drop, which would be wasted if there were no divergent extension of the nozzle, is utilized in the divergent portion to give additional velocity to the escaping stream. This velocity is given in a definite and useful direction, whereas if there were no divergent extension of the nozzle the fluid, after leaving the nozzle, would expand laterally and its parts would acquire velocity in directions such that no use could be made of the kinetic energy so acquired.

Consider what happens with an orifice or nozzle that has no divergent extension. Fluid is expanding through the nozzle from a source where the pressure is P_1 into a space where the pressure is P_2. Assume the back-pressure P_2 to be less than P_t as calculated by Eq. (8). In that case the pressure in the jet, where it leaves the nozzle, will be P_t and the further drop of pressure to P_2 will occur through scattering of the stream. The discharge is determined by Eq. (11). *It is not increased by any lowering of the back-pressure* P_2, because any lowering of P_2 does not affect the final pressure in the nozzle, which remains equal to P_t. Osborne Reynolds[1] explained the apparent anomaly by pointing out that the stream is then leaving the nozzle with a velocity equal to that with which sound (or any wave of extension and compression) is propagated in the fluid, and consequently any reduction of the pressure P_2 cannot be communicated back against the stream: its effects are not felt at any point within the nozzle. The pressure in the stream at the orifice therefore cannot become less, however much lower the back-pressure P_2 may be. But if P_2 is increased so as to exceed P_t, the lateral scattering close to the orifice ceases, the velocity is reduced, the pressure at the orifice then becomes equal to P_2, the discharge is reduced, and its amount is to be calculated by writing P_2 for P in Eq. (7) or (7 *a*).

In applying these results to a nozzle of any form the least

[1] *Phil. Mag.* March, 1886; *Collected Papers*, vol. II, p. 311.

section is to be regarded as the throat: if there is a divergent extension beyond the least section the amount of the discharge is not affected, though the final velocity of the stream is increased. Taking a nozzle of any form, and a constant initial pressure P_1, if we reduce the back-pressure P_2 from a value which to begin with is just less than P_1, the discharge increases until P_2 reaches the value $P_1\left(\frac{2}{m+1}\right)^{\frac{m}{m-1}}$. After that, any further reduction of P_2 does not increase the discharge. But the velocity which the fluid acquires before it leaves the nozzle may then be augmented by lowering P_2 and adding to the divergent portion of the nozzle. The nozzle will be rightly designed when it provides for just enough expansion to make the final pressure equal to the back-pressure; the jet then escapes as a smooth stream, and the energy of expansion is utilized to the full. If the nozzle does not carry expansion far enough; if in other words the final pressure exceeds the back-pressure, energy will be wasted by scattering. If on the other hand the back-pressure is too high for the nozzle, so that the nozzle provides for more expansion than can properly take place, vibrations are set up which cause some waste. We shall now consider the application of these general results to air and to steam.

133. Application to Air. In applying the above formulas to any permanent gas, such as air, the index m is γ, the ratio of the two specific heats (§ 40). Its value for air may be taken as 1·40. Substituting this number in Eq. (8) we have, for a jet of air expanding under adiabatic conditions,

$$\frac{P_t}{P_1} = \left(\frac{2}{2{\cdot}4}\right)^{3{\cdot}5} = 0{\cdot}528.$$

Hence if the jet is being delivered against a back-pressure less than $0{\cdot}528P_1$ a divergent extension of the nozzle is required to give the greatest possible velocity to the issuing stream, though the quantity delivered will be the same as that which would be delivered against a back-pressure of $0{\cdot}528P_1$. If the back-pressure be increased it must exceed $0{\cdot}528P_1$ before there is any diminution in the discharge.

As a numerical example, suppose that air, with an initial pressure of 300 lb. per sq. inch, is discharged through a convergent-divergent nozzle into the atmosphere, or against a back-pressure of say 15 lb. per sq. inch. The pressure at the throat is 158·4 and, since the final ratio of pressure is one to twenty, the ratio of the final cross-section to the cross-section of the throat should, by Eq. (12), be

$$\frac{X_f}{X_t} = \frac{\sqrt{0{\cdot}4019 - 0{\cdot}3349}}{\sqrt{0{\cdot}01385 - 0{\cdot}00588}} = 2{\cdot}90.$$

This is for the ideal case of adiabatic expansion. Effects of friction are for the present disregarded; they will be considered later.

134. Application to Steam. In applying the general equations to the formation of a steam jet we have to bear in mind an important distinction between the kind of adiabatic expansion which occurs in a jet, and the kind already treated of in §§ 68, 86 and 90. It was assumed there that at each stage in the expansion the fluid was in thermal equilibrium: on beginning to expand it consequently at once became wet (if it was initially saturated), and at every stage it consisted of a mixture of saturated steam with the proportion of water necessary to keep the entropy of the mixture constant. That kind of expansion may be distinguished as the equilibrium type of adiabatic expansion.

But in the adiabatic expansion of a steam jet the action is too sudden for thermal equilibrium to be maintained. The steam becomes *supersaturated*: in the earlier stages of the expansion—up to the throat and for some way beyond it—there is practically no water present: the steam remains almost if not quite dry, and even when it leaves the nozzle it contains much less water than would correspond to the equilibrium condition. The proper amount of condensation has not had time to take place. Ultimately part of the expanded vapour does condense, for the supersaturated state is not a stable one: in the absence of suitable nuclei round which drops may form there is slow condensation on the sides of the containing vessel.

When the expanded vapour passes adiabatically from the supersaturated to the equilibrium state, the condensation of a part of it causes heat to be developed; the temperature accordingly rises, and the entropy of the fluid as a whole is increased. This change of condition is irreversible and dissipates energy: thus one effect of supersaturation in a steam jet is to add to the quasi-frictional losses. So long as it remains supersaturated the vapour is also *supercooled*, that is to say, its temperature at any pressure is lower than the temperature of saturation.

It follows from this retarded condensation, as Callendar has pointed out[1], that when these equations are applied to a steam jet (assumed to be formed under ideal frictionless conditions), the value which should be assigned to the index m is 1·3. This is the index for "steam gas," that is to say, for steam which remains completely dry during the expansion to which the calculation refers. It applies not only to superheated steam but to steam which, though its equilibrium condition would be one of wetness, has temporarily escaped becoming wet during expansion by remaining supersaturated.

If the steam be superheated to begin with, it behaves like a gas in the initial stages of the expansion, and its equilibrium continues stable until its temperature has fallen to the value that corresponds to saturation at the pressure then reached. It is only in further expansion, beyond this stage, that supersaturation occurs. If the steam be saturated to begin with there is some supersaturation as soon as the expansion begins.

According to Callendar's equations, the adiabatic expansion of superheated steam follows the law

$$P\,(V - b)^{1\cdot 3} = \text{constant},$$

where b is a small term representing the volume of water at 0° C. or 0·016 cub. ft. per pound. This is relatively so small (except at very high pressure) that it may as a rule be neglected and the formula be taken as $PV^{1\cdot 3}$ = constant. The same formula may be taken as applying in the supersaturated state. Using this index in Eq. (8), we find for steam

$$\frac{P_t}{P_1} = 0{\cdot}545,$$

and then by Eq. (11)

$$\frac{M}{X_t} = \left(\frac{2}{2{\cdot}3}\right)^{\frac{10}{3}} \sqrt{\frac{2 \times 32{\cdot}2 \times 1{\cdot}3}{2{\cdot}3}} \sqrt{\frac{P_1}{V_1}} = 3{\cdot}786 \sqrt{\frac{P_1}{V_1}}$$

with pounds and feet as units. With the units more commonly employed this gives

$$\frac{M \text{ in lb. per sec.}}{X_t \text{ in sq. inches}} = 0{\cdot}3155 \sqrt{\frac{P_1 \text{ in lb. per sq. inch}}{V_1 \text{ in cub. ft. per lb.}}}$$

as a formula for calculating the size of the throat in a nozzle

[1] "On the steady flow of steam through a nozzle or throttle." *Proc. Inst. Mech. Eng.* Feb. 1915.

supplied with saturated or superheated steam, or conversely the discharge through a given nozzle. The result however is subject to a correction for friction, the effect of which is slightly to reduce the discharge. No divergent extension of the nozzle is required unless the pressure drops through a greater ratio than 1 to 0·545.

Before it was recognized that a jet of initially saturated steam is necessarily supersaturated when it passes the throat, it was usual to calculate the throat pressure by using for the index m a value appropriate to the equilibrium type of adiabatic expansion. The index 1·135 was generally taken as applicable, with the result of making $P_t = 0{\cdot}577P_1$, and of making the numerical factor 0·3003 instead of 0·3155 in the above formula for M/X_t. The equilibrium theory gave too high a value for the pressure in the throat, and consequently the calculated discharge for a given size of throat was too small. Experiments on the flow of steam through nozzles were found to give a discharge which was actually greater than that which had been calculated for the ideal case of no friction, although the effect of friction would be to reduce the experimental discharge below its ideal amount. When, however, account was taken of the fact of supersaturation, by using 1·3 as the index, the calculated discharge was brought into harmony with the results of experiment. The revised theory gave a calculated discharge slightly greater than the actual discharge, but with no more difference than could properly be ascribed to frictional effects. The loss of energy through friction and supersaturation in nozzles has been the subject of an experimental investigation by a Committee of the Institution of Mechanical Engineers[1]. Their first Report (1923) describes the methods of experiment and discusses earlier work. Later Reports give results for various nozzles by stating the "velocity coefficient" and the "energy coefficient" or efficiency of the nozzle. The velocity coefficient is the ratio of the measured velocity of the jet to the velocity calculated from the adiabatic heat-drop, and the energy-coefficient is the square of that ratio. A nozzle of the kind used in impulse turbines, where the drop in pressure at each stage is not so great as to require any divergent extension, was found to have a practically constant efficiency of 0·891 for velocities ranging from 900 to 1500 feet per second, and somewhat higher efficiencies both below and above that range. A "reaction" nozzle of the kind

[1] *Proc. Inst. Mech. Eng.* January and March, 1923; May and October, 1924; May, 1925.

used in Parsons' turbines had an efficiency ranging from ·98 to ·90 throughout the working range of 280 to 700 feet per second.

135. Effects of frictional losses in a turbine. We shall next consider as a whole the process of expansion in a steam turbine from admission to exhaust. In general a turbine is multi-compound; the heat-drop occurs in a series of steps or stages as the steam passes successive rings of blades; at each step it suffers a fall of pressure and thereby acquires momentum which is used to drive the ring. The progressive drops in pressure imply increases in volume and in the effective area of the channels to be provided for the passage of the steam, which is finally discharged to the condenser at a very low pressure, making the whole heat-drop (from the initial state) as great as possible, and with a small velocity, so that the leaving loss—or unutilized kinetic energy—shall be unimportant.

The stages may be comparatively few, or they may be so many as to make the expansion approximate closely to a continuous process. It is instructive to regard the process as continuous and to compare the actual relation of pressure to volume with the relation that would hold in adiabatic expansion of the equilibrium type.

The term "frictional losses" is used here in a broad sense to cover the losses from any irreversible actions that occur during the passage of the steam through the turbine, including friction against nozzles and blades, windage of discs or drums, eddies caused by leakage over blade tips, etc. Among the contributing items are also effects of supersaturation, for the expansions take place too quickly to allow thermal equilibrium to be maintained when the steam has lost its superheat.

At each stage the volume is greater than it would be in adiabatic expansion, because the effect of the frictional losses is to generate heat. At any assigned pressure the fluid is warmer, if it is a dry gas, or drier, if it is a wet mixture; in either case its volume is increased. Hence if we take the pressure-volume diagram for the expansion as a whole the effect of frictional losses is to give the actual expansion curve a form such as BC' (fig. 57), the curve for adiabatic expansion under the condition of equilibrium at every stage being BC. But though this suggests a gain of work there is really a loss. The pressure-volume diagram does not measure an actual output of work but an artificial quantity which we may call the "gross apparent work." Of this gross apparent work a part is reconverted into heat as the expansion proceeds, namely, a quantity sufficient to supply enough heat at each stage to bring the expansion curve out from *BC* to *BC'*.

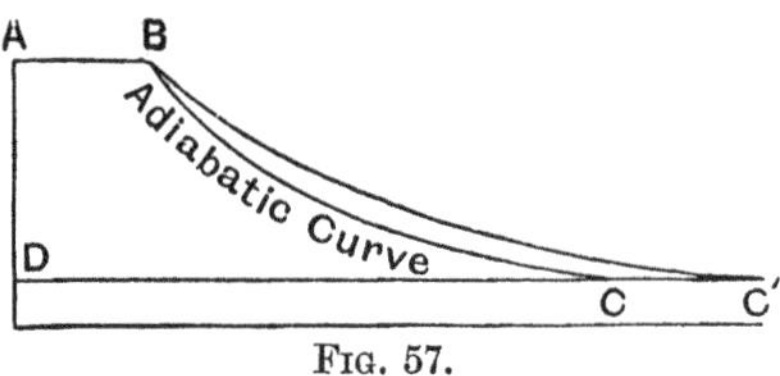

Fig. 57.

The area *ABCD* represents the adiabatic heat-drop, and measures the utmost amount of work ideally obtainable in any method of utilizing the energy of the steam. It is the quantity with which the actual output is to be compared in finding the "efficiency ratio" of any engine or turbine. The actual output is necessarily less in all cases than the area *ABCD*.

At the end of the operation shown by the curve *BC'* the net amount of work which has been obtained, far from being greater than the adiabatic area *ABCD*, is less than that area by the equivalent of $I_2' - I_2$, where I_2' is the total heat at *C'* and I_2 is the total heat at *C*. In other words it is less by the quantity of heat which would change the condition of the expanded fluid from *C* to *C'* under constant pressure.

Turning to the entropy-temperature diagram (fig. 58 or fig. 59) the ideal case without frictional losses is represented by *ABCD*, an area which (omitting the small feed-pump term, § 94) is the thermal equivalent of the work area *ABCD* of fig. 57. Frictional losses give the expansion curve some such form as *BC'* in which the entropy increases progressively as the expansion proceeds. The area *MBC'M'* represents the heat produced by frictional or quasi-frictional actions: it is the heat required to give the expansion curve its actual form, and since no heat comes from outside sources it is supplied at the expense of the output which the steam would give under ideal adiabatic conditions. The gross apparent work is represented by the area *DABC'*, and the actual output is less by the area *MBC'M'*. Thus we have for the net amount of work obtained from the steam

Area *DABC'* – Area *MBC'M'*

or Area *DABC* – Area *MCC'M'*.

The area *DABC* is the adiabatic heat-drop $I_1 - I_2$, and the net amount of heat which the turbine converts into useful mechanical effect is less than the adiabatic heat-drop by the area $MCC'M'$, which is $I_2' - I_2$.

In fig. 58 the steam is initially saturated; in fig. 59 it is super-

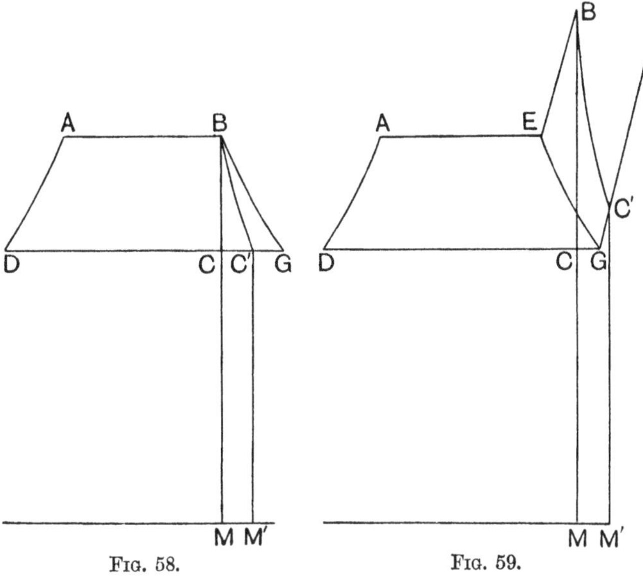

FIG. 58. FIG. 59.

heated. EB and GC' are constant pressure lines for the initial and final states respectively.

For practical purposes it is better to represent the effects of friction on the Mollier diagram of entropy and total heat. Thus, in fig. 60, B is the initial state (in this example there is some superheat: the broken line is the boundary curve), BC represents an ideal adiabatic equilibrium process of expansion, and BC' the actual process. I_1 is the initial total heat, I_2 the total heat that would be left in the steam after adiabatic expansion to the final pressure P_2, and I_2' the total heat actually left in the steam after working

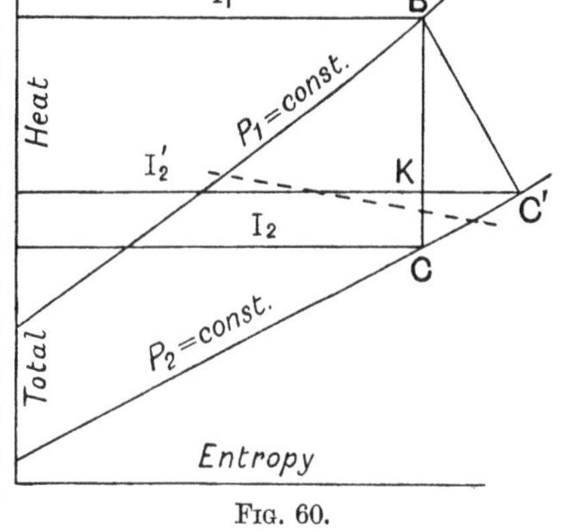

FIG. 60.

down to that pressure. The actual heat-drop, to which the net amount of work done is equivalent, is BK or $I_1 - I_2'$ and the net loss is KC or $I_2' - I_2$.

The student may find it useful to express the effect of friction thus. When there is no friction, and the expansion is adiabatic,

$$dI = VdP,$$

where dI represents (in units of work) the drop of total heat which takes place while the pressure drops by dP. When there is friction

$$dI' = V'dP - dQ,$$

where dI' represents the drop of total heat as affected by friction, and V' the volume as affected by friction, dQ being the quantity of heat generated by friction at the expense of the gross apparent work and restored to the fluid as heat. Hence

$$dI - dI' = dQ - (V' - V)\, dP.$$

Integrating between limits corresponding to P_1 and P_2,

$$I_1 - I_2 - (I_1' - I_2') = Q - \int (V' - V)\, dP,$$

where Q is the whole quantity of heat generated by friction. Since I_1 and I_1' are the same, this gives

$$I_1' - I_2 = Q - \text{Area } BCC' \text{ of fig. 57},$$

which expresses the fact that in consequence of friction the net loss of mechanical effect is equal to the heat generated, less the work that is recovered through the augmentation of volume which friction brings about.

136. Theoretical efficiency ratio. Whether the stages in a steam turbine are many or few, provided no heat escapes to the outside by conduction or by leakage of steam, and provided the kinetic energy of the current of steam is negligible on its exit from the turbine, the actual heat-drop $I_1 - I_2'$ is all represented by work done upon the rotor. Let η_t stand for the ratio of the actual heat-drop to the adiabatic heat-drop,

$$\eta_t = \frac{I_1 - I_2'}{I_1 - I_2}.$$

Under the conditions stated above this fraction expresses the efficiency ratio of the turbine as a whole, namely, the ratio of the work done on the rotor to the work ideally obtainable by adiabatic expansion through the same range. The whole adiabatic heat-drop

$I_1 - I_2$ would be converted into work only if the turbine were reversible and therefore thermodynamically perfect. Owing to internal irreversibility the work converted into heat is less, apart from any loss of heat by conduction.

We may call η_t the *theoretical efficiency ratio.* It is what the efficiency ratio would be if the whole *actual* heat-drop $I_1 - I_2'$ were converted into work, which would be the case if there were no loss of heat to the outside, and if the steam had no more than a negligible leaving velocity.

137. Action in successive stages. If the proportion is known beforehand of the frictional loss KC (fig. 60) to the ideally available heat-drop BC we can mark the point K on the adiabatic line through B, and draw a horizontal line through K to find C'. When there are experimental data for estimating the frictional losses in expansion down to various intermediate temperatures this construction can be applied to trace the actual expansion curve BC' in a series of steps. The method is applied to compound steam turbines as a means of determining the state of the steam after each of a series of stages in the passage of steam through the turbine. Fig. 61 illustrates this application of the Mollier diagram

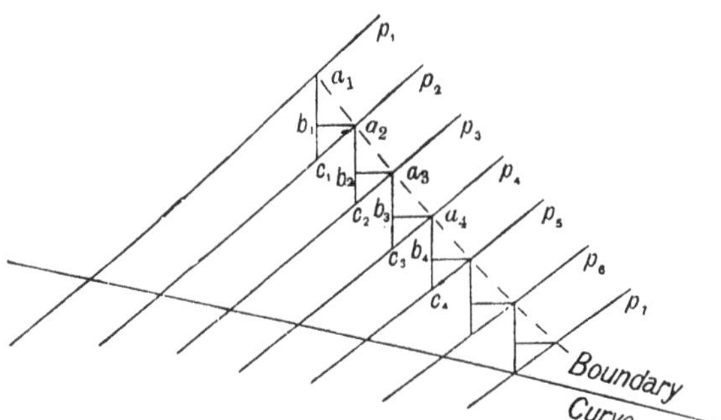

FIG. 61. Mollier diagram for successive stages.

by showing some of the early stages in a turbine using superheated steam. Beginning with the initial pressure, a series of constant-pressure lines are drawn, p_1, p_2, p_3, etc., corresponding to the pressures at which the steam enters the successive stages. In the first stage the pressure drops from p_1 to p_2, in the second stage from p_2 to p_3, and so on. In the first stage, adiabatic expansion from p_1 to p_2 would be represented by a_1c_1, and the length of that line would be a measure of the adiabatic heat-drop; but the actual heat-drop is the smaller quantity a_1b_1, and a_1b_1 is the heat converted into work while the steam passes through the first stage. The condition of the steam at the end of the first stage, and beginning of the second, is represented by the point a_2, which is found by drawing a line of constant total heat through b_1 to meet the constant-pressure curve p_2. In the second stage adiabatic expansion would give the line a_2c_2. The actual heat-drop, which also measures the work done, is a_2b_2, and the condition of the steam as it passes on to the third stage is represented by a_3. Similarly in the third stage the work done is a_3b_3, the steam passes to the fourth stage in the condition a_4, and so on. The diagram shows the process of expansion by stages down to the boundary curve; it is readily extended into the wet region. In each stage the fraction ab/ac measures the ratio of the work done to the adiabatic heat-drop for that stage. The points a_1, a_2, a_3, etc., lie on what is called the "curve of condition," a curve showing what the condition of the steam would be as it passes from stage to stage if the assumption were correct that no heat is lost to the outside. The curve of condition consequently corresponds to the outer curve BC' of fig. 57 or fig. 60. The total work done on the rotor is the sum of the work done in the successive stages, namely, Σab.

138. Stage efficiency and Reheat Factor. Taking any stage of a compound turbine, the ratio of the work done on the rotor to the adiabatic heat-drop, in that stage, may be called the stage efficiency and denoted by η_s; thus

$$\eta_s = \frac{ab}{ac}.$$

The total work done on the rotor

$$\Sigma ab = \Sigma\eta_s\,(ac),$$

and if η_s can be treated as constant from stage to stage,

$$\Sigma ab = \eta_s\Sigma ac.$$

The quantity Σac is called by some writers the "cumulative heat-drop." This quantity is greater than the whole adiabatic heat-drop $I_1 - I_2$, between the initial and final pressures, to an extent that depends upon the stage efficiency. The ratio

$$R = \frac{\Sigma ac}{I_1 - I_2}$$

is called the *Reheat Factor*. It can be determined from the diagram when the stage efficiency is known, and its value is relatively high when the stage efficiency is low, or, in other words, when there is much loss through irreversible action within each stage.

Treating η_s as constant we have

$$\eta_s R = \frac{\eta_s \Sigma ac}{I_1 - I_2} = \frac{\text{work done on rotor}}{\text{adiabatic heat-drop}} = \eta_t,$$

under the conditions postulated, which make the actual heat-drop a measure of the work done on the rotor.

From the equation

$$\eta_t = \eta_s R$$

it will be seen that in a compound turbine η_t is greater than the stage efficiency η_s, since R is greater than unity.

We might have defined the reheat factor by reference to fig. 57 as

$$R = \frac{\text{area } ABC'D}{\text{area } ABCD},$$

for in a compound turbine of many stages the curve of condition is represented by BC' and the area $ABC'D$, which was called the "gross apparent work" in § 135, is the mechanical equivalent of the "cumulative heat-drop" Σac. The work done on the rotor is $\eta_s \times$ area $ABC'D$, and is less than the area $ABCD$, the efficiency ratio being

$$\eta_t = \frac{\eta_s \times \text{area } ABC'D}{\text{area } ABCD}.$$

139. Real efficiency ratio. The foregoing expressions involve the proviso that there is no leakage of heat and that the leaving velocity of the steam is negligible. But in any real turbine there is some leakage of heat, and there is also appreciable kinetic energy in the current of steam at its discharge from the last ring of blades to the condenser; hence the actual heat-drop $I_1 - I_2'$ includes a quantity representing the loss due to these causes, in addition to the work done on the rotor. Let that loss be expressed as a fraction of the adiabatic heat-drop, namely,

$$x(I_1 - I_2).$$

Then

$$I_1 - I_2' - x(I_1 - I_2)$$

is that part of the actual heat-drop which is converted into work on the rotor.

Hence, allowing for this loss, the net or real efficiency ratio of the turbine becomes

$$\frac{I_1 - I_2' - x(I - I_2)}{I_1 - I_2} = \eta_t - x,$$

since η_t is, by definition (§ 136), the ratio of the actual heat-drop $I_1 - I_2'$ to the adiabatic heat-drop.

The amount of work obtained from the steam is therefore

$$(\eta_t - x)(I_1 - I_2).$$

Writing η_r for the real efficiency ratio, its relation to the other quantities is given by the equation

$$\eta_r = \eta_t - x = \eta_s R - x.$$

In the process of designing a turbine a value is estimated for the stage efficiency η_s; then the curve of condition is deduced, which allows the reheat factor to be found and also the probable volume and velocity of the steam at each stage. In this way data are obtained for determining the dimensions and form of the steam passages.

140. Importance of division into stages. The whole available heat-drop, from such pressures as are usual in steam-supply down to the vacuum obtainable in a good condenser, may easily be as much as 200 pound-calories with saturated steam or even 250 when the steam is considerably superheated. If the whole of this drop were employed to give velocity to the steam in a single operation, the jet would leave the nozzle at a speed of about 4240 or 4740 feet per second, since $v^2 = 2gJ(I_1 - I_2)$, from which $v = 300{\cdot}2\sqrt{I_1 - I_2}$. Making allowance for frictional losses we have still to reckon on velocities of 4000 feet per second or more. If the kinetic energy of the jet is to be extracted by making it impinge on the blades or buckets of a simple turbine wheel, the condition for efficiency is that the velocity of the blades should be not far short of one-half the velocity of the jet. To get the best return we should therefore require to have a speed of something like 2000 feet per second at the periphery of the wheel which carries the blades. This is an impracticable speed: there are no materials of construction fitted to withstand the forces which it would involve. De Laval's turbine does indeed run at very high speeds, but they are far short of this and involve the sacrifice of a considerable part of the available energy of the jet. There is in addition a loss through friction, which becomes important when the jet rubs against the blade surfaces at a very high speed. Another point to be considered is that steam impinging with great velocity on turbine blades exercises a cutting action, especially when it contains particles of water in suspension, an action which is practically absent at lower speeds: consequently low or moderate steam velocities are not only better from the point of view of efficiency, but tend to secure greater durability in the machine.

On these grounds it will be clear that there is much advantage in dividing the heat-drop into stages. It is essential to thermodynamic efficiency and is in fact done in all turbines designed for large output. Though the inventions of De Laval have secured for his single-wheel turbine a degree of efficiency which for a one-stage action is remarkably good, it is only in small sizes that his construction can compete with other types, for in large sizes the consumption of steam in multi-stage turbines can be brought down to a much lower figure.

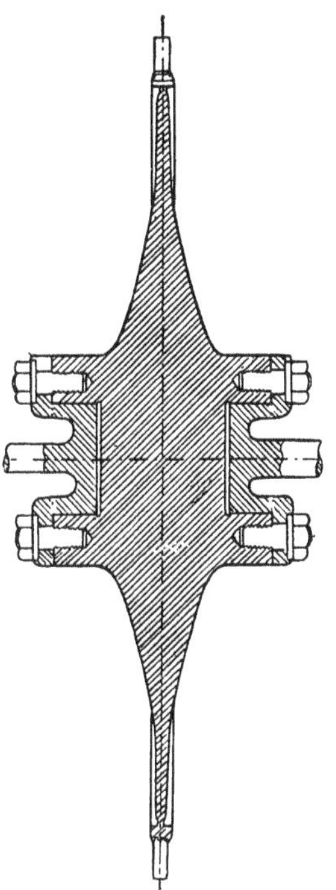

Fig. 62.
De Laval turbine wheel.

141. De Laval's turbine. In De Laval's turbine, which dates from 1889, the steam expands in a single step from the initial to the final pressure in a divergent nozzle of the type already described. The jet thus formed is directed against a ring of blades which are set all round the circumference of a wheel designed to be capable of very rapid rotation. With this object the wheel is much thickened in the neighbourhood of the axis to provide strength there for the stresses to which rotation gives rise. This is seen in fig. 62, which is a section through the wheel showing the thickened centre and the blades projecting at the circumference. The wheel, in this example, has a solid centre, to avoid the increased centrifugal stresses that arise when there is an axial hole, and the shaft, which is in two parts, is attached by bolts on the sides. Further, the shaft is made so thin as to be flexible, in order that the period of transverse vibration of the shaft, loaded with the mass of the wheel, shall be much longer than the time taken to complete a single revolution. In other words, the speed of revolution is much above the "critical" speed at which the loaded shaft is unstable in respect of transverse oscillations. This device enables the wheel to run with great steadiness and protects the shaft itself and the bearings and foundations from the vibration that would be caused by any want of balance if the shaft were rigid. In starting the turbine, a completely steady state is reached only when the speed has passed the critical value.

The position of a steam nozzle in relation to the blades is shown in fig. 63, where the arrow on the left shows the direction of rotation of the wheel and the other arrow shows how the steam escapes to exhaust after having its course diverted by acting on the blades, only a few of which are shown in the figure. Steam is delivered against one side of the ring of blades and passes to exhaust from the other side. In all except the smallest of these turbines there are several nozzles placed at intervals round one side of the wheel, each delivering an independent jet, and the output of power is regulated by opening more or fewer of the nozzles as may be required. In a small De Laval turbine developing about 5 horse-power the wheel makes some 30,000 revolutions per minute; in a turbine developing 600 horse-power the wheel has a mean diameter of 37 inches and makes 9500 revolutions per minute, with a peripheral speed of 1550 feet per second. The turbine shaft is geared, by means of double-helical gear-wheels carrying teeth of specially fine pitch, to a second-motion shaft running at about one-tenth of the speed, which is connected to the dynamo or other machine the turbine has to drive. Trials of De Laval turbines fitted with condensers have shown an average consumption of about 20 lb. per hour per brake horse-power in a 60 horse-power turbine, and 15 or 16 lb. in a 300 horse-power turbine.

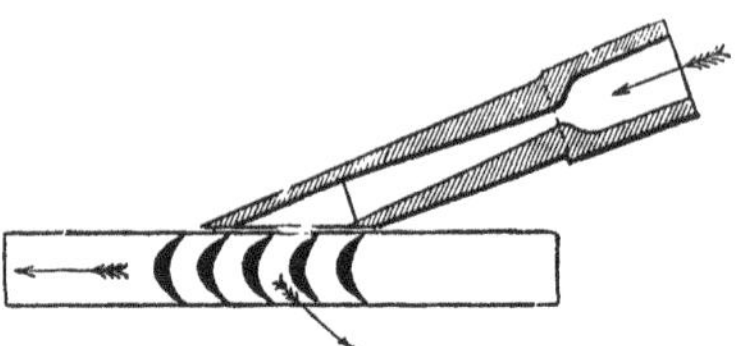

FIG. 63. De Laval nozzle and blading.

De Laval's is the only non-compound turbine; it is also interesting as the first turbine to run above the critical speed, and the first to use reduction gearing, thereby making the turbine speed independent of limits imposed by the mechanism to be driven. As we shall see later, the use of gearing is now common in many large turbines and is universal in marine turbines of modern design.

142. Types of Turbines. De Laval's turbine acts purely by impulse and the steam is discharged after a single passage through the ring of moving blades. It is therefore to be classed as a *simple impulse turbine.*

The word *impulse* is here applied, as one applies it in speaking of a Pelton water-wheel, to describe an action in which the fluid does not suffer any reduction of pressure during its passage through the ring of moving blades or buckets. The drop in pressure which is required to give velocity to the fluid has already occurred in the fixed nozzle. The stream of fluid then passes through the wheel, and it exerts a driving force on the moving blades only because the direction of its motion relative to them becomes changed in the passage.

In describing turbines, the word *reaction* is reserved for those in which the steam suffers a drop in pressure while it is passing through the ring of moving blades. There is then a reaction on the blades, due to the fresh momentum which the stream acquires as a consequence of that drop in pressure.

Turbines in which the whole heat-drop is divided into a number of successive stages, in each of which the stream acquires a moderate velocity, are called *compound.* Compound working, in this sense, was first introduced by Sir Charles Parsons in 1884, and is a feature in all practical steam turbines with the single exception of De Laval's.

A very extensively used modern type is the *compound impulse turbine,* in each stage of which the steam passes through fixed nozzles, or blade passages acting as nozzles, in which it undergoes a moderate heat-drop and acquires a corresponding velocity, and then acts by impulse on a ring of moving blades. The turbine-case is divided by a series of fixed transverse diaphragms into chambers, each of which contains a revolving disc, which has a central shaft and carries at its circumference a ring of radial blades. The steam passes from one chamber to the next through channels in the diaphragm between, formed by fixed blades so shaped as to make the channels sufficiently convergent to act as nozzles; these give the steam a suitable velocity and also direct it at a suitable angle against the moving blades of the disc.

One set of fixed and moving channels together constitute a *stage*. In the fixed channels the steam suffers heat-drop and acquires momentum: in the moving channels of an impulse turbine there is no further heat-drop, but the momentum is given up, almost completely, in driving the mechanism before the steam goes on to the next stage. This type of turbine lends itself well to the economical production of power on a large scale. Its successful development (from 1898 onwards) owes much to Rateau, and the compound impulse type is often associated with his name.

A modified form of impulse turbine, due to Curtis, and developed about 1900, is characterized by the use of what is called a *velocity compounded wheel*. In this device the steam, having acquired momentum in passing through fixed nozzle channels, gives it up to two or more parallel rings of impulse blading between which there are fixed guide-blades whose function is to alter the direction of the stream. They do this without giving it any additional kinetic energy, for no expansion takes place in them. The device allows a large part of the energy of the stream to be taken up by the wheel without requiring nearly so high a circumferential speed as is demanded in a single row of moving blades. Thus in a Curtis wheel with two rows of moving blades, between which there is a row of fixed guide-blades, the speed of best efficiency will be only about half that which would be required in a single row. In early Curtis turbines as many as four rows of moving blades were used, but the frictional losses were found to be so considerable that in modern practice the principle of "velocity compounding" is usually limited to two or at most three rows.

For small powers turbines have been constructed in which the whole heat-drop is utilized by one Curtis wheel, but Curtis wheels may advantageously be compounded in the ordinary (pressure) sense, so that the heat-drop is divided between two or more of them, through which the steam passes in series. Turbines so arranged are *compound Curtis turbines*. In each Curtis wheel it is intended that the steam should not undergo any change of pressure as it passes the moving blades and the guide-blades fixed between them; the pressure-drop occurs only in the nozzles that admit steam to the wheel. The nozzles and the wheel together constitute a "stage."

A *pure reaction turbine* would be one where the nozzles in which the steam expands are themselves the moving part and are driven by the reaction which results from the fact that the steam is acquiring relative velocity while it passes through them. An ancient toy described by Hero of Alexandria, in which a pivoted boiler with projecting arms is caused to revolve by discharging steam from tangentially set nozzles at the ends of the arms, is an example of a pure reaction turbine. This type has not come into use: it would require an enormous speed of recoil to work efficiently. But a mixture of reaction with impulse characterizes the action of the turbine which is historically the most important of all—that of Parsons—the design which first made steam turbines practicable. The whole modern development of the steam turbine may be said to have sprung from Parsons' invention of a compound turbine with many successive stages, which dates from 1884. The use of such stages allows the blade velocity to be made relatively low. Each stage comprises a ring of fixed blades projecting radially inwards from the case, and a ring of moving blades projecting radially outwards from a drum-shaped rotor. The fixed blades are so shaped as to make up a ring of convergent passages which act as nozzles, but only about half the heat-drop of the stage occurs in them, for the moving blades are of the same convergent shape, and in passing through these the steam undergoes the other half of the heat-drop, thereby acquiring fresh momentum and exerting a reaction effect. The rings of fixed and moving blades alternate from end to end throughout the turbine, becoming progressively larger to provide for the increasing volume of the steam. In general, the fixed and moving blades for each stage are alike, and the expansion of the stage is divided equally or almost equally between them.

Turbines of the Parsons type are usually spoken of as *reaction turbines*. The name is not fully descriptive, for the steam acts on the moving blades partly by impulse, but it serves as a convenient means of distinguishing turbines of this type from purely impulse turbines such as those of De Laval or Rateau or Curtis.

In one of the early forms of Parsons' turbine the steam, instead of flowing parallel to the axis, flowed radially outwards, the fixed blades being attached to a fixed disc and the moving blades to a parallel disc, which revolved about an axis through the centre of the fixed disc. An interesting and highly ingenious modification of this arrangement was made in 1910 by B. and F. Ljungström, who let both discs revolve, but in opposite directions. In the Ljungström turbine, which is also compounded of many stages but all within one pair of discs, there are therefore no fixed blades: both sets of blades are equally driven, and a doubled relative velocity is obtained for a given frequency of revolution. Such a turbine is described as a *radial flow* turbine to distinguish it in this respect from the much more common *longitudinal flow* type.

In compound turbines, whether of the impulse or reaction types, there are generally so many stages that the nozzles, or blade passages which act like nozzles, are not of the kind described in § 131, but are only convergent, for the drop of pressure in each stage does not involve expansion beyond the "throat." We saw in § 134 that the pressure of steam in a nozzle might drop by expansion to 0·545 times its initial value without requiring the nozzle to have a divergent expansion. Ten such steps would suffice to bring the pressure down from, say 200 lb. per sq. inch to the lowest pressure attainable in a good condenser, and usually the number of stages is greater than ten—in reaction turbines it is as a rule much greater. With a Curtis wheel however the drop of pressure in one stage may be so considerable that the nozzles for it have to be divergent (compare § 148).

In each stage the area of the passages must of course be designed with reference to the volume to which the steam has then expanded, a volume which becomes relatively enormous in the final stages when the pressure approaches that of the condenser. The necessary increase of area is obtained by lengthening the blades or increasing the diameter of the drums or discs, or in both ways, and, further, in the latest stages the inclination of the blades is increased so as to give a larger axial component to the velocity of discharge, with the result that a greater volume of steam passes per second through the annular area over which the flow takes place, the volume of flow per second being the product of the axial velocity into that area. In some instances an increased area is obtained by providing two channels in the later stages of the expansion: this is done by providing a separate cylinder for the low-pressure stages, and letting the steam enter that at the middle of its length and expand towards both ends. Examples of this construction will be found in §§ 155 and 156.

Many turbines are properly described as *combination turbines* because they are made up by combining more than one type. Thus it is by no means unusual to have a Curtis velocity-compounded wheel (often called, for brevity, a *velocity wheel*) as the first stage in a turbine which is otherwise of the Rateau, or of the Parsons, type.

Other terms of classification which may be mentioned here relate to the character of the steam-supply. An *exhaust steam turbine* is one which takes steam at a low pressure after it has done some other duty, generally in a non-condensing engine, and expands it down to the much lower pressure of the condenser. A *mixed pressure turbine* also does this, but is designed, in addition, to use high-pressure steam when that is required to supplement a (possibly intermittent) low-pressure supply. A *reducing* or *pass-out turbine* extracts a part of the energy from high-pressure steam before letting it pass out at a lower pressure to serve some other industrial purpose, as for instance in paper manufacture, often under conditions of a varying demand. Such turbines may be arranged to admit of extraction of steam at more than one stage in the expansion. In large power plants the device called "bleeding," which means the extraction of a small part of the steam, often at more than one stage, is employed for the purpose of heating the condensate, or discharge from the condenser, on its way back (as feed-water) to the boiler, with the effect of increasing the thermodynamic efficiency of the cycle as a whole. This device will be referred to more fully later. A reducing turbine may pass out all its steam at a lower pressure, keeping none for further expansion, and so dispensing with a condenser: in that case it is often called a *back-pressure* turbine.

143. Action of the steam on the blades. Velocity diagram for impulse blading. We have now to consider the dynamics of the action on the moving blades of a turbine, and shall in the first instance deal with a single row of impulse blading, forming, along with the nozzles which deliver steam to it, a "stage" in a compound impulse turbine. The general arrangement of the nozzles and blades of such a stage will be clear from the right-hand side of fig. 64, which relates to a turbine by the British Thomson-Houston Company. In that turbine the first stage is a two-row Curtis wheel (shown on the left), which is followed by single-row stages. Of these only the first two appear in the figure; later stages must of course have nozzles and blades of progressively greater radial length to accommodate the increasing volume of the steam. Each row of blades has, secured to the tips, a ring of "shrouding" which gives the available channel a definite area, and limits any radial spreading of the jet as it passes through the blade channels. The whole effective channel, measured in the plane of the wheel, is an annulus whose area is the blade height multiplied by the mean periphery, less a small deduction for the thickness of the

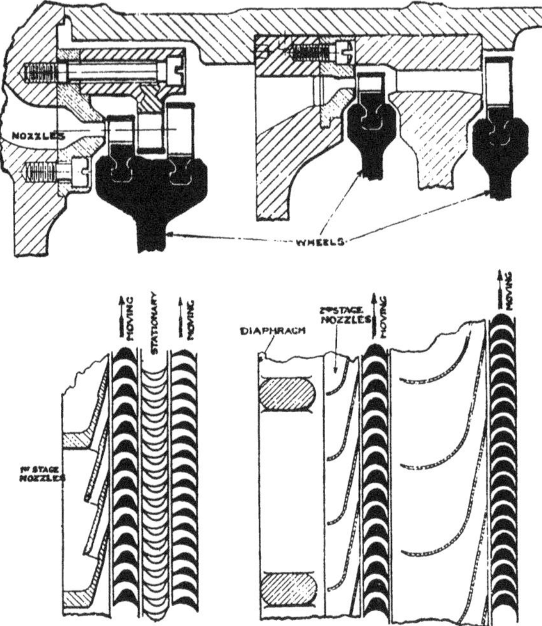

Fig. 64. Impulse blading. Curtis wheel and Rateau stages.

edges: in general the edges are so sharp that this deduction is unimportant. Viewed in the direction of the axis, the channel between any two neighbouring blades is a nearly rectangular figure, whose height is the radial length of the blades, and whose width is nearly equal to their pitch, the neighbouring rectangles being separated only by the small thickness of the sharp edges.

In pure impulse blading the pressure is the same at entrance and exit. The blade channels do not run full; that is to say, they are not completely filled by the stream as nozzle channels necessarily are. When the jet has entered the blade channel, its direction is deflected to follow the curvature of the blade: this, by centrifugal action, tends to make it spread laterally, but the shrouding and the bottom of the channel limit the spreading. On the exit side the channel may be nearly or just filled by the escaping steam: this depends, as will be seen later, on the extent to which the axial component of its velocity has been reduced during the passage. Where the channel does not run full the space not occupied by the stream forms a sort of dead-water of steam in which there are eddies but no stream-line flow.

The nozzles of course run full and are convergent, that is to say, their sectional area measured at right angles to the direction of

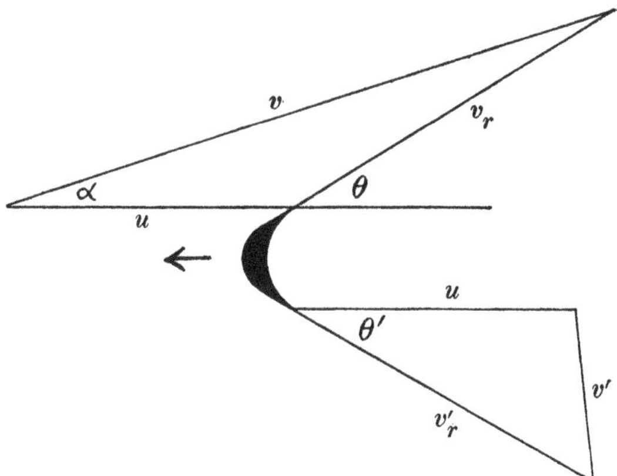

FIG. 65. Velocity diagram for impulse blade.

motion of the steam contracts towards the delivery end. The form of their section is nearly rectangular. The nozzles are shaped, as will be seen in the figure, to direct their jets at a suitable angle to the plane of the wheel. Usually this angle is about 20°. In the latest stages it is often more; on the other hand in earlier stages it may be as little as 16° or even 12°. The nozzles are less closely pitched than the moving blades. Their function is (1) to convert the heat-drop for the stage into the kinetic energy of the jets, which is done subject to the loss referred to in § 134, and (2) to give the jets the desired direction.

Let a row of impulse blades, moving with velocity u (fig. 65) be acted on by jets with velocity v inclined at an angle α to the direction of motion of the blades. With radial blades, as in the ordinary longitudinal type of turbine, u is to be taken as the mean velocity, namely, the peripheral velocity at the middle of the height. Then v_r, the resultant of v and u, is the relative velocity with which the steam enters the blade channels; call its inclination θ. If the blade face has an entrance angle θ as shown in the figure, the steam will glide on to it without any sudden change of direction, or as is generally said "without shock." In passing through the blade channel it gains no relative velocity, for there is no expansion due to drop in pressure. Were it not for frictional or quasi-frictional loss its relative velocity at exit v_r' would be equal to v_r. Actually friction reduces the relative velocity, making $v_r' = kv_r$, where k is a coefficient, less than unity, to be deduced from experiment. The direction of v_r' is determined by the exit angle of the blades, θ', which may or may not be equal to θ. Combining v_r' with the

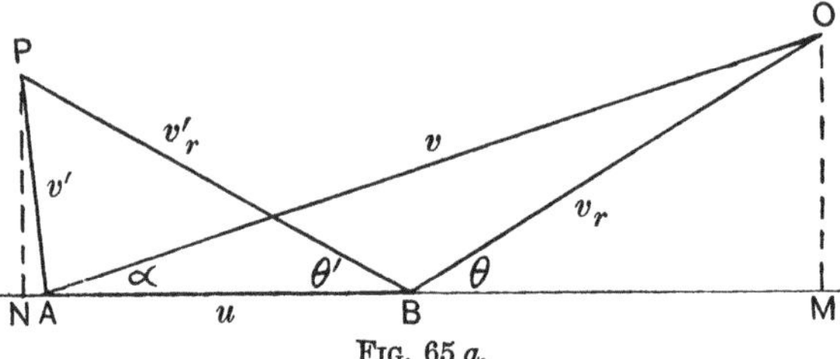

FIG. 65 *a*.

blade velocity u we find v', which gives, in magnitude and direction, the absolute velocity which the steam has when it leaves the blades.

The two triangles of velocities may conveniently be placed together on the same base AB as in fig. 65 *a*. OM and PN are drawn at right angles to AB produced. In moving over the blade surface the steam has changed its absolute velocity from OA to PA: the component of its velocity tangential to the wheel has therefore changed by the total amount MN, as a consequence of the force between the steam and the blade. Hence MN measures, in kinetic units, the momentum communicated to the blades, in the direction of their motion, per lb. of steam that has passed; and the work done on the blades, per lb. of steam, is uMN, which may be written

$$uMN = u\,(v_r \cos\theta + v_r' \cos\theta') = uv_r \cos\theta \left(1 + k\,\frac{\cos\theta'}{\cos\theta}\right).$$

To express the work in foot-pounds we must of course divide by g, but for the present we shall continue to use kinetic units.

The kinetic energy of the jet is $\frac{1}{2}v^2$ per lb. of steam. Hence the fraction of this which has been utilized as work on the blades, or what is called the *diagram efficiency* η_d, is

$$\eta_d = \frac{2uMN}{v^2}.$$

The ratio u/v of blade velocity to jet velocity is a quantity of much importance in this connexion. Calling it r, we have

$$\eta_d = \frac{2rMN}{v}.$$

Further, since

$$MN = (v_r \cos\theta + v_r' \cos\theta') = v_r \cos\theta \left(1 + k\frac{\cos\theta'}{\cos\theta}\right),$$

and $$v_r \cos\theta = v\cos\alpha - u,$$

$$\eta_d = 2r(\cos\alpha - r)\left(1 + k\frac{\cos\theta'}{\cos\theta}\right).$$

It will be seen that when r and α are assigned, the blade-entrance angle θ is determined; it is a function of these quantities, the relation being

$$\tan\theta = \frac{\sin\alpha}{\cos\alpha - r}.$$

The blade-exit angle θ' is independent of r and α, and after these quantities have been assigned its adjustment is within the choice of the designer to meet other requirements.

If, however, the blading be symmetrical ($\theta' = \theta$), or if a constant ratio be maintained between $\cos\theta'$ and $\cos\theta$, then the quantity $\left(1 + k\frac{\cos\theta'}{\cos\theta}\right)$ becomes a constant factor independent of α and r, and in that case a very simple rule holds for the relation between the jet angle α and the speed ratio r which will give maximum diagram efficiency, when one or the other of these two variables is assigned. For, taking the jet angle as assigned, and the speed ratio as variable, η_d will then be a maximum when $\frac{d\eta_d}{dr} = 0$, that is to say, when $r = \frac{1}{2}\cos\alpha$, with the value

$$\max.\ \eta_d = \tfrac{1}{2}\cos^2\alpha\left(1 + k\frac{\cos\theta'}{\cos\theta}\right).$$

MN is then $u\left(1 + k\frac{\cos\theta'}{\cos\theta}\right)$, and the work done on the blades, per lb. of steam, is $u^2\left(1 + k\frac{\cos\theta'}{\cos\theta}\right)$.

This very simple condition for maximum diagram efficiency, namely, that the speed ratio u/v should be $\frac{1}{2}\cos\alpha$, is approximated to in actual cases and is a useful guide in turbine design. It would give a speed ratio of 0·48 for a jet angle of 16°, 0·47 for 20°, and 0·44 for 28°. But any moderate departure from the speed ratio so found reduces the diagram efficiency only slightly, as may be seen by drawing a curve of efficiency in relation to u/v for any assumed jet angle. Further, and this is a point of much importance, the diagram efficiency is by no means a complete criterion of the most favourable conditions of working, for other sources of loss have to be taken into account, some of which are more or less dependent on the speed ratio. In practice it is usual to make the speed ratio somewhat smaller than that which would give maximum diagram efficiency, with a view to other considerations.

144. Axial component of velocity. In a velocity diagram drawn to suit the condition $u/v = \frac{1}{2}\cos\alpha$ (fig. 66), we should have

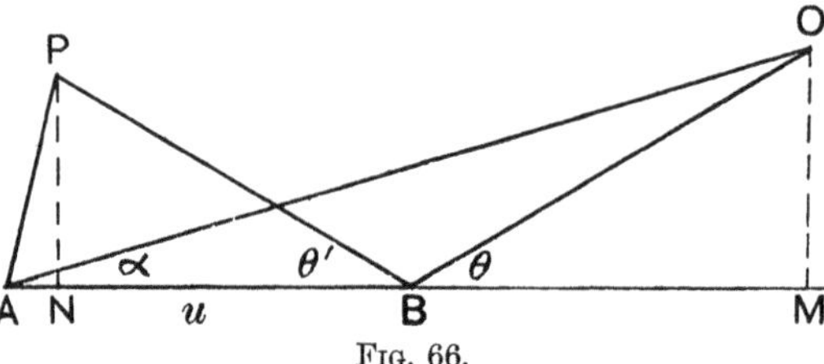

FIG. 66.

$BM = u$ and $\tan\theta = 2\tan\alpha$, so that a jet angle of 16° would require the-blade entrance angle θ to be nearly 30°, a jet angle of 20° would require θ to be about 36° and so on. On account of friction, BP is substantially less than BO, perhaps 10 or 15 per cent. less. It will be obvious that under these conditions the leaving velocity PA could not be perpendicular to the plane of the wheel, and also that its axial component PN would be considerably less than OM unless we were to make PA more oblique by increasing θ'. We might on the other hand lessen the obliquity of PA somewhat by reducing θ' but with the result of making PN still smaller. At each stage of a compound impulse turbine it is desirable that PA should not be far from normal to the wheel and also that PN should not be greatly less than OM. For OM represents the axial velocity with which the steam leaves the annulus of nozzles, and PN represents the axial velocity with which it leaves the annulus of blades. Its volume is the same in both cases, and the volume is the product of the axial velocity into the effective area of the annulus. Hence, even if the blades were running full at exit, the area of the blade annulus on the exit side multiplied by PN must not be less than the area of the nozzle annulus multiplied by OM. This means that the blade height must be greater than the nozzle height in at least the ratio OM to PN. A reference to fig. 64 will show that in each stage the blade height is greater than the nozzle height, but a large difference would have disadvantages.

In practice the speed ratio u/v is usually made less than $\frac{1}{2}\cos\alpha$, with the effect (fig. 66 *a*) of making BM greater than u. This increases OB and BP so that PA stands more nearly at right angles to AB. It also increases the axial velocity for the same values of u and α. In fig. 66 the speed ratio is 0·48. Fig. 66 *a* shows the effect of reducing it to 0·45 while α and θ' are unchanged:

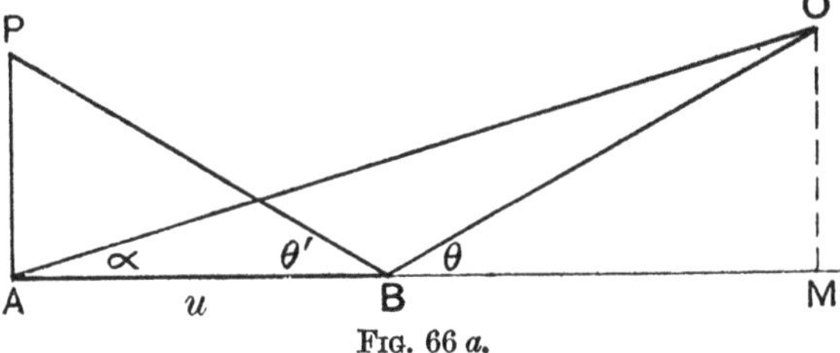

Fig. 66 *a*.

N then practically coincides with A. The friction coefficient k is taken as 0·9 in both figures.

One effect of having PN less than OM is that a dynamical thrust is exerted on the ring of blades in the direction of the axis. For each lb. of steam loses axial momentum to the extent $OM - PN$ by passing through the ring, and is therefore pushed backwards with a force $(OM - PN)\,g$, in pounds: in other words, it exerts on the ring an axial thrust of that magnitude in the direction of the flow, although there is no difference of steam pressure on the two sides. The aggregate amount of this axial thrust in an impulse turbine is small, and it is met by the thrust collar which serves to keep the rotor from longitudinal displacement.

In the latest stages of a turbine, when it becomes difficult to provide enough annular area of passage way without making the blades excessively long, the axial component of velocity may with advantage be increased by using larger jet angles and blade angles. Blades designed for these conditions are called *wing blades*. Their diagram efficiency is somewhat less than that of the blading used throughout the earlier stages.

The exit angle in wing blades may be 40° or more. In the blading of earlier stages the exit angle usually lies between 20° and 30°. By making it too small we should reduce the ratio of PN to OM so much as to make the length of the blades excessive in comparison with that of the nozzles. It will be obvious that the axial velocity is important in design because on it depends how much steam will pass through a given turbine. Let X be the effective area of the annulus, at any stage, which may be taken as nearly equal to πDh, where D is the diameter measured to the middle of the blades, and h is the blade height. Let the axial velocity be v_a. Then $v_a X$ is the volume of steam passing per second, and the number of pounds passing per second is

$$M = \frac{v_a X}{V},$$

where V is the volume of the steam per lb. in the condition it has then reached.

To estimate the dimensions suitable for a turbine which is to develop a given amount of power, it is easy to determine from the known heat-drop and probable efficiency ratio how many pounds of steam should operate per second: from that the product $v_a X$ is found for each stage, having regard to the specific volume V. The efficiency, as we have seen, depends on the velocity ratio: the higher u is made, the higher (for the same efficiency) is v and also v_a. Hence by increasing the blade speed a given area of annulus will give increased power; or to produce the same power with the same efficiency a smaller area will suffice.

145. Examples of velocity diagrams. To illustrate these and other points a velocity diagram is drawn in fig. 67 from data in a paper by Mr Baumann[1]. The diagram relates to part of the final blading in a large turbine of the compound impulse type. The jet velocity is 1840 feet per second, and the jet angle α is nearly 24°. As calculated from the figures given for the velocities it is 23° 33′. For the reason above indicated the blading is designed to give a

[1] K. Baumann, "Some recent developments in large steam turbine practice." *Jour. Inst. Elect. Eng.* vol. LIX. 1921, p. 651.

fairly high axial component of velocity. The blade velocity u is 732 feet per second, hence the speed ratio u/v is 0·40. These data fix the blade-entrance angle θ as 37° 36′. The blade-exit angle is 40°, and k is close on 0·9.

This makes the leaving velocity 702 feet per second, and MN, which measures the whole change of tangential velocity, is 1781 feet per second, showing that the tangential force exerted on the blades, for each lb. of steam passing per second, is $1781/g$ or 55·3 pounds, and that the work done per lb., namely $u.MN/g$, is 40490 foot-pounds. Since the energy of the jet is $(1840)^2/2g$ or 52570 foot-pounds for each lb. of steam, it follows that the diagram efficiency η_d is 0·77. The same figure would be found by applying the equation for η_d in § 143.

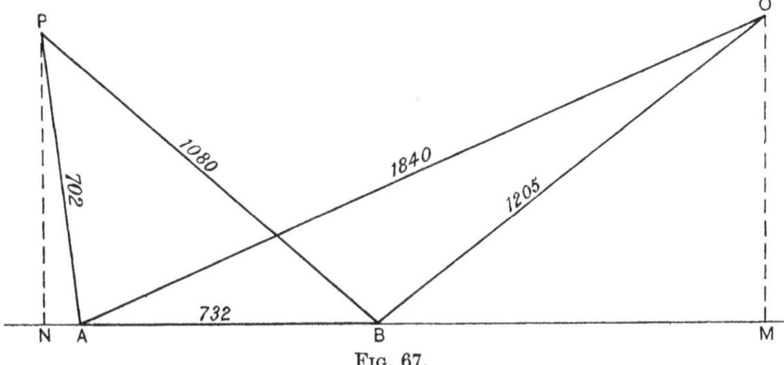

FIG. 67.

The difference between the energy of the jet and the work done on the blades is made up of two items, (1) the "friction loss" which is $(OB^2 - BP^2)/2g$ or $OB^2(1 - k^2)/2g$, and (2) the "leaving loss" which is $PA^2/2g$. From the diagram we can find these and draw up an energy balance-sheet in foot-pounds per lb. of steam:—

Energy supplied in the jets	52570	Work done on the blades	40490
		Friction loss	4430
		Leaving loss	7650
			52570

In a diagram relating to any earlier stage the quantity $PA^2/2g$ is only lost as regards the individual row of blades: it is carried over and may be more or less completely utilized in the succeeding stage. At the worst it will contribute to the total heat available for subsequent expansion. But in the final stage (as here) it represents a real leaving loss: it goes to the condenser along with the remaining heat-content of the steam.

Since there are 1400 foot-pounds in a pound-calory the leaving loss in this example is equivalent to 5·45 calories, which is about $2\frac{1}{2}$ per cent. of the total heat-drop. The axial components of velocity (OM and PN) are 735 at entrance to the blades and 702 at exit. The corresponding longitudinal thrust $(OM - PN)/g$ is 1·68 pounds for each lb. of steam passing per second.

Here a main object in the design has been to make the fullest use of the area of the annulus in passing steam to the condenser. This implies that on the exit side the channels should be completely filled. In practice there may even be a slight departure from pure impulse working, with the same object: that is to say, the blade channels may run full with a small amount of expansion in them. On this point Mr Baumann makes the following general remark[1]:—

"It is necessary that the steam should fill completely the blade passages in the last row of moving blades to ensure that the leaving losses are a minimum for a given leaving area. This condition is met by using a small amount of reaction, *i.e.* the pressure between the last row of guide-blades and the last row of moving blades should be slightly higher than the pressure in the exhaust, a condition which automatically obtains when a turbine is designed without reaction in the last stage, but is run at a higher vacuum than that for which it is designed."

In earlier stages of a compound impulse turbine it is usual to guard against differences of pressure on the two sides of a blade wheel, such as might arise through sudden changes in the working conditions, by providing openings in the wheels to ensure that the pressure on both sides shall be the same.

Another example, for an early stage in a compound impulse turbine, is shown in fig. 67 *a*. Here the jet angle is 16°, the jet velocity 1206 feet per second and the blade velocity 534, so that u/v is 0·443. This makes $\theta = 28°$. The blades are symmetrical, the exit angle being also 28°. The friction coefficient k is taken as 0·87, with the result that PA is almost normal to AB. MN is 1169. Numerical values of the velocities are given in the diagram,

[1] Baumann, *loc. cit.*, p. 588. The practice of some makers of impulse turbines is to introduce a little reaction into several of the latest stages, letting their channels run full.

from which it will be found that the jet energy per lb. of steam is 22580 foot-pounds, of which 19390 are converted into work, 1890 are lost in friction, and 1300 are carried over to the next stage. The diagram efficiency is therefore 0·86.

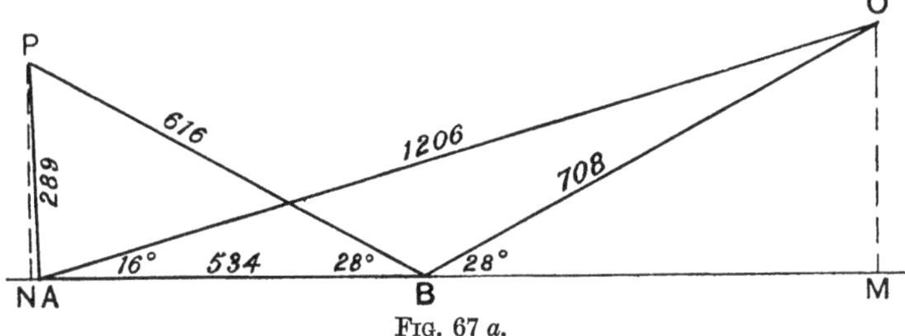

FIG. 67 *a*.

146. Section of impulse blading. A section of symmetrical impulse blade suitable for these angles is shown in fig. 68; it is drawn so that the blade has the same angle on its back or convex side as on its face or concave side, at each of the two edges. Fig. 68 *a* illustrates an unsymmetrical blade, in which the exit angle θ' is made 6° less than the entrance angle θ. The angle of the face at the inlet edge is often made a little larger than the angle of the back, to give the inlet edge greater strength, and it is not found that this exposes the steam to any sensible "shock" on entry. This feature is also shown in fig. 68 *a*, where it will be seen that while the back of the blade at the inlet edge is inclined at θ, namely 28°, the face is inclined there at 31° or $\theta + 3°$.

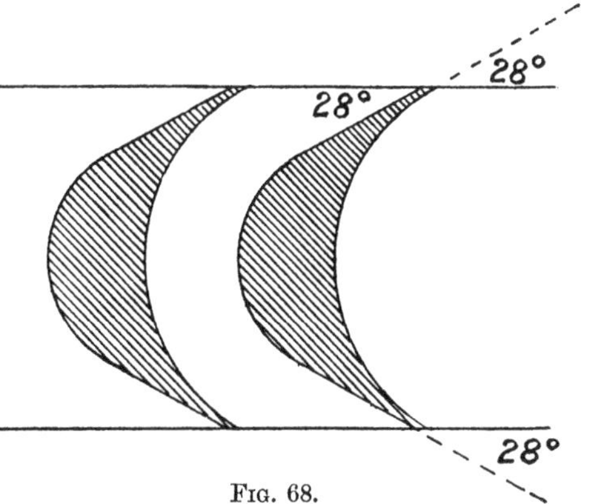

FIG. 68.

The face as seen in section is generally a circular arc with a radius nearly equal to the pitch, or circumferential distance from blade to blade, and the pitch is rather more than half the axial width of the blade ring. The back is formed by a circular arc connecting straight lines inclined at θ and θ' respectively. These lines are drawn so that the blade has a small finite thickness at each edge. With this construction the channel between neighbouring blades has a cross-section which in symmetrical blading may be uniform from exit to entrance, and in all cases is nearly uniform. The cross-section of the blade itself gives it the strength and rigidity needed to stand the bending moment due to the impulse of the steam, and the radial tension due to the rotation. In long blades such as are found in the low-pressure stages of a large turbine a further consideration is of particular importance: the ring must have sufficient lateral rigidity to escape the kinds of vibration to which a flexible disc or other flat structure is liable when rotated at a high speed; this in practice limits the dimensions of the blade ring in relation to the speed[1].

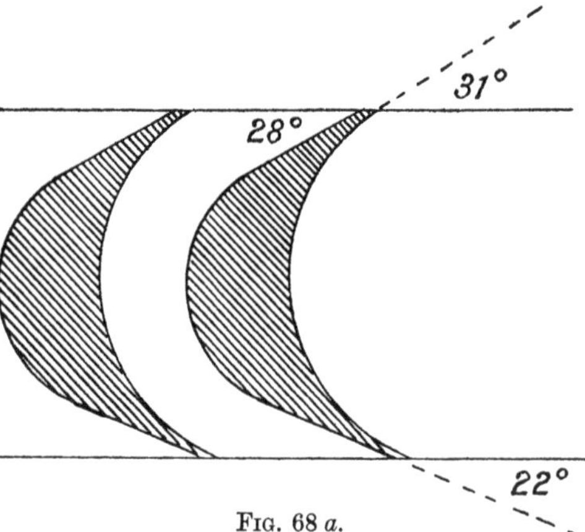

Fig. 68 *a*.

147. Diagram efficiency and Stage efficiency. There is much uncertainty as to the proper value to be assigned, in any given case, to the friction coefficient k. This makes the diagram to some extent conjectural, especially as regards the leaving velocity and the efficiency. A design based on the method of the diagram has in general to be adjusted empirically before the best results are obtained in actual working. It will further be obvious that the velocity diagram, important though it is as an aid to design, does not by itself determine the conditions of economic performance. The larger idea of stage efficiency, already referred to in § 138, includes elements which diagram efficiency leaves out of account, such as losses due to friction in the nozzles, to leakage of steam past the blades or through glands, to thermal conduction, and to windage or friction between the blade wheel and the

[1] See a paper by W. Campbell, "The protection of steam turbine disc wheels from axial vibration," which gives examples of failure arising from this cause. *American Soc. of Mech. Eng.* May, 1924.

atmosphere of steam in which it revolves. Accordingly the final ratio, in the action as a whole, between the work done on the rotor and the adiabatic heat-drop is considerably less than might perhaps be expected from a study of diagram efficiencies alone.

148. Velocity diagram for a Curtis wheel. Fig. 69 illustrates how the method of the velocity diagram may be applied to a "velocity compounded" or Curtis wheel, with, in this case, two rows of moving blades. The general arrangement of such a

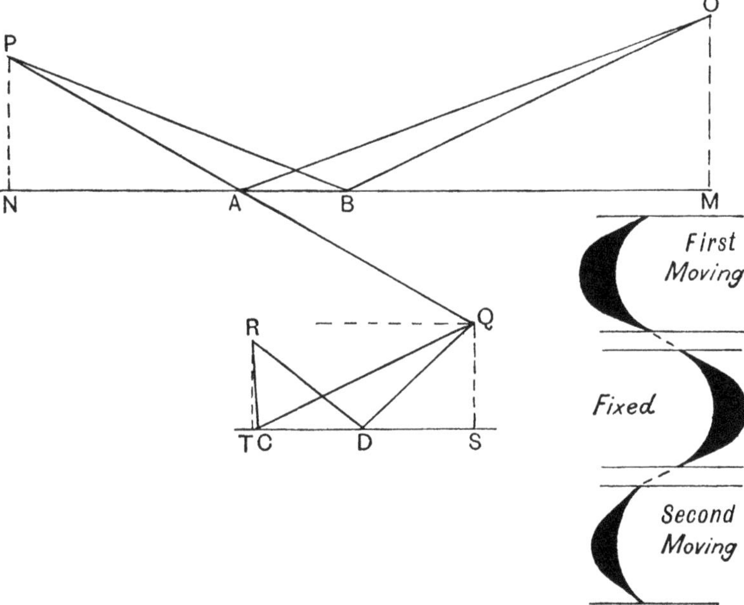

FIG. 69. Velocity diagrams and blade sections for a two-row Curtis wheel.

wheel was shown on the left-hand side of fig. 64, and the blading is more clearly seen in fig. 70. It will be noted that the heat-drop for the wheel is large enough to require the nozzles to have some divergent extension. As was pointed out in § 142, a Curtis wheel is frequently used as the first stage not only in impulse turbines, but in turbines which are otherwise of the reaction type.

In fig. 69 the velocity diagram $OABP$ relates to the first row of moving blades in a wheel of this type, and $QCDR$ to the second row. The function of the fixed blades between is to alter the direction of the stream from PA to QC. The blade speeds AB

and CD are the same for both rows. The sketch alongside shows appropriate blade sections for the fixed as well as the two moving rows.

Steam enters with the jet velocity OA which, when compounded with AB as in an ordinary impulse pair, determines OB, the relative velocity of entry to the first row, and the entrance angle OBM. BP is the relative velocity at exit from the first row; its direction is fixed by the given exit angle ABP, and its magnitude is $k.OB$. Join PA and produce it, making $AQ = PA$. This represents the velocity with which the steam enters the fixed blades, and determines the proper entrance angle. It leaves them with the reduced velocity QC which is equal to $k.PA$, in a direction determined by the given exit angle of the fixed blades. QC is the velocity with which the steam reaches the second row: it is combined with the blade speed CD to give the relative velocity QD, which is reduced by friction in the second row to DR. The direction of DR is fixed by the given exit angle of the second row. Finally, joining R with C we find the velocity RC with which the steam passes on to the next stage.

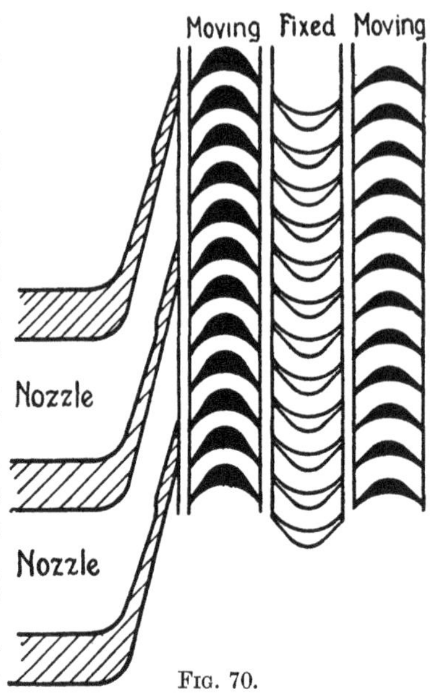

Fig. 70.

It will be obvious that the diagrams for the two rows might be superposed, by making CD coincide with AB: they are kept separate here for the sake of greater clearness.

The work done is $u.MN$ on the first moving row and $u.ST$ on the second moving row, showing that the work done on the first row is fully three-fourths of the whole. The friction loss is $\frac{1}{2}(OB^2 - BP^2)$ on the first moving row, $\frac{1}{2}(PA^2 - QC^2)$ on the fixed guide-blades, and $\frac{1}{2}(QD^2 - DR^2)$ on the second moving row. The carry over energy is $\frac{1}{2}RC^2$. These items together account for the jet energy $\frac{1}{2}OA^2$. All the quantities are in kinetic units.

The longitudinal thrust is $OM + QS - (PN + RT)$. Assuming

that each of the blade channels is full on the exit side, the blade heights at the exit edge in (1) the nozzles, (2) the first moving row, (3) the fixed row, and (4) the second moving row should be inversely proportional to the axial velocities, namely to OM, PN, QS and RT respectively. Thus the ratio OM/RT may be used to fix what is called the height ratio of the wheel, namely the ratio in which the height has to be increased from the nozzle to the exit edge of the second row. In some examples the blades of each row are shaped so that the height increases from the entrance to the exit side, the shroud rings being conical; in others the height is constant from one to the other side in each row.

It should be observed that in a two-row Curtis wheel the jet gives up a tangential component of velocity approximately equal to $4u$, as against $2u$ in a simple impulse pair. Thus the condition for efficient working is to be reached by making u/v approximate to $\frac{1}{4}\cos\alpha$ instead of $\frac{1}{2}\cos\alpha$. Similarly in a three-row wheel u/v should be about $\frac{1}{6}\cos\alpha$.

The diagram illustrates how the principle of velocity compounding enables a wheel with moderate blade speed to take up the energy corresponding to a relatively large heat-drop. A Curtis wheel with two rows of moving blades is equivalent in this respect to four single impulse wheels running at the same speed, for these would require the jet velocity to be halved and each would therefore take only one-quarter of the heat-drop that is taken by the two-row wheel. Similarly a Curtis wheel with three moving rows would take up the same heat-drop as nine single wheels: the frictional losses, however, would make it much less efficient.

In the example for which fig. 69 is drawn the jet velocity is 2000 ft. per second, and the jet angle α is 20°. The blade speed is 420, making $u/v = 0{\cdot}21$. The other data are k, which is taken as 0·9, and the three exit angles. The values of these are given below, and also the approximate values of the entrance angles as found by drawing the diagram.

	Entrance Angles (found from the diagram)	Exit Angles (data)
First moving blades	25°	21°
Fixed guide-blades	29°	25°
Second moving blades	42°	38°

The moving and fixed blade sections shown in the figure are drawn with these angles.

With these data it will be found (according to the diagram) that of the whole jet energy, which is 62100 foot-pounds per lb., 48450 foot-pounds are converted into work on the wheel, 11800 are lost through friction, and 1850 are carried over. The height ratio OM/RT is 2, which means that on the exit side of the second moving row the blades should have a height twice that of the nozzles by which steam enters the wheel. The axial thrust is 7·3 pounds for each lb. of steam passing per second.

The relatively large loss through friction will be noted. Experience shows that the realized efficiency in a velocity wheel is even less than these figures would suggest. The results are much affected by uncertainty as to the value of k for each part of the action.

In the practice of some makers there is a departure from strict impulse working; the blade channels are running full, and the steam is expanding slightly in them. The chief part of the pressure-drop occurs in the nozzles, but there is a little further drop in both moving and fixed blade channels. Consequently, though the driving force on the blades is mainly produced by impulse, there is a small additional driving force due to reaction. When this is the case a diagram may be drawn on the basis of assuming blade heights to begin with, and selecting values of PN, QS, and RT so that the reciprocals of these are proportional to the assumed heights. In this procedure we ignore k but determine the length of BP from the value of PN and the given exit angle. If BP is then found to be more nearly equal to OB than friction would warrant, the result means that there is some expansion, giving an increase of velocity—or rather reducing the loss of velocity that is caused by friction—and using up a small heat-drop in supplement to that which has been used in the nozzles. The same procedure is to be followed in the fixed row and in the second moving row. In the absence of an exact knowledge of k the results do not enable us to distinguish between the effect of friction on the one hand and that of supplementary heat-drop on the other, in accounting for the difference which is found in the diagram between BP and OB, between QC and AQ or PA, and between DR and QD. Whenever there is any expansion in the blade channels it will more or less counteract in the diagram the difference ascribable to friction[1].

[1] For a diagram drawn in this way see Prof. Goudie's book on *Steam Turbines*, Edition of 1922, p. 306.

149. Reaction blades and their velocity diagrams. In a Parsons reaction turbine the fixed and moving blades are alike (fig. 71). They are set in alternate rings which succeed each other from end to end of the turbine. A ring of fixed blades and the next ring of moving blades constitute a "stage." Each ring of moving blades takes up the energy of jets directed against it by the preceding ring of fixed blades, and to that extent the turbine acts by impulse. But in passing through the channels of the moving

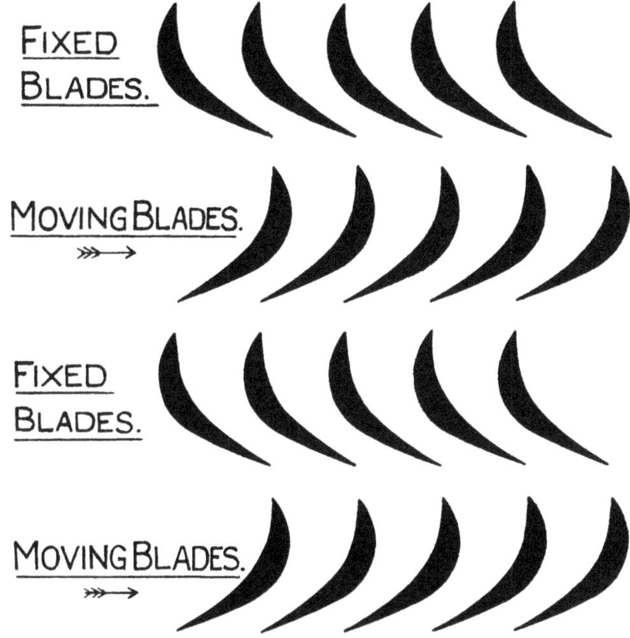

FIG. 71. Typical Parsons reaction blading.

blades the steam becomes accelerated. Its velocity relatively to the blades is increased as well as altered in direction, and it consequently exerts on the blades, by reaction, a greater driving force than it would exert if it passed over them without increase of relative speed. There is a drop of pressure in the moving blade channels as well as in the fixed blade channels. Parsons makes the action in both alike, so that the steam loses pressure and acquires relative velocity to the same or nearly the same extent in a row of moving blades as it does in the preceding row of fixed blades. The heat-drop of the stage is accordingly shared between the two rows. The blades are in an annular space between the stator and rotor, that is to say, between the "cylinder," or enclosing case, and the rotating drum. The fixed blades are secured in grooves in the cylinder and project inwards, nearly touching the surface of the rotor drum: the moving blades are secured in grooves on the rotor and project outwards, nearly touching the inner surface of the cylinder. There used to be no shroud rings over the blade tips, but now, for blades in the high-pressure portion, Parsons uses a device called "end tightening" (see next section) which requires both fixed and moving blades to be shrouded. The cylinder is stepped so that the annular space becomes progressively larger towards the low-pressure end, to give longer blades and so provide a greater area of flow for the steam as its volume increases. The rotor drum also is generally stepped in a few places, to give a higher velocity to the low-pressure blades and thereby allow the steam velocity to be greater at the low-pressure end, so that less enlargement of the area of flow may suffice, for otherwise the blades would have to be made unduly long.

The number of stages depends on what peripheral speed it is convenient to use, and is often (especially in the older marine turbines designed for use without gearing) much larger than the number of stages in an impulse turbine, the peripheral speed being less. Modern reaction turbines, however, often use fairly high blade speeds with a comparatively small number of stages.

In the ideal turbine there should be a continuous enlargement of the steam passage way from row to row of the blades, corresponding to the continuous increase in volume which is undergone by the steam as it passes from row to row. Recent designs in fact tend towards this form, but in many of the older turbines, where the stages were more numerous, the changes of blade length were, for convenience in construction, confined to a limited number of steps, and the radial height of the blades was the same for several rows in each step; in this way a sufficiently good approximation to the ideal form was obtained. In a few of the last rows the axial velocity of the steam is increased by the use of "semi-wing" and "wing" blades, with blade angles larger than those used throughout the earlier stages.

Fig. 72 shows a full-size section of the blades in an early stage of a 6000 kw. Parsons turbine, along with the velocity diagram. Here the blade length is $1\frac{1}{16}$ in. The peripheral speed u is 236 ft.

per second, the mean diameter of the blade ring being $22\frac{1}{2}$ ins., and the revolutions 2400 per minute. Treating this as a moving row, steam is delivered from the preceding fixed row at an angle α of 20° to the direction of motion, and with a velocity of 315 ft. per second.

Accordingly the relative velocity of the steam at entry is OB, and the blades are shaped to receive steam at that angle without shock. The channels are convergent, and there is a drop of pressure, so that the steam is accelerated in passing through them. Its velocity relative to the blades changes from OB to $B'P'$, and by drawing $B'A'$ equal to u we find the absolute velocity $A'P'$ with which it enters the next fixed row. In this example the exit angle

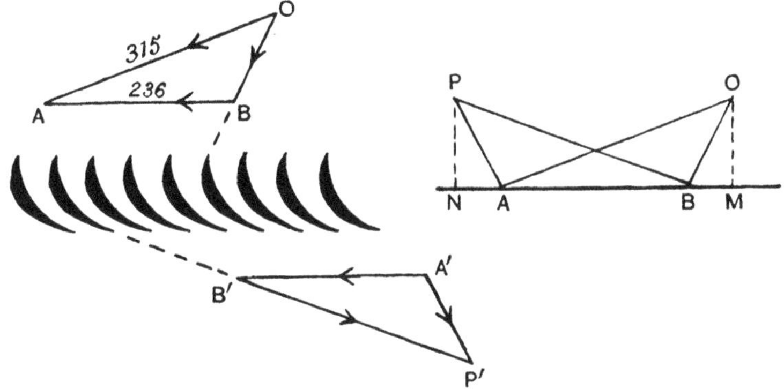

FIG. 72. Parsons blading with velocity diagrams.

is 20°, and, as there is as much heat-drop in the moving channels as in the fixed, $B'P'$ is equal to OA. The two diagrams may conveniently be superposed, as is shown on the right of the figure, where the triangle PAB takes the place of $P'A'B'$.

For efficient working in a series of such rows OB, PA, etc. should be not far from normal to the direction of motion of the blades. This requires that the velocity ratio u/v should approximate to $\cos\alpha$, instead of to $\frac{1}{2}\cos\alpha$ as in impulse blading (§ 143). With $\alpha = 20°$ this would make $u/v = 0{\cdot}94$; in practice it is usually less, as in this example, where it is just over 0·75.

The work done on the rotor per pound of steam in passing through the pair of rows which form a complete stage is $u.MN$, and is obtained at the expense of two heat-drops, which in this instance are equal, in the two elements of the pair.

Fig. 72 *a* shows (also in full-size section) blades of the final row in the same turbine with their velocity diagram. Here the mean

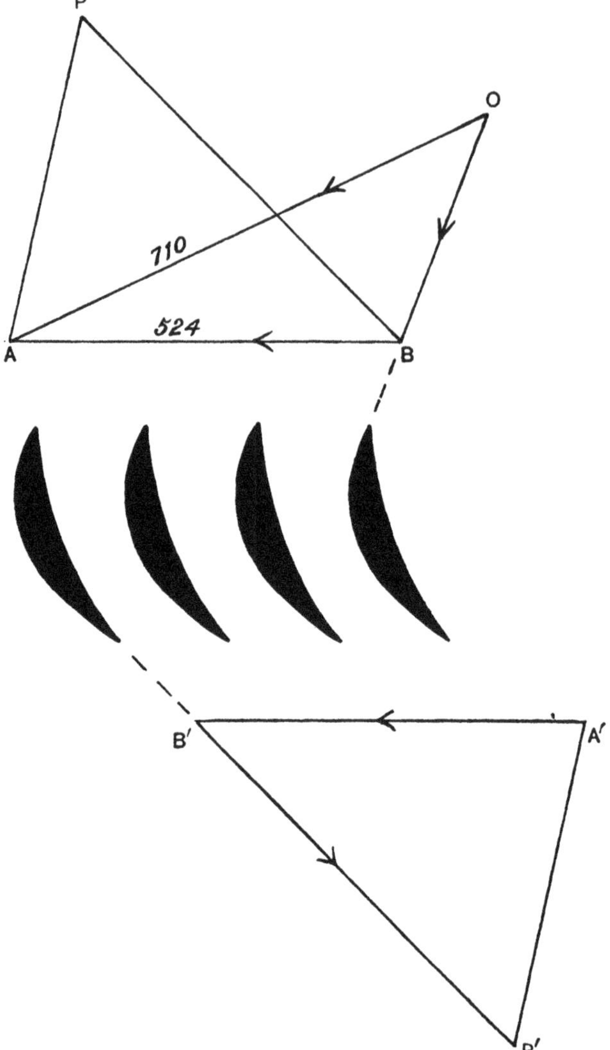

FIG. 72 *a*. Parsons wing blading with velocity diagrams.

diameter of the blade ring is 50 inches, and consequently u is 524 ft. per second. These are "wing" blades with an exit angle of 45°, designed to give the steam a relatively large amount of axial velocity;

the jet velocity from the preceding fixed row is 710 and the jet angle is 25°. Steam enters the blades with the relative velocity OB, and leaves them, after acceleration to a relative velocity $B'P'$, with an absolute velocity $A'P'$, which is not far from axial in direction, and is large enough to let a moderate blade length ($10\frac{1}{2}$ inches) deal with the great volume of the expanded steam.

The diagrams are superposed in the upper part of the figure. In this example $PA^2/2g$ measures the leaving loss per pound of steam.

150. Forms of reaction blading. End tightening. In the later stages of a Parsons turbine the leakage of steam over the blade tips, due to the difference of pressure on the two sides of the row (whether moving or fixed), is not a serious source of loss, for the clearance over the tips is very small compared with the length of the blades. Enough clearance must be left to allow for such slight bending as the rotor may undergo through inequalities of temperature or centrifugal strain, and for greater safety the tips are sharpened to a knife edge so that a casual grazing contact between the ends of the moving blades and the inner surface of the casing, or between the ends of fixed blades and the surface of the rotor, will only grind the tips down and not do serious damage. With short blades, such as those in the early stages, the loss through leakage over the tips was more considerable, but Parsons now adopts the plan of shrouding these by a ring with a projecting sharp edge which is adjusted to run nearly in contact with a continuous flat surface formed by one side of the projecting bases of blades in the neighbouring row. Any leakage therefore takes place not over the tips but between the sharp edge of the shrouding and a flat surface with which it can be brought into close juxtaposition by longitudinal adjustment of the position of the rotor in the case. This "end tightening" device is shown in figs. 73 and 74. Fig. 73 gives a general view with a portion of the blading removed: fig. 74 is a drawing of a pair of shorter blades, fixed and moving. It will be seen from both figures that the fixed blades carry a shroud ring with a sharp edge projecting so that it is nearly in contact with the flat side B of the base of the adjoining moving blades, and similarly they carry a shroud ring whose sharp edge is nearly in contact with a corresponding flat side B on the base of the fixed blades. The amount of axial clearance at the sharp edges of the shrouding is controlled by a fine screw adjustment of a double-acting thrust collar which holds the rotor shaft from longitudinal displacement. With this device there is no restriction as to the amount of radial clearance that may be allowed. End tightening, the use of which dates from 1912, was at first applied to a few rows

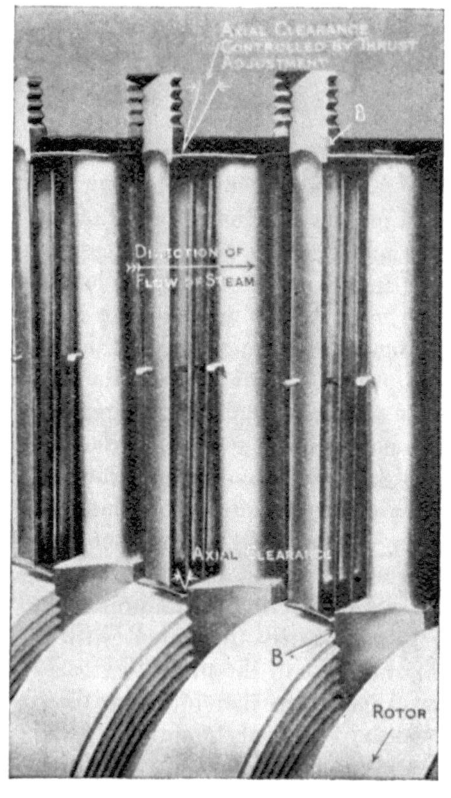

Fig. 73.

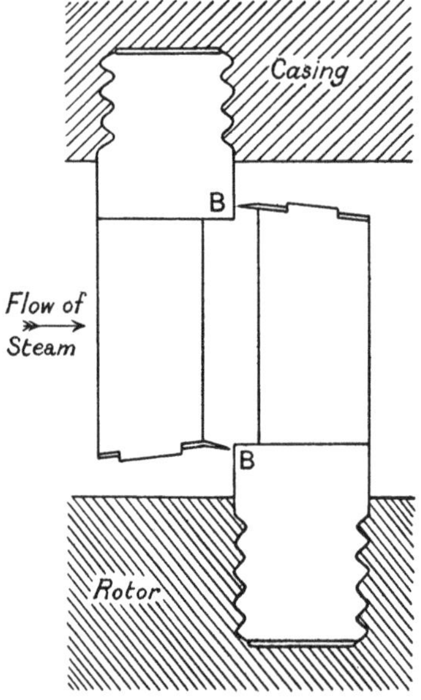

Fig. 74.
End tightening in reaction blading.

at the high-pressure end. It is now often extended to all except the low-pressure stages, and has done much to contribute to the efficient use, by reaction blading, of the high pressures which have become common in steam turbine practice. Its full advantage is only secured by a nice adjustment of the longitudinal position of the rotor to suit conditions of temperature which vary somewhat with changes in the output.

Comparing the forms of reaction with those of impulse blading it should be noted that a lighter type of blading serves in the reaction turbine, especially towards the high-pressure end. The comparatively small velocity of the steam in a reaction turbine makes it unnecessary to give the blades nearly so rigid a section as is required at corresponding stages in the impulse type.

Fig. 75 shows long blades of a Parsons turbine, at a stage near the exhaust end, and also illustrates how the blades are inserted.

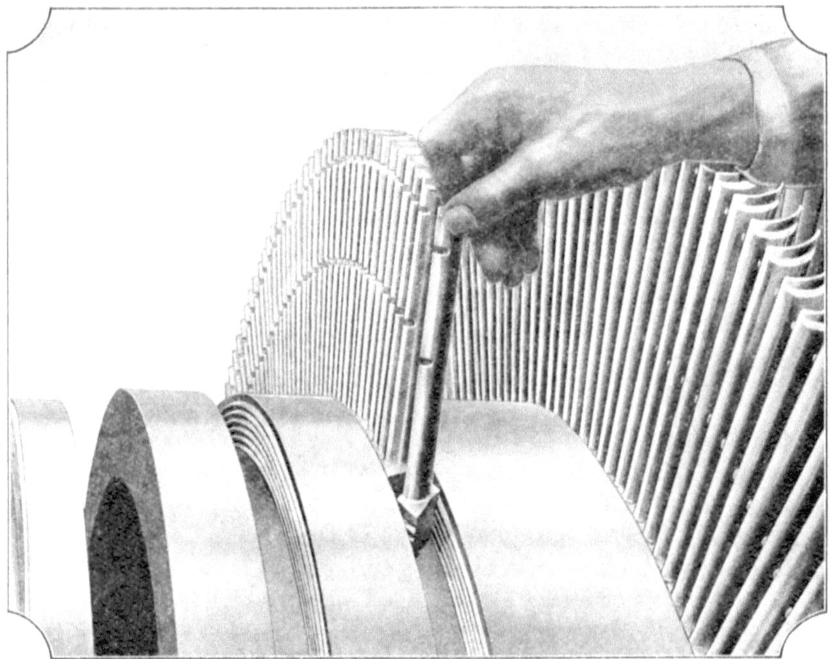

Fig. 75.

They are held firm by serrations on two sides of a lozenge-shaped root or base which is made integral with the blade itself. Each blade after insertion in the groove is twisted round by hand until the serrations gear with those in the groove, and is then knocked up tight. The fixed blades project inwards from similar serrated grooves in the casing. These blades, both fixed and moving, carry no shrouds but are sharpened at the tips, and leakage is prevented from being excessive by making the radial clearance small. The blades are stiffened by being brazed to rings of binding wire on the entrance side, one near the middle of the height and another near the tip. With very long blades three or four such rings of wire are used. Since the wires are on the entrance side, where the relative steam velocity is small, they offer very little obstruction to the flow.

The same method of fixing is used for end-tightened blades (figs. 73 and 74), but there the roots project beyond the groove for a sufficient distance to furnish the flat side faces *B*. The blades of fig. 73 are stiffened by a wire at the middle of their length as well as by the shroud ring.

The sectional forms of Parsons blading will be seen from figs. 71, 72 and 72 *a*, and also from fig. 76, which shows, to full size,

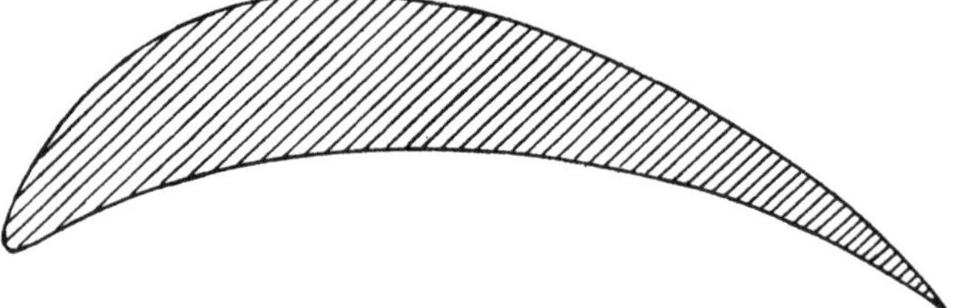

FIG. 76. Full-size section of Parsons blading in a large turbine.

the section of a blade from the last ring of a 50,000 kw. steam turbine. Each of the blades of fig. 76 is 40 ins. long and weighs 21 lb. The blunt round edge on the admission side, together with the finely tapered tail, gives blades of this type a "stream-line" form, which is effective in avoiding turbulence. It has been arrived at by trials of efficiency rather than by conforming strictly to the lines of a velocity diagram. The comparatively low velocity with which steam enters the blade channels of a reaction turbine makes a rounded admission edge more suitable than it would be in an impulse turbine.

151. Comparison of turbines. Parsons coefficient K. In most of Parsons' turbines, as was mentioned above, the blade heights do not increase regularly from stage to stage, but in a number of steps, each of which includes a group of stages in which the blade height is constant. The steps are formed by abrupt changes in the diameter of the drum or of the enclosing cylinder or both. Each group of stages in which the diameter and blade height are constant is called by Parsons an "expansion." Thus in each "expansion" there are several successive stages with the same blade height, mean diameter, and blade speed; and the whole heat-drop of the "expansion" is divided nearly equally among them. The next "expansion" will have a greater blade height, a greater diameter, and consequently a greater blade speed.

This arrangement was structurally convenient when the number of stages was great. But in more modern designs of reaction turbines, by Parsons himself and by other makers, blade speeds are made relatively high, with the result that fewer stages suffice, and in such cases the blade height may increase progressively from one stage to the next. The need for very many stages has disappeared through the introduction of gearing. So long as the turbine shaft was directly connected to a mechanism (such as the screw-propeller of a steamship) which could not be efficiently driven at a high speed, there was a practical limit to the number of revolutions and therefore to the blade speed; consequently the stages had to be very numerous. With gearing, on the other hand, the angular speed of the turbine is no longer limited by that of the driven mechanism: each can be set to run at its own best speed, and with freedom to use a high speed for the turbine shaft a comparatively small number of stages will give a thoroughly efficient design. The efficiency depends upon the blade speeds and the number of stages in a way which has now to be considered.

Writing D for the diameter in inches (to the middle of the blade) and N for the number of revolutions of the turbine shaft per minute, the blade velocity in feet per second, at any stage, is

$$u = \frac{\pi DN}{12 \times 60} = \frac{DN}{229{\cdot}2}.$$

If δH represent the effective heat-drop *of the stage*, half of this, in a reaction turbine, is used in the fixed blades to give the steam its jet velocity v; hence

$$v = \sqrt{2gJ\delta H/2} = 300{\cdot}2\sqrt{\delta H/2} = 212{\cdot}3\sqrt{\delta H},$$

when the thermal unit is the pound-calory. Hence

$$\left(\frac{u}{v}\right)^2 = 0{\cdot}4225 \times 10^{-9}.\frac{D^2N^2}{\delta H}.$$

This is for a single stage; but in the design of a turbine the quantity u/v, on which the diagram efficiency of each stage depends, is kept nearly constant from stage to stage. Treating it as constant and taking the action as a whole we may therefore write

$$\left(\frac{u}{v}\right)^2 = \frac{0{\cdot}4225 \times 10^{-9}\Sigma D^2N^2}{H},$$

where H is the whole effective heat-drop from admission to exhaust, or $\Sigma\delta H$, and ΣD^2N^2 is summed for all the rows of moving blades.

In a Parsons turbine of the multi-stage type, if D_1, D_2, D_3, etc. be the mid-blade diameters, and n_1, n_2, n_3, etc. be the number of stages in the successive "expansions,"

$$\Sigma D^2N^2 = (n_1D_1{}^2 + n_2D_2{}^2 + n_3D_3{}^2 + \ldots)\,N^2,$$

when (as is usual) N is the same for all. In some instances there may be separate shafts for the high and low-pressure parts of the turbine, with different values of N, and the terms for the different shafts will have to be separated[1]. In any case, whether the stages in a turbine be many or few, ΣD^2N^2 is readily calculated when the diameters and revolutions are known, or are provisionally assumed for the purposes of design.

The quantity $10^{-9}\Sigma D^2N^2$ is commonly called the Parsons Coefficient, and is denoted by K. It serves as a useful basis of comparison between designs, since, for a given heat-drop, any arrangements for which K has equal values will give the same value of u/v, or rather of its mean square, and may therefore be expected to give nearly the same overall efficiency. Using the symbol K in this sense, we accordingly have

$$\frac{u}{v} = 0{\cdot}65\sqrt{\frac{K}{H}}.$$

This is when heat is measured in pound-calories: when British Thermal Units are used the formula becomes

$$\frac{u}{v} = 0{\cdot}87\sqrt{\frac{K}{H}}.$$

The same method of comparison is obviously applicable to impulse turbines. In each stage of an impulse turbine, however,

[1] As for example in the geared turbines of H.M.S. "Hood," where the high-pressure shaft runs at 1500 revolutions per minute and the low-pressure shaft at 1100.

the heat-drop is not halved, and hence $v = 300{\cdot}2\sqrt{\delta H}$, with the result of making

$$\frac{u}{v} = 0{\cdot}46\sqrt{\frac{K}{H}}$$

when H is in pound-calories, and

$$\frac{u}{v} = 0{\cdot}62\sqrt{\frac{K}{H}}$$

when H is in British Thermal Units, K standing for $10^{-9}\Sigma D^2N^2$ as before.

In reckoning K for a turbine with one or more Curtis velocity wheels, the equivalent number of simple impulse rows should be substituted, namely, four for a two-row Curtis wheel, and nine for a three-row wheel (§ 148).

As an example of K in a modern reaction turbine, reference may be made to a 12,000 kw. Parsons turbo-generator built for the Colenso power station and exhibited at Wembley in 1924. With a speed of 3000 revolutions per minute this machine has thirty-eight stages, in which D ranges from 24 to 48 ins. The supply pressure is 250 lb. by gauge, the steam is superheated to 371° C. (700° F.), and expands to a vacuum of 28½ ins., making the adiabatic heat-drop equal to 247 calories (444 B.T.U.). The coefficient K defined as above[1] is 363. Hence if we were to take the whole adiabatic heat-drop as effective, in the above formula, we should have

$$\frac{u}{v} = 0{\cdot}65\sqrt{\frac{363}{247}} = 0{\cdot}79.$$

But this allows nothing for losses which make the heat-drop that is effective for giving velocity less than the adiabatic heat-drop.

In a modern impulse turbine using steam under similar conditions the value of K is from 200 to 250, with the result that u/v, on the same basis, is about 0·42 to 0·47. It will be observed that to secure efficient values for the speed ratio, K should be greater in a reaction than in an impulse turbine.

In estimating the effective heat-drop for use in the above formulas, to determine a mean value of u/v, it is evidently proper to deduct from the total adiabatic heat-drop the leaving loss

[1] *Engineering*, June 6, 1924. The writer there follows a usage according to which K is defined as $10^{-6}\Sigma D^2N^2$, and so gives it 1000 times the numerical value stated in the text.

and also any losses that occur through thermal leakage during the passage of the steam. Further, at any one stage v is not in fact the full velocity due to the heat-drop of the stage but a velocity reduced by nozzle friction. On the other hand, the effect of the reheat factor (§ 138) is to increase the heat-drop available at any stage and therefore to increase the velocity. Some designers multiply the whole heat-drop by the reheat factor, then deduct the leaving loss, and use the quantity so found for H in the formula. This results in giving to v at each stage a value greater than the real velocity—what may be called an ideal value in which the reduction due to nozzle friction is ignored, for v then becomes the velocity "theoretically" due to the pressure-drop. For purposes of comparison between different turbines this usage serves conveniently enough, but one should bear in mind that it gives u/v a conventional meaning, making it numerically less than the real u/v of the velocity diagram.

152. Balance of longitudinal forces. Dummies. Labyrinth packing. Since in reaction blading there is a greater steam pressure on the admission side of each row of moving blades than on the other side, there is a resultant longitudinal thrust on the rotor pushing it towards the exhaust end. In Parsons' earliest turbine this was balanced by admitting steam at the middle of the length and making the construction symmetrical with an exhaust at each end. In modern construction, a symmetrical double-ended form is often given to the low-pressure portion of a turbine when that is contained in a separate cylinder, distinct from the high-pressure portion, the chief motive of the arrangement being to secure a larger total area for the flow of steam in the latest stages. Usually, however, admission to a turbine takes place at one end and exhaust at the other, and to balance the steam thrust which would result, in the reaction type, Parsons attaches to the rotor a number of stepped "dummy" rings on which the steam presses the other way. These are stepped so that they correspond in diameter to the several portions of the rotor, measured in each instance to the middle of the blade height, and steam passages are provided which secure that the same pressure shall act on the dummy ring, forcing it away from the exhaust end, as acts on the corresponding portion of the rotor, forcing it towards the exhaust end. The arrangement will be seen in the sectional drawings of Parsons turbines (figs. 87–89 below). Each dummy ring has half the corresponding blade area, because only half the steam thrust towards the exhaust is borne by the rotor blades, the other half being borne by the fixed blades. The dummy increases by a series of steps, at each of which there is a steam connexion with the appropriate place in the main turbine channel, to make the pressure equal. The back of the largest dummy is in communication with the exhaust. This secures that a longitudinal balance will be preserved, notwithstanding variations in the admission or exhaust pressure of the steam.

No steam-tight packing is permissible between the dummy rings and the cylinder, but escape of steam there is minimized by a device introduced by Parsons, usually called a "labyrinth." The labyrinth is a group of fine edges running as nearly as possible in contact with a number of surfaces, the object being to reduce leakage between the fixed and moving parts, without actual contact, by causing any steam that passes to suffer a succession of throttlings alternated with expansions. The result is to offer a greatly increased resistance to the flow. The labyrinth device is applicable not only to dummy rings but to other parts of turbines where leakage is to be minimized without close packing, as for instance in the clearance between the rotor shaft and the fixed diaphragms of an impulse turbine. It is often used also in the glands through which the shaft passes out of the casing at each end. Figs. 77 and 77 *a* illustrate two labyrinth arrangements of Parsons' design. In Fig. 77 the clearance is axial; in Fig. 77 *a* it is both radial and axial, and the axial clearance is adjustable by giving a small longitudinal displacement of the thrust collar in the rotor bearings. In impulse turbines there is no endways force on the rotor due to steam pressure, but there are, as we have seen, small forces due to changes in the axial component of the steam's velocity.

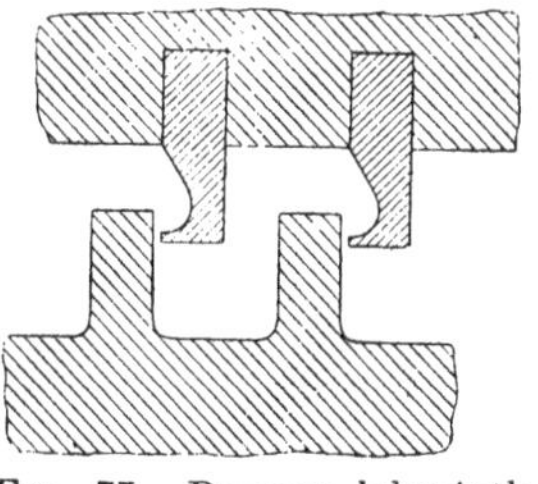

FIG. 77. Parsons labyrinth packing. Dummy gland rings with axial clearance.

With large turbines of whatever type it is usual to control the longitudinal position of the rotor by means of a thrust collar, capable of sustaining any unbalanced force that may tend to force the rotor either way. For this purpose a double-acting Michell thrust block is perfectly effective: in it the thrust is borne by a number of small flat metal pads working against the flat sides of a ring projecting from the shaft. Each pad has a pivot on its

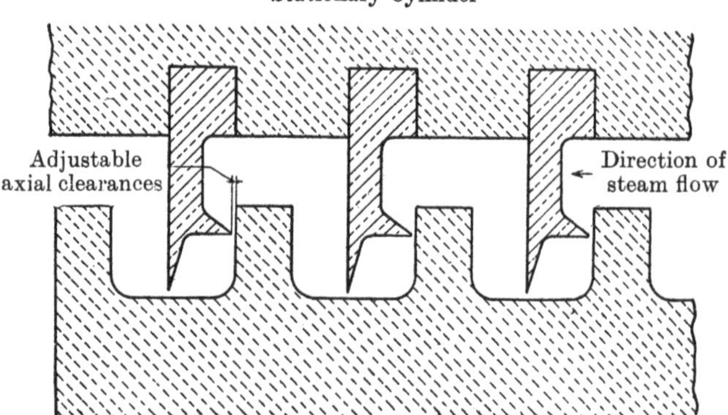

FIG. 77 *a*. Parsons labyrinth packing. Dummy gland rings with axial and radial clearance.

back near the middle, so that it may rock slightly, with the result that the oil which is supplied to it and is drawn in by viscous friction, takes the form of a wedged shaped stream, entirely

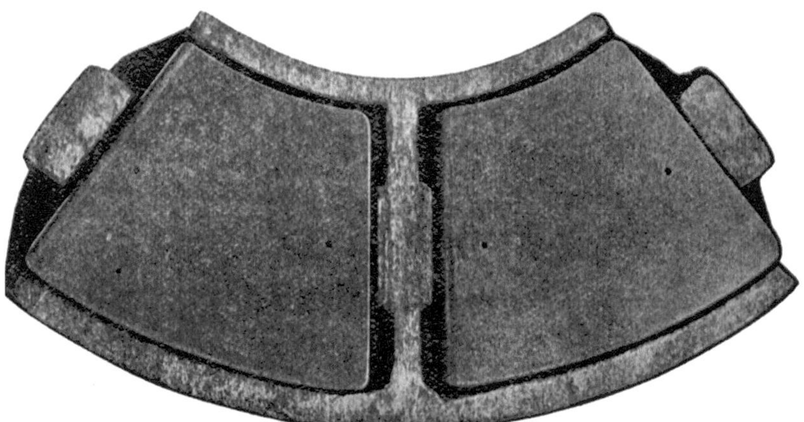

FIG. 78. Pads of Michell thrust block.

separating metal from metal, and tilting the pad so that the thin end of the oil wedge points in the direction of motion. The thrust is transmitted through the film of oil, which is able, on account of its motion, to stand a great pressure[1]. The device is illustrated in fig. 78, which shows two rocking pads and a portion of the containing ring, which is held fast in the fixed casing. The flat collar on the shaft moves past the faces of the pads in a counter-clockwise direction, drawing in a film of oil on the left where the pads have a rounded edge.

153. Lubrication and Governing. A continuous circulation of oil under moderate pressure is maintained for the supply of the bearings and also for actuating the servo-motor or relay through which the speed of the turbine is governed. A worm on the end of the shaft gears with a worm wheel whose spindle serves two purposes: it drives the oil-pump and also some form of centrifugal speed governor. A common form of oil-pump is the "gear-pump" shown in fig. 79, which delivers oil under a moderate pressure by catching it in the spaces between the teeth of geared wheels running with very little clearance in a closed case. The oil is fed to the upper side of the bearings: the film becomes thinner as it approaches the bottom where it sustains the weight which the journal has to carry, according to the principle referred to in § 152. It returns to a reservoir which supplies the pump and is circulated over and over again, passing on its way through a cooler made up of water-cooled pipes. The same circuit supplies oil for the thrust block and for the teeth of the gear-wheels in a geared turbine.

Fig. 79. Oil-pump.

An important difference between turbines and reciprocating engines is that in a turbine there is no mixture of oil with the

[1] This type of thrust bearing, invented by an Australian engineer, Mr A. G. Michell, is a beautiful application of the theory of lubrication enunciated by Osborne Reynolds in 1886 (*Collected Papers*, vol. II, p. 228), and founded on experiments by Beauchamp Tower (*Proc. Inst. Mech. Eng.* 1885). In America the device is known as a Kingsbury thrust bearing.

working steam. Lubrication has to be provided only on surfaces which are out of contact with steam, and it becomes possible to keep the oil and steam completely apart. This has a great advantage in securing a clean condensate, ready for return to the boiler, and in making it easy to maintain a high vacuum, on which the thermodynamic efficiency of the turbine very largely depends.

The usual method of governing is by throttling the steam before admission, and the throttle-valve is connected to the centrifugal

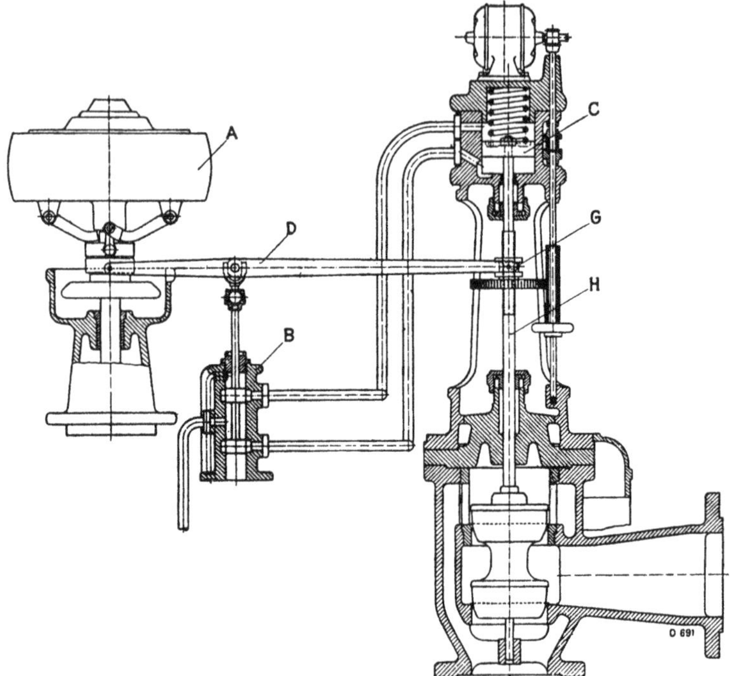

Fig. 80. Governing apparatus.

governor through an oil-worked servo-motor or relay, thereby securing a much greater nicety of regulation than would be possible if the governor had directly to do the work of moving the valve. A typical arrangement for the purpose is shown in fig. 80 (as applied by Messrs Escher Wyss and Co. to their "Zölly" turbines). The centrifugal governor A lifts or lowers the lever D about G as a fulcrum and so causes oil (under pressure), which is supplied to the middle of the control valve B, to be admitted to one or other side of the servo-motor piston C. If the speed falls, it is to the under side of C that oil is admitted. Thus C rises and with it the spindle H of the throttle-valve, admitting more steam to the turbine. The throttle continues to open until G has risen enough to bring D again to a central position. A spring above the piston C is compressed when it rises; this makes the movement slightly stable. A hand wheel on the right-hand side allows the height of G on the spindle H to be adjusted, so as to vary within narrow limits the speed which the governor will maintain.

In many impulse turbines the nozzles which deliver steam to the first blade wheel do not extend all round the circumference, but are grouped in separate arcs which may be under the control of separate admission valves. This is called *partial admission*. It allows the regulation of speed to be mainly effected by opening one after another of the groups, so that the supply may be adapted to the load. In such cases the opening of the groups may be made automatic by causing the speed governor to actuate, through a servo-motor, a shaft provided with cams which successively open one after another of the admission valves when the load increases, and close them one after another when the load falls.

An essential element in the speed control of a turbine is an emergency or "runaway" governor which stops the machine if, through any cause, the speed rises beyond a safe limit. A separate governor, which acts only when the safe limit is passed, with a trip stop-valve, is commonly used for this purpose; and, where the steam capacity is great, a device is added for spoiling the vacuum should the emergency governor come into action. This is necessary, for instance, in turbines where the steam is reheated at an intermediate stage of its expansion, for the reheating pipes add so large a volume of only partially expanded steam that if the load were suddenly removed a dangerous speed might be reached before the effects of stopping the admission could be felt.

Turbines are often called upon to take, temporarily, a considerable overload—as much, say, as one-third more than the load in normal working. They can do so, at some sacrifice of efficiency, by having the boiler steam admitted at a later stage than the first, and for this purpose a by-pass is provided through which steam may enter so as to short-circuit the first two or three or more stages. In normal working the by-pass is closed by a stop-valve, which may be opened by hand, or automatically through the governor relay, when the speed falls as a result of overloading.

In impulse turbines with partial admission the overload is usually provided for by opening additional nozzles.

154. Compound impulse turbines. Some features of compound impulse working will be seen in fig. 81, which is a half section through the cylinder of a turbine for moderate output by the Metropolitan-Vickers Company. There the first stage is a two-row Curtis wheel with partial admission, its nozzles being grouped in three sets, one sufficing for half load, two for full load, and the third coming into use for overload. The first stage brings the pressure down from 175 lb. to about 35 lb., and the remainder

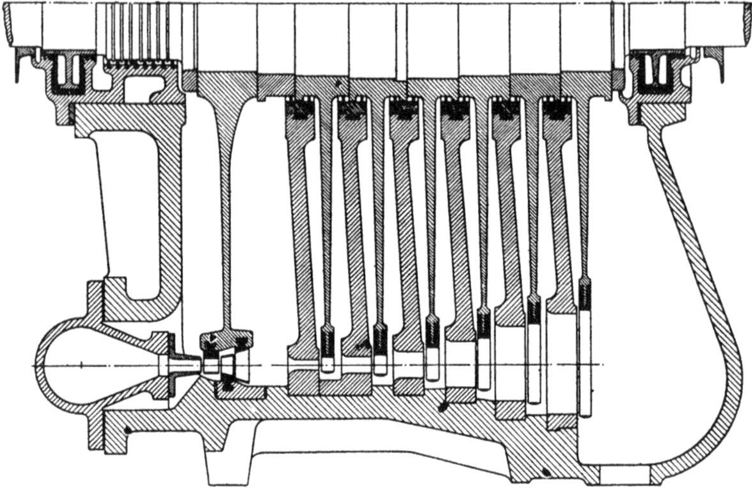

Fig. 81. Compound impulse turbine.

of the expansion to condenser pressure is done in six Rateau stages. An incidental advantage of the Curtis wheel as the initial stage is that although the steam is supplied to the nozzles in a highly superheated state, its temperature has greatly fallen before it comes into contact with the blades and with the main body of the turbine case. This is specially important when the case is of cast iron, for that metal is found to undergo a slow growth when exposed to high temperature steam, and it is consequently necessary to use steel for any parts of a turbine that are exposed to temperatures such as are now usual in superheating. The moving blades are carried by steel discs which are forced on to the shaft. The blades are milled to the required shape out of solid metal, usually steel; in the Curtis wheel they are secured by having T-shaped roots; in the other wheels the base is a fork which fits over and is riveted to the rim of the disc. The diaphragms fit steam-tight in the cylindrical case, and leakage of steam between them and the rotor is restricted by labyrinth rings without actual contact. At the high-pressure end the shaft passes out of the case through a labyrinth gland and beyond that there is, in this example, a "water seal." The water seal, which is also used at the exhaust end to keep air from entering, is shown to a larger scale in fig. 81 *a*. It consists of a wheel or "impeller" which revolves in a deep groove in the casing. Water is supplied to the groove, under a sufficient head, and is thrown by centrifugal force to the

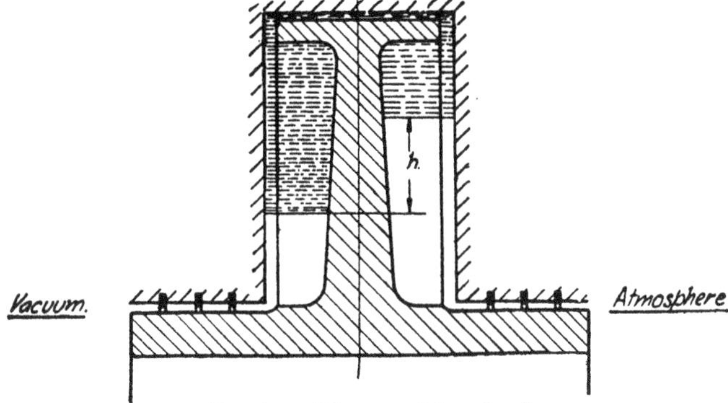

FIG. 81 *a*. Water seal for gland.

circumference. If there were no difference of pressure on the two sides of the impeller, the water would stand at the same height on both: as it is, the difference of height h (fig. 81 *a*) balances the excess of air pressure, and a complete seal is maintained. The water gradually boils off in consequence of the churning it undergoes, and the steam from it passes to the condenser. At the high-pressure end there is a connexion to the condenser between the labyrinth gland and the water-seal device; accordingly both water seals work under the same conditions. After the turbine has acquired a speed sufficient to set the water seals into action, there is no leakage of steam outwards nor of air inwards. Alternative methods of gland packing will be seen later in other examples: most makers prefer to use graphite segments rather than the water seal.

A distinctive feature of Metropolitan-Vickers design, known as *multiple exhaust blading*, is illustrated in fig. 82. Here again the first stage is a Curtis two-row wheel with partial admission from nozzle boxes A and nozzles B. Then follow the ordinary single wheels C, bringing the steam to the diaphragm D from which it passes to the wheel H. There is a division ring E in that diaphragm, between the parts G and F, and a corresponding ring in the wheel, which separates the inner annulus H from the outer annulus I. The blading of F and I is so designed that the part of the steam which goes through them thereby completes its heat-drop and is ready to pass direct to the condenser. The other part of the steam has traversed G and the inner annulus H; the blading there leaves it with some further power of expansion. It divides between L and K; the part that goes through K does work on the outer

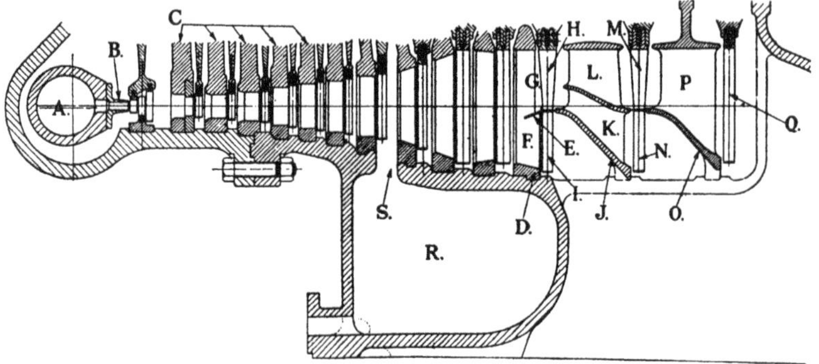

FIG. 82. Impulse turbine with multiple exhaust blading.

annulus of the next wheel, namely N; and the part that goes through L passes through the inner annulus M of that wheel to another set of guide-blades P in which it undergoes a final expansion and acquires a velocity that is taken up by the last wheel of all, namely Q. The effect of this device, invented by Mr K. Baumann, is to secure larger areas for the final stages of the expansion than could be provided in a single-cylinder turbine, at the same speed, without unduly increasing the length of the blades. The whole area available for the final expansion is that of the complete wheel Q, plus the outer annulus N and the outer annulus E of the two preceding wheels.

A feature that will be noticed in the sketch is the chamber R, to which steam has access through S from a middle stage in the expansion. This is a common feature of many large turbines: the object, which was mentioned in § 142 and will be discussed in § 161, below, is to allow the feed-water to be heated by "bleeding," that is to say, by removing steam for the purpose. Steam is often bled in this way at more than one stage.

Many compound impulse turbines are of the pure Rateau type, each wheel having only a single row of blades. From the point of view of stage efficiency Rateau wheels are preferable to Curtis wheels: on the other hand, the practical advantage already mentioned of the Curtis wheel in securing a large drop of pressure and temperature before the steam comes into contact with the blades and casing frequently leads designers to include it as the first element in the series, especially under the conditions of high pressure and high superheat that are now usual. The practice of partial admission has conveniences which make it common, especially when the first element is a Curtis wheel, although it involves loss of energy through the churning of steam by the inactive portions of the wheel.

Fig. 83 is a section of a typical single-cylinder impulse turbine designed by the English Electric Company to develop 15,000 kw. at 1500 revolutions per minute. Steam is supplied at a gauge pressure of 235 lb. per sq. inch and a temperature of 385° C. (725° F.), and is expanded to a vacuum of 29 ins. by a two-row Curtis wheel followed by eleven simple impulse stages. Admission to the Curtis wheel is partial, and overload is dealt with by the automatic opening of supplementary nozzles. The left-hand portion of the cylinder case, which encloses the Curtis wheel and the first three simple stages, is of cast steel, to stand steam of high temperature and pressure, and is bolted to the right-hand portion, which is of cast iron.

The mean diameter of the wheels ranges from $83\frac{1}{2}$ to 98 ins., giving a greatest blade-tip speed of about 770 ft. per second, with a greatest stress at the hub of the wheels of 24,500 and at the root of the blades of 14,500 lb. per sq. inch. The wheels, which are of forged steel, are fixed to the shaft not directly but by being pressed over an intervening ring of other metal at each side (to avoid molecular seizing), in the manner shown in fig. 83*a*, which also illustrates the labyrinth packing of the diaphragm glands. The main shaft glands are also of the labyrinth type and are (as is usual in all such cases) provided with an inner steam passage to which steam may be admitted when necessary, and an outer passage through which a little steam is allowed to blow off, to secure that air will not be drawn into the turbine. The drawing shows that all but the last four wheels are "ventilated," that is to say, they have holes to secure equality of steam pressure on the two sides. In the last four the holes are omitted; this gives them greater strength, and also suits the practice, already mentioned, of letting their blade channels run full with a small amount of reaction effect.

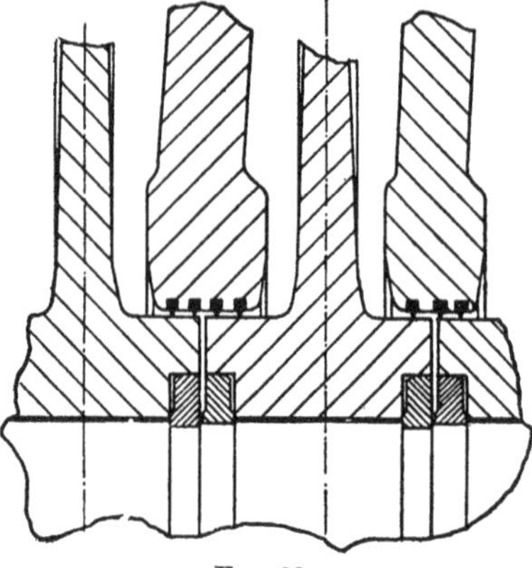

FIG. 83 *a*.
Diaphragm glands and wheel fixing.

In this example, as in other large turbines, the speed of revolution is well below the critical speed for the shaft. Many small compound impulse turbines, however, for use with gearing, run above the critical speed. Thus in a 1000 kw. turbine by the same makers the speed is 6000 revolutions per minute: there is no Curtis wheel, but eight simple stages with a mean diameter of about 27½ ins. and a blade speed of 800 ft. per second. There is complete admission to the whole circumference of the first wheel, and a by-pass to the third wheel deals with overloads. The critical number of revolutions per minute is under 4000. The firm of C. A. Parsons and Co. make high-speed impulse turbines for small powers, an example of which is shown in fig. 84. This turbine, which is designed for a normal load of 250 kw. at a speed of 9000 revolutions per minute (a speed much above the critical), comprises a Curtis wheel, with partial admission, and five simple stages, all carried by discs which are turned solid on the steel forging which forms the shaft. The diaphragm glands are of the labyrinth type, but each of the shaft glands has four packing rings of graphite which are made to press lightly against the revolving surface by means of garter springs.

A single-cylinder turbine by the British Thomson-Houston Company for the Rotherham power station, with a continuous output of 30,000 kw. has a two-row Curtis wheel followed by 13 simple stages. The diameter of the largest wheel is 12 ft. to the tips, and the speed is 1500 revolutions per minute.

Single-cylinder impulse turbines have been built by the General

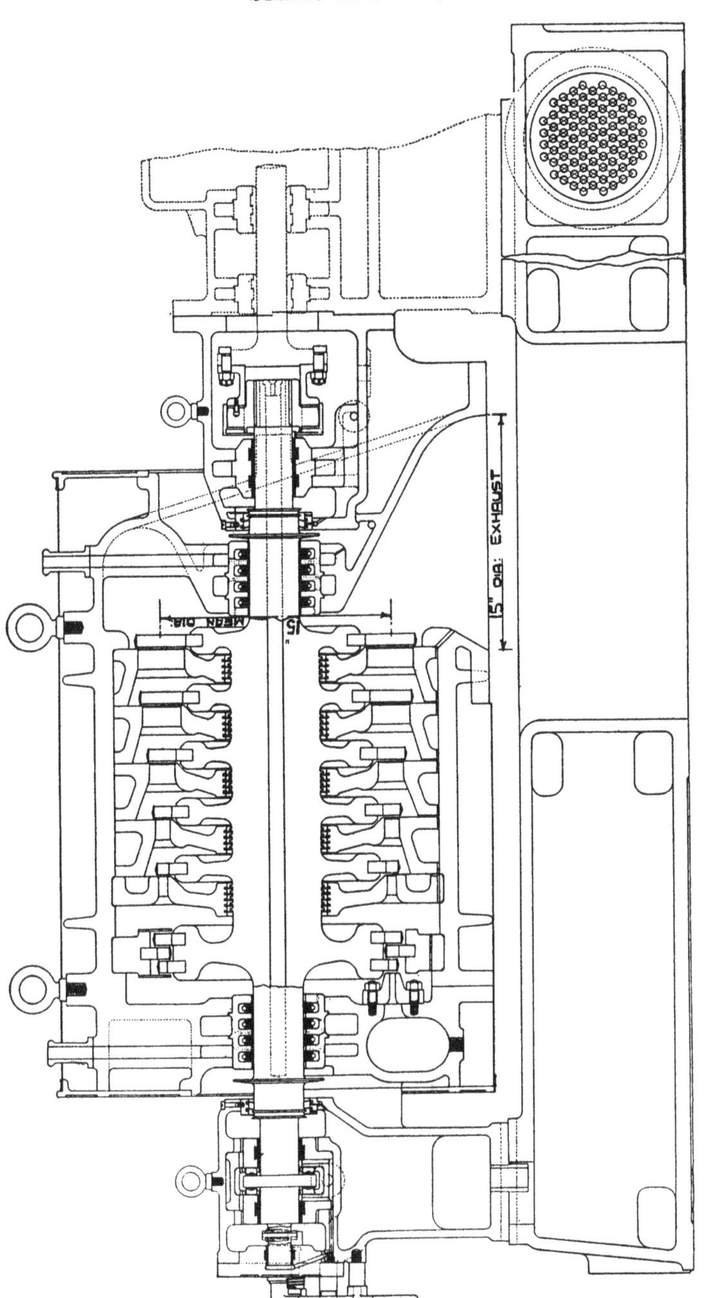

Fig. 84. High-speed impulse turbine for small output (Parsons).

Electric Company of Schenectady for outputs of as much as 60,000 kw. One of these machines, designed for steam at an initial pressure of 350 lb. per sq. inch by gauge, superheated to 700° F., is shown in fig. 85. It runs at 1500 revolutions per minute and has 20 simple impulse stages, expanding to a vacuum of 29 inches. There is complete admission to the first row of blades, with a by-pass to the ninth row for overload. Steam is bled from three points for heating the feed-water. Although the rated capacity of this turbine is 60,000 kw. the most economical load is approximately 40,000 kw.

Another single-cylinder design by the General Electric Company is shown in fig. 86. Here the steam has a much higher initial pressure, namely 600 lb. per sq. inch by gauge, and is superheated to 725° F. The machine runs at 1800 revolutions per minute (for 60 cycles per second) and has 19 stages, expanding as before to a vacuum of 29 inches. Its special feature, which is unusual in a single-cylinder turbine, is that the steam is reheated at an intermediate stage of the expansion. The figure shows the provision which is made in the turbine casing for withdrawing the steam after the seventh stage, when the pressure has fallen to about 167 lb. absolute: it is then taken to the boiler room and re-superheated, again to 725° F. It returns at a slightly reduced pressure, re-entering the turbine at the eighth stage. Provision is also made for bleeding at three stages, and in this way the feed-water is heated to about 250° F. Several of these machines were installed during 1925, with a rated output for each unit of 40,000 kw.[1]

In machines which approach this power, however, it is more usual to divide the heat-drop between two, or in some case three, cylinders, the low-pressure cylinder often having central admission with exhaust at both ends to provide large passage ways in the latest stages. The cylinders may be arranged in tandem on the same shaft, but it is more common in turbines of great power to make them "cross-compound," each with its own shaft and its own alternator, synchronous running being secured by connecting the alternators electrically. With steam pressures up to 500 lb. and

[1] For a general account of American practice in turbines of large output, up to the year 1924, see a paper by Mr F. Hodgkinson, *Trans. of the First World Power Conference*, vol. II, p. 1516. Several of the turbines referred to in the text are of later date.

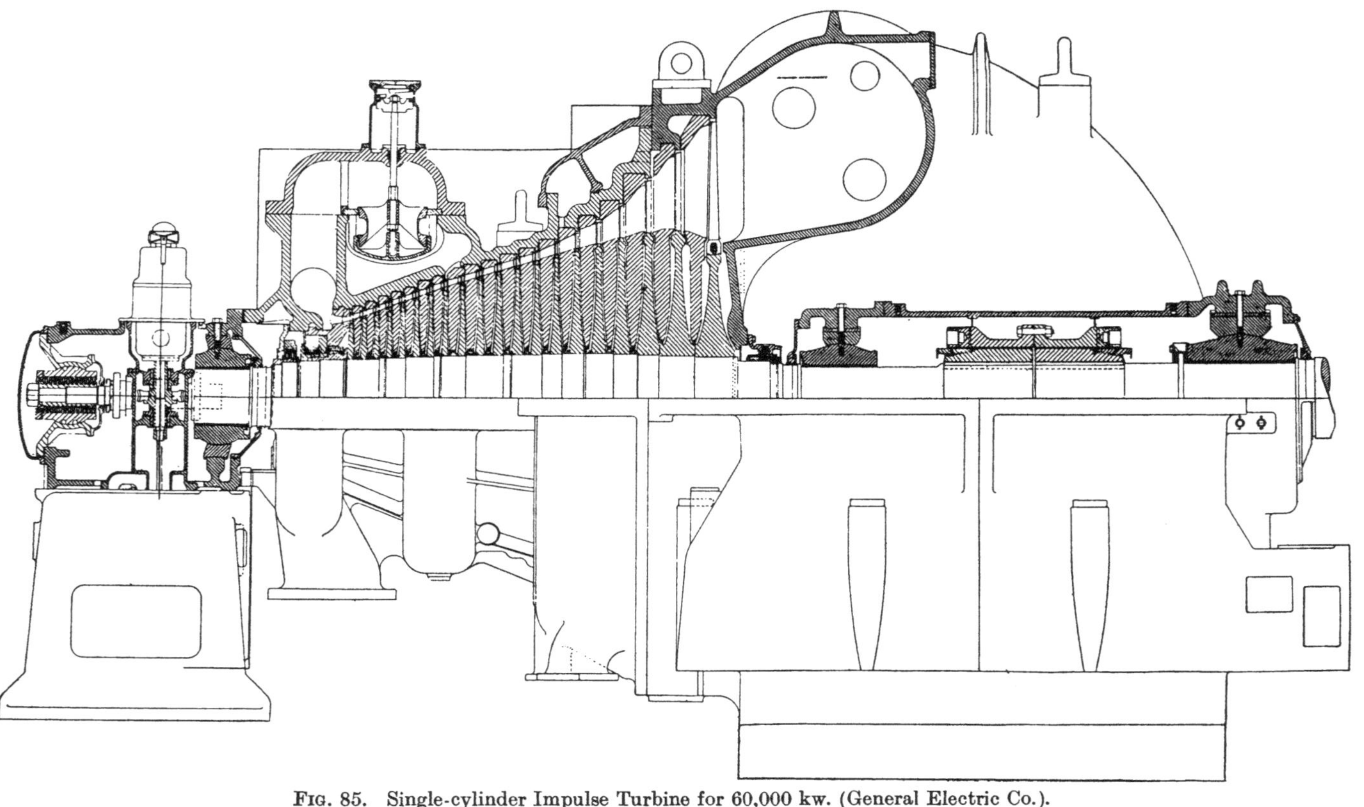

FIG. 85. Single-cylinder Impulse Turbine for 60,000 kw. (General Electric Co.).

over, which are now common, a division of the action between two or even three cylinders has advantages which probably outweigh such increase as it may cause in the cost of construction of a large machine. A similar division, as we shall see later, is usual in large reaction turbines; it was introduced by Parsons for land turbines in 1899, and he has made it a feature of marine practice ever since the first application of the turbine to the propulsion of ships.

Among large turbines of the impulse class built by the General Electric Company, the tandem arrangement is exemplified in several machines placed in service in 1925 and 1926 to develop 50,000 kw., with steam supplied at 375 lb. per sq. inch by gauge and superheated to 675° F. They run at 1800 revolutions per minute and have 20 stages in all, 18 of which are in the high-pressure part. The low-pressure part consists of two stages only, but these are arranged with double flow from the middle towards both ends. The casing of the turbine, which is continuous for the two parts, provides a very large connecting channel between them. Steam is bled at four points for feed-heating. The cross-compound design, which is more usual in turbines of great power, is used by the same makers in their 60,000 kw. and 77,000 kw. machines built for the Commonwealth Edison Company of Chicago. In the 60,000 kw. turbine the high-pressure portion comprises 10 stages and runs at 1800 revolutions per minute, taking steam with a stop-valve pressure of 550 lb. per sq. inch by gauge, and a temperature of 725° F., and expanding it to about 120 lb. absolute. At this pressure the steam is returned to re-superheating boilers and has its temperature again raised to 725° F. before entering the low-pressure portion, where there are 14 stages and the speed is 1200 revolutions per minute. The exhaust pressure corresponds to a 29 inch vacuum. Feed-heating is effected by bleeding at three points. This turbine, which is designed to have maximum efficiency at its full load, was put into service in 1924 at the Crawford Avenue Power Station, where, as will be seen in the next section, a large Parsons turbine of the reaction type has also been installed. The 77,000 kw. turbine, the most economical load of which is 65,000 kw., has 14 stages in the high-pressure portion, where the rotor speed is 1800 revolutions per minute, and 10 stages in the low-pressure portion, where the speed is 1200 revolutions

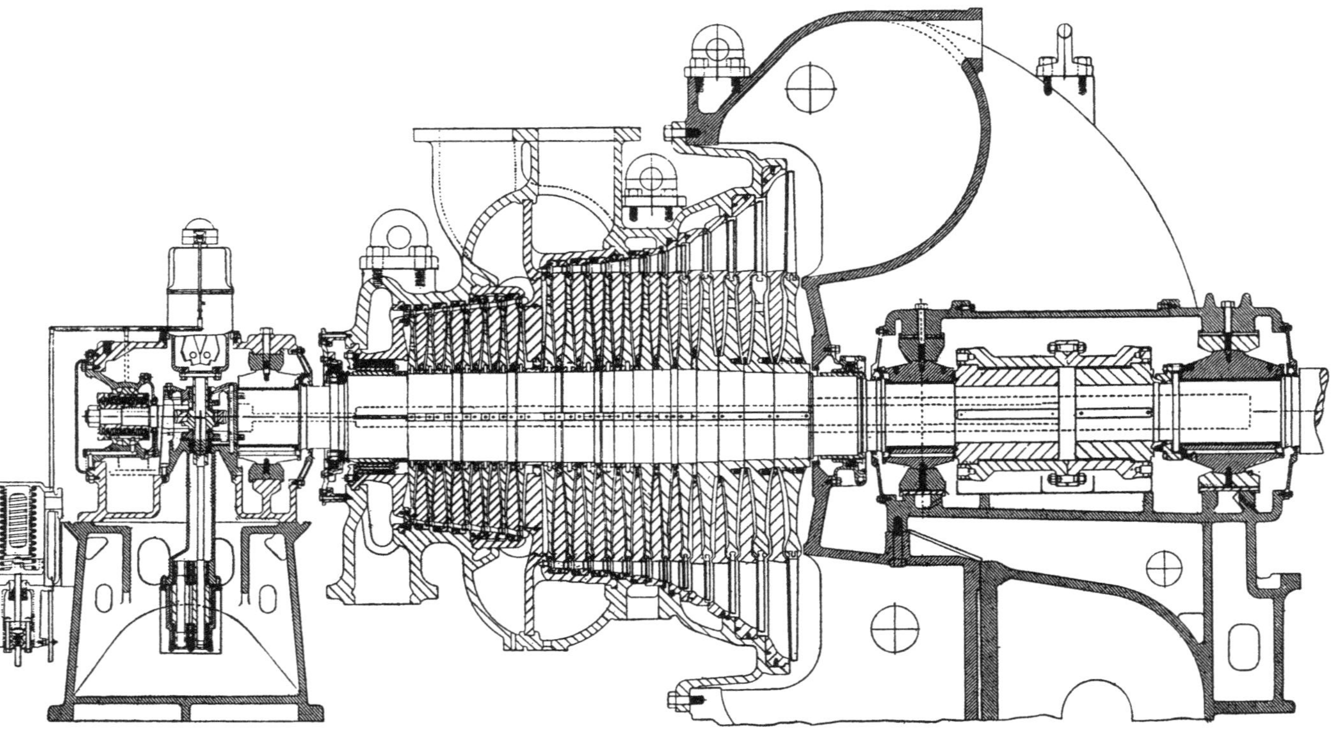

Fig. 86. Single-cylinder Impulse Turbine with provision for Reheating (General Electric Co.).

per minute. There is no re-superheating, but by bleeding steam at four points the feed-water has its temperature raised to about 340° F.

An interesting new departure by the same makers is a high-pressure turbine which was put into operation in 1925 by the Edison Electric Illuminating Company of Boston. It takes steam from a special boiler at an initial pressure of 1200 lb. per sq. inch. The rotor runs at 3600 revolutions per minute, giving 60 periods per second by direct coupling to a two-pole alternator, and comprises 20 simple impulse stages. These expand the steam down to a pressure of 365 lb., at which it passes, through a re-heater, into the general steam-supply from other boilers for the main portion of the station plant. The machine develops about 3000 kw.: its dimensions are small, the rotor being only about a foot in diameter and two feet long in the bladed part. The blade wheels are machined integral with the shaft. This experiment in the use of steam with an ultra-high pressure is important, but it may be added that the chief problems in such an experiment relate rather to boiler design: the turbine side of the question presents comparatively little difficulty.

155. Parsons reaction turbines. The single-cylinder reaction turbine shown in fig. 87 and the tandem form in fig. 88 exhibit features to which reference has already been made and are representative of Parsons' 1925 practice. Both run at 3000 revolutions per minute and carry the blades on solid forged steel rotors: the last six rings, however, of fig. 87 are on disc wheels which are shrunk on to a turned-down portion of the shaft, and are in contact with one another at the circumference as well as the hub, forming a very rigid structure. Each shaft is bored through from end to end to test the soundness of the metal. To the left in fig. 87, and in the high-pressure part of fig. 88, three dummies will be seen, corresponding in diameter (more or less) to groups in the blading. There is labyrinth packing between them and the case, with pressure-equalizing steam passages from corresponding places in the turbine. In the turbine of fig. 87 there are in all forty-one stages, beginning with three "expansions" of six each, then seventeen stages on a conical portion of the rotor with progressively increasing blade length, and finally six on the disc wheels, also

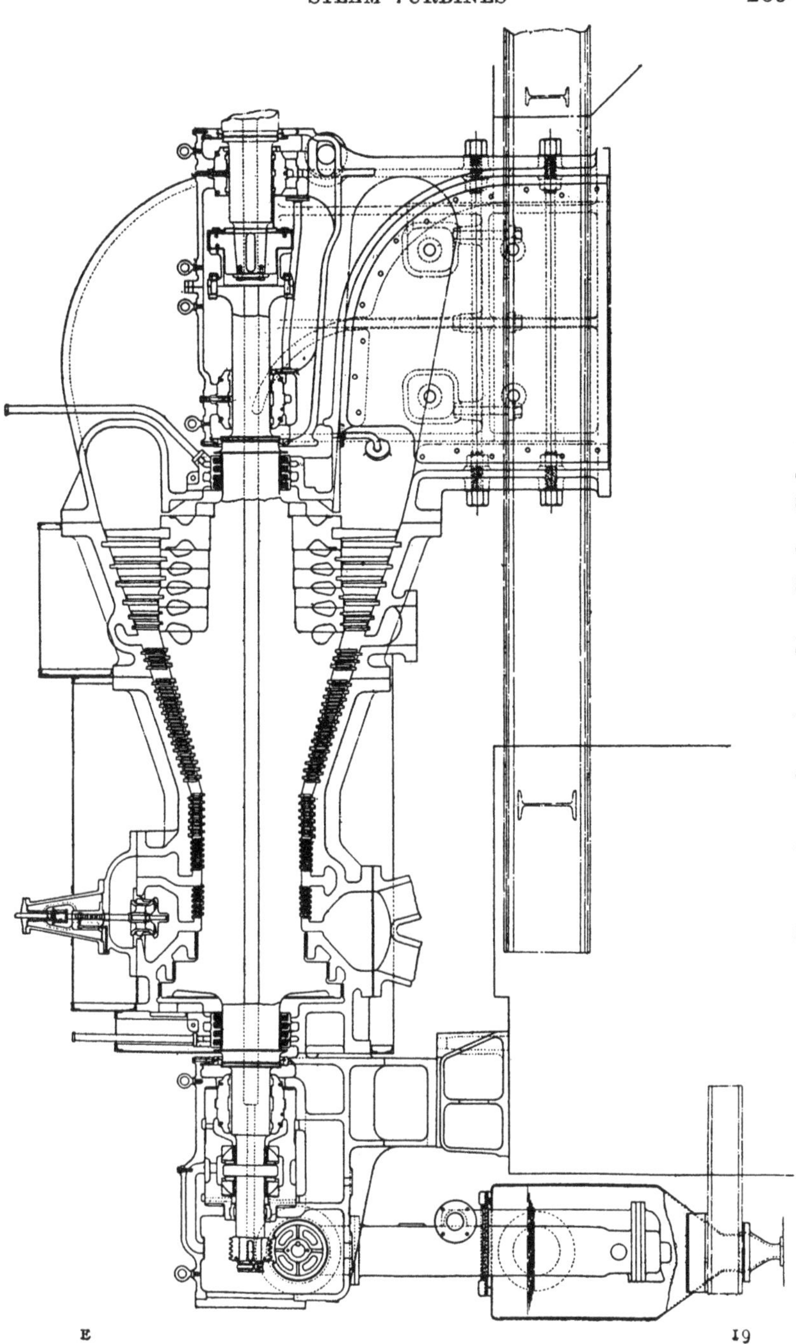

FIG. 87. Parsons' reaction turbine. Single-cylinder type.

increasing in blade length from stage to stage. All the blading except in the last six stages is of the end-tightened type. Steam (at a stop-valve pressure of 200 lb. by gauge and temperature of 528° F.) enters at the bottom through a multiple steam-pipe, the multiple form being adopted because its much smaller rigidity reduces straining through change of temperature. It expands to a vacuum of 29 ins. The maximum output in continuous running is 8000 kw. and the economical output, for most efficient running, is 6000 kw. The valve on top is a by-pass which admits live steam for overload to a steam belt after the first six stages. Up to the thirty-second stage the casing is of steel. After the thirty-fifth there is a provision for "bleeding" steam for the purpose of feed heating. The glands between the casing and the shaft are each packed by four segmental rings of graphite. To the left of the journal at the high-pressure end is seen the Michell thrust collar, and on the extreme left is the worm gear which drives the oil-pump and governor. To the right of the other journal is a flexible "claw" coupling between the shaft of the turbine and that of the alternator or other driven mechanism. The condenser, not shown in the drawing, is attached to the exhaust-pipe below the floor, and the free opening of the exhaust-pipe to it measures 7 ft. by 5 ft.

The greatest mean blade speed in this turbine is 677 ft. per second, and the value of K is 343. The heat-drop, expressed in British Thermal Units, is 408·4. If we were to take that quantity, without correction for leaving loss and reheat factor, in the expression

$$\frac{u}{v} = 0{\cdot}87\sqrt{\frac{K}{H}},$$

we should find u/v to be 0·80.

The tandem turbine of fig. 88 develops a maximum continuous output of 20,000 kw. and a normal or economical output of 15,000 kw. It has a double-ended low-pressure cylinder, to the middle of which steam is led after expanding through thirty-six stages in the high-pressure cylinder, to a pressure which in normal working is about 1 lb. above that of the atmosphere. Either of the alternative paths in the low-pressure cylinder provides eight more stages, making forty-four in all. The whole of the high-pressure cylinder is of steel, and all the blading in it is end-tightened: in the low-pressure cylinder the blades have sharpened tips with radial clearance. Each rotor is a solid steel forging, lightened in the low-pressure cylinder by having five deep grooves turned out. The highest mean blade speed is again 677 ft. per second. The by-pass valve admits steam for overload after the first "expansion" of ten stages. Each of the rotors has its own Michell thrust collar, and there is a flexible coupling between the two shafts, as well as one on the shaft of the alternator. The condenser is in two parts, with a short and wide exhaust-pipe leading to each. The glands are packed by graphite rings.

This turbine is designed for a steam-supply at a stop-valve pressure of 350 lb. and a temperature of 700° F. Here the coefficient K is 401, to suit the greater heat-drop of about 488 B.T.U.; on the same basis as before this makes $u/v = 0{\cdot}79$.

As a further example of a Parsons reaction design, reference may be made to a turbine of 50,000 kw. normal capacity supplied by him in 1925 to the Crawford Avenue Power Station, Chicago. This plant has three cylinders through which the steam passes in series, each with its own alternator. Steam is generated at 600 lb. per sq. inch and reaches the stop-valve at 550 lb. and a temperature of 400° C. (750° F.). It expands through thirty-six stages in the high-pressure turbine to a gauge pressure of 100 lb.; it is then passed through a re-heater in the boiler house and returns for delivery to the intermediate-pressure turbine at 700° F., where it expands through twenty-three stages to an absolute pressure of about 2 lb., and goes on to the low-pressure cylinder for expansion through five more stages, making sixty-four in all, down to a vacuum of 29¼ ins. The blade lengths range from 2½ ins. in the first ring to 40 ins. in the last, where they have a section already shown in fig. 76 (§ 150). The H.P. and I.P. shafts run at 1800 revolutions per minute and drive four-pole alternators of 16,000 and 29,000 kw. respectively; the L.P. shaft runs at 720 revolutions per minute and drives a ten-pole alternator of 6000 kw., making a total output of 51,000 kw. The highest blade-tip velocity is 760 ft. per second. The L.P. cylinder is set coaxially with the I.P., and the steam passes from one to the other through a conical connecting passage, with the least possible disturbance of the flow, so that its carry-over energy may be conserved. The arrangement of this passage in relation to the I.P. and L.P. turbines and the intermediate bearings will be clear from fig. 89. The alternator of the I.P. turbine is to the left of the drawing: that of the L.P. turbine is to the right. From the L.P. cylinder the steam is discharged very directly into a pair of large surface-condensers with vertical tubes, one on each side of the shaft, with a total condensing surface of 56,000 sq. feet. Steam is bled from the turbines at three places for heating the condensate on its way back to the boiler, namely at the exhaust of the high-pressure turbine and at points where the (absolute) pressure is 20 lb. and 5 lb. The quantities of steam drawn off per hour for this purpose at the three points

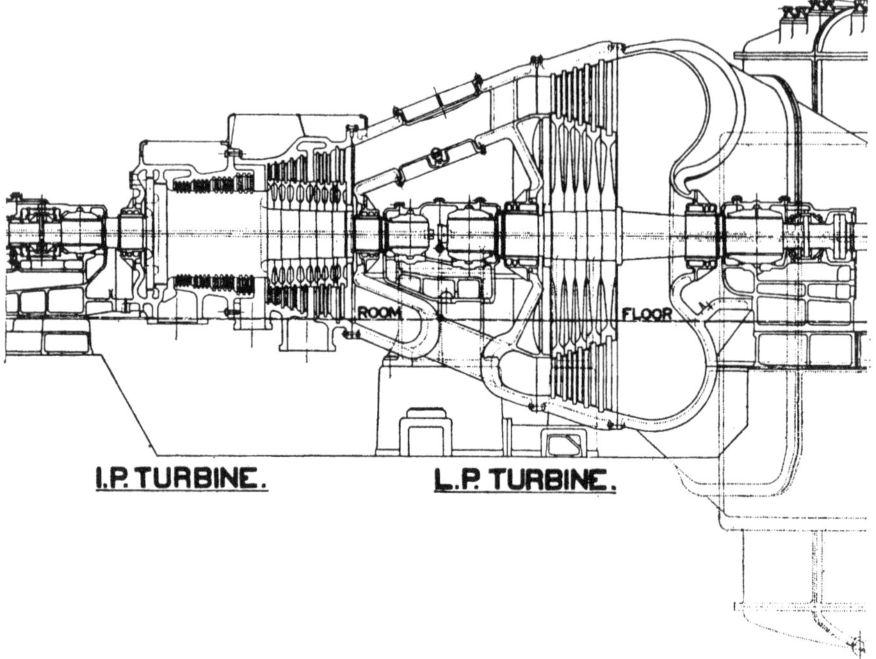

Fig. 89. Parsons' reaction turbine (Crawford Avenue). Intermediate and low-pressure cylinders.

are about 45,500 lb., 22,000 lb., and 25,000 lb. respectively, out of a total admitted of about 420,000 lb. per hour. Steam from the drains and high-pressure glands is also utilized to give some preliminary heating. The feed-water is thus heated from about 18° C. to 157° C. It is heated further by economizers, and the flue gases also preheat the air supplied to the furnaces. With this plant the full normal output of electrical energy should represent about 33 per cent. of the heat supplied to the steam and 27·8 per cent. of the energy of the fuel. Parsons estimates that by raising the pressure of supply to 1500 lb. the overall efficiency, from fuel to electricity, would be increased to 30·2 per cent.[1]

The value of K for this turbine as a whole is 487·2, divided as follows:—

High-pressure turbine	134·7
Intermediate-pressure	286·5
Low-pressure ,,	66

FIG. 90. Brown Boveri reaction turbine with Curtis wheel.

156. Combined Impulse and Reaction Turbines. The firm of Brown Boveri and Co., of Baden, who undertook to manufacture Parsons' turbine on the continent of Europe, modified his design by substituting an impulse wheel, generally of the two-row Curtis type, for several of the earliest high-pressure stages. This was before the introduction of end-tightened blading, when the loss

[1] See his paper on Steam Turbines, *Trans. of the First World Power Conference*, 1924, vol. II, p. 1477, which includes an important general discussion of the conditions that determine efficiency. Further particulars of this large turbine set, with many illustrations, will be found in *Engineering* of March 5, 1926.

through blade-tip leakage was relatively large in the short high-pressure blades: the change had the advantage of avoiding part of that loss, and also of shortening the rotor by substantially reducing the number of stages. The Curtis wheel was included in the same casing with the drum which carried the reaction rows, but had a larger diameter, sometimes much larger, and consequently a higher blade velocity, which enabled it to deal with a considerable fraction of the whole heat-drop. This combination of a Curtis wheel for the first stage with reaction blading for the remainder of the turbine has been and still is largely used. It has been adopted by Parsons himself for turbines of small output, that is, of 1000 kw. and under. An example by Brown Boveri is shown in the photograph, fig. 90, where the Curtis wheel and the dummy are seen on the right.

Later Messrs Brown Boveri have developed other combined types in which a series of impulse wheels is followed by a series of reaction wheels in the same cylinder. An example is given in fig. 91, where it will be seen that there are seven impulse wheels of the simple Rateau type, followed by seven reaction stages. The reaction blades are carried by separate steel wheels instead of a drum, and the fixed blades in the reaction portion have shrouded ends. The impulse and reaction stages form together a continuous annular channel enlarging without any sudden break. To the left of them will be seen a small dummy which balances the end thrust. The steam which leaks through the labyrinth packing of the dummy is conveyed, by an external pipe, to a point at which the reaction blading begins, and does work on that. This turbine has an output of 30,000 to 40,000 kw., and runs at 1500 revolutions per minute. It is designed for an ordinary working pressure of 300 lb. or less.

To adapt their turbines to much higher pressures, such as 80 to 100 atmospheres, the same firm supplement the design by adding what they call a "primary" turbine, which is a small impulse turbine adapted to take the first part of the expansion, running at a much higher speed, and connected by gearing to the shaft of the main turbine. Usually the primary turbine is in two parts through which the steam passes in series, one part on each of two small shafts carrying pinions which gear with a wheel on the main shaft. A characteristic of each part of the primary turbine is that it is overhung on its pinion shaft: the turbine case encloses the end of the shaft beyond the bearing, and the shaft is supported

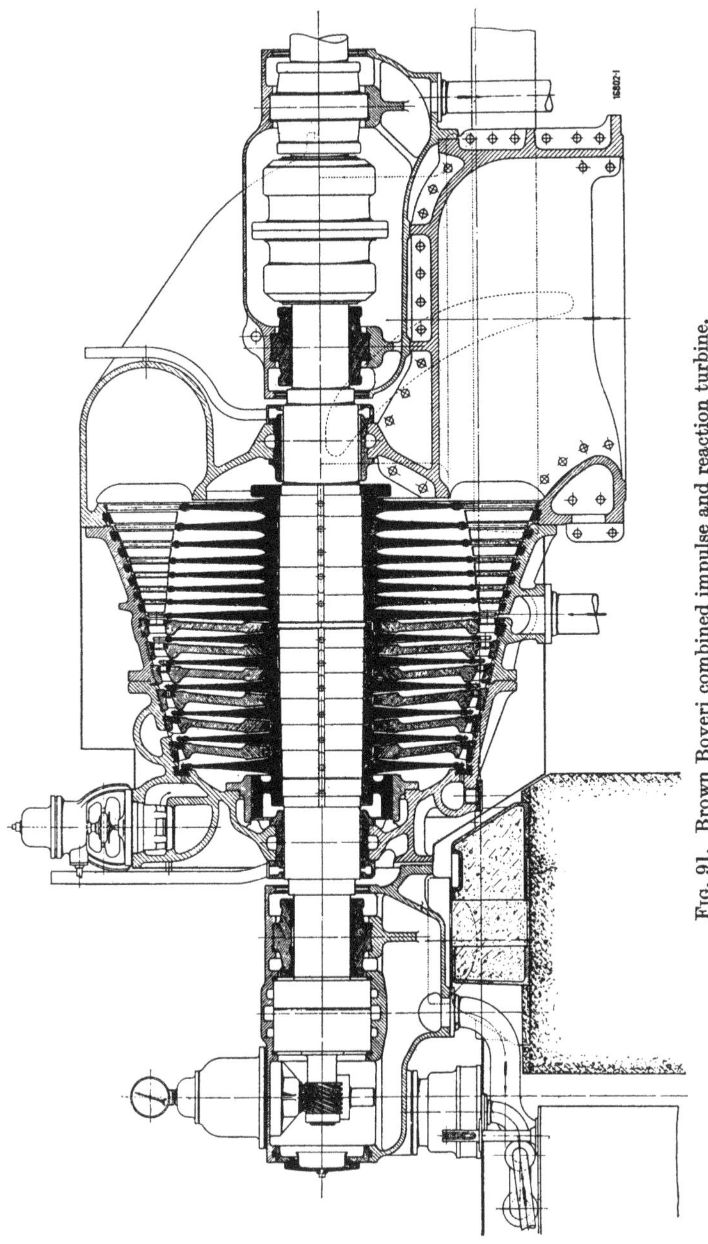

FIG. 91. Brown Boveri combined impulse and reaction turbine.

on one side only. It may run at about 8000 revolutions per minute and carry two small impulse wheels, so that the whole of the primary process involves four stages. By this supplement to the machine the steam is expanded down to a pressure suitable for its admission to the main turbine. In some of the firm's designs ultra-high-pressure turbines of this class are overhung on the pinion shafts of gearing which drives an electrical generator, and works in series with a turbine of more normal type driving another generator in electrical connexion with the first.

Fig. 92 shows a two-cylinder tandem turbine by the English Electric Company resembling the turbine of fig. 88, but with reaction blading in the low-pressure cylinder only. The blading of the other cylinder consists of twelve impulse wheels of the Rateau type in which the steam expands to nearly atmospheric pressure. It then passes to the middle of the low-pressure cylinder which is double-ended with nine reaction stages between admission and exhaust. Under normal conditions of load, with admission at 200 lb. by gauge and 310° C., about 57 per cent. of the whole work is developed in the low-pressure cylinder; the steam passes on to it at a gauge pressure of about 1 lb. and with still some 10° C. of superheat. It may of course be reheated between the cylinders. Each rotor consists of a steel forging, the discs which carry the impulse blading being turned out of the solid, an arrangement which reduces the centrifugal stress and shortens the rotor by narrowing the axial distance from stage to stage. The machine shown is for an output of 20,000 kw. at 3000 revolutions per minute. The blade diameters are so moderate that the centrifugal stress at this speed nowhere exceeds 14,000 lb. per sq. inch. The shafts run below their critical speed. There is admission round the whole circumference of the first ring, and overload is dealt with by an automatic by-pass to a belt between the fourth and fifth stages. The high-pressure casing is of steel up to the eighth stage, after that of cast iron. Except that there is a labyrinth gland at the high-pressure end, the arrangement of glands and thrust collars is the same as in fig. 88.

This design is based on the consideration that while reaction blading is intrinsically more efficient, impulse working offers advantages for steam whose pressure and temperature are high which are held to justify its being preferred in the first cylinder; further, a separate double-ended cylinder for the reaction portion not only gives a balance of longitudinal forces but also makes it easy, without excessive blade speed, to provide enough passage way for expansion to the highest attainable vacuum.

157. Ljungström Turbine. One more type of turbine remains to be described, the radial-flow reaction turbine of the Ljungström brothers, in which alternate rows of blades are attached to discs revolving in opposite directions, with the advantage that the relative speed is twice that of either row. In the design of this turbine there are many ingenious features which contribute to give it an exceptionally high thermodynamic efficiency, at the expense, however, of making its construction somewhat complex. It is not built in very large sizes. A Ljungström turbine by the Brush Company, with an output of 5000 kw., supplied with steam at 166 lb. and 662° F., has been found on trial to consume only 10·36 lb. per kw. hour, a result not surpassed except in the very best performance of much larger turbines of other types.

The distinctive features will be seen by reference to fig. 93, which is a half section showing the blade system of a 1500 kw. Brush Ljungström turbine. The turbine stands between two alternators, electrically coupled, and each of the two turbine rotors is overhung from the shaft of the alternator which it drives. The rotors h and h_1, which may be described as discs, revolve in opposite directions, each at 3000 revolutions per minute. The bottom of the figure is on the axis of rotation. The turbine spindles p and p_1, which respectively carry h and h_1, enter the casing through labyrinth glands ww, whose construction is shown to a larger scale in fig. 94. Attached to the case are fixed discs f and f_1, and between them and discs which revolve with the turbine is another labyrinth packing, shown in detail in fig. 94 a. These labyrinth discs oppose leakage of steam to the condenser; and they also serve as dummies to balance the longitudinal thrust with which the steam tends to force the turbine discs apart. Steam enters the system at n on each side from the steam chest e, and expands radially through the blading which is carried, in alternate rows, by the discs h and h_1. At each stage the blades are held in a stiff blade ring which is supported from h (or from h_1) in the manner shown in fig. 95, namely, by an expansion ring of dumb-bell section which prevents any temperature straining, and lateral leakage of steam is restricted by a sharp-edged packing strip on either side which is nearly in

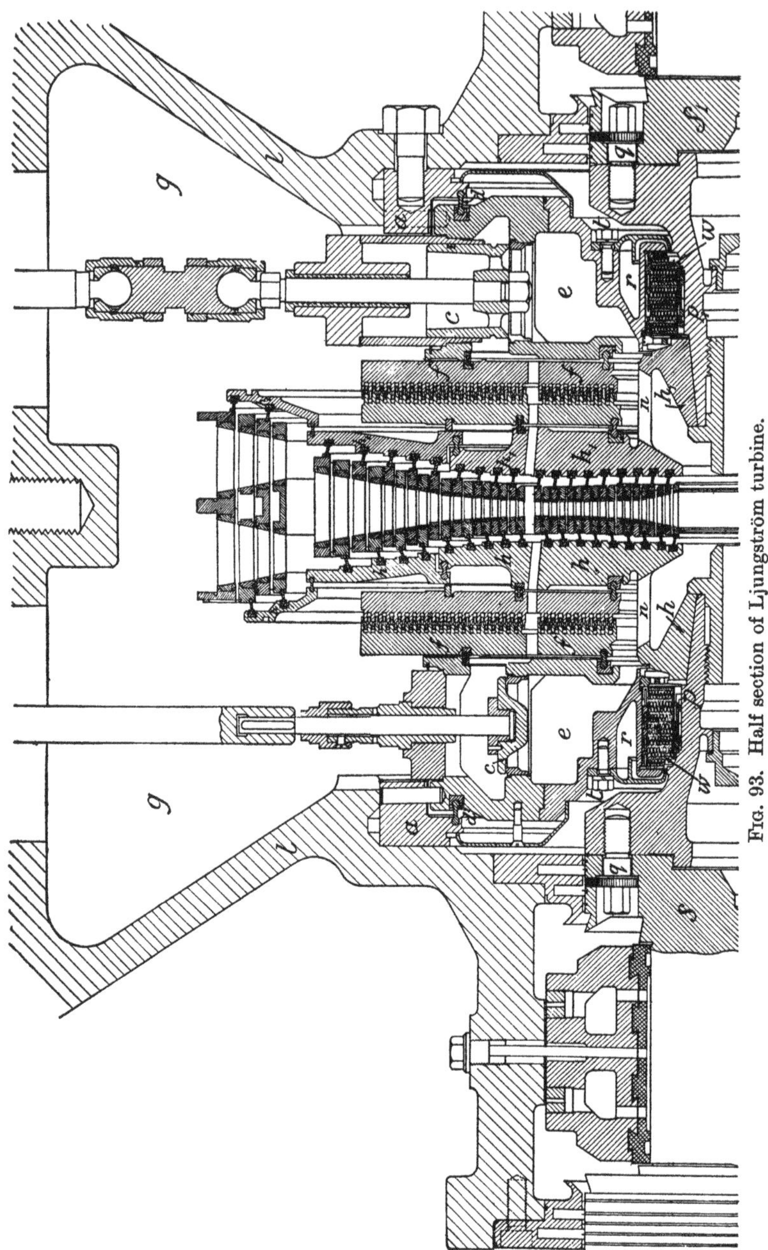

FIG. 93. Half section of Ljungström turbine.

contact with the under side of the next blade ring. Starting from the axis and going out radially it will be observed that the blades at first become shorter; this is because the enlargement of the circumference is more than sufficient to provide the necessary increase of passage way for the steam. In later stages the increased volume of steam requires the blades to be made longer, and near the circumference they are so long as to need division into

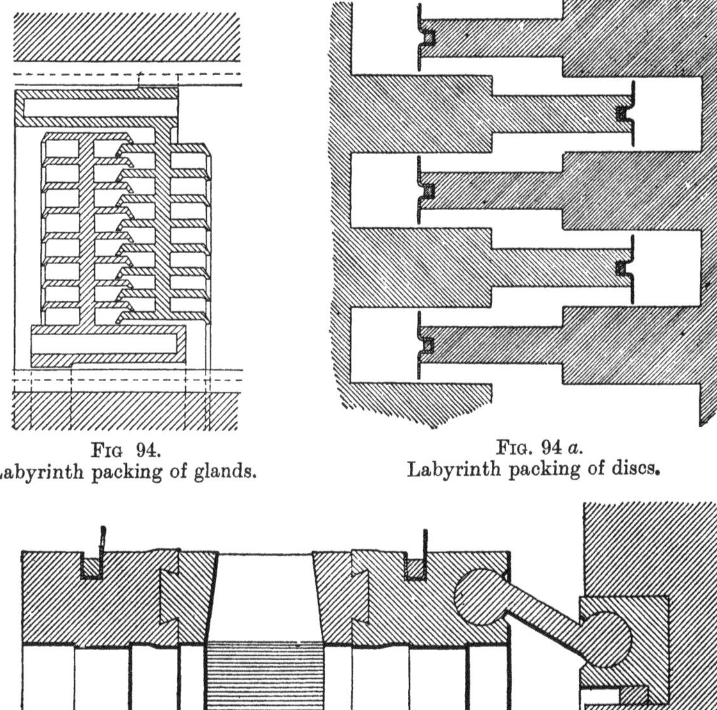

Fig 94.
Labyrinth packing of glands.

Fig. 94 *a*.
Labyrinth packing of discs.

Fig. 95. Blade rings and expansion rings.

parts with a stiffening rib between. In the turbine of 5000 kw. the final stage of expansion is performed through a set of radially placed blades on either side, in order to give it a sufficiently large passage way. From the exhaust *g* it passes to the condenser. Overloads are dealt with by the by-pass valves *cc*, one of which is automatic and the other operated by hand. There are joint rings with dumb-bell section between the blade discs and the moving element of the labyrinth packing at *f*. The shafts revolve much below their critical speed. The efficient working of this turbine is to be ascribed in part to the advantage of high relative velocity and in part to excellence of the labyrinth packing, which can be maintained in fine adjustment on account of the comparative absence of temperature strains.

158. Applications of the Steam Turbine. Marine use. Gearing. From the first, a chief use of the steam turbine has been to drive dynamo-electric machinery. This has been developed on an immense scale with the growth of central stations for distributing electric power. In their early days such stations were generally equipped with reciprocating engines, but the turbine soon showed itself to be more economical of fuel as well as better adapted for driving machines in which a high speed of rotation was required, with the result that it has completely displaced, for this purpose, steam-engines of other types. In most large-scale examples it is directly coupled to alternators, which are successfully designed for efficient working at the high speeds the turbine demands. A usual frequency for the alternating currents sent out from a power station is 50 periods per second: a turbine running at 3000 revolutions per minute will generate these when directly coupled to a two-pole alternator, or one running at 1500 revolutions when directly coupled to a four-pole alternator[1]. Such are very usual conditions in the working of large machines; the aggregate electrical horse-power which is thus devèloped in central stations amounts to many millions. Direct-coupled turbo-alternators of 20,000 kw. are in common use, and some units have an output of 50,000 kw. or even more. Small turbines often run at speeds much exceeding 3000 revolutions: when these are employed to drive alternators it is necessary to reduce the speed by gearing. In the driving of continuous current dynamos gearing is generally resorted to, for in that way a better efficiency is obtained in both turbine and dynamo than is practicable with direct coupling. The loss of power in transmission through gear wheels with well cut and properly lubricated teeth is barely two per cent.; this is more

[1] In American power stations the frequency of alternating currents is often 60 periods per second and accordingly many direct-coupled turbines there are designed to run at 1800 revolutions per minute with a four-pole alternator, or at 1200 with a six-pole alternator.

than made up for by the advantage that comes of being able to select independent speeds which are best suited for the performance of the turbine and the driven machine respectively.

A minor application of turbines in which direct coupling continues to be used is to drive centrifugal blowers and air compressors for the delivery of air in large volume under moderate pressure.

The successful introduction of gearing for the transmission of large amounts of power has greatly widened the field in which the turbine is applied, making it serve as a convenient and economical prime mover for rolling mills, paper mills, textile factories and so forth. But the most conspicuous effect of gearing has been on ship propulsion. In all modern designs of marine turbine a single or double set of gear wheels is interposed between the turbine shaft and the propeller shaft. The speed of rotation natural to a steam turbine is so much higher than that natural to a screw propeller, if both are to work under favourable conditions, that to connect them through a reducing gear adds much to the efficiency of the combination even in the fastest ships. It has also brought the advantage of the turbine within reach of ships of any speed, however low.

The change, however, from direct coupling to the use of gears did not take place until after the turbine had established itself as a highly efficient means of propelling ocean liners, warships, and other vessels of high speed. By his experimental yacht "Turbinia," tested in 1897, Parsons demonstrated the practicability of applying directly coupled turbines to drive a ship at what was then an unprecedented speed[1]. The "Turbinia" had three shafts driven respectively by a high, intermediate and low-pressure turbine, and each carried a small high-speed screw propeller. Parsons showed that the phenomenon which he called cavitation (namely the formation of vacuous spaces behind the propeller blades) imposed a limit of speed on the directly coupled shaft. But this limit did not prevent direct-coupled turbines from being successfully applied to the propulsion of fast vessels of all sizes. Such turbines were characterized by low-shaft speed, large drums to carry the blades, and a great number of stages, usually with reaction blades throughout. They came quickly into use for cross-channel and other passenger vessels: they were adopted in all new ships of the British Navy

[1] See Parsons, *Trans. Inst. Nav. Arch.* 1903, p. 284. An appendix to that paper (p. 293) gives a report of trials made in 1897 by the present writer.

and were selected for the "Lusitania" and "Mauretania," ships whose exceptional size and speed—requiring about 70,000 H.P.—led (in 1904) to a special inquiry as to best means of propulsion. It was then recognized that the turbine offered possibilities for the concentration of power in a single vessel to an extent not previously attempted. By 1918, when direct coupling became obsolete in the designs of marine turbines, some fifteen millions of horse-power had been supplied in turbines of this class for the propulsion of ships by Parsons and his licencees. There was in addition a considerable use of Curtis turbines, especially in ships of the U.S. Navy, also with direct-coupled propeller shafts.

In the Parsons marine turbines of that period the coefficient K rarely exceeded 100 on account of the limitation of speed and the need for avoiding excessive weight and size in the cylinders and drums. Usually three or four shafts were fitted: thus in the "Lusitania" and her sister ship there were four shafts arranged in two compound units with a high and a low-pressure turbine on each side of the centre line. Among the last examples of direct coupling on a large scale were the turbines of the "Aquitania," where 60,000 H.P. were developed on four shafts at 155 revolutions per minute. The wing shafts carried high and intermediate-pressure turbines respectively: the two middle shafts carried a pair of low-pressure turbines taking steam in parallel. Thus the steam passed in series through three cylinders, and K had a somewhat higher value, namely 122. For going astern or manœuvring each wing shaft carried a separate high-pressure turbine element which ran reversed and these worked in series with corresponding low-pressure reverse elements in the two low-pressure turbines. The battle-cruisers "Lion" and "Princess Royal" with turbines of 70,000 H.P. offer other late examples of the now obsolete system of direct coupling to the screw propeller.

At one time an effort was made to obtain improved efficiency in marine propulsion by using a triple expansion reciprocating engine in series with a steam turbine, which enabled the expansion to be usefully carried further than is practicable under a piston. This combination was employed in the "Britannic" and other ships of the White Star Line: it slightly extended the lower limit of speed within which the direct-coupled steam turbine could be advantageously applied.

The change to gearing was brought about by Parsons' improvements in the cutting of accurate gears and by his large-scale experiments in 1910 with the "Vespasian," an old ship originally driven by triple-expansion engines. These were removed, after exhaustive trials, and a set of geared turbines was substituted, in which a pair of 5-inch pinions on the high and low-pressure turbine shafts geared, by double helical teeth, with a wheel 99½ inches in diameter on the propeller shaft.

The result showed a reduction of 15 per cent. in the consumption of coal: it also proved that transmission of the driving power through gearing of this type, under the conditions that apply on board ship, was completely satisfactory[1]. After 1914 the use of geared marine turbines made a rapid advance, and by the end of 1925 the number of horse-power supplied in that form exceeded sixteen millions. A large-scale example is furnished by the turbines of H.M.S. "Hood," completed in 1920, which aggregate 144,000 H.P. on four propeller shafts. In many turbine-driven ships single-reduction gears are used, with a speed ratio that may be as high as 20, but more recent practice for slow speed vessels prefers double-reduction gearing in which a total speed ratio of 40 or so is reached in two steps through the use of short intermediate gear shafts[2].

Hydraulic and electric methods of transmission have also been employed, but the greater simplicity of mechanical gearing, along with its high efficiency, gives it the first place. More than 20,000 H.P. has been transmitted through a single pinion. In Naval use the introduction of gearing has a special advantage in the great economy it effects when the ships steam, as they ordinarily do, at reduced or "cruising" speed[3].

A typical example of double-reduction gearing as applied by the Parsons firm to marine propulsion is shown in fig. 96. Here the turbine is in two parts, on separate shafts, running at somewhat

[1] See Parsons, *Trans. Inst. Nav. Arch.* 1910, p. 168, and 1911, p. 29.

[2] See Walker and Cook, "Mechanical Gears of Double Reduction for Merchant Ships," *Trans. Inst. Nav. Arch.* 1921.

[3] On the subject of mechanical gearing and the proper selection and best treatment of the steel to avoid stripping of teeth, see Parsons, Cook and Duncan, *Trans. Inst. Nav. Arch.* 1923, p. 60. On the use of geared turbines in Warships, see Tostevin, *Trans. Inst. Nav. Arch.* 1920, p. 129. On electric transmission, see Emmett, *Trans. Inst. Nav. Arch.* 1923, p. 91. Also E. Berg, *Engineering*, February 26, 1926, who gives a list of vessels using electric transmission. This includes two U.S. airplane carriers under construction in 1926, each with turbines designed to develop 180,000 S.H.P.

different speeds. Pinions on these shafts gear with wheels on a pair of short intermediate shafts which carry "second reduction pinions," and these in their turn gear with large wheels on the propeller shaft. In this example the propeller shaft runs at 80 R.P.M., the intermediate shafts at 425, and the turbine shafts at 3300 for the high-

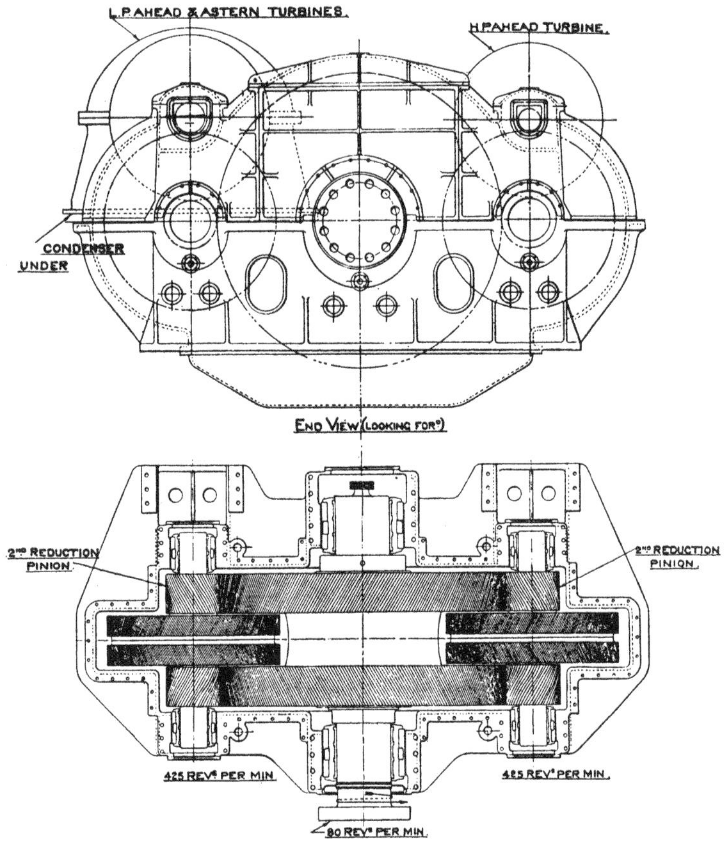

FIG. 96. Double-reduction gearing for marine use (Parsons).

pressure and 2500 for the low-pressure, their pinions being of different diameter. In the low-pressure casing there is incorporated a supplementary turbine for going astern.

In a projected design of a double-geared turbine of 5000 horse-power, it is proposed to supply steam at 500 lb. per sq. inch, superheated to 700° F., and to expand this in a three-cylinder turbine (high, intermediate and low), the three shafts of which run at 5000, 3000 and 1700 revolutions respectively, and carry pinions that gear with two short second-motion shafts driving a central propeller shaft at 90 revolutions, giving the ship a speed of 14 knots.

As an example of usual marine practice the turbine engines of the twin-screw Orient liner "Orama" may be cited, which give a speed of 20 knots with 19,800 S.H.P. Each of the two screws has its independent set of Parsons turbines in three cylinders, the shafts of which run at 1370 R.P.M. and are connected through single-reduction gearing to a propeller shaft running at 95 R.P.M. Steam is supplied at 215 lb. and superheated to 281° C. It expands first through a two-row Curtis wheel and then through 89 reaction stages in all. On each propeller shaft there is a double-helical gear wheel 13 ft. in diameter which the high, intermediate, and low-pressure turbine shafts drive by three separate pinions[1].

159. Exhaust and pass-out turbines. Thermal accumulators. The fact that a steam turbine can make effective use of low-pressure steam leads to the employment of what are called exhaust turbines, which take their supply from the exhaust of other engines or utilize steam that has already served some other purpose. Such turbines may act as auxiliaries to non-condensing plant, and may be applied to engines which work intermittently, like rolling-mill engines in steel-works or winding engines in collieries, by adding a device for storing the thermal energy of the exhaust steam so that the turbine may receive a sufficiently constant supply. Rateau has designed various thermal accumulators of this kind, generally in the form of a horizontal cylindrical boiler-shell kept about two-thirds full of water and provided with pipes through which the irregular supply of steam is admitted below the water level in such a manner as to set up a vigorous circulation in the process of being condensed. About 250 lb. of water are allowed for each lb. of steam, but the quantity of course depends on the variability of the supply and the latitude permissible in the demand[2].

[1] Drawings and particulars of these engines will be found in *Engineering* of Oct. 31, 1924.

[2] For results obtained by the use of steam accumulators, see J. Ruths, *Trans. of the First World Power Conference*, 1924, vol. II, p. 1265.

The facility with which steam may be drawn from or supplied to a turbine at any stage in its expansion has led to the development of special types. In factories where steam is used not only for the production of power but for other purposes, as in paper-making, sugar-refining, brewing and various chemical processes, or for the heating of rooms, it is often advantageous to use a high boiler pressure, in conjunction with a turbine which will first expand the whole of the supply down to a suitable pressure, at which a part is drawn off for the process in question. Thus the early stages of the turbine act as a reducing valve, but with this difference, that they take useful work out of the steam. "Pass-out" turbines are designed to work in this way between the extreme limits (1) when all the steam is being passed out at an intermediate pressure or pressures, and none reaches the condenser, and (2) when none is being passed out and the action is that of an ordinary condensing turbine. When the demand for low-pressure steam varies largely, it may be arranged to pass out all the steam from the turbine, and to return what is not required in the outside process to a point from which it completes its expansion.

The readiness with which steam may be extracted at any point of the expansion is taken advantage of in the process of feed-heating in steps by "bleeding." This process, already mentioned in describing large turbines, is discussed in § 161.

160. Use of a high vacuum. The main thermodynamic advantage of the turbine over the reciprocating engine is its ability to continue the extraction of work from steam which has already expanded to a pressure below what would be effective under a piston (§ 94). To take full advantage of this characteristic the best possible vacuum must be maintained in the condenser, and, as we have already seen, the exhaust steam must have unrestricted freedom in passing to the condenser from the final row of blades.

The thermodynamic advantage that is secured in a turbine by reversible or nearly reversible expansion to a very low pressure will be clear from the entropy-temperature diagram, fig. 97. There *ABCEF* represents the (theoretical) cycle for a reciprocating engine supplied with saturated steam, in which release takes place at *E*, and the steam is condensed mainly along the constant-volume line *EF* (see § 96), the temperature of the condenser being at the level *AF*. The height of *E* corresponds to a pressure such as is often found at release in a piston engine. The area *ABCEFA* represents the heat available for conversion under these conditions. But if expansion be continued to the condenser pressure, as it may be in a steam turbine, the area representing heat available for conversion is the whole adiabatic heat-drop *ABCDA*, the area *EDFE* representing that part of the heat-drop which the piston engine loses through incomplete expansion and the turbine saves. It will also be evident that with incomplete expansion under a piston there is but little to be gained by improving the vacuum in the condenser, whereas in a turbine an improved vacuum gives a relatively important

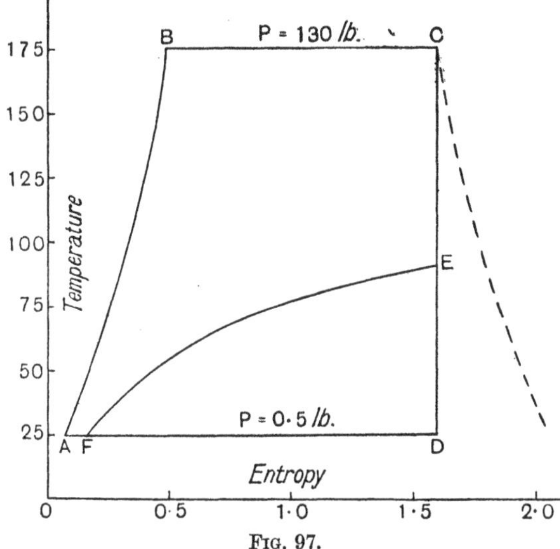

FIG. 97.

increase of the work done, by lowering not simply *AF* but the whole line *AD*.

The advantage of lowering *AD* is seen by comparing values of the heat-drop for the same conditions of admission and for different final pressures. Thus with steam at say 200 lb. (absolute), superheated through 125° C., the heat-drop for a vacuum of 27 inches is 204·5 calories and for a vacuum of 29 inches it is 235·8, or fully 15 per cent. more. A corresponding difference in the output of a turbine is attainable by improving the vacuum from 27 inches, which was a usual enough figure when reciprocating engines only had to be considered, to 29 inches, which represents good turbine practice.

These considerations led Parsons to invent (about 1903) an arrangement called a vacuum augmenter which is shown in fig. 98. There the condensate drains from the bottom of a surface-condenser into a pipe bent to form a water seal. The air mixed with residual vapour is separately extracted from a higher level in the condenser by means of a steam-jet pump which delivers into a chamber called the "augmenter condenser," where the steam of the jet pump is condensed. Since the steam jet delivers the air against pressure its effect is to maintain in the main condenser a vacuum higher than that in the augmenter condenser, and therefore

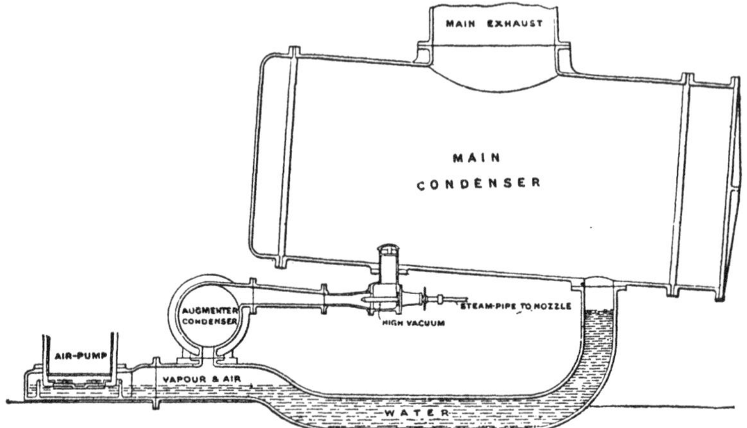

FIG. 98. Parsons' Vacuum Augmenter.

higher than the "air-pump," which here removes the air and condensate together, would be competent to produce if acting alone. This is done at the expense of a small quantity of steam used in the jet, but the beneficial effect of a high vacuum on the efficiency of the turbine is so great that there is a net saving.

The device, though not now made in precisely this form, embodies a principle that is applied in many recent condensers. It led to the practice of separately extracting the air from a condenser by means of a steam-jet pump. In modern usage, this is most commonly done by two steam jets acting in series, forming together a two-stage ejector which discharges the air against atmospheric pressure. A single steam-jet ejector will serve to give a moderate vacuum of 26 inches or so, but the high vacua required for large turbines make two stages necessary. The condensate is independently dealt with by a small centrifugal pump which delivers it to a feed-tank from which it is returned by the feed-pump to the boiler. Part of the condensate is used to condense the steam which has been supplied for the steam jets: the heat of that steam is thereby returned to the feed-water. This combination of jet ejectors for the air with a separate pump for discharging the condensate has taken the place of the old-fashioned "air-pump," which, in earlier practice, removed both air and condensate together. It secures an augmented vacuum in substantially the same manner as Parsons' original device, and it enables a very close approach to be made to the theoretical limit which is imposed by the temperature of the available cooling water. With circulating water capable of cooling the condensate to 79° F. (26° C.) this limit would be 29 inches[1]. Actually 29 and even 29·25 inches are reached in well-designed condensing plants when the conditions as to cooling water are favourable. For large turbines surface-condensers are used with a tube surface of one square foot for every 6 lb. or so of steam per hour, and the cooling water is caused to circulate through the tubes, often by a centrifugal pump driven by an electric motor. The same motor serves also to drive a smaller centrifugal pump which discharges the condensate to the "hot well" or feed-tank. In Messrs Brown Boveri's practice a water-jet ejector is usually substituted for the first of the two steam-jet ejectors in extracting the air, one steam jet still acting in series with it to reach the full effect in producing a high vacuum. In jet ejector devices the fluid which is to be extracted has velocity communicated to it by mixing with the jet, and then loses its momentum by passing through an expanding nozzle, in which process it acquires the pressure necessary for discharge.

161. Thermodynamic performance of turbines. Improvement of the cycle. The striking advance in thermodynamic performance which has been made in recent years depends partly on improvement in the efficiency ratio; in other words, a modern turbine converts into work a larger fraction of the adiabatic heat-drop. It also depends on the fact that the adiabatic heat-drop has been itself increased, by increasing the effective range of tem-

[1] When a vacuum is stated in inches of mercury a standard barometer-pressure of 30 inches is assumed. The vapour-pressure of water at the temperature named corresponds to one inch of mercury.

perature through which the drop occurs. Further, in the most recent practice there has been a change in the cyclic process as a whole which has brought it nearer to the cycle of Carnot.

The efficiency ratio is greater because turbines are bigger, with better blade design and reduced leakage, and run at speeds which have brought the relation of the blade velocity to the steam velocity into closer agreement with the value which secures a maximum of hydraulic efficiency. It has also been improved through the indirect influence of high superheat, for superheating, apart from its effect in augmenting the effective range of temperature, reduces those losses within the turbine which arise from friction and from supersaturation.

The effective range of temperature has been increased (1) at the lower end of the range, by improving the vacuum and so lowering the temperature at which heat is rejected. The advantage of a high vacuum has already been emphasized. And (2), at the upper end, by the use of higher pressure and also by superheating. Both of these changes raise the average temperature at which heat is received by the working substance.

In 1907 what was then regarded as a remarkably good result was recorded in tests of a Parsons turbo-alternator developing 5000 kw., supplied with steam at 200 lb. by gauge, superheated 67° C. to a temperature of 264·7° C., the vacuum being 29·04 inches (barometer 30 inches). Under these conditions the consumption was 13·19 lb. per kilowatt hour. At the pressure and temperature of supply the total heat I was 709 units and the adiabatic heat-drop to the stated vacuum was 224·4 units[1]. Taking one kilowatt hour as equivalent to 1896 heat units, each lb. of steam yielded an amount of electrical energy equivalent to $\frac{1896}{13.19}$, or 143·7 units. The electrical output therefore corresponded to $\frac{143 \cdot 7}{224 \cdot 4}$, or 64 per cent. of the adiabatic heat-drop. The heat supplied per lb. of steam was 683 units, and of this 21 per cent. was converted into electrical energy.

In 1918 another Parsons machine of 25,000 kw. capacity, tested after four years' service, was found to consume only 10·84 lb. per kilowatt hour, with steam at an absolute pressure of 217 lb. superheated to 287° C., and with a vacuum of 28·87 inches. Here the electrical energy developed per lb. of steam was equivalent to

[1] The unit is the pound-calory, equal to $\frac{9}{5}$ of a B.T.U.

175 units, the adiabatic heat-drop was 227 units, and the ratio to it of the electrical energy was 0·77.

Allowing for loss in the alternator the mechanical output in this example must have been fully 80 per cent. of the adiabatic heat-drop.

In another official test, at Carville "B" Power Station in 1916, a Parsons turbo-alternator developing 10,000 kw. used only 10·03 lb. of steam per kw. hour with an absolute pressure of 251 lb., temperature 378° C., and vacuum 29·03 inches. The electrical energy developed per lb. of steam was therefore 189 units, the adiabatic heat-drop 258 units, and their ratio was 0·73[1].

Although this last test gives a smaller efficiency ratio than the preceding one, the actual thermodynamic efficiency is somewhat greater, when the heat converted into electrical energy is expressed as a fraction of the heat taken in by the steam. This is because of the higher pressure and higher superheat. Allowing for the heat returned to the boiler by the feed-water at a temperature corresponding to the condenser pressure, the heat taken in by the steam in the last test was 742 units per lb., and of that 25·5 per cent. was converted into electrical energy. In the preceding test the fraction converted was 25·2 per cent.

The higher pressures and higher superheats now employed, and the practice of reheating, have substantially improved these figures, bringing the thermal efficiency near to 30 per cent. as a matter of estimate. Authoritative tests, however, of very large turbines at full output are rarely made, and experimental data have still to be obtained. A still better thermal efficiency—probably approximating to 33 per cent.—is within reach when, in addition to using high pressure, high superheat, and reheating of the steam at an intermediate stage of its expansion, the system is resorted to of heating the feed-water in a series of steps by bleeding steam from several points in the turbine. The effect of this device is to depart from the Rankine cycle and make the thermodynamic process as a whole approach more closely to the cycle of Carnot.

The thermodynamic defect of the Rankine cycle is that the cold feed-water has its temperature raised to that of the boiler by an *irreversible* application of heat: it is this that makes the Rankine

[1] For results of other tests of Parsons' turbines at the same station, see § 178.

cycle less efficient than the completely reversible cycle of Carnot. Suppose, however, that in the passage of the condensate back to the boiler it is gradually heated in a large number of steps by being brought, at each step, into thermal contact with steam taken from the turbine at a temperature only very slightly higher than that which the feed-water has reached at the step in question. By sufficiently increasing the number of such steps, so that in each of them the water takes in only a little heat from a source scarcely

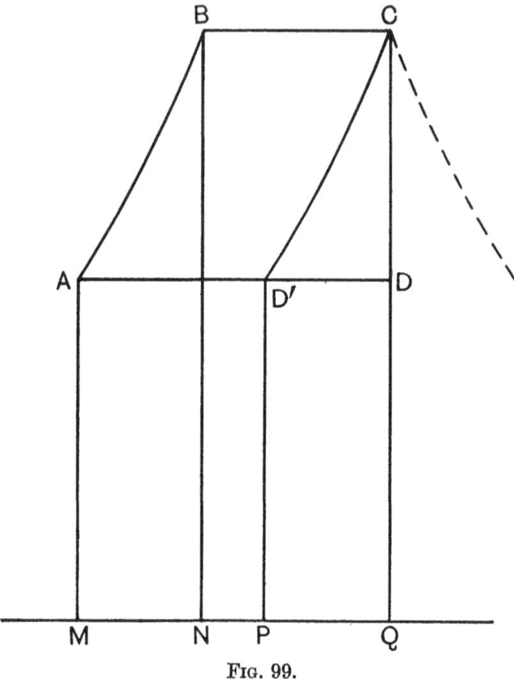

FIG. 99.

higher in temperature than itself, the process of heating may be made very nearly reversible, and consequently the Rankine cycle is converted into what is substantially a Carnot cycle.

To make this point clear, suppose for simplicity that there is no superheating and consider how regenerative heating of the feed-water would affect the entropy-temperature diagram. Without it, the process would be represented by the Rankine cycle *ABCD* (fig. 99). The heat which the feed-water takes up before steam begins to form is the area *MABN*. In complete regenerative feed-heating that heat is taken reversibly during the expansion from steam which is bled from the turbine to supply it; consequently the amount of steam to be bled at each stage must be such that the process of expansion is represented by the curve CD', parallel at each level of temperature to BA, so that the area under every part of the curve CD' shall be equal to the area under the corresponding part of BA. The whole area $PD'CQ$, which is equal to $MABN$, represents heat transferred from the steam to the feed-water. The heat taken in from the hot source by each pound of steam now becomes $NBCQ$, and the heat rejected to the condenser becomes $MAD'P$. The ratio of these two quantities is that of the absolute temperatures BN and AM. Thus (under the assumed condition of no superheat) a strict Carnot cycle would be obtained in the ideal case where the number of steps in the process of regenerative feed-heating is made indefinitely large.

In practice a fair approximation to reversible feed-heating, with material improvement in the efficiency of the cycle, is made by bleeding steam for the purpose at three or four stages. This method, which is now usual in large installations, formed a conspicuous feature in the design of the North Tees Power Station, planned in 1917 by Messrs Merz and McLellan. There the boiler pressure is 475 lb., and there are three main feed-heaters, grouped in "cascade." The condensate on its way back to the boiler first picks up what heat is available from drains and from the steam that has escaped through the turbine glands; it then passes through the first main heater where it receives heat from steam bled at a pressure of 5 lb. (absolute) from the low-pressure turbine, then through one at intermediate pressure, and finally through the third at 65 lb., which is supplied by steam bled from the exhaust of the high-pressure turbine cylinder. The water leaves this third heater at a temperature of 300° F. By three such steps something like two-thirds of the advantage is practically secured that would result from a complete regenerative heating, that is to say, from a substitution of the Carnot for the Rankine cycle[1].

Another feature of modern practice which the same installation illustrates is reheating between the high-pressure and low-pressure

[1] See *Engineering*, June 13 and July 11, 1924, for particulars of the station and for entropy-temperature diagrams drawn to illustrate cascade feed-heating and reheating of the steam.

cylinders. From a stop-valve pressure of 450 lb. and temperature of 650° F. the steam is first expanded to 65 lb. in the high-pressure turbine; it then returns to the boiler house and is reheated to 500° F. before expansion in the low-pressure turbine to a vacuum of 29¼ inches. Although this means the taking in of supplementary heat at a temperature on the whole somewhat lower than that at which the main supply is received, there is a considerable advantage as regards efficiency in keeping the steam much more nearly dry in the later stages of its expansion. As a general effect of these changes, namely, higher initial pressure, reheating of the steam, and cascade heating of the feed-water, it is computed that the ratio of the electrical output to the heat supplied in the steam, which was 0·255 in the Carville "B" tests, has been raised to fully 0·3.

Cascade feed-heating involves a change in boiler practice. It prevents the feed-water from serving to take up, by means of an economizer, heat which is left in the furnace gases after they have acted on the heating surface of the boiler, and the waste heat of the gases has therefore to be utilized in another way. This is done by passing the gases through an air heater in which their heat is extracted by air on its way to maintain combustion in the furnaces. Preheating the air by means of the furnace gases is not a new device, but its use has been extended and its value increased by the introduction of cascade feed-heating. Its effect is to improve what is the ultimate measure of thermodynamic performance, the ratio of the power developed to the thermal energy latent in the fuel. This overall thermal efficiency, from fuel to electricity, is the real figure of merit of a power station: to reckon it completely account must be taken of the supply to auxiliary plant. In a station such as North Tees the overall efficiency is nearly 21 per cent.

Further improvement in the steam turbine cannot be looked for by any lowering of the condenser pressure; nor by any great raising of the upper limit of temperature, because it appears that a limit not much above 400° C. is imposed by considerations of the safety of materials under stress. But at present most of the heat is taken in at a temperature much below that limit, namely, at the temperature of saturation corresponding to the boiler pressure. With a pressure of 500 lb. per sq. inch absolute the temperature of saturation is only 243° C.; with the pressure raised to 1500 lb.

it would be 313° C., and at 2000 lb. it would be 336° C. Hence some thermodynamic improvement is possible by raising the pressure to figures much above 500 lb., and it is in this direction that experimental advances are being made[1].

Several large power stations in the United States use a boiler pressure of 650 lb. per sq. inch, with steam superheated to 750° F. (400° C.). At the Langerbrugge (Belgium) station of the Flanders Electricity Company the boiler pressure is 800 lb. These figures are much exceeded by the experimental plant of the Edison Electric Illuminating Company of Boston, already referred to in § 154 as working with a boiler pressure of 1200 lb. In all these installations there is initial superheating, and in most of them there is also reheating, usually to about 700° F.[2]

However high the pressure be raised, initial superheating is still desirable, but the amount of it is necessarily restricted. With very high pressures the need for reheating becomes increased, not only because of the greater total range of expansion but because the initial superheat that may be given is less. There may even be (in theory) an advantage in reheating more than once, though this is not likely to be realized if the steam has to be taken to the boiler house for the purpose: an independently fired re-superheater close to the turbine is perhaps not impracticable.

So long as the pressure of supply does not much exceed 500 lb. per sq. inch one reheating is sufficient. This will be seen from fig. 100, which is an entropy-temperature diagram drawn to scale for the (theoretical) action when steam at 500 lb. absolute is initially superheated to 400° C., then adiabatically expanded to the saturation line, when its pressure is approximately 70 lb. and temperature 150° C. It is then reheated, again to 400° C., and finally expanded adiabatically to the condenser temperature, taken here as 30° C. Except in the very last stages it remains dry. In the action of a real turbine the reheat factor (§ 138) would operate to keep the steam drier at the finish than it is in the adiabatic conditions which are here assumed.

[1] See Sir Charles Parsons' paper on Steam Turbines, *Trans. of the First World Power Conference*, 1924, vol. II, p. 1477.

[2] For particulars, see papers by D. S. Jacobus and by W. S. Monroe, *Trans. of the First World Power Conference*, vol. II, pp. 1405 and 1441; also Sir James Kemnal, *ibid.* p. 1367, and *Inst. of Engineers and Shipbuilders in Scotland*, February, 1926.

It should be borne in mind that the thermodynamic advantages of reheating are apt to be seriously lessened in practice by losses of heat and pressure in the pipes by which the partially expanded steam is conveyed to the reheating flues and back. Data are scanty as to the extent to which reheating as it is now carried out actually improves the efficiency of power plants in station working.

On the general subject of the thermodynamic cycle in turbines mention should be made of the use of mercury as a substitute for steam in the upper part of the temperature range, by means of a

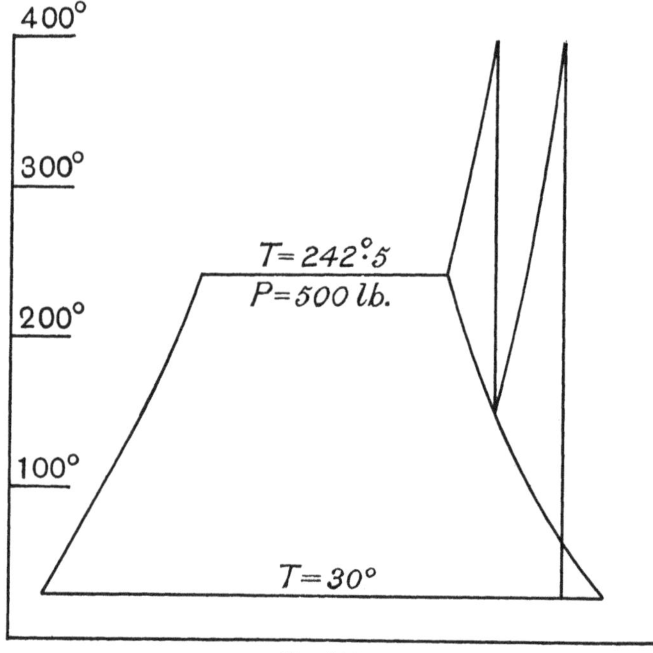

Fig. 100.

binary vapour combination in which mercury and water-vapour act in thermal series, the mercury-vapour being formed in a high-temperature boiler, at a moderate pressure, then expanded in the first turbine to a low pressure, at which however its temperature is still so high that in being condensed it forms steam at a pressure appropriate for use in a second turbine. A plant working on this binary cycle, designed by Mr W. L. R. Emmet, has been built by the General Electric Company of Schenectady, and experimentally installed by the Electric Light Company of Hartford, Conn.

Mercury is selected for the first part of the action in order that a high temperature for the reception of heat (during the conversion of the liquid mercury into vapour) should be associated into only a very moderate pressure, the object being effectively to widen the total thermodynamic range by raising T_1 while the final temperature of rejection remains as usual conditioned only by the temperature of the water that is available for condensing steam. The thermodynamic advantages of this binary working are fully discussed by Mr W. J. Kearton in a paper where particulars are contributed by Mr Emmet and others as to the construction and working of the plant[1]. The use of mercury for such a purpose is open to grave objections. Its latent heat is comparatively so small that something like nine times as much mercury as water has to pass through the cyclic process. Mr Kearton states on the authority of Mr Emmet that the weight of mercury in the system was 7 lb. for each kilowatt of output capacity. This makes the cost of the mercury a formidable part of the whole capital cost. For that reason, as well as on grounds of safety to health, no leakage of mercury can be tolerated, a condition not easy to maintain at the high temperature of working. The experiment is a bold and interesting one, but it appears to the writer that the estimated advantages of high-temperature reception through this use of mercury are not so much greater than those which can be secured by means of the single fluid, steam, on the lines already indicated, as to encourage the idea that binary working with its obvious drawbacks will find general application.

[1] *Proc. Inst. Mech. Eng.* November, 1923. See also Mr Emmet's paper in *Trans. of the First World Power Conference*, vol. II, p. 1391.

CHAPTER IX

THE TESTING OF HEAT-ENGINES

162. Indicated power and brake power. Mechanical Efficiency. In the testing of an engine of any kind, as a means of developing power, we are concerned with (*a*) the work done within the engine by the working substance, and (*b*) the effective output, in the form of work that is available outside the engine for driving other mechanism. Of these quantities the second is always less than the first, for some of the work done by the working substance is consumed within the engine itself in overcoming friction and in driving auxiliary parts (such as pumps) which are essential to the operation. In an engine of the piston and cylinder type the power developed by the working substance may be measured from the indicator diagram; hence (*a*) is called the indicated power. In an engine of any type the output (*b*) may be found by testing with a brake, or by measuring the torque transmitted by the shaft: it is therefore called the brake or shaft power. The ratio of the brake or shaft power to the indicated power is called the mechanical efficiency.

In reciprocating steam-engines the mechanical efficiency is about 0·85 in favourable cases: that is to say, about 15 per cent. of the indicated work is ineffective for any useful purpose outside the engine, being spent in overcoming the friction of the piston and piston-rods, valves, guide-blocks, pins, journals, etc., and in driving pumps. Occasionally the mechanical efficiency may approach 0·9; in general it is a good deal less. In oil-engines of the Diesel type, to be described later, which are now largely employed for marine propulsion, it is found on trial to range from about 0·71 to 0·77 in continuous working. With any engine it is of course the brake or shaft power that is the real criterion of useful performance, and the consumption of fuel per shaft horse-power-hour is a matter of much more practical importance than the consumption per indicated horse-power-hour.

In steam turbines no measurement is possible of indicated power: the output can be determined only in the form of shaft power. In marine propulsion this is done by measuring the torsion of the propeller shaft, or sometimes its longitudinal thrust. When a turbine drives a dynamo the output is measured electrically, and as it is possible to make a fairly accurate estimate of the efficiency of the dynamo[1] the shaft power may be readily inferred. The shaft power bears a much higher ratio to the work done by the steam in a turbine than in a reciprocating engine, the frictional losses in a turbine being relatively small.

163. The Indicator. In any reciprocating engine the indicator serves a double purpose. Besides measuring the work done within the cylinder, it exhibits in detail what is going on and gives the engineer a means of critically examining every part of the action. In a steam-engine it shows the time and manner of the four events of the stroke, namely, the admission, cut-off, release and compression, which together make up what is called the "distribution" of the steam; it detects faults in the setting or in the working of the valves and suggests changes by which the distribution may be improved. When the information which it gives is supplemented by a knowledge of how much steam is passing through the cylinder per stroke a complete analysis of the action becomes possible; the wetness of the steam at any stage may then be determined, as well as the exchanges of heat that take place between the steam and the cylinder walls.

Similarly, in the internal combustion engine, the indicator is invaluable, apart from its use to measure power, in showing whether the events are properly timed and in supplying material for a study of each phase of the cycle.

The indicator, invented by Watt and improved by M'Naught, Richards and others, consists in its simplest form of a small steam-cylinder, fitted with a piston which slides easily within it and is pressed down by a spiral spring of steel wire. The cylinder of the indicator is connected by a pipe below this piston to one or other end of the cylinder of the engine, so that steam from the engine cylinder has free access, and the piston of the indicator consequently rises and falls in response to the fluctuations of pressure which occur in the engine cylinder. The indicator piston actuates a pencil, which rises and falls with it and traces the diagram on a sheet of paper fixed to a drum that is caused to turn back and forth about its axis through a certain angle, in unison with the motion of the engine piston. In M'Naught's indicator the pencil

[1] In large alternators the dynamo efficiency may be as much as 0·97.

was directly attached to the indicator piston, in Richards' the pencil is moved by means of a system of links so that it copies the motion of the indicator piston on a magnified scale. This has the advantage that an equally large diagram is drawn with much less movement of the indicator piston, and errors which are caused by the piston's inertia are consequently reduced. With engines even of moderate speed these errors may be serious unless pains are taken to avoid effects of inertia in the indicator piston and the moving parts connected with it. In Richards' indicator the linkage employed to multiply the indicator piston's motion is an arrangement similar to the parallel motion which was introduced by Watt as a means of guiding the piston-rod in beam engines. In several forms of indicator lighter linkages are adopted, and other changes have been made with the object of fitting the instrument better for high-speed work. One of the best of these modified forms of Richards' indicator is that made by the Crosby Company, which is shown in figs. 101 and 102. The pressure of steam in the engine cylinder raises the piston *F* (which is shown in section in fig. 102), compressing the spring above it and causing the pencil to rise in a nearly straight line through a distance proportional, on a magnified scale, to the compression of the spring, and therefore to the pressure of the steam. At the same time the drum *D*, which carries the paper, receives motion through the cord *C* from the cross-head of the engine. Inside this drum there is a spiral spring *E* which becomes wound up when the cord is pulled, and serves to turn the drum in the reverse direction during the back stroke when the cord is relaxed. The cap *A* of the indicator cylinder has holes in it which admit air freely to the top of the piston, keeping the upper

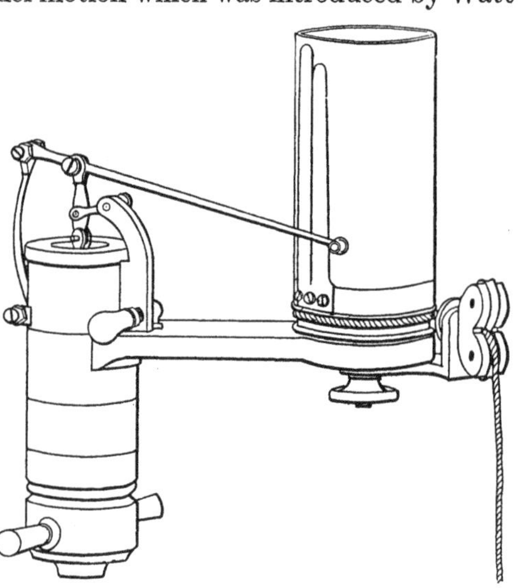

FIG. 101. Crosby Indicator.

side at the pressure of the atmosphere, and the piston has room to descend, extending its spring, when the pressure of the steam is less than that of the atmosphere. The spring is easily taken out and replaced by a stiffer or less stiff one when higher or lower pressures have to be dealt with. Springs adapted to various ranges of pressure are supplied with the indicator and are marked with a number which states the pressure, in lb. per sq. inch, which will raise the pencil through a distance of one inch on the paper. The accuracy of this number should be verified by testing the indicator under steam against a standard pressure-gauge, or against a mercury column. Tests made by applying water under pressure are not suitable, unless a proper allowance be made for the change of elasticity of the spring through change of temperature. The spring is stiffer (generally by two or three per cent.) when cold than at the temperature (about 100° C.) which it takes up when in use. In testing an indicator the pressure of the steam should be raised slowly or by steps, and the test should be made with rising and also with falling pressure, to see that there is no material friction error, which would show itself by causing the indicator pencil to stand higher during the fall than during the rise when the steam had the same pressure.

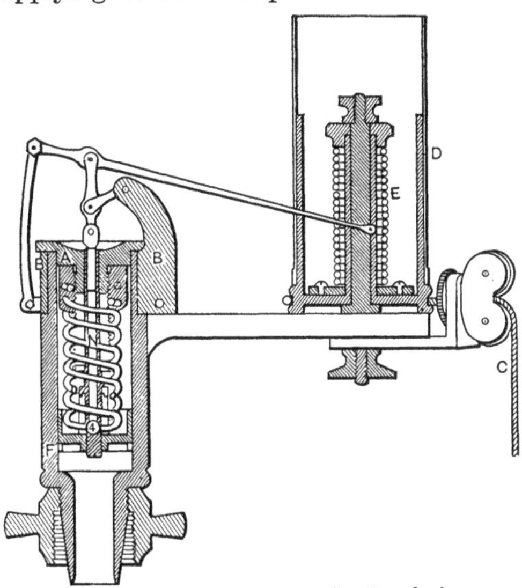

FIG. 102. Crosby Indicator. Sectional view.

A tap, placed just below the indicator but not shown in the diagram, allows the communication with the cylinder to be closed at pleasure, and also puts the space below the indicator piston into communication with the atmosphere, thereby allowing the "atmospheric line" to be drawn on the diagram.

The pencil is withdrawn from the paper by turning back the piece *BB* which is separate from the rest of the indicator cylinder.

The small handle shown in fig. 101 is provided for this purpose, and a stop behind the handle prevents the pressure of the pencil against the paper from exceeding a regulated amount.

In other examples the spring which controls the movement of the piston is fixed outside the indicator cylinder, so that its elasticity will not be affected by changes of temperature.

An indicator of this kind with light moving parts may be made to work satisfactorily enough for speeds up to about 300 or 400 revolutions per minute. But in modern engines, especially of the internal combustion type, speeds have to be dealt with at which it would wholly fail. What is essential is that the natural period of oscillation of the indicating system should be much shorter than the time of revolution of the engine on which the indicator is to be used. This consideration compels, for engines whose speed much exceeds that limit, a radical departure from the old form. The indicator piston is allowed a very small range of movement, being under the control of a very stiff spring, or in some examples a stiffly controlled diaphragm is substituted for a piston. Since any large mechanical magnification of the movement would re-introduce inertia, one or other of two plans is followed: either (1) an optical magnification is used instead, or (2) the diagram takes a miniature form, to be magnified afterwards by photography or examined under a microscope. Forms of indicator embodying these alternative methods are described in the next two sections. The satisfactory working of such instruments depends on the care that is taken in their construction to avoid backlash as well as effects of inertia.

164. Optical Indicator. In some forms of optical indicator an elastic diaphragm has been substituted for the piston, the elasticity of the diaphragm itself furnishing the control without a separate spring. But it appears on the whole preferable to retain the use of a piston with a separate stiff spring, which can be changed at will to suit the different ranges of pressure. With this there is no difficulty in securing a uniform scale, and it has the further advantage that it escapes the variations of temperature, and therefore of elastic stiffness to which a diaphragm is subject in use. A form of optical indicator with a piston designed by B. Hopkinson for gas-engine researches is shown in figs. 103 and 103 *a*. The block *A* which is screwed into the indicator hole of the engine is bored to receive a piston F, made hollow for the sake of lightness and furnished with three grooves to reduce leakage. Round the outside of A fits a frame B abutting against a shoulder at the top and pressed up against the shoulder by a spiral spring in compression. The shoulder is furnished with a ball-race to make the frame B turn readily about the axis of the piston. The spring which controls the motion of the indicator piston is a flat bar D clamped to the frame B by screws EE. The part between the

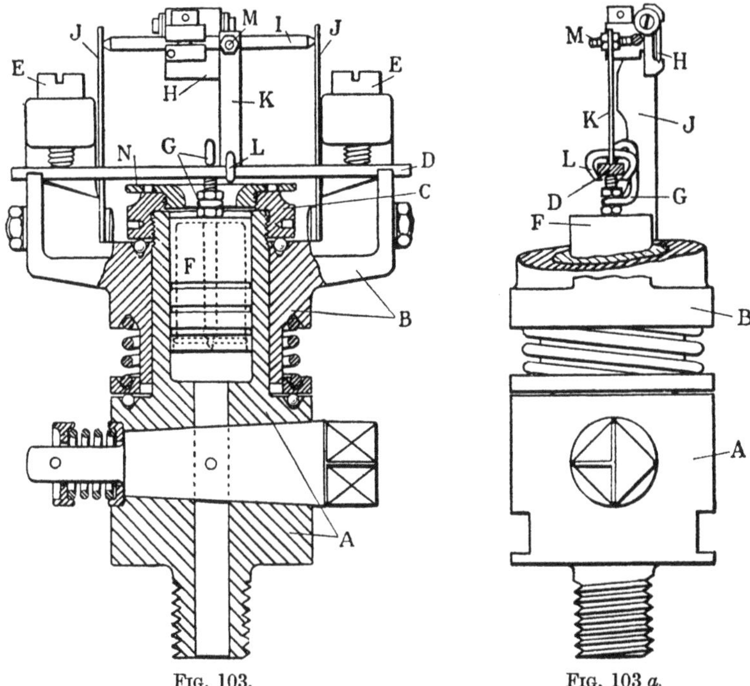

FIG. 103. FIG. 103 *a*.

screws is free to bend as a beam and is connected with the piston by the hook G. The movement of the piston and spring is communicated to a small mirror M, pivoted by an axle I between flexible side supports JJ, through a flexible arm K which makes the mirror tilt by an amount that is proportional to the small movement of the piston. A beam of light reflected from the mirror is accordingly displaced in proportion to the changes of pressure in the engine cylinder. At the same time the mirror receives another motion, at right angles to the first, proportional to the displacement of the piston of the engine. This is done by making the frame B rock about its axis, by connecting it with an excentric which causes this rocking motion to correspond exactly with the motion of the main piston. The beam of light consequently traces out an indicator diagram. The diagram is made visible by the optical device shown in fig. 104. Light falls on the indicator

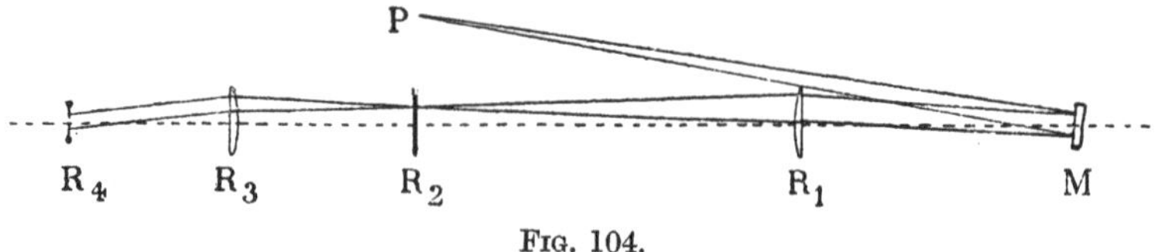

Fig. 104.

mirror M, from a source P, consisting of a small hole illuminated by an incandescent lamp. The reflected beam is brought to a focus at R_2 by the lens R_1. This lens being 4 inches in diameter is big enough to include the largest displacements of the beam. A transparent screen is placed at R_2 and a second lens R_3 is interposed between it and the eye, which is placed at R_4, the principal focus of R_3. The transparent screen is engraved with horizontal and vertical lines on which the diagram is seen projected. As an alternative the indicator may have a light camera fitted to it to give a photographic record as in fig. 105[1].

Fig. 105. Optical indicator with camera.

165. Micro-Indicator. In the micro-indicator of the Cambridge Instrument Company the diagrams are approximately 3 mm. long and $2\frac{1}{2}$ mm. high. They are inscribed by a fine tracing point on a surface of celluloid, a method of recording which the inventor, Mr W. G. Collins, has successfully applied to instruments of

[1] For a fuller account of this indicator, with examples of diagrams given by it, see Hopkinson's paper on the "Indicated Power and Mechanical Efficiency of the Gas-Engine" (*Proc. Inst Mech. Eng.* Oct. 1907).

various kinds. A group of ten indicator diagrams (for a high-speed internal-combustion engine), taken successively on a celluloid disc which is moved a step forward between one diagram and the next, are shown to double the actual size in fig. 106, and one diagram is photographically enlarged from the celluloid record in fig. 107. The indicator piston, which has a travel of little more than 1 mm., is shaped as a portion of a sphere where it touches the cylinder so that no jamming can occur. It is controlled by having the top of its rod held against a straight flat spring the projecting end of which gives motion to the stylus. The celluloid disc is carried on an arm projecting from a post which turns back and forth through a distance proportional to the motion of the engine piston, and an

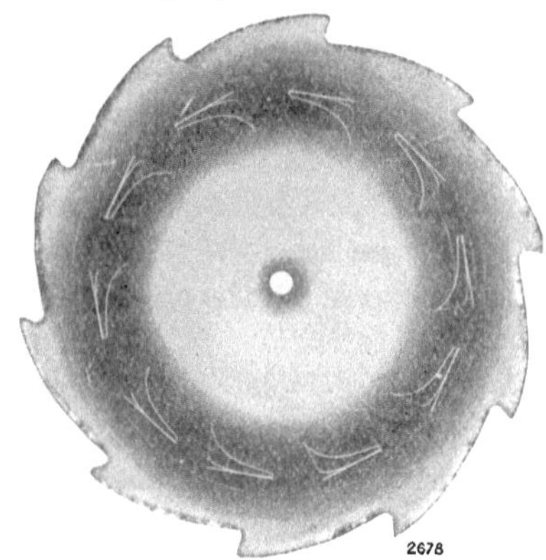

FIG. 106.
Micro-indicator diagrams.

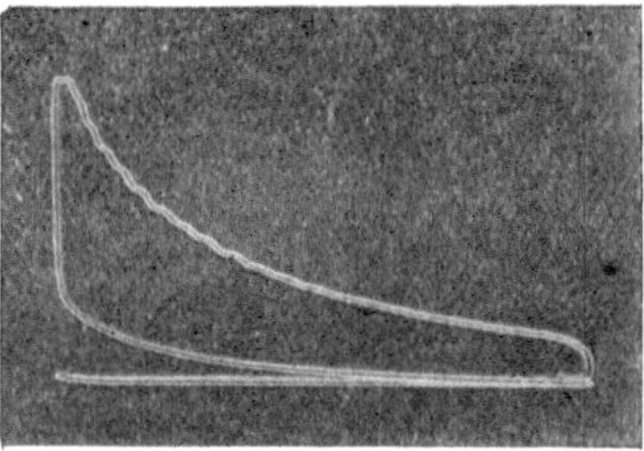

FIG. 107.
Micro-indicator diagram enlarged.

electromagnetic connexion enables the diagram to be drawn during a single cycle only, and the celluloid to be advanced a step before the next record is made.

The scribing point acts with remarkably little friction. It slightly indents the surface, heaping up the material on both sides of the indentation, with the effect that the enlarged record appears as a dark line between two white ones. The diagram of fig. 107, which records a single cycle, shows this triple feature in the line: measurements can be made with great accuracy to either edge, or to the central dark portion.

166. Indicating on a time-base. Some of the difficulties of high-speed indicating are removed when the variations of pressure are recorded not in relation to the displacement of the piston, but in relation to the angular movement of the engine-shaft. That is to say, the paper or other surface which receives the record takes a uniform motion from the shaft instead of a reciprocating motion from the piston, and a continuous curve is drawn upon it which exhibits the changes of pressure in relation to what is essentially a time-base, so long as the engine is running steadily. Provided the dead-points of the stroke are marked on this continuous record, it is a matter of simple geometry to draw, from it, a cyclical diagram of the usual form and to determine the mean effective pressure and the indicated power. The continuous diagram has a great advantage, especially in application to fast internal-combustion engines; it not only reduces inertia troubles but spreads out the record of what is the most important phase in the action—the ignition and combustion which occur while the piston is near its dead-point. It also shows in a very useful way any variation from one cycle to the next.

An example of an indicator of the usual type, but recording on a continuously running paper, is shown in fig. 108. The same principle is applied to optical indicators and to the Cambridge miniature form.

A variant of this arrangement is obtained by placing the paper round a drum which makes one revolution for each complete cycle of the engine's action. In that case the diagrams are still on a time-base, but they are superposed for successive cycles.

167. Balanced Indicators. This name is given to a class in which the pressure record is taken at isolated points, without a general movement of the indicator piston, by a device which, at each such point, balances the momentary pressure on the piston against another force. Thus, if the other force is set to have a definite value, the instant at which the pressure on the indicator piston becomes equal to that may be recorded with no more than an infinitesimal movement of the piston. Successive points may be marked in this way in successive cycles, until (provided the working is uniform) a complete and consistent diagram is built up, the record being taken on a surface which is either reciprocating with the piston or revolving so that it makes one turn for each cycle.

An important example of this class is the "Farnboro" electric indicator developed by the British Air Force, in which the balancing force against the piston is supplied by an air pressure which is caused to vary slowly between the extremes that have to be recorded, in a time which includes many revolutions of the engine. At any instant when the cylinder pressure of the engine balances the air pressure, a valve or disc which is free to move between

FIG. 108. Indicator giving a continuous diagram on a time base.

stops about 0·01 inch apart moves from one to the other stop, with the effect that it momentarily breaks an electric circuit, and causes a spark to pass from the recording point to the paper, on which the spark makes a fine mark. The position of the recording point is controlled by the air pressure (acting against a spring), and the paper is carried by a drum geared to the shaft so that it makes one turn for each cycle. The diagram is accordingly recorded point by point, one for each cycle, the air pressure being slightly changed from one cycle to the next. With a high-speed engine, running regularly, the whole process takes only a fraction of a minute. This indicator has been successfully applied to aeroplane engines in flight, at speeds over 2000 revolutions per minute[1].

168. Conditions of accurate working. The following notes and those of § 169 relate primarily to the indicating of steam-engines under conditions such that an indicator of the Crosby or a similar kind will serve. In part, however, they will be seen to apply to indicators of other types. To register correctly, an indicator must satisfy two conditions: (1) the motion of the piston must be proportional to the change of pressure in the engine cylinder; and (2) the motion of the drum must be proportional to that of the engine piston.

The first of these conditions requires that the pipe which connects the indicator with the cylinder shall be short and of sufficient bore, and that it shall open in the cylinder at a place where the pressure in it will not be affected by the kinetic action of the steam during admission or exhaust. Frequently pipes are led from both ends of the cylinder to a central position where the indicator is set, so that diagrams may be taken from either end without shifting the instrument. This arrangement is convenient and shows the double action prettily; but except with small cylinders it makes the connecting pipes so long as to give rise to serious errors. In large engines it is therefore not admissible: a pair of indicators should rather be used, each fixed with the shortest possible connecting pipe, or the diagrams should be taken successively from the two ends of the cylinder with a single instrument set first at one end and then at the other. The general effect of an insufficiently free connexion between the indicator and the engine cylinder is to make the area of the diagram too small.

The first condition of correct working is also invalidated to some extent by the friction of the indicator piston, of the joints in the linkage, and of the pencil on the paper. The piston must slide very freely; nothing of the nature of packing is permissible, and any

[1] For a fuller description of the Cambridge, Farnboro, and other indicators, with a discussion of the general problem, see *Proc. Inst. Mech. Eng.* Jan. and Feb. 1923, pp. 95–197.

steam that leaks past it must have a free exit through the cover. The pencil pressure must not exceed the minimum which is necessary for clear marking. By careful use of a well-made instrument the error due to friction in the piston and connected parts may be kept so small as not to be serious. The disturbing effects of inertia are seen by oscillations being set up when the indicator piston suffers a comparatively sudden displacement. These oscillations, superposed upon the legitimate motions of the piston, give a wavy outline to parts of the diagram, especially when the speed is great and when the last-named source of error (the friction) is small. When they appear on the diagram a continuous curve may be sketched by hand midway between the crests and hollows of the undulations. To keep the oscillations within reasonable compass in high-speed work a stiff spring must be used and an indicator with light parts has to be selected. Still another possible source of error is backlash through too great looseness in the joints. The strain of the spring must of course be well within the limit of elasticity, so that the rise of the pencil from the datum-line may be proportional to the excess of pressure above that of the atmosphere.

With regard to the motion of the drum it is, in the first place, necessary to have a reducing mechanism which will give a sufficiently accurate copy, on a small scale, of the engine piston's stroke. Some contrivances used for this purpose have aimed at a rigorous geometrical solution of the problem—as for instance by adapting some form of pantagraph—but the multiplicity of joints in such mechanisms is apt to give rise, through backlash, to greater errors than would occur in simpler forms of gear designed to give no more than a close approximation. Of these simpler gears a usual form is that shown in fig. 109. *AB* is a long pendulum rod, pivoted on a fixed centre at *A*. *BC* is a short link connecting *B* with the cross-head *C*. The indicator drum receives its

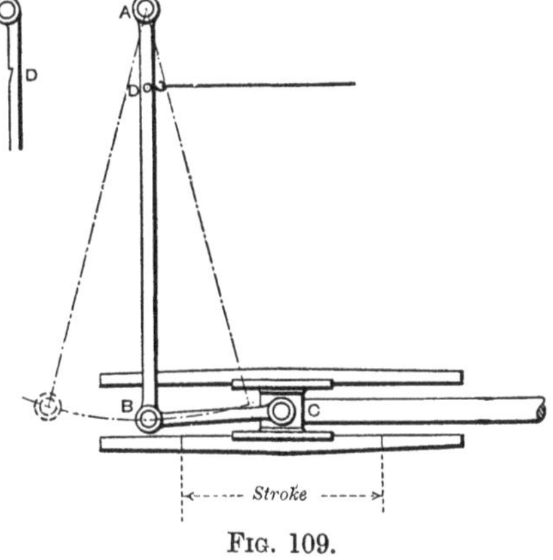

FIG. 109.

motion from any suitable point *D* on the rod *AB* by a cord which is led over pulleys if necessary. A convenient arrangement, shown by the separate sketch at the side, is to have a notch in the rod AB at D in which a hook at the end of the cord will engage when the hook is slipped along from A: this gives a ready means of throwing the indicator drum in or out of gear. To make the mechanism fairly accurate the length of the pendulum rod AB should be considerably greater than that of the stroke. Another form of indicator gear is obtained by omitting BC, joining AB directly to the cross-head and setting the pin A on a sliding block in fixed guides which allow it to slide towards or away from the piston as the rod oscillates. This is geometrically a better form, but it requires careful construction to escape backlash at A.

Even when the cord which is to move the indicator drum is connected to the piston-rod in such a way as to copy its motion correctly, the motion of the drum itself may become incorrect because the length of the cord is not strictly constant. The varying pull causes varying amounts of extension, and when the cord is a long one the error which this involves may be serious. The tension in the cord varies from three causes: (1) the varying resistance of the drum spring, (2) the varying acceleration of the drum, and (3) the friction of the drum and of the guide-pulleys, if there are any. Causes (1) and (2) may be arranged to counteract each other, but the friction of the drum necessarily tends to make the cord longer during the forward motion of the drum than during the backward motion. Hence it is important to see that the drum and pulleys turn readily with little friction, and still more important to make the cord short. Where there is a long distance between the indicator and the point from which motion is taken—as will generally be the case in large engines—cord should be used only at places where flexibility is required and stout wire should as far as possible be substituted. Even in comparatively small engines wire may be used with advantage. Of the errors to which indicator diagrams are liable perhaps none are so often neglected as those that come from the stretching of long driving cords[1].

169. Directions for taking indicator diagrams. In taking indicator diagrams the following practical hints may be found

[1] For an important discussion of the errors of the indicator, see papers by Osborne Reynolds and Brightmore (*Min. Proc. Inst. C. E.* 1885; Reynolds' *Collected Papers*, vol. II).

useful:—Before attaching the indicator to the engine, see that the piston moves freely; that the joints of the lever and links are oiled and are sufficiently slack to avoid friction, but not so slack as to allow the pencil to shake; that the pencil point is sharp, and that it is adjusted to press lightly upon the paper drum; and that the paper drum turns freely without shaking.

Select a spring appropriate to the pressure within the cylinder and to the speed of the engine. With the Crosby indicator the diagram should not be more than $1\frac{3}{4}$ inches high; thus a 50 spring should not be used if the range of pressure to be indicated exceeds 87 lb. per sq. inch. When the engine runs fast it is necessary to use a still stiffer spring, to prevent the diagram from showing an inconvenient amount of oscillation. If large oscillations occur the process of smoothing the diagram by sketching a line midway between the crests and hollows is unsatisfactory, and a new diagram must be taken with a stiffer spring.

In putting a spring in and screwing the parts together, try whether there is any backlash or shake between the spring and the indicator piston. If there is any it is to be taken up (in the Crosby instrument) by means of the set screw under the piston.

Screw the indicator cock to the pipe on the engine cylinder, and couple up the indicator, taking care to tighten up the coupling collar in such a position that it leaves the handle of the cock free to turn. See that the cord from the drum has a clear course to the oscillating lever, and that its mean position during the oscillation is about perpendicular to the lever. Adjust the length of the cord and the amount of its motion so that when the cord is in gear the drum turns backwards and forwards without coming up against a stop at either end of its travel. If it touches one stop or the other the cord is too long or too short: if it touches both stops the travel of the drum is too great and a point nearer the fulcrum of the oscillating lever must be taken for the attachment of the driving cord.

Do not keep the indicator drum moving except while diagrams are being taken. Stop the drum by disconnecting the cord from the oscillating lever before attempting to put a paper on the drum. In putting on the paper see that it is taut and clear of wrinkles, and fold down the projecting edges so that they may not touch the lever which carries the marking pencil.

Turn on steam to the indicator for a minute or so before taking the diagram. Then make the pencil touch the paper lightly, keeping it on long enough to complete a single diagram. Withdraw the pencil. Shut the cock leading to one end of the cylinder and open the cock from the other end (if pipes from both ends come to the same indicator). Make the pencil touch the paper again to take the other diagram. Withdraw it and shut the indicator cock. Touch the pencil again to the paper to draw the atmospheric line. Stop the drum by disconnecting the cord. Remove the paper and mark the diagrams to show which end of the cylinder each refers to. Note the scale number of the spring, and the speed of the engine, with the date and hour and any other particulars that may be wanted.

170. Calculation of the indicated horse-power. By measuring the mean height of the diagram between the top and bottom lines we find the *mean effective pressure*, which when multiplied by the area of the piston and the length of the stroke gives the work done per stroke.

The mean height of the diagram is most accurately found by measuring the area of the diagram with a planimeter or otherwise (as for instance by applying Simpson's rule) and dividing that area by the length of the base, namely, the distance between lines drawn perpendicular to the atmospheric line and touching the diagram at its extremities. Often the mean height is found by dividing the base into ten or twelve equal parts, drawing a perpendicular to the atmospheric line through the middle of each of these parts, measuring the lengths of the perpendiculars between the top and bottom lines and taking the mean of these lengths. The lengths of the perpendiculars are most conveniently measured by applying to each line in succession a scale graduated in inches and tenths of an inch: with a little practice it is easy to estimate to hundredths of an inch. The mean height in inches is multiplied by the scale number of the spring to find the mean effective pressure in lb. per sq. inch. The mean effective pressure for the other side of the piston is found from the other diagram in the same way. Calling these mean effective pressures p_m and p_m', in lb. per sq. inch, and the net areas of the corresponding sides of the piston a and a' in sq. inches, and the length of the stroke l in feet, the work done by the steam per revolution is

$$l\,(p_m a + p_m' a')$$

in foot-pounds.

The work done per minute is

$$nl\,(p_m a + p_m' a'),$$

n being the number of revolutions per minute; and the indicated horse-power is

$$\frac{nl\,(p_m a + p_m' a')}{33{,}000}.$$

In general a and a' are nearly equal. The sum of them may then be taken and multiplied by the mean pressure $\frac{1}{2}(p_m + p_m')$, as a substitute for the quantity within brackets. And it is convenient when many diagrams are to be worked out for one engine to express the quantity $\frac{l\,(a + a')}{33000}$ as a single constant factor which has only to be multiplied by the mean pressure $\frac{1}{2}(p_m + p_m')$ and by n to find the indicated power.

171. Thermodynamic tests. Measurement of the supply of steam by means of the feed. When engine trials are to serve as tests of thermodynamic performance, either the heat supplied or the heat rejected has to be measured, for comparison with the work done. Measurements of the supply of heat are most usual. Sometimes, however, this method of testing may be impracticable and a test by means of the rejected heat may be easy. In any case a measurement of the rejected heat furnishes a valuable check on the accuracy of the other method, and the most satisfactory trials are made by measuring the heat supplied as well as the heat rejected; this allows a balance-sheet to be drawn up in which the heat given to the engine is more or less completely accounted for.

To determine the supply of heat the quantity of steam used by the engine is measured. Except when the engine has a surface-condenser, this has to be done by measuring the amount of feed-water that is required to keep the level of water in the boiler constant during a prolonged run. A somewhat long run is necessary in a trial of this kind because the level of water in the boiler cannot be read very exactly and the whole consumption of feed-water should be so great that any error due to this cause will become negligible. With a Cornish or Lancashire boiler a run of six or eight hours may be required: on the other hand, if the engine is getting its steam from a tubular boiler working hard under forced draught, or from a water-tube boiler, the evaporation may be so rapid, and the water surface so comparatively small, that a single hour or even less will suffice. Care should be taken to have all the conditions of the experiment as closely as possible the same at the end as at the beginning of the trial: if for instance the feed-pump is working at the beginning it should be working at the end at the same rate, and the pressure in the boiler should be the same. In these circumstances the quantity of water in the boiler, for a given reading in the gauge-glass, may be taken to be the same at the end as at the beginning of the run, and the quantity of feed-water that has been supplied in the interval is therefore equal to the quantity of steam that has left the boiler. If there has been no leakage and no blowing off at the safety-valve or otherwise, this quantity of steam has been delivered to the engine.

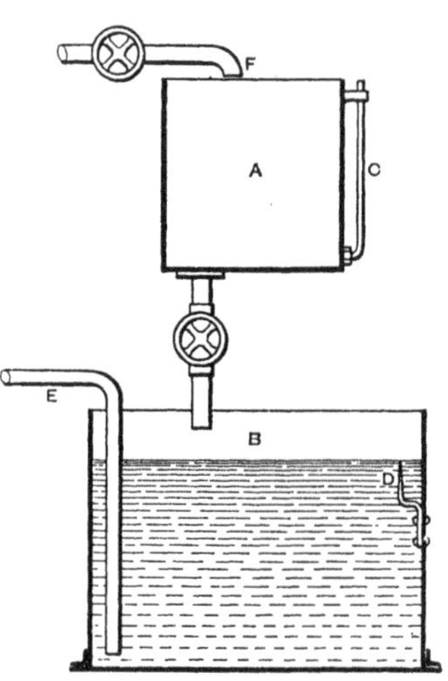

Fig. 110. Arrangement of tanks for measuring feed-water.

To measure the feed-water a convenient plan is to have a small tank (A, fig. 110) set above another (B) so that it may drain into B. The weight of water contained by A when full must be accurately known, and it should be furnished with a gauge-glass C to let fractions of the whole contents be read. B must have a float or a point gauge D or other mark in it to indicate when the water reaches some one standard level. The feed-pump draws water from B by the pipe E; fresh water can be run into A at pleasure from the supply pipe F, and there is a stop-cock between A and B. At the beginning of the test it has to be seen that the water in B is at the standard level and that the stop-cock between the tanks is shut. Water is supplied during the test by completely filling A as often as may be necessary, letting its contents drain completely into B each time and noting the hour and minute at which each fill of A is emptied into B. At the end of the trial, after A has been filled for the last time, just enough of its contents is passed into B to bring the level of water in B up to the standard level, the gauge-glass of A showing what fraction has been passed. In a long run it is useful to check the work by dividing the whole period into a series of parts, in each of which the supply of feed-water is separately noted. The boiler pressure, the speed, and all other conditions of working must of course be kept as nearly uniform as may be throughout and should all be noted at regular intervals during the trial. It is useful to exhibit the log of the trial graphically by plotting all the observed quantities on section-paper with time as the base.

The engine should work for some time under the prescribed conditions as to speed, pressure, and load before the period of the test begins, in order that it may get thoroughly warmed up and that a uniform action may be established. During the trial indicator diagrams are taken from time to time and the times are noted. Where there is a mechanical counter the whole number of revolutions made during the period of trial is found by reading the counter at the beginning and at the end.

172. Measurement of the supply of steam by means of the condensed water. In engines which work with a surface-condenser the amount of steam passing through the engine in a given time is readily determined by measuring or weighing the water discharged from the condenser. An important advantage of this method is that a satisfactory trial of the engine can be made in much less time than is necessary when the steam used is to be determined from the feed-water. Provided the engine has been running long enough for the action to become uniform before the trial begins, the discharge need not be collected during more than ten or fifteen minutes, and thus a series of distinct trials under different conditions can be made in a single day.

If the engine has steam-jackets the water condensed in them must be measured in addition to the water discharged from the condenser; and even when the whole supply of steam is inferred from the feed it may be desirable to determine separately the amount that is used in the jackets. This is done by draining them into a tank or tanks so that the condensed water may be weighed. The water must escape freely enough to prevent its accumulating in the jacket and yet not so freely as to let steam blow through, which may be secured by fitting a gauge-glass to a vertical part of the jacket drain, with a throttle-valve on the drain below it. By adjusting this valve the escape of water condensed in the jackets can be regulated so that the surface of the water will show itself in the glass at a constant height; the water is then passing off just as fast as it is condensed. To prevent evaporation of the discharged water the continuation of the drain may pass in the form of a bend or worm through a tank of cold water so that the jacket-water may be cooled before it reaches the vessel in which it is to be measured.

173. Comparison of feed-water with discharged water. In many trials the quantity of steam used by the engine is measured by both of the means that have been described, namely, by finding on the one hand how much feed is supplied to the boiler, and on the other hand how much is discharged from the condenser and the jacket drains. In most cases some discrepancy is observed: the feed-water may be as much as five per cent. more than the discharged water. This apparent loss of substance is due in part to water-vapour being discharged from the air-pump without being included in the measurement, but it is mainly due to leakage. Steam may escape at joints in small quantities showing little trace of its presence, and there is often some leakage into the fire-box, as for instance at the ends of tubes. In most cases the measurement of the water discharged by an engine gives a fairer test of its performance than is given by measuring the feed.

174. Calculation of heat supplied and heat rejected. Knowing the amount of steam supplied to the cylinder and jackets we may go on to calculate the amount of heat which the working substance takes up. In modern practice the steam is usually more or less superheated, but if there is no superheating the supply may be somewhat wet. In the absence of information[1] as to wetness the assumption that the steam is dry is the only safe one, though this may do some injustice to the engine by over-estimating the supply of heat. The pressure and temperature of the steam and

[1] Attempts have been made to measure wetness in a steam-supply by a device called a "throttling calorimeter," in which a sample drawn from the supply pipe is passed through a reducing valve and the amount of superheating thereby produced is observed. But the difficulty of ensuring that the steam so treated shall be in a condition representative of the whole supply, when it reaches the reducing valve, makes this method quite untrustworthy.

the temperature of the feed having been noted, the quantity of heat taken up by each lb. may be treated (§ 64) as sensibly equal to

$$I' - I_{w_0},$$

where I' is the total heat of steam under the conditions of supply (assuming for greater generality that there may be superheat) and I_{w_0} is the total heat of water at the temperature at which the feed is delivered to the boiler or to the economizer if there is one. From the thermodynamic point of view an economizer in which the feed is heated by the furnace gases is to be regarded as part of the boiler.

The heat rejected is measured by observing the quantity of the condensing water and the amount by which its temperature rises as it passes through the condenser. With small engines the quantity

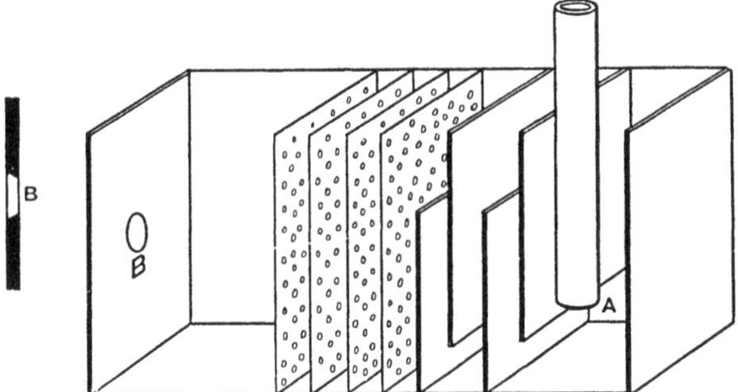

FIG. 111. Weir-box with circular orifice.

may be found by direct weighing or measuring in a large tank, or by the use of a pair of measuring tanks arranged so that one fills while the other empties. But in general the quantity of condensing water is too great to be easily treated in this way, and it has rather to be gauged as a stream, by observing the *head* under which it flows through an orifice of known size, or over a weir. This gauging is generally done after the water leaves the condenser, in which case, if the condenser is of the injection type, the quantity that is measured is the sum of the cooling water and the condensed steam, and the amount of the cooling water alone can be inferred by deducting from the whole a measured or estimated allowance to represent the feed.

When the stream to be gauged is large an open weir with a rectangular or V-shaped notch will be found most convenient: but for small streams a submerged circular orifice has the advantage that the accuracy of the result is less affected by any small error that may be made in measuring the head. The stream to be gauged enters at A (fig. 111) a box containing baffle plates and perforated diaphragms (sheets of perforated zinc or gauze will do well), so that turbulence disappears before it reaches the orifice B. This is a circular hole in a flat plate, and is bevelled to a sharp edge with the bevel outside. The head of water in the chamber close to the orifice is to be observed by means of a device such as a float and scale, or a point-gauge, which is not shown in the diagram. If h is the head in feet, measured from the surface to the centre of the hole, and s is the area of the hole in square feet, the discharge Q in cubic feet per second is given by the formula

$$Q = cs\sqrt{2gh},$$

where c is a "coefficient of discharge," the ordinary value of which for a circular hole in a large flat surface is 0·62, when the head is sufficient to bring the surface of the water to a considerable height above the hole[1].

A compact and simple weir-box may easily be made by using a tall rectangular sheet-iron tank, a foot or so square in horizontal section and 3 or 4 feet high. The pipe bringing in the water is brought down inside the tank, in a corner, and opens close to the bottom. The lower part of the tank is filled by sheets of gauze serving as baffle plates. The water rises steadily above these, and escapes by one or more sharp-edged circular orifices cut in the side of the tank about a foot from the top.

175. Wetness of the steam during expansion. In § 117 it was explained how to find the proportion of water present in the cylinder at any stage of the expansion and to represent the results

[1] When an open rectangular notch with sharp edges in a vertical plate is used for a weir, h is to be measured from the bottom of the notch to the free level of the surface, at a distance far enough back to give practically still water; then

$$Q = 3{\cdot}33\,(b - 0{\cdot}2h)h^{\frac{3}{2}},$$

where b is the breadth of the notch.

With a triangular notch cut so that the breadth is twice the depth $Q = 2{\cdot}54h^{\frac{5}{2}}$, where h is the depth of the bottom of the notch below the still-water surface level. For the justification of these formulas reference must be made to books on hydraulics or to papers by James Thomson, *Rep. Brit. Assoc.* 1858, 1861, and 1876, p. 243.

of the calculation graphically by means of a "saturation curve" upon the indicator diagram—namely, a curve which represents the volume which the steam in the cylinder should fill at any pressure if it were dry throughout. To draw this curve for either side of the piston we should in strictness know how the whole amount of the cylinder feed is shared by the two ends of the cylinder—a matter which the test does not determine. But in general the action in the two ends is so nearly symmetrical that practically correct results may be obtained by combining the indicator diagrams for the two into a mean diagram, taking for clearance the mean of the two actual clearances, and taking half the cylinder feed per revolution as the quantity of steam that enters the cylinder per stroke. The diagram shown in § 117 is in fact a combination diagram drawn in that way. To determine the wetness of the steam during expansion is an important part of an engine test, and it is useful to exhibit the results, so far as this particular is concerned, by showing the saturation curve alongside of the actual curve of pressure and volume. In dealing with compound engines a saturation curve may be drawn separately for each cylinder, or the diagrams for the several cylinders may be combined into one by means of a device which will be described in the next chapter.

The process of estimating the water present during expansion by comparing the saturation volume with the volume actually filled by the working substance depends on the assumption that the whole quantity of substance does not change from the time at which cut-off is complete until release begins. Any leakage of steam, in or out, through the valve or past the piston, will invalidate the calculation.

Having determined what proportion of the working substance is steam and what is water throughout the expansion we may go on to calculate how much heat is taken from or given to the walls of the cylinder and piston during any stage of its action. This analysis of the transfers of heat, introduced by Hirn and developed by his pupils and followers, has been pursued at great length in some engine tests[1]. It proceeds on the basis that if any two points in the action be taken and the internal energy calculated for each, the heat taken in or given out between the two is measured by

[1] See Dwelshauvers-Dery, "*Étude Calorimétrique de la Machine à Vapeur.*" Also J. G. Mair, *Min. Proc. Inst. C. E.* vols. LXX and LXXIX.

taking account of the change of internal energy together with the work done by or spent upon the substance between the two points, which is determined by reference to the pressure-volume curve of the indicator diagram. The data are not in general very exact, and the analysis is not particularly instructive. As regards the heat stored and restored during expansion, this is readily found from the entropy-temperature diagram in a manner which has already been sufficiently explained.

176. Tests of mechanical efficiency. Measurement of brake horse-power. In tests of mechanical efficiency the engine is commonly set to work against some form of friction brake arranged to serve as an absorption dynamometer. For engines of small power no form is so simple or so easy of application as a rope or band brake of the type shown in fig. 112. Two, three, or more parallel turns of rope with a few wood blocks to hold them apart (the number of ropes depending on the quantity of power that is to be absorbed) are made to clasp the fly-wheel in the manner sketched; the slack end is attached to a spring balance and the other end is loaded with weights, either directly, or through a lever if the amount of load is inconveniently great. A little grease applied to the surface of the metal makes the brake work quietly and steadily. The wheel may be kept cool by water; a wheel with internal flanges on the rim, forming an internal channel to which cold water may be supplied, is convenient. The resistance is adjusted by varying the amount of the weight T_1. A platform or stop fixed a little way below this weight allows the brake to be applied and removed by pulling or slacking away the rope by which the spring balance is suspended; by pulling the rope which is shown at the top of the figure the weights are lifted off their platform and the brake comes into action. When the brake is in action the pull T_2 indicated by the spring balance is noted from time to time. The effective resistance is $T_1 - T_2$, and the work done against the

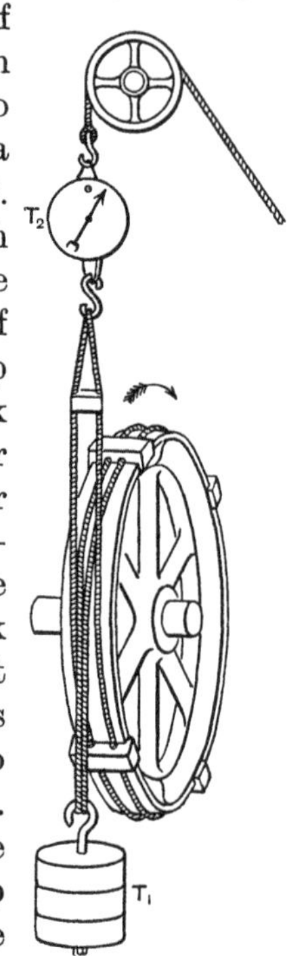

FIG. 112. Rope brake.

brake per revolution is $2\pi r\,(T_1 - T_2)$, where r is the radius measured from the axis of rotation to the middle of the rope's thickness. Hence the brake horse-power is

$$\frac{2\pi nr\,(T_1 - T_2)}{33{,}000}.$$

The mechanical efficiency is the fraction $\frac{\text{B.H.P.}}{\text{I.H.P.}}$; or, without reducing to horse-power, it is the ratio of the work done on the brake per revolution to the work done by the steam per revolution, namely,

$$\frac{2\pi r\,(T_1 - T_2)}{l\,(p_m a + p_m' a')},$$

in the notation of § 170.

A flexible band made by using a few strips of cotton listing has the advantage as compared with rope of working smoothly and silently without any lubrication and is to be preferred to rope for small engines. When only two or three horse-power have to be measured a single strip of listing will be found to make an excellent brake.

In dealing with large powers the most effective and accurate absorption dynamometer is the Froude type, in which the work of the engine is spent in churning water by turning a species of turbine wheel in a casing through which water is continuously passed. The casing is held from turning by applying weights to a lever arm, and this measures the moment exerted by the engine-shaft, on which the turbine wheel is fixed. Osborne Reynolds designed for his experiments a very perfect brake of this kind and applied it not only in engine trials but also, as was mentioned in Chapter II, to measure the mechanical equivalent of heat by observing, along with the work done, the quantity of water which in passing through the brake had its temperature raised from 0° C. to 100° C.[1]

In testing engines employed for pumping, the useful output may be determined by measuring the volume of water delivered and the pressure against which it is delivered, the product of these quantities giving the effective work got from the pump. This however is less than the brake horse-power by an amount which depends on the losses that occur in the pump itself. Similarly, in

[1] *Phil. Trans. R. S.* 1897; *Min. Proc. Inst. C. E.* 1889; *Collected Papers*, vol. II, pp. 336 and 601.

testing engines which generate electricity, the useful output of the dynamo is readily measured by electrical appliances, but here again the quantity measured is less than the brake horse-power by an amount which represents the loss in converting mechanical into electrical energy. In either case the efficiency determined, when the effective output is divided by the indicated power of the engine, represents a joint efficiency of engine and pump or engine and dynamo, and is necessarily less than the mechanical efficiency of the engine alone.

177. Torsion dynamometers. Various forms of dynamometer have been devised for measuring the work done by an engine by finding the torque or twisting moment transmitted through the shaft in the working of the engine. The need for such an instrument arose when the steam turbine was applied to the propulsion of ships, and the power could no longer be found by "indicating" the engine. A portion of the length of the shaft is taken, and by static experiments or by calculation from the known dimensions and known modulus of rigidity of the material a relation is determined beforehand between the torque and the angle of twist. Then when the engine is running the angle of twist is observed or recorded, the corresponding torque is inferred, and the work done per revolution is calculated as $2\pi M$, where M is the torque. Several devices have been employed for measuring the angle of twist, by finding the relative angular displacement of two wheels or collars which are clamped to the shaft at a suitable distance apart. In one of the earliest instruments of the kind, the Denny and Johnson torsionmeter, an electrical appliance was used for measuring the relative displacement of the collars, and a length between them of 15 or 20 feet was the least that would give satisfactory results. With such torsional strains as occur in propeller shafting the displacement under full load is only about one two-thousandth part of the distance between the collars. Refined means of measurement are therefore necessary if a reasonable percentage of accuracy is to be attained on a short length of shaft. In Hopkinson and Thring's torsionmeter a length of three or four feet is enough; the relative displacement of the two collars is measured by the tilting of a mirror carried by one of them in such a manner that an arm projecting from the other tilts it through a small angle which is proportional to the torque. The amount of the tilt is observed by using a scale and lamp at the side. The arrangement is illustrated in fig. 113. Collars are clamped to the shaft at A and B, the collar clamped at B having a long sleeve which extends over the intervening length and is an easy fit over the collar

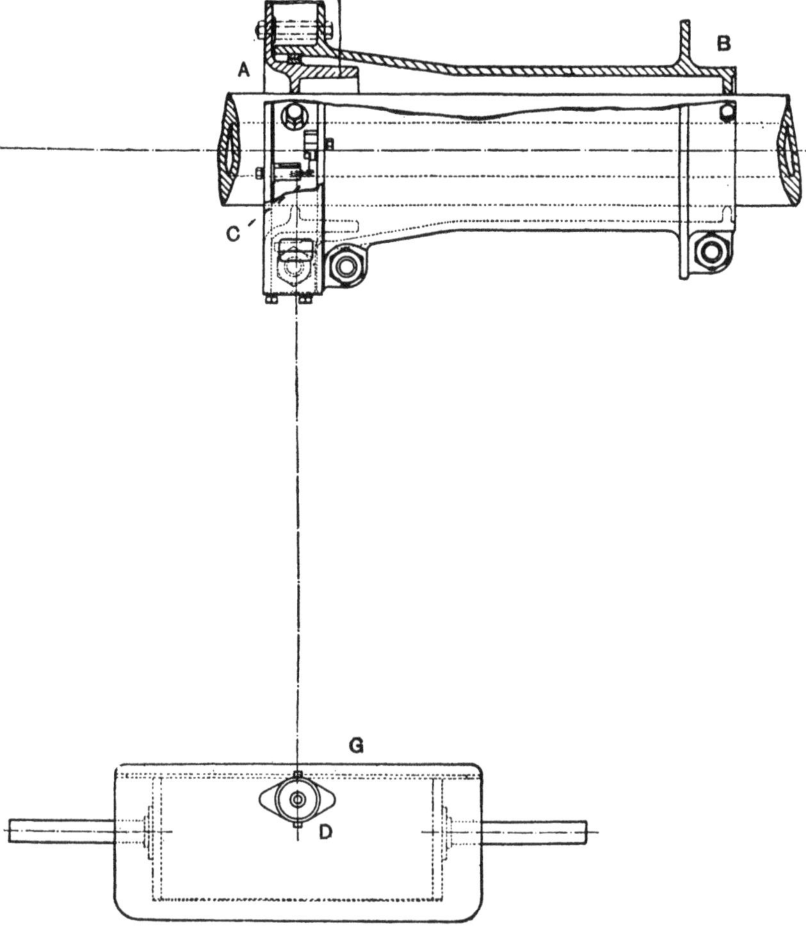

FIG. 113. Hopkinson-Thring torsionmeter.

clamped at A. There is provision for a small amount of relative turning which has the effect of tilting the little mirror C and thereby deflecting a beam of light which comes from the lamp D and is reflected to the fixed scale G. The collar also carries a fixed mirror, reflecting another spot of light from the same lamp: the two spots are adjusted to coincide on the scale when the torque is zero, and the torque is therefore measured by the distance between them. It should be noticed that what is measured is the torque at the instant, in each revolution, when the reflected beam passes the screen. If the torque varies throughout the revolution, as it will do in a shaft driven by reciprocating engines, an approximate mean may be found by using three or more mirrors spaced regularly round the shaft[1].

Among other devices for the same purpose is the Denny-Edgecumbe torsionmeter, depending on direct vision[2]; also the Ford electrical torsionmeter by Siemens Brothers, where the displacement between the two collars clamped to the shaft is measured by its effect on a pair of air-gaps in the magnetic circuits of two small transformers mounted on them and connected to primary and secondary electric circuits outside by means of ring and brush contacts. The two air-gaps are equal where there is no torque; in that condition the transformers are balanced and there is no secondary electromotive force. When torque is produced one of the gaps becomes narrower and the other wider, with the result that a secondary current is produced which is proportional to the torque and may be read at any convenient place in the ship[3]. In another electrical form of torsionmeter by Mr E. B. Moullin, the relative displacement of the two collars alters the air-gaps of a pair of choking coils and thereby varies an alternating current whose amplitude is recorded by continuously photographing the indications of a Duddell oscillograph. The photographic record shows how the torque varies during each revolution when the shaft is piston driven, especially when it is driven by an internal-combustion engine. It also shows the wider variations that arise when the shaft runs through a critical speed in which the periodic fluctuations of turning moment synchronize with its natural frequency of torsional oscillation[4].

[1] For details of the instrument see a paper by Hopkinson, *Trans. Inst. Nav. Arch.* 1910, p. 184.

[2] *Engineering*, March 20, 1925.

[3] *Engineering*, Feb. 13, 1925.

[4] The Moullin torsionmeter was used by a joint committee appointed by the Institution of Mechanical Engineers and the Institution of Naval Architects to make trials of marine oil-engines. A description of the instrument with specimens of the records will be found in their Reports, *Proc. Inst. Mech. Eng.* Nov. 1924 and March 1925; *Proc. Inst. Nav. Arch.* 1925, p. 264.

178. Trials of an engine under various amounts of load. The Willans Line. Although some engines are required to work always under the same or nearly the same conditions as to load, more commonly the load is liable to variation, and it may be as important to examine the performance under light loads as to make trials at full power. In power stations, for example, much of the work is done at less than full load and the efficiency under such conditions is a matter of real interest. To be complete a trial should include a series of tests made at various grades of output from the largest down to a limit when the engine merely drives itself without doing external work. Such results may be represented graphically by drawing a curve in which the ordinates are the

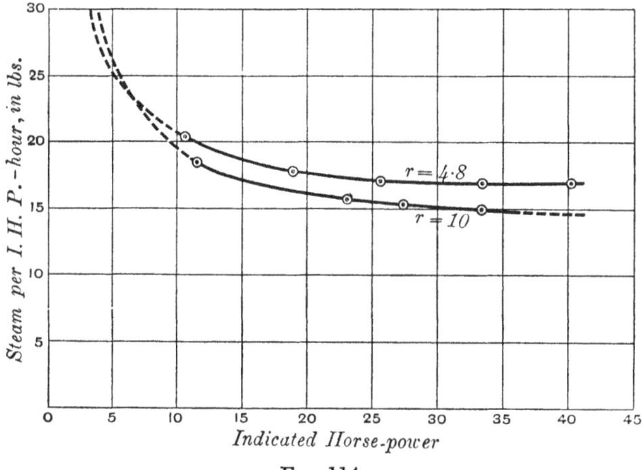

FIG. 114.

numbers of pounds of steam consumed per horse-power-hour, and the abscissæ are the rates of output in horse-power.

Two curves of this kind are shown in fig. 114, relating to an old series of tests by Willans of one of his compound high-speed single-acting engines, using a condenser[1]. One curve is for a set of trials in which the ratio of expansion was kept constant at 4·8, and the output was varied by altering the pressure of supply. The other curve is for a similar trial in which the ratio of expansion was kept constant at 10.

Another useful way of showing the performance at all powers is to plot the whole quantity of steam consumed per hour in

[1] *Min. Proc. Inst. C. E.* vol. CXIV, 1893.

relation to the horse-power. Curves of this kind were first used by Willans and are often called Willans Lines: examples of them are given in fig. 115 relating to the same two sets of trials as fig. 114. Under the conditions of these trials Willans found that the curve of total steam consumption in relation to power (fig. 115) was sensibly a straight line. With variable cut-off and constant pressure, on the other hand, the Willans line as found in other trials was very slightly curved, having a somewhat steeper gradient at high powers.

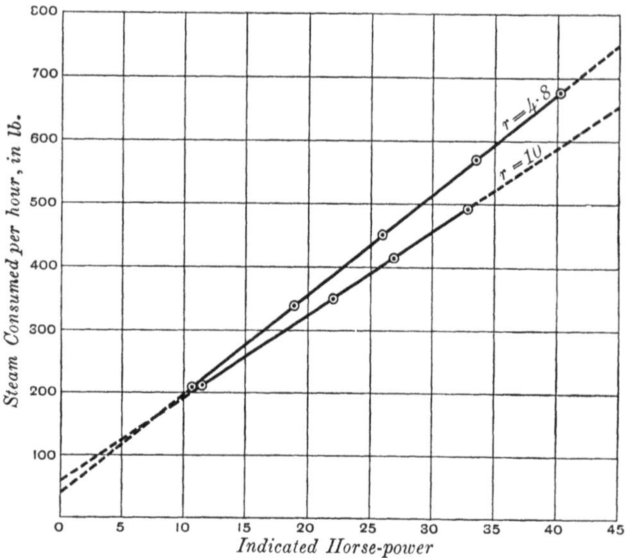

FIG. 115. Willans Lines.

The same types of diagram are useful in representing the consumption of steam in relation to brake horse-power, pump horse-power, electrical output in kilowatts, etc. They exhibit clearly what the efficiency will be under the various conditions as to output that may arise in practice.

When the Willans line is straight the whole consumption of steam at any load may be regarded as made up of two parts—the constant unproductive consumption that takes place without doing work and a further consumption that is simply proportional to the power developed. The whole consumption is equal to

$$a\,(n + b),$$

where n is the number of horse-power and a is the rate at which steam is taken per horse-power after the unproductive supply ab has been furnished. Such a formula is also applicable when the output of a turbine is expressed electrically, n being then the number of kilowatts, and a the rate at which the consumption of steam rises, per kilowatt of effective output, when the load is increased.

A Willans line may of course be drawn with the external or brake horse-power as the base; ab in the formula then represents the steam that is used in making the engine drive itself; and b is what may be called the "idle-work," a quantity which is somewhat greater than the indicated work done in overcoming engine friction.

A comparison of Willans lines relating to indicated and brake power respectively serves to show how far the work spent on engine friction remains constant for all outputs. In general it may be expected that this quantity will be greater at high output since the forces at the joints of the mechanism become on the whole increased.

Two further examples of Willans lines are given in figs. 116 and 117. Fig. 116 relates to tests of a Schmidt engine of 300 horse-power, using highly superheated steam. The following are par ticulars of the trials:—

Indicated horse-power	312	239	175	47
Steam per hour, lb.	2925	2150	1670	975
Steam per I.H.P.-hour, lb.	9·37	9·00	9·54	20·7
Boiler pressure (by gauge) in lb. per sq. inch	141	141	142	140
Temperature of steam on leaving superheater, ° C.	425	420	430	420
Vacuum, inches	27¼	28	28	29

In this case the line showing the total quantity of steam used per hour is distinctly curved. A line is also drawn to show the steam used per horse-power-hour. It passes a minimum at about 250 horse-power, the consumption then being only 9 lb. per indicated horse-power-hour. This result has been already referred to in Chapter VII, as an example of the economy attainable with high superheat, even in small engines.

The other example (fig. 117) relates to official tests of a Parsons turbine at the Carville "B" station of the Newcastle-on-Tyne Electric Supply Company (1920). Here the output was measured electrically as useful power delivered to the mains, and the Willans line is accordingly drawn on a kilowatt base, for loads which ranged from 5000 to 11,000 kw. Reduced to standard conditions, namely, with a stop-valve pressure of 250 lb. per sq. inch by gauge, a

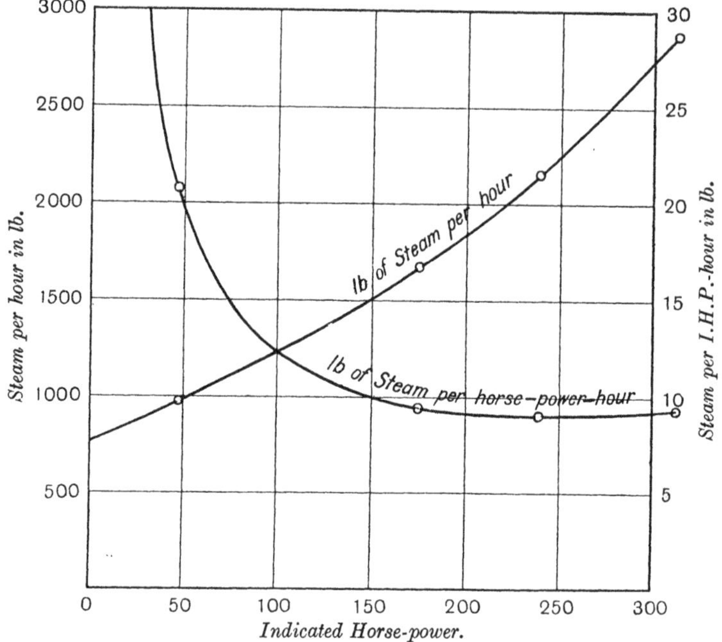

FIG. 116. Willans Line for tests of a Schmidt engine.

stop-valve temperature of 650° F., and vacuum 29 inches, the tests gave the following results:—

Load in kilowatts	5060	7760	10,220	11,070
Steam per hour, lb.	57,200	82,900	106,500	115,000
Steam per kw.-hour, lb.	11·31	10·68	10·42	10·39

It will be observed that the four points fall well on a straight Willans line, which has the equation

$$\text{Steam per hour, in lb.} = 9{\cdot}6\,(n + 900),$$

where n is the load in kilowatts.

In the tests of this turbine it was found that with steam superheated rather more highly, namely to 696° F. (369° C.), the absolute pressure being 252 lb. per sq. inch and vacuum 29·04 inches, the consumption was only 10·04 lb. per kw.-hour, when the load was 10,000 kw. Under these conditions the adiabatic heat-drop was 256·7 calories. Since 1 kw.-hour is equivalent to 1895·6 pound-

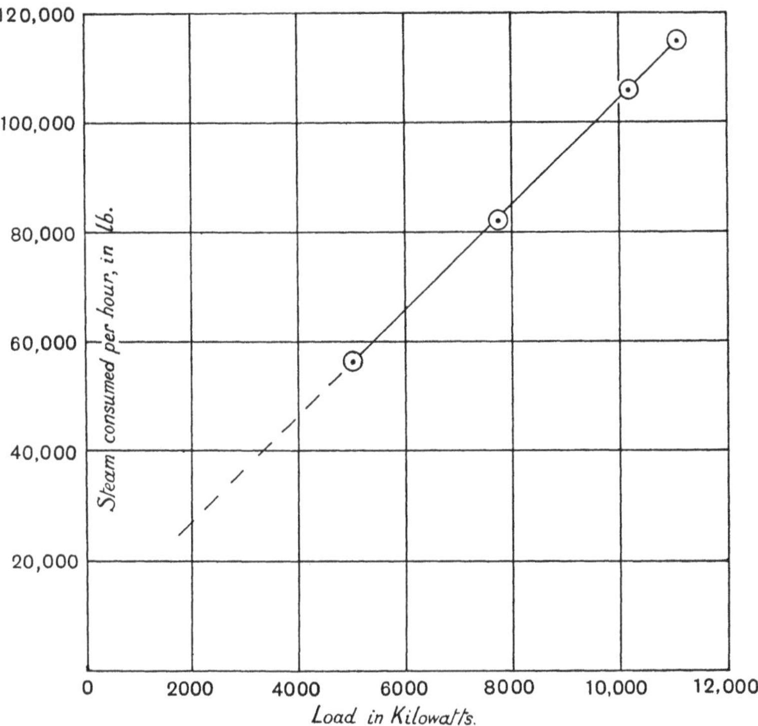

FIG. 117. Willans Line for tests of a Parsons Turbine.

calories, the quantity of heat converted into useful electrical work per lb. of steam was $\frac{1895 \cdot 6}{10 \cdot 04}$ or 188·8 calories. This represents $\frac{188 \cdot 8}{256 \cdot 7}$ or 0·736 of the adiabatic heat-drop.

We may also compare the useful output with the heat taken in, which is equal to the total heat in the steam as supplied minus the heat present in the feed-water. Here the feed-water, taken direct from the condensate, brings back to the boiler 25·3 calories per lb. The total heat of the steam as supplied was 763·1 calories.

Hence the heat taken in per lb. was 737·8 calories. Of this the amount converted into electrical energy was 188·8 calories, or 25·6 per cent.

179. Curve of expansion to be assumed in estimating the probable indicated horse-power of steam-engines. Although, in a reciprocating engine, the exchanges of heat between the steam and the cylinder largely affect the consumption of steam, it will be seen by examining indicator diagrams that their influence on the form of the expansion curve is but slight. In practical cases the curve is never very different from a rectangular hyperbola. The simple supposition that the pressure during expansion varies inversely with the volume will answer sufficiently well for the purpose of estimating the probable power to be exerted by an engine of given size, when the speed, the initial pressure, the back-pressure and the ratio of expansion are assigned.

If there were no clearance, if the pressure of admission, p_1, were maintained up to the point of cut-off, if the expansion continued to the very end of the stroke, and if there were a sharp release and a uniform back-pressure p_b, without compression, then the assumption that the curve of expansion may be treated as a common hyperbola would make the mean effective pressure equal to

$$\frac{p_1 (1 + \log_\epsilon r)}{r} - p_b,$$

where r is the ratio of expansion, namely, the ratio which the whole volume of the stroke bears to the volume that is swept through up to the point of cut-off. To show this, we may express the area of the indicator diagram as

$$p_1 v_1 + \int_{v_1}^{v_2} p dv - p_b v_2,$$

where v_1 and v_2 are the volumes at cut-off and release respectively, so that $r = \frac{v_2}{v_1}$.

Since by assumption pv at any point in the expansion $= p_1 v_1$, the area is

$$p_1 v_1 \left(1 + \int_{v_1}^{v_2} \frac{dv}{v}\right) - p_b v_2$$
$$= p_1 v_1 (1 + \log_\epsilon r) - p_b v_2.$$

The mean effective pressure is found by dividing the area by v_2, whence the formula given above is obtained. The same formula would be applicable to compound expansion, r being interpreted as the final ratio, if the further condition were satisfied that there should be no loss of pressure during the transfer of the steam from one cylinder to the next.

In practice, of course, these conditions are not fulfilled, and the general result is to make the actual mean effective pressure p_m less than the pressure calculated in this manner, in a proportion which is sometimes stated by the use of a coefficient e, thus:—

$$p_m = e\left\{\frac{p_1(1 + \log_\epsilon r)}{r} - p_b\right\}.$$

The diagram factor, as Professor Unwin has called e, is a number less than unity, which may be estimated as a matter of experience from the results given by other engines of like types, working under more or less similar conditions.

CHAPTER X

COMPOUND EXPANSION

180. Woolf engines. An engine of the piston and cylinder type is said to be compound when the expansion is begun in one cylinder and continued in another, or in more than one other. The steam may be made to pass directly from one cylinder to the next, or it may pass from the first cylinder into an intermediate chamber, called a *receiver*, where it is temporarily stored, and from which the second cylinder draws its supply. An advantage of the latter plan is that it does not require the reception of steam by the second cylinder to be simultaneous with the rejection of steam by the first. This allows the cranks to be set at any angle: it also allows the distribution of the expansion between the two cylinders to be readily adjusted. Compound engines are rarely used with immediate transfer of steam from one cylinder to the other. The connecting passages between the cylinders have a certain capacity and form a receiver even in cases where no special chamber is added for the purpose.

The original form of compound engine invented by Hornblower and revived by Woolf had no receiver. Steam passed directly from the high to the low-pressure cylinder, entering the low-pressure cylinder during the whole period of exhaust from the high-pressure cylinder. Such an action can occur only when the high and low-pressure pistons begin and end their strokes together, that is to say, when their movements either coincide in phase or differ by half a revolution. This was the case in many of the early compound engines, which were beam engines with a high-pressure cylinder and a low-pressure cylinder standing side by side, the pistons rising and falling together. Engines whose high and low-pressure cylinders are placed side by side, and act either on the same crank or on cranks set at 180° apart, may discharge steam directly from one to the other cylinder. This kind of working is also possible in tandem engines, which have the high and low-pressure cylinders in one line, with one piston-rod common to both pistons. The name "Woolf engine" has been applied to those compound engines which discharge steam directly from the high to the low-pressure cylinder without the use of an intermediate receiver.

181. Receiver engines. An intermediate receiver becomes necessary when the phases of the pistons in a compound engine do not agree. With two cranks at right angles, for example, a portion of the discharge from the high-pressure cylinder occurs at a time when the low-pressure cylinder cannot properly receive steam. The receiver may be an independent vessel connected to the cylinders by steam-pipes; generally, however, a sufficient amount of receiver volume is afforded by the steam chests containing the valves, together with the steam-pipe which connects the cylinders. In marine engines for example the receiver consists simply of the steam chests and connecting pipe, the capacity which these passages necessarily have in order to let the steam pass freely being sufficient for the amount of storage that is required.

Even when the phases of the pistons do agree, as in tandem engines, it is generally preferred to treat the connecting passages as a receiver rather than transfer the steam direct from one cylinder to the other. This is partly because their use as a receiver makes it easy to adjust the proportion of work done by each cylinder.

When a receiver is used the low-pressure cylinder, as well as the high-pressure cylinder, has a definite point of cut-off. During part of the stroke the receiver is storing steam, and the steam which has been admitted to the low-pressure cylinder is expanding independently in that cylinder. By varying the position in the low-pressure stroke at which cut-off takes place the distribution of work between the two cylinders may be adjusted at will.

As a rule matters are regulated so as to make both cylinders do equal, or nearly equal, amounts of work. If the pistons act on separate cranks this has the effect of giving the same value to the mean turning moment for both cranks. The adjustment of the work will be discussed in § 183 below.

182. Drop in the Receiver. Compound diagrams. Care must be taken that there is no large amount of unresisted expansion into the receiver; in other words, the pressure in it should not be greatly below that in the high-pressure cylinder at the moment of release. Any drop of pressure as the steam passes from the high-pressure cylinder to the receiver will show itself in an indicator diagram by a sudden fall at the end of the high-pressure expansion. From the thermodynamic point of view it is irreversible, and therefore wasteful. Practically some small amount of drop is desirable for the same reasons which make a rather incomplete expansion preferable to complete expansion in the working of a single cylinder. Drop can be avoided or reduced to any desired extent, as we shall presently see, by selecting a proper point of cut-off in the low-pressure cylinder. When there is no drop the expansion that occurs in a compound engine has precisely the same effect in doing work as the same amount of expansion would have in a simple engine, provided the law of expansion be the same in both and the waste of energy which occurs by the friction of ports and passages in the transfer of steam from one to the other cylinder be negligible. The work done in either case depends merely on the relation of pressure to volume throughout the process: and so long as that relation is unchanged it is a matter of indifference whether the expansion be performed in one vessel or in more than one. It has, however, been explained in Chapter VII that in practice

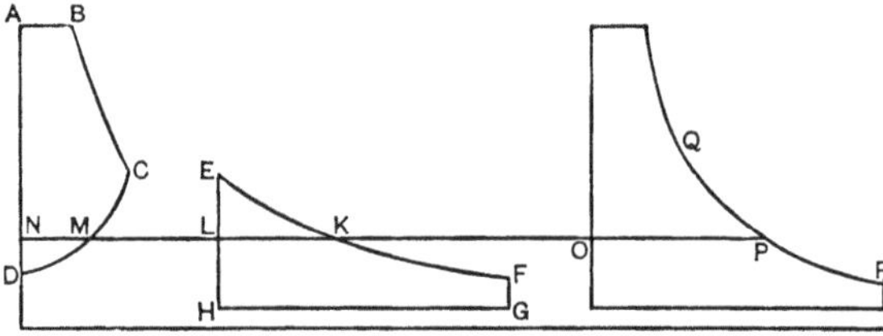

FIG. 118. Compound Diagrams: Woolf type.

a compound engine has a thermodynamic advantage over a simple engine using the same pressure and the same expansion, inasmuch as it reduces the exchange of heat between the working substance and the cylinder walls and so makes the process of expansion more nearly adiabatic.

The ultimate ratio of expansion in any compound engine is the ratio of the volume of the low-pressure cylinder to the volume of steam present in the high-pressure cylinder at the point of cut-off. Fig. 118 illustrates the combined action of the two cylinders in a hypothetical compound engine of the Woolf type, in which for simplicity the effect of clearance is neglected and also the loss of pressure which the steam undergoes in transfer from one to the other cylinder. *ABCD* is the indicator diagram of the high-pressure cylinder. The exhaust line *CD* shows a falling pressure in consequence of the increase of volume which the steam is then undergoing through the advance of the low-pressure piston. *EFGH* is the diagram of the low-pressure cylinder, and is drawn alongside of the other for convenience in the construction which follows. It has no point of cut-off; its admission line is the continuous curve of expansion EF, at each point of which the pressure is the same as at the corresponding point in the high-pressure exhaust line CD. At any point K, the actual volume of the steam is $KL + MN$. By drawing OP equal to $KL + MN$, so that OP represents the whole volume, and repeating the same construction at other points of the diagram, we may set out the curve QPR, the upper part of which is identical with BC, and so complete a single diagram which exhibits the equivalent expansion in a single cylinder. The area of the figure so drawn is equal to the sum of the areas of the high-pressure and low-pressure diagrams.

In a tandem compound engine of the receiver type the diagrams resemble those shown in fig. 119. During CD (which corresponds to FG) expansion is taking place into the large or low-pressure cylinder. D and G mark the point of cut-off in the large cylinder, after which GH shows the independent expansion of the steam now shut within the large cylinder, and DE shows the compression of steam by continued discharge from the small cylinder into the receiver. At the end of the stroke the receiver pressure is OE, and this must be the same as the pressure at C, if there is to be no drop of pressure at release. In the diagram sketched it is assumed that there is none. The case of drop would be illustrated if we were to cut off the corner at C by a vertical line drawn from some earlier point in BC to meet the curve CD; this would of course also imply a shortened high-pressure stroke. Diagrams of a similar kind may be sketched without difficulty for the case of a receiver engine with any assigned phase-relation between the pistons. It may be noted in passing that a receiver has the thermodynamic advantage (over the Woolf type) that it reduces the range of temperature in the high-pressure cylinder, and so helps to prevent initial condensation of the steam.

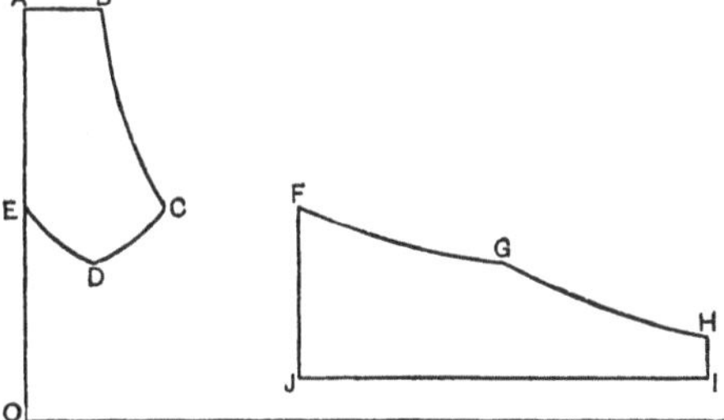

FIG. 119. Compound Diagrams: Receiver type.

183. Adjustment of the division of work between the cylinders, and of the drop. Graphic method. By making the cut-off take place earlier in the large cylinder we increase the mean pressure in the receiver; the work done in the small cylinder is consequently diminished. The work done in the large cylinder is correspondingly increased, for the total work (depending as it does almost wholly on the initial pressure and the total ratio of expansion) is unaffected or scarcely affected by the change. Hence we have the apparently anomalous result that a shorter admission to the low-pressure cylinder causes it to do a larger share of the whole work.

Further, the same adjustment—namely, hastening the cut-off in the low-pressure cylinder—serves, in case there is drop of pressure

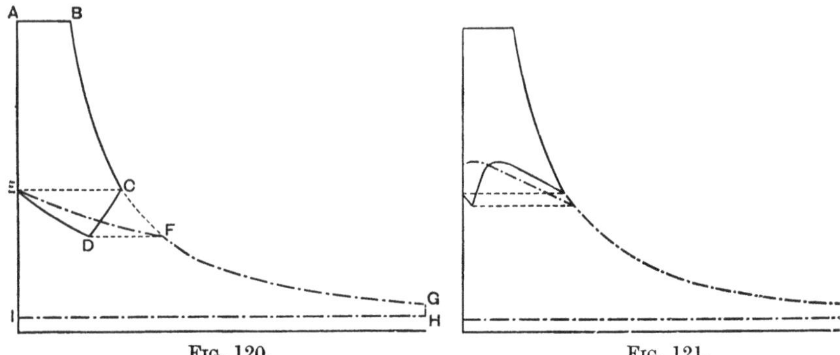

FIG. 120. FIG. 121.

FIGS. 120 and 121. Determination of the point of cut-off in the low-pressure cylinder of a compound engine.

at release, to remove it, for it raises the pressure in the receiver. By selecting suitable values of the ratio of cylinder volumes to one another and to the volume of the receiver, and also by choosing a proper point for the low-pressure cut-off, it is possible to secure absence of drop along with equality in the division of the work between the two cylinders.

To determine beforehand that point of cut-off in the low-pressure cylinder which will prevent drop when the ratio of cylinder and receiver volumes is assigned is a problem most easily solved, or approximately solved, by a graphic process. The process consists in drawing the curve of pressure during admission to the low-pressure cylinder until it meets the curve of expansion which is common to both cylinders. In fig. 120 (where for the sake of simplicity the effects of clearance are neglected) AB represents the admission line and BC the expansion line in the small cylinder. Release occurs at C, and from C to D steam is being taken by the large cylinder. D corresponds to the cut-off in the large cylinder, which is the point to be found. From D to E steam is being compressed into the receiver. If it is desired to avoid drop the receiver pressure at E is to be the same as the pressure at C. E is therefore known, and may be employed as the starting-point in drawing a curve EF which is the admission line of the low-pressure diagram $EFGHI$. This line is drawn by considering at each point in the low-pressure piston's stroke what is then the whole volume of the steam. The place at which EF intersects the continuous expansion curve BCG determines the point of cut-off which will give no drop. The sketch (fig. 120) refers to the case of a tandem receiver engine; but the process may also be applied to an engine with any assumed phase-relation between the cranks. Fig. 121 shows a pair of theoretical indicator diagrams determined in the same way for an engine with cranks at right angles, the low-pressure crank leading. In these examples the volume of the receiver has been taken equal to the volume of the high-pressure cylinder. With a larger receiver the variations of pressure during the back stroke of the high-pressure piston would be less conspicuous. In using the graphic method any form may be assigned to the curve of expansion. Generally this curve may be treated without serious inaccuracy as a common hyperbola, in which the pressure varies inversely as the volume. The construction may obviously be applied to triple and quadruple expansion engines. For an accurate solution it would be necessary to take the effect of clearance into account and also to allow for some loss of pressure in the passage from one vessel to another. The figures given here omit these complications, and treat the expansion as hyperbolic.

184. Algebraic method. When this simple relation between pressure and volume is assumed, it is not difficult to find algebraically the low-pressure cut-off which will give no drop, with assigned ratios of cylinder and receiver volumes. Taking the simplest case—that of a tandem engine, or of an engine with parallel cylinders whose pistons move together or in opposition—we may proceed thus. Since the point of cut-off to be determined depends on volume ratios we may for brevity treat the volume of the small cylinder as unity. Let R be the volume ratio of the receiver to the small cylinder, and L the volume ratio of the large to the small cylinder. Let x be the required fraction of the stroke at which cut-off is to occur in the large cylinder; and let p be the pressure at release from the small cylinder. If there is to be no drop, p is also the pressure in the receiver at the beginning of admission to the large cylinder. During that admission the volume changes from $1 + R$ to $1 - x + R + xL$, and the pressure at cut-off is therefore $\frac{p(1+R)}{1-x+R+xL}$. The steam that remains is now compressed into the receiver, from volume $1 - x + R$ to volume R. Its pressure therefore rises to

$$\frac{p(1+R)}{1-x+R+xL} \cdot \frac{(1-x+R)}{R},$$

and this, by assumption, is to be equal to p. We therefore have

$$(1+R)(1-x+R) = R(1-x+R+xL),$$

whence $$x = \frac{R+1}{RL+1}.$$

Thus, with $R = 1$ and $L = 3$, cut-off should occur in the large cylinder at half-stroke (which is the case illustrated by the diagram of fig. 119); with a greater cylinder ratio the cut-off in the large cylinder should be earlier, as it is, for instance, in fig. 120.

The above expression may be written $x = \dfrac{1+\frac{1}{R}}{L+\frac{1}{R}}$, from which it will be seen that when the receiver is very large x approximates in value to $\frac{1}{L}$.

A similar calculation for a compound engine whose cranks are at right angles, and in which cut-off occurs in the large cylinder before half-stroke, shows that the condition of no drop is secured when

$$2R(xL - 1) = 1 - 2\sqrt{x(1-x)}.$$

In some compound engines a pair of high-pressure cylinders discharge into a common receiver; in some a pair of low-pressure cylinders are fed from a receiver which takes steam from one high-pressure cylinder, or in some instances from two. With these arrangements the pressure in the receiver may be kept much more nearly constant than is possible with the ordinary two-cylinder type.

185. Ratio of cylinder volumes. The size of the low-pressure cylinder in a compound engine is fixed by reference to the power the engine is intended to develop, when the speed, the boiler pressure, and the total ratio of expansion are given. But the size of the high-pressure cylinder remains a matter of choice when all these things are settled. Say that the total ratio of expansion is to be r; we may choose any ratio L less than r for the volume ratio of the large to the small cylinder. It will then be necessary to make the cut-off in the small cylinder happen at a fraction of the stroke equal to $\frac{L}{r}$ in order that the final volume of the steam, when it fills the whole of the large cylinder, may be r times its initial volume up to the point of cut-off in the small cylinder. Thus an earlier or later adjustment of the cut-off in the high-pressure cylinder will allow the whole ratio of expansion to take whatever value may be wanted, no matter what may be the ratio of the cylinder volumes.

Again, as we have seen above, by varying the cut-off in the large cylinder we can adjust matters so that equal amounts of work are done in both cylinders, irrespective of their sizes.

But it is only when a suitable ratio of volumes has been selected that this adjustment to equalize the work will also secure a reasonable absence of drop—or that an adjustment of the low-pressure cut-off to avoid drop will not too seriously disturb the balance of work.

This consideration serves to fix in a general way the proper proportion of the volumes. No hard and fast rule is followed; an exact balance in the work is not essential, and a complete absence of drop is not even desirable. The same practical considerations which make it undesirable in a simple engine to have complete expansion apply in regard to compound engines: unless there is some little drop the last part of the stroke is ineffective.

In a two-cylinder compound condensing engine, using steam of about 100 lb. pressure, the large cylinder may have from four to four and a half times the volume of the small cylinder. In triple-expansion engines of the mercantile marine, with pressures usually ranging from 150 to 220 lb., the third cylinder has about 7 times the capacity of the first cylinder, and the second cylinder from 2·6 to 3 times that of the first. In land engines of this type, where an earlier cut-off may be resorted to without inconvenience, the first cylinder may be relatively rather larger. In quadruple expansion engines the proportion of volumes in the successive cylinders is usually something like 1 : 2 : 4 : 8, though in some examples the fourth cylinder may have $8\frac{1}{2}$ or even 9 times the volume of the first, the boiler pressure ranging generally from 200 to 250 lb. per sq. inch.

In many large compound engines, especially in marine practice, the last stage of the expansion is carried out in two cylinders instead of one, the steam being divided between the two. In this way the use of a single excessively large cylinder is avoided, and the arrangement often facilitates balancing, as will be shown in Chapter XIII. A type of marine engine that has been much used is one with triple expansion, but with four cylinders, two of which jointly serve to carry out the last stage of the compound action.

186. Thermodynamic advantage of compound expansion. The thermodynamic advantage of compound expansion has already been pointed out. It allows high-pressure steam to be used without the excessive waste which would occur if a high grade of expansion were attempted in a single cylinder. So long as the boiler pressure does not much exceed 100 lb. the advantage is sufficiently secured by dividing the expansion into two stages. Beyond this triple expansion becomes advisable, and when the expansion is divided into three stages it is advantageous to make pressure considerably higher: thus with triple engines a pressure of 160 to 200 lb. is usual. Quadruple expansion has little if any advantage when the pressure is under 200 lb.; up to this pressure and perhaps beyond it the thermodynamic benefit of a fourth stage is scarcely sufficient to justify the mechanical complication it involves. With water-tube boilers it is easy to go to pressures which justify quadruple expansion.

187. Mechanical advantage of compound expansion. Uniformity of effort in a compound engine. A simple engine using high-pressure steam with an early cut-off has the drawback, from the mechanical point of view, that the thrust of the steam on the piston during the early part of the stroke is very great in comparison with the mean thrust. The initial pressure of the steam acts on the full area of a piston whose size is determined by reference to the mean pressure. The piston and connecting-rod, the framing and other parts of the machine must be made strong enough for this relatively great initial thrust, and for steady motion a large fly-wheel becomes necessary.

The compound engine avoids the extreme thrust and pull which would have to be borne by the piston-rod of a single-cylinder engine working at the same power with the same initial pressure and the same ratio of expansion. If all the expansion took place in the low-pressure cylinder, the piston at the beginning of the stroke would be exposed to a thrust greater even than the sum of the thrusts on the two pistons of a compound engine of equal power. Thus in the tandem-engine diagrams of fig. 118 the greatest sum of the thrusts will be found to amount to less than two-thirds of the thrust which the large piston would be subjected to if the engine were simple. The mean thrust throughout the stroke is of course not affected by compounding; only the range of variation in the thrust is reduced. The effort on the crank-pin is consequently made more uniform, the strength of the parts may be reduced, and the friction and wear at joints lessened. Thus even in a tandem compound engine there is mechanically some advantage, and the benefit of compounding in this respect is greater when the cylinders are placed side by side, instead of tandem, and work on cranks at right angles. As a set-off to its advantage in giving a more uniform effort, the compound engine has the drawback of requiring more working parts than a simple engine with one cylinder. But in many instances—as in marine engines—two cranks and two cylinders are in any case almost indispensable, to give a tolerably uniform effort and to get over the dead-points without the aid of a heavy fly-wheel; and the comparison should then be made between a pair of simple cylinders and a pair of compounded cylinders. Another point in favour of the compound working is that, although the whole ratio of expansion is great, there need not be a very early cut-off in any cylinder; hence the common slide-valve, which is unsuited to give an early cut-off, may be used. The mechanical advantages of compound working were recognized sooner than its thermodynamic economy, and did much to bring it into favour before, indeed, the practice had grown up of using steam high enough in pressure to make compounding distinctly advantageous from the thermodynamic point of view.

Again, apart from its improved economy in fuel consumption, the mechanical merits of the triple engine contributed much to bring it to the position it has long held in marine practice. The advantage of three cranks over two in giving uniform effort and comparatively little friction and wear is conspicuous.

It also gives a better balance as regards the effects of inertia, and in this respect a great further gain may be secured by using four cranks, whether with quadruple expansion or, as is more usual, with triple expansion and a division of the third stage between two cylinders.

188. Examples of indicator diagrams from compound engines. A few representative indicator diagrams taken from compound engines are shown in the following figures. Fig. 122 gives a pair of diagrams from the two cylinders of a Woolf engine,

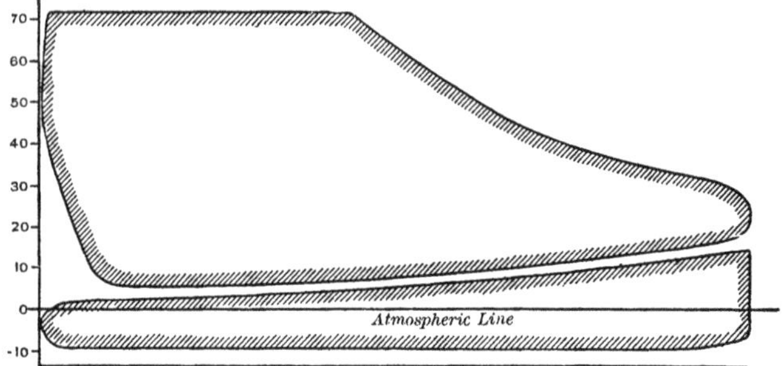

FIG. 122. Indicator diagrams of a Woolf engine.

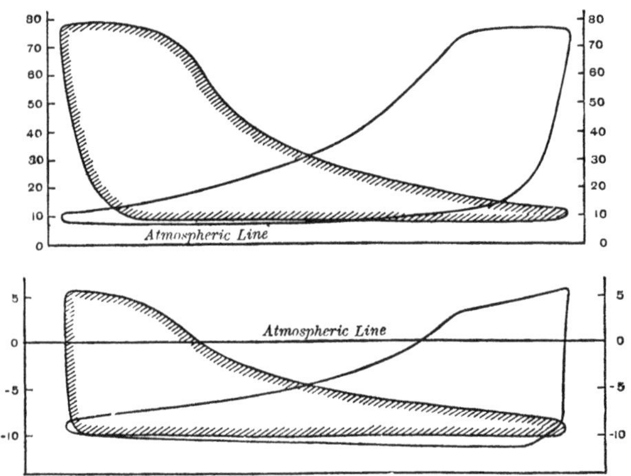

FIG. 123. Diagrams of a tandem receiver engine.

in which the steam passes as directly as possible from the small to the large cylinder. Both pistons have the same length of stroke. The diagrams are drawn to the same scale of stroke and therefore to different scales of volume, and the low-pressure diagram is turned round so that it may fit into the space below the high-pressure diagram. There is some drop at the high-pressure release, and frictional losses in the passages cause the admission line of the large cylinder to lie slightly lower than the exhaust line of the small cylinder. The transfer of steam goes on throughout nearly the whole of the back stroke until compression begins in the small cylinder. The steam then present in the large cylinder continues expanding for the small part of the stroke that is left until the point of release is reached.

Fig. 123 shows the diagrams of a tandem compound engine of the receiver type with cylinders 30 and 52 inches in diameter and 6 ft. stroke (volume ratio 1 to 3), taking steam at an initial pressure of 80 lb. by gauge. With this proportion of volumes and with the somewhat early cut-off shown by the diagrams there is a complete absence of any objectionable drop and a nearly equal division of work between the cylinders. Expansion valves (see Chapter XI) were used to produce this early cut-off. The exhaust line of the small cylinder dips in the middle, as in fig. 119 or 120, but much less, for here the receiver is more capacious. When the cranks are set at right angles this line rises towards the middle, as fig. 121 indicates.

Fig. 124 shows a set of triple-expansion diagrams, from trials of the steamship "Iona," by a Committee of the Institution of Mechanical Engineers. The cylinder diameters were 21·9 in., 34 in. and 57 in., giving a volume ratio of 1 : 2·4 : 6·8, and the stroke was 39 in. The engines made 61 revolutions per minute and developed 208 I.H.P. in the first cylinder, 217 in the second, and 220 in the third, with a consumption of 13·35 lb. of steam per I.H.P.-hour. A simple slide-valve was used on each cylinder[1].

189. Combination of the indicator diagrams in compound expansion. The indicator diagrams of a compound engine may be combined in such a way that the pressures and volumes in the several cylinders are displayed in proper relation to one another,

[1] Report of Research Committee on Marine Engine Trials, *Proc. Inst. Mech. Eng.* April 1891.

by the use of a single scale of pressures and a single scale of volumes. Some care, however, is necessary in the interpretation of such combined diagrams, and the construction to be adopted will depend on the object aimed at.

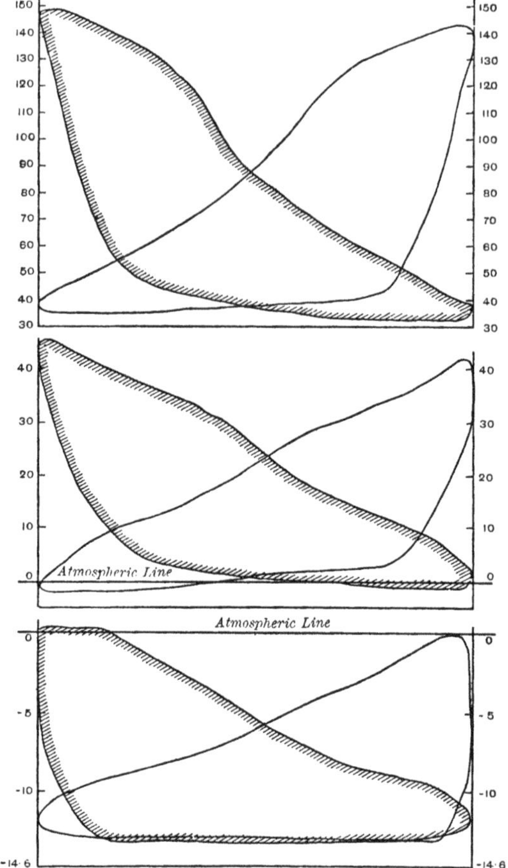

FIG. 124. Indicator diagrams of a triple-expansion engine.

A common practice is to set out each diagram from the line of no volume through a distance which represents the clearance in the corresponding cylinder. This is illustrated in fig. 125, where each of the two diagrams is a mean for the two sides of the piston, and the distance of each from the line OY is the mean clearance in the corresponding cylinder. Diagrams drawn in this way are not without their uses, but it must be remembered that the amount of substance which is taking part in the expansion is different in the two parts of the combination, and consequently a single adiabatic curve or a single saturation curve cannot properly be drawn to apply to both. The line SS is the saturation curve for the first stage of expansion, and the line $S'S'$ for the second stage, drawn in the manner which was described in § 117. In this example the cylinder feed per single stroke was 0·0498 lb., and the cushion steam was 0·0074 lb. in the small cylinder and 0·0022 lb.

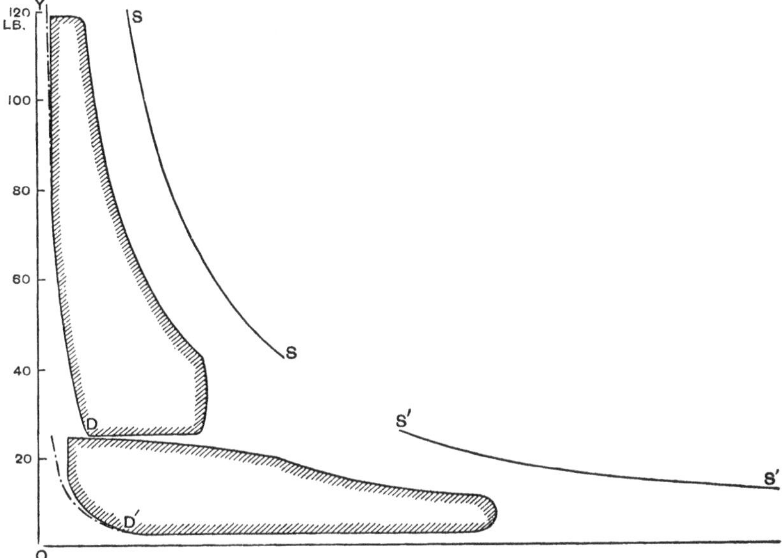

FIG. 125. Combination of compound diagrams.

in the large cylinder. The saturation curve SS is accordingly drawn for 0·0572 lb. and $S'S'$ for 0·052 lb.

The amount of the substance present in the cylinder is in general different in the successive stages because of differences in the amount of cushion steam in the several cylinders: the cylinder feed is the same throughout. If therefore we modify the diagram in such a way as to eliminate the cushion steam, leaving the cylinder feed only, we may draw a single saturation curve which will serve for the whole of the expansion.

This is done in fig. 126, which represents the same pair of diagrams, transformed by the following device. From points D, D'

(fig. 125) taken at the places where compression has begun and the exhaust is complete, saturation curves are drawn for the cushion steam in the respective cylinders. These curves are indicated by broken lines in the figure: the one that relates to the small cylinder is scarcely distinguishable from the compression curve of the indicator diagram. The diagrams are then redrawn as in fig. 126, using horizontal distances from these curves as abscissæ. This is equivalent to subtracting from the actual volumes throughout the diagram a quantity which represents the volume the cushion steam would occupy if it were saturated at all pressures. The result is

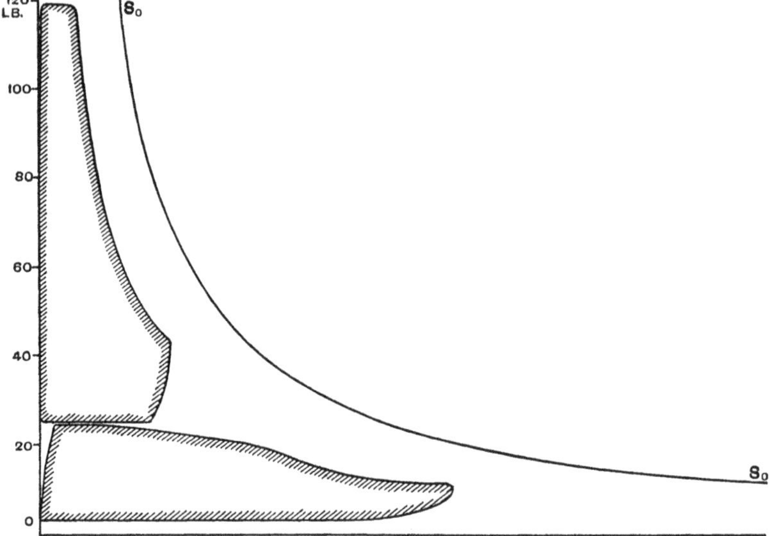

FIG. 126. Combined diagrams with cushion steam eliminated.

that the area of the diagram remains unaltered: its area is still a true measure of the work. But a single saturation curve S_0S_0 may now be drawn—namely, for a quantity of steam equal to the cylinder feed—which will apply equally to both (or all) stages of the compound expansion. The horizontal distance at any pressure between the expansion curve in fig. 126 and the saturation curve S_0S_0 is the same as the horizontal distance at that pressure between the expansion curve in fig. 125 and its corresponding saturation curve. It still represents the volume which has disappeared by condensation and leakage or what is often called the "missing quantity." The chief advantage of this construction is that it makes a single saturation curve possible, and so allows the changes in the amount of water present to be readily exhibited as the steam passes through the whole course of its expansion.

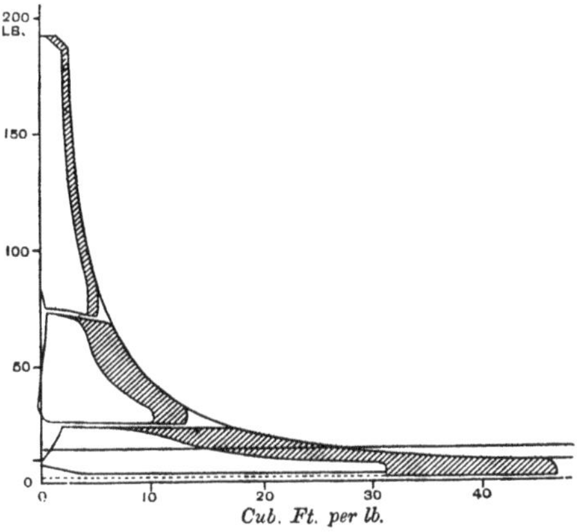

Fig. 127. Combined diagrams for a triple-expansion engine.

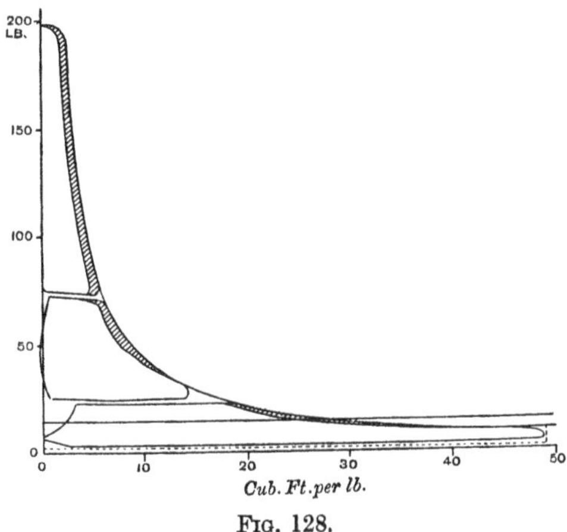

Fig. 128.

This will be apparent from figs. 127 and 128, which are copied from Osborne Reynolds' account of trials of a triple-expansion engine[1]. Here the cushion steam has been eliminated in the manner just described and a single saturation curve has been drawn for the cylinder feed. The horizontal width of shaded space between

[1] *Min. Proc. Inst. C. E.* vol. XC; *Collected Papers*, vol. II.

the actual expansion curves and this line measures the missing quantity at each stage in the expansion. Fig. 127 refers to a test made without steam in the steam-jackets, and fig. 128 to a test when all the jackets were supplied with steam at the full boiler pressure of 190 lb. The drying influence of the jacket is conspicuous: in fig. 128 there is scarcely any condensation in the third cylinder.

These diagrams relate to an engine built for experimental use in which the three pistons could move independently, at different speeds, and the speeds were in fact different. Hence to prepare the diagrams for combination a further device was employed: the common scale of length of the diagrams was chosen so that the volumes represented in each are reckoned per lb. of cylinder feed. The scale of volume is accordingly divided in the figures to show cubic feet per lb. of water passing through the engine. This method of graduation might be followed with advantage even in ordinary cases, where it is not rendered necessary by the pistons having independent speeds, for it facilitates comparison between various trials.

CHAPTER XI

VALVES AND VALVE-GEARS

190. The slide-valve. In early steam-engines the distribution of steam was effected by means of conical lift-valves, rising and falling on conical seats, and worked by tappets from a rod which hung from the beam. The slide-valve, the invention of which is credited to Murdoch, an assistant of Watt, came into general use with the introduction of locomotives, and is still, in one form or another, the most usual means of controlling the admission and exhaust of the steam in engines of the piston and cylinder type.

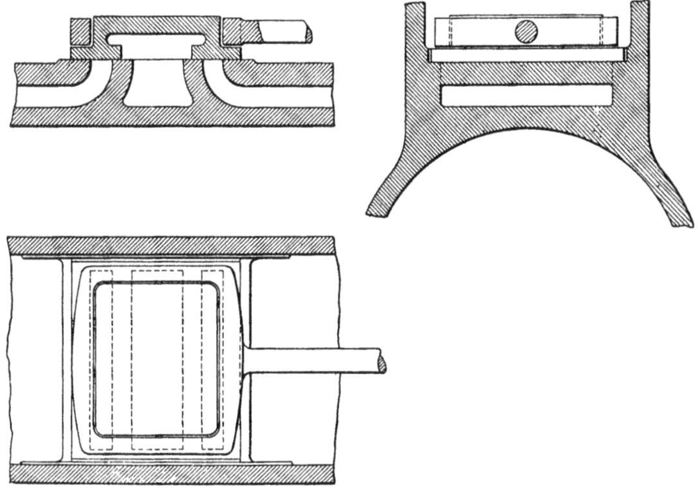

FIG. 129. Common Slide-Valve.

The common or D-shaped slide-valve is illustrated in fig. 129, which shows a sectional side and end elevation and a plan. The seat, or surface on which the valve slides, is a plane surface formed on or fixed to one side of the cylinder, with three openings called ports, which extend across the greater part of the cylinder's width. The ports are shown in the plan by dotted lines. The central opening is the exhaust-port through which the steam escapes; the others, or steam-ports, which are narrower, lead to the two ends of the cylinder respectively. The valve is a box-shaped cover which slides upon the seat, and the whole is enclosed in a chamber called the valve-chest, to which steam from the boiler is admitted. The valve is pulled backwards and forwards across the ports by means of a valve-rod which passes out of the valve-chest through a steam-tight stuffing-box. The valve is attached to the valve-rod, not rigidly but in such a way that while it has no longitudinal freedom to slide along the rod it is free to take a close bearing on the seat, under the pressure exerted by the steam on its back. In its middle position the valve covers both steam-ports completely, but when it is moved a sufficient distance to either side of the middle position, it allows fresh steam to enter one end of the cylinder from the valve-chest, and allows the steam which has done its work to escape from the other end of the cylinder through the cavity of the valve into the exhaust-port. The valve-rod is generally moved by an eccentric on the engine-shaft, which is mechanically equivalent to a crank whose radius is equal to the eccentricity, or distance of the centre of the shaft from the centre of the eccentric disc or sheave. The sheave is encircled by a strap to which the eccentric-rod is fixed, and the rod is connected by a pin-joint to the valve-rod outside of the valve-chest. The eccentric-rod is generally so long that the motion of the valve is sensibly the same as that which it would receive were the rod infinitely long. Thus if a circle (fig. 130) be drawn to represent the path of the eccentric-centre during a revolution of the engine, and a perpendicular PM be drawn from any point P on a diameter AB, the distance CM is the displacement of the valve from its middle position at the time when the eccentric-centre is at P. AB is the whole travel of the valve.

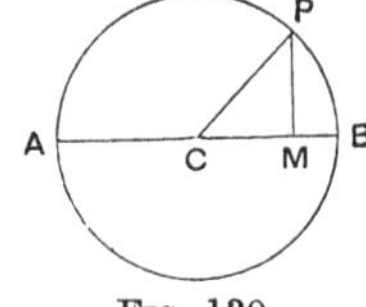

FIG. 130.

191. Lap, lead, and angular advance. If the valve were formed so that when in its middle position it did not overlap the steam-ports (fig. 131), any movement to the right or the left would admit steam, and the admission would continue until the valve had returned to its middle position, or, in other words, for half a revolution of the engine. Such a valve would not serve for expansive working; it would admit steam to one end of the cylinder during all the stroke, and at the same time would exhaust steam from the other end during all the stroke. As regards the relative position of the crank and eccentric it would have to be set so that its middle position was coincident in point of time with the extreme position of the piston; in other words, the eccentric radius would have to be set 90° in advance of the crank.

To make expansive working possible the valve must be able to keep the cylinder ports closed during some part of the stroke. For this purpose it must have what is called *lap*, that is to say, its edges must project beyond the ports as in fig. 132, where e is the *outside lap* or *steam lap* and i is the *inside lap* or *exhaust lap*. Admission of steam to either end of the cylinder now begins only when the displacement of the valve from its middle position is equal to the outside lap, and continues only until the valve returns

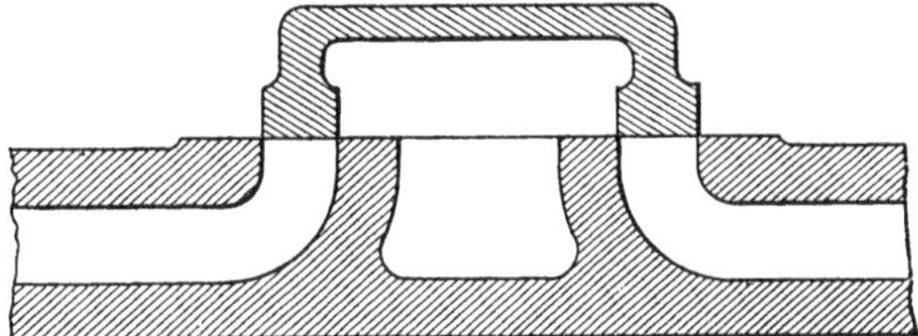

FIG. 131. Slide-Valve without Lap.

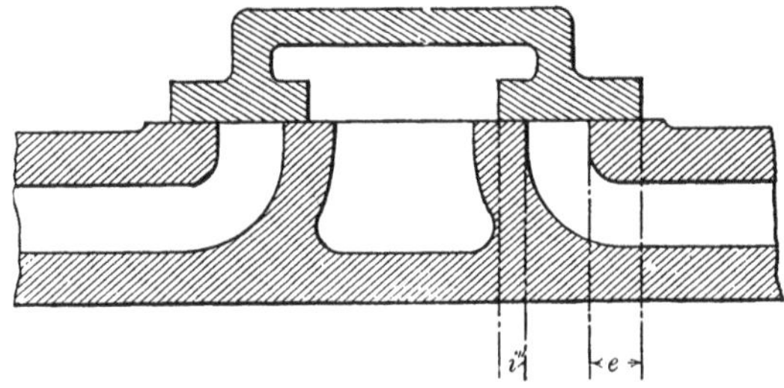

FIG. 132. Slide-Valve with Lap.

to the same distance from its middle position. Further, exhaust begins only when the valve has moved past the middle position by a distance equal to the inside lap and continues until the valve has again returned to this distance from its middle position. Thus let a circle (fig. 133) be drawn to represent the path of the eccentric-centre, on a diameter fg which is the whole travel of the valve, let om be set off equal to the outside lap e and on equal to the inside lap i, and let perpendiculars amb and cnd be drawn at these distances from the centre. The points a, b, c and d then mark the positions of the eccentric-centre at which the four events of admission, cut-off, release and compression respectively occur for one end of the cylinder. As to the other end the four events are determined in the same way of setting off the corresponding outside lap to the left of *o* and the inside lap to the right of *o*. The laps may or may not be equal for the two ends of the cylinder. For the sake of clearness we may for the present confine our attention to one of the two ends. Of the whole revolution the part from *a* to *b* is the arc of admission; in other words, the port is open to steam while the shaft turns through an angle equal to *aob*. Similarly *bc* is the arc of expansion, *cd* that of exhaust, and *da* that of compression.

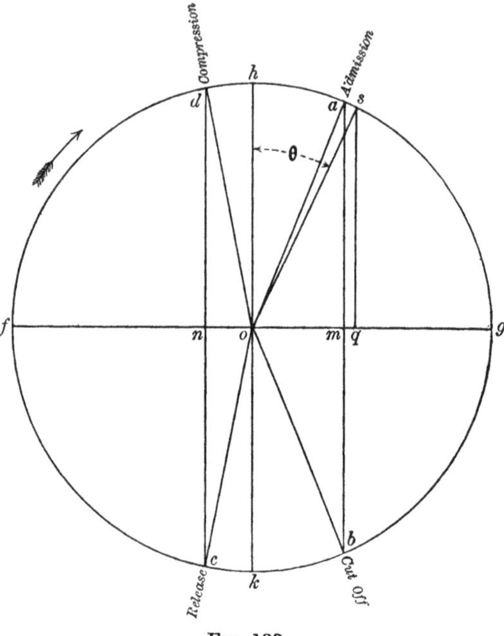

FIG. 133.

The relation of these events to the piston's position is still undefined. If the eccentric were set in advance of the crank by an angle equal to *foa*, the valve would be just beginning to open as the piston stroke begins. It is, however, desirable, in order to allow the steam free entry, that the valve should be already some way open when the piston stroke begins, and hence the eccentric is set at a rather greater angular distance in advance of the crank. Thus if the angular position of the eccentric be *os* while the crank is at the dead-point (on the line *of*) the valve is already open by the distance mq, which is called the *lead*. The angle θ by which the whole angle between the crank and the eccentric exceeds a right angle is called the *angular advance*, this being the angle by which the eccentric is set in advance of the position it would hold if the primitive arrangement without lap were adopted. The lap e, the lead l, the angular advance θ, and the half-travel or throw of the eccentric r are connected by the equation

$$e + l = r \sin \theta.$$

An effect of lead is to cause *preadmission*, that is to say, the lead allows steam to enter before the back stroke is quite completed, and this increases the mechanical effect of the compression in "cushioning" the piston during the reversal of its motion.

The greatest amount by which the valve is ever open during the admission of steam is the distance mg. The width of the steam-port is made at least equal to this distance, and is often greater in order that advantage may be taken of the wider opening nf which occurs during exhaust.

192. Graphic method of examining the distribution of steam given by a slide-valve. Let the circle APB (fig. 134)

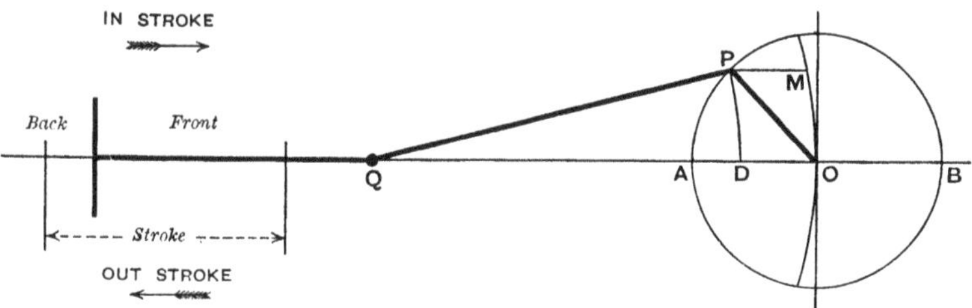

FIG. 134.

represent the path of the crank-pin about the centre O, the stroke being AB. When the crank is at any point P the position of the piston may be found by projecting the point P on AB by drawing a circular arc PD with the length of the connecting-rod PQ as radius and the cross-head Q as centre. Then DO represents the displacement of the piston from its middle position, and AD and DB represent its distance from the two ends of the stroke. Another construction equivalent to this is to draw through O the arc OM with the length of the connecting-rod as radius, and draw PM parallel to AB. PM, being equal to DO, measures the displacement of the piston from its position at mid-stroke. In speaking of the two ends of the cylinder we shall distinguish the one nearer the crank as the front end and the other as the back end. The stroke towards the crank may be called the in-stroke and the other the out-stroke, as marked in fig. 134.

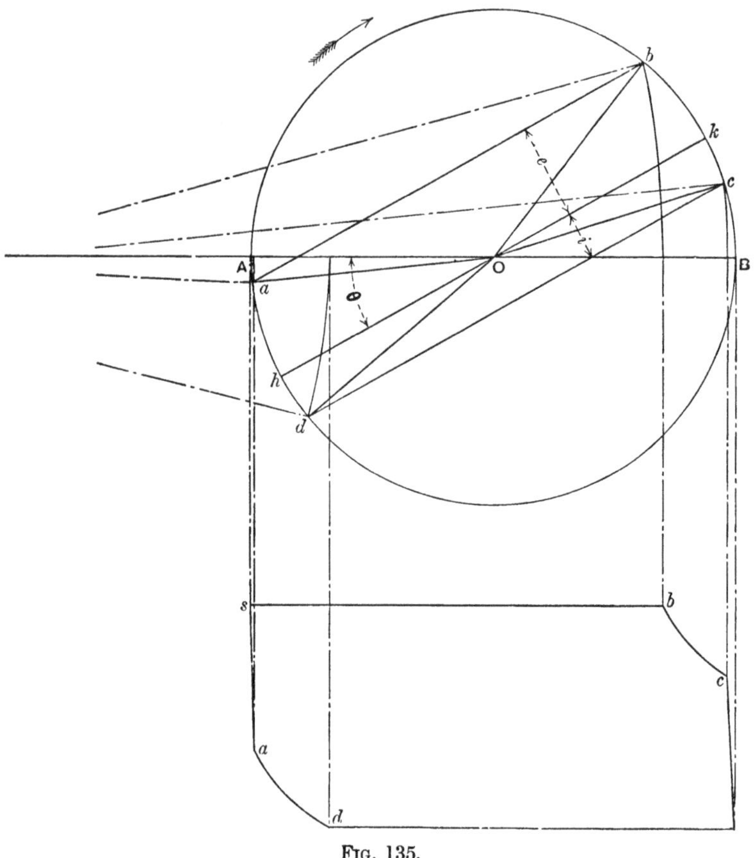

FIG. 135.

To find the position of the piston at each of the four events we have to make a construction which is equivalent to transferring from fig. 133 the four positions of the crank which correspond to the positions a, b, c, d of the eccentric. This is most readily done by drawing a single circle (fig. 135) to represent the motion of the crank-pin on one scale and the motion of the eccentric-centre on another scale. Taking the diameter AB to represent the piston stroke, draw another diameter hk to represent the line hk of fig. 133 turned back through an angle of $90° + \theta$, so that the angle AOh (fig. 135) is equal to θ. Draw ab and cd parallel to this line at distances from it equal to the outside and inside laps respectively. The effect is that each of the points a, b, c and d is turned back, in fig. 135, through an angle equal to $90° + \theta$ as compared with its position in fig. 133. Consequently these points in fig. 135 show the positions which the crank has at the four events. And the corresponding positions of the piston may be found by projecting the points a, b, c, d on AB by means of circular arcs. This is shown

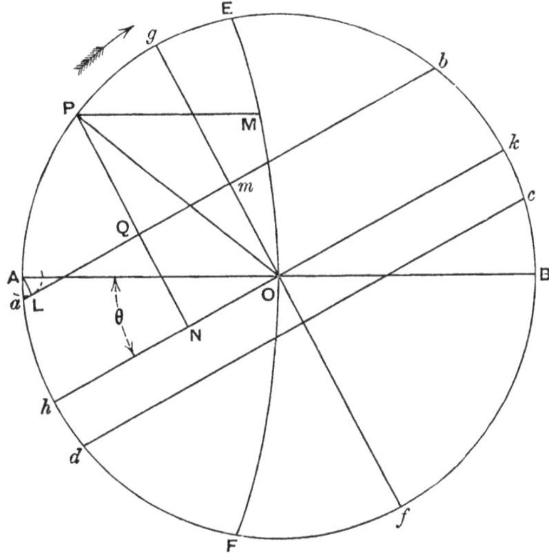

FIG. 136.

in fig. 135, and the indicator diagram is also sketched by reference to the positions projected on AB.

The following is an equivalent and rather more convenient construction. Let a circle (fig. 136) be drawn as before to represent the eccentric's motion on one scale and the crank's on another, and let AB be the piston stroke. Draw hk as before so that the angle $AOh = \theta$, the angular advance. Taking a centre on OA produced, draw the arc EOF through the centre O with radius equal to the length of the connecting-rod. Then when the crank has any position OP the displacement of the valve from its middle position is PN (drawn perpendicular to hk) and the displacement of the piston from mid-stroke is PM. Also, if ab and cd be drawn as before at distances from hk equal to the laps, the four events happen at a, b, c and d, and PQ is the extent to which the valve is open when the crank is at P. Similarly AL is the extent to which the valve is open at the beginning of the stroke, that is the *lead*. The port has its maximum opening when the crank is at Og during admission and at Of during exhaust, unless its width is so small that it has become completely uncovered with a smaller displacement of the valve.

The diagram shown in fig. 136, which is a modified form of one due to Reuleaux, may readily be applied to determine the characteristics which a slide-valve must have to give a stated distribution of steam. Suppose for instance that the travel of the valve, the lead, and the position of cut-off are assigned. Having marked b, the position of the crank-pin at the given point of cut-off in relation to the stroke AB, draw a circle with centre A and radius AL equal to the lead. Then draw a line through b tangent to this circle. This will be the line ba of the diagram. Its inclination to BA determines the angular advance, and a perpendicular on it from O gives Om, which is the outside lap. The inside lap becomes determinate when either the point of release c or that of compression d is assigned, and it is found by drawing a line through c or d parallel to ab, and measuring the distance of this line from O.

193. Inequality of the distribution on the two sides of the piston. So far we have dealt only with the events corresponding to one end of the cylinder, namely (in the diagram) the back end. This has been done only to avoid complicating the diagram with too many lines. In fig. 137 the construction of fig. 136 is repeated with the outside-lap lines ab and $a'b'$ drawn for both ends, and also the inside-lap lines cd and $c'd'$, and the corresponding events are marked. The construction lines relating to the front end of the cylinder are distinguished by being dotted and their reference letters are accented. The laps have been taken equal for the two ends, and an obvious result is that the cut-off is considerably later at the back than at the front. Compare bm with $b'm'$, these being the distances by which the piston has passed mid-stroke when the cut-off occurs at the back and front respectively. This want of symmetry, proceeding as it does from the obliquity of the connecting-rod, is slight when the rod is many times longer than the crank but becomes important when the rod is short. In the sketches the length of the rod is taken to be three times that of the crank.

This inequality may be remedied by making the outside laps unequal, giving less lap to the front end of the valve. When this is done, however, the amounts of lead (which are equal with equal laps) become unequal. But in general it is better to sacrifice equality of lead and to secure at least approximate symmetry in the positions of the two points of cut-off, when the admission of steam is controlled by a simple slide-valve. When a separate expansion valve is used (§§ 207–208) the cut-off is determined by it,

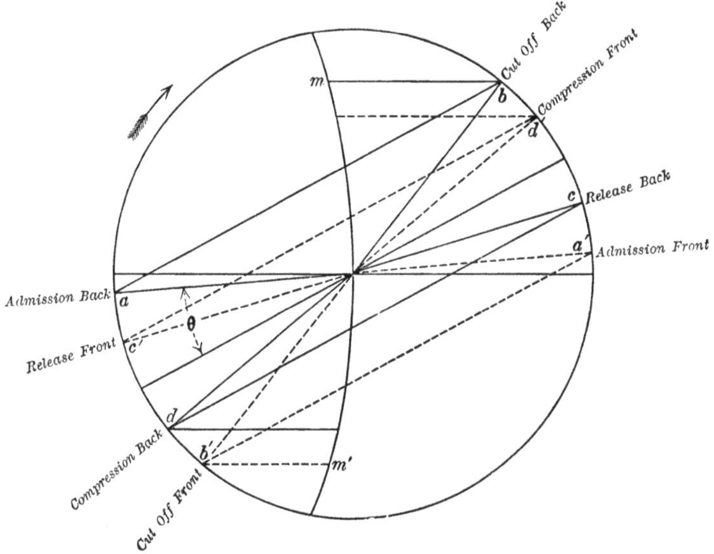

FIG. 137.

and not by the main slide-valve, and in that case the amounts of lead may properly be made equal. In certain cases a somewhat unequal distribution of steam is to be preferred, as in the ordinary vertical marine engine, where the work done by the steam against the front or bottom end of the piston is partly spent in raising the piston and rods and consequently should be greater than the work done against the back or top end which is supplemented by the descent of these heavy weights.

In cases where the eccentric-rod is itself so short that its obliquity should be taken account of, this is readily done in Reuleaux's diagram (fig. 136 or 137) by using circular arcs in place of the straight lines *ab*, *hk*, *cd*, these arcs being described with a radius which represents the length of the eccentric-rod on the same scale as that on which the diameter *AB* represents the travel of the valve, from centres on *Of* produced beyond *f*. Except in rare cases it leads to no appreciable error to treat the eccentric-rod as infinitely long.

Fig. 138 illustrates how a symmetrical distribution is secured by reducing the outside lap at the front end. There *ab* is the outside-lap line for the back end and *a′b′* is the corresponding line for the front end. These lines are drawn so that the cut-off occurs

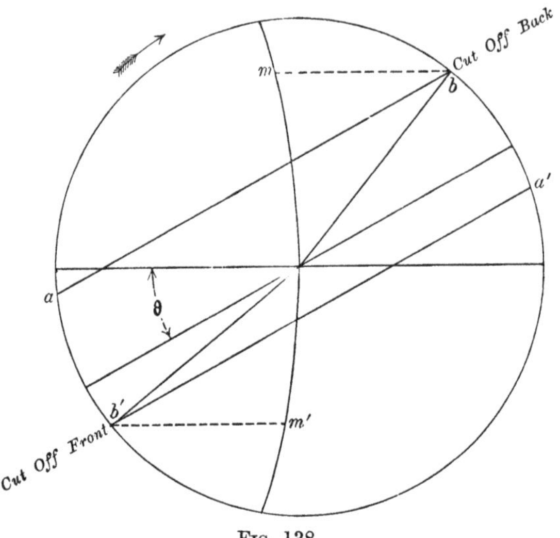

FIG. 138.

at the same percentage of the stroke at both ends: *bm* and *b′m′* are equal. The inside laps may also be adjusted in the same way to give equal amounts of compression on both strokes (or, alternatively, to give symmetrical points of release). The amounts of lead, of course, are no longer equal: the lead at the front end has been considerably increased by the reduction of the lap.

194. Zeuner's valve diagram. A graphic construction much used by slide-valve designers was published by Zeuner in 1856[1]. On the line *AB* (fig. 139), which represents the travel of the valve, let a pair of circles (called valve-circles) be drawn, each with

[1] G. Zeuner, *Treatise on Valve-Gears*, transl. by M. Müller, 1868.

diameter equal to the half-travel. If a radius CP be drawn in the direction of the eccentric centre at any instant, it is cut by one of the circles at a point Q such that CQ represents the corresponding displacement of the valve from its middle position. That this is so will be seen by drawing PM and joining QB, when it is obvious that the triangles CPM and CBQ are equal in all respects and $CQ = CM$, which is the displacement of the valve. The line AB with the circles on it may now be turned back through an angle of $90° + \theta$ (θ being the angular advance), so that the valve-circles take the position shown to a larger scale in fig. 140. This makes the direction of CQP (the eccentric) coincide on the paper with the simultaneous direction of the crank, and hence to find the displacement of the valve at any position of the crank we have only to draw the line CQP in fig. 140 parallel to the direction which the crank has at the instant under consideration, when CQ represents the displacement of the valve to the scale on which the diameter of each valve-circle represents the half-travel of the valve. CL is the valve's displacement at the beginning of the stroke indicated by the arrow. Draw circular arcs EF and IJ with C as centre and with radii equal to the outside lap and the inside lap respectively. CE is the position of the crank at which pre-admission occurs. The lead is LM. The greatest steam opening during admission is GB, and the greatest opening to exhaust is the whole width of the port, namely KH. Intercepts on the radii within the shaded areas give the steam and exhaust openings for any angular positions of the crank. The cut-off occurs when the crank has the direction CF. CI is the position of the crank at release, and CJ marks the end of the exhaust, or the beginning of compression.

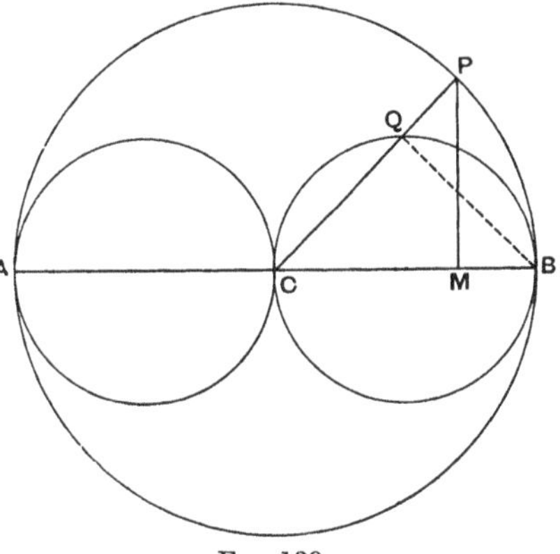

FIG. 139.

In the diagram given in fig. 140 radii drawn from C mark the angular positions of the crank, and their intercepts by the valve-circles determine the corresponding displacements of the valve. It remains to find the corresponding displacements of the piston. For this Zeuner employs a supplementary graphic construction, shown in fig. 141. Here ab or a_1b_1 represents the connecting-rod, and bc or b_1c the crank. With centre c and radius ac a circle ap is drawn, and with centre b and radius ab another circle aq. Then for any position of the crank, as cb_1, the intercept pq between the circles

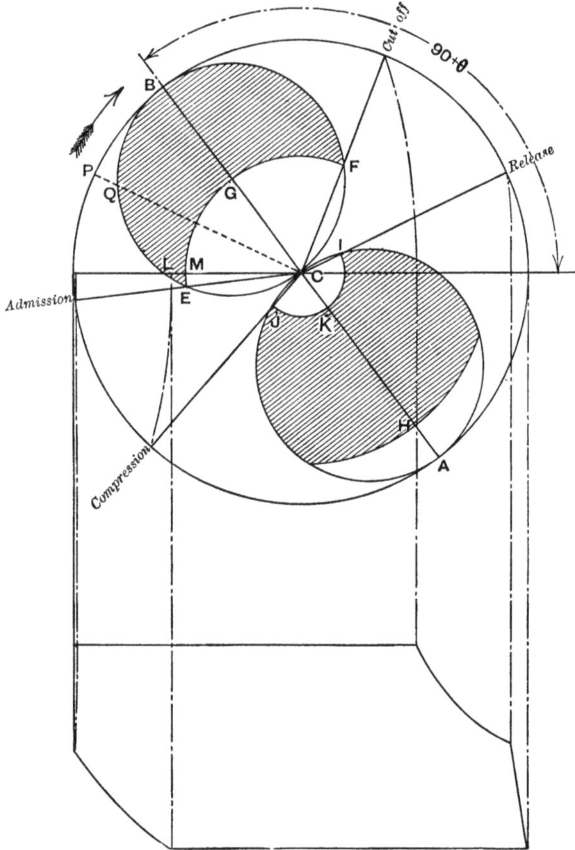

FIG. 140. Zeuner's Slide-Valve Diagram.

is equal to aa_1, and is therefore the distance by which the piston has moved from the extreme position which it had at the beginning of the stroke. In practice this diagram may be combined with that of fig. 140, by drawing both about the centre and using different scales for valve and piston travel. A radial line drawn from the centre parallel to the crank in any position then shows the valve's displacement from its middle position by the intercept CQ of fig. 140, and the simultaneous displacement of the piston

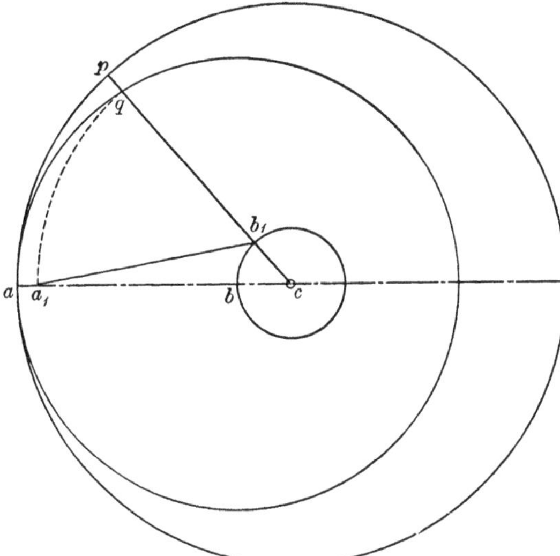

FIG. 141. Zeuner's construction to find the displacement of the piston.

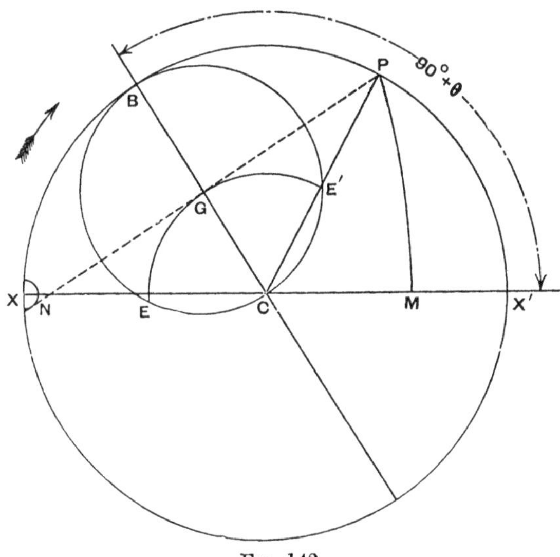

FIG. 142.

from the beginning of its motion by the intercept pq of fig. 141. As an alternative to this the piston's displacement may be found in Zeuner's diagram by the construction used in Reuleaux's, which was described in connexion with figs. 134–137.

To exemplify the use of Zeuner's diagram take the same problem as before, namely, to find the outside lap and angular advance when the point of cut-off and the lead for the corresponding side of the piston are assigned, as well as the travel of the valve. The solution is shown in fig. 142. On the base-line XX' mark the point M to represent the required point of cut-off and project this on the circle XPX' to find CP, which is the angular position of the crank at cut-off. With X as centre draw a circle with radius XN to represent the given lead. From P draw PN tangent to this circle. Then CGB drawn perpendicular to PN defines the diameter of the valve-circle. The angle $X'CB$ is the angular advance, plus 90°, and CG is the required outside lap.

So far as the simpler problems of the slide-valve are concerned Zeuner's diagram has no marked advantage over Reuleaux's. It is however more readily applicable to devices such as the Meyer expansion gear which will be described later.

195. Bilgram's valve diagram. This is an alternative construction shown in fig. 144 and arrived at as follows. In fig. 143

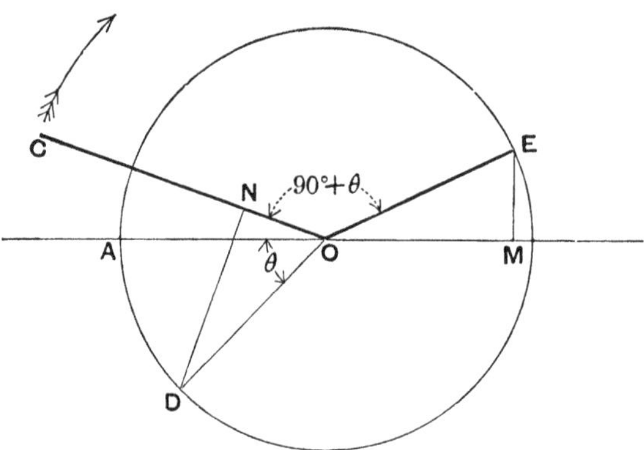

FIG. 143.

let AED be a circle representing the path of the eccentric-centre E, and let C be the angular position of the crank when the eccentric is at E, so that COE is $90° + \theta$. Let OD be drawn making the angle AOD equal to the angular advance θ. Draw DN perpendicular to the direction of the crank OC. Then the triangles ODN and

OEM are equal, and *DN*, being equal to *OM*, measures the displacement of the valve from its middle position when the crank is at *C*.

Hence the position of the crank when admission takes place is determined by the fact that the perpendicular from *D* on *OC* must then be equal to the outside lap: in other words, *OC* will then be tangent to a circle described about *D* as centre with the outside lap as radius. Further, when the crank travels round so that *CO* produced is tangent to the same circle on the other side it has reached the position of cut-off. Similarly, if a second circle be drawn about *D* with the inside lap as radius, the positions in which

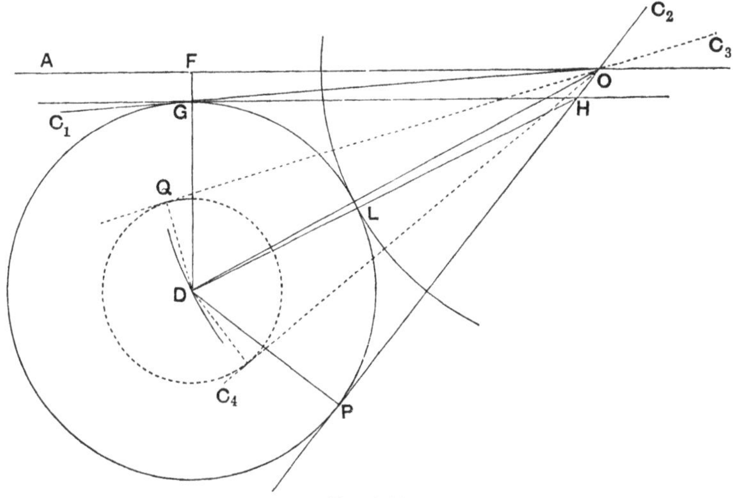

FIG. 144.

OC is tangent to it on its two sides are those of compression and release.

The complete construction, for one end of the cylinder, is shown in fig. 144. There *AOD* is the angular advance, *OD* the half-travel of the valve, *DP* the outside lap, and *DQ* the inside lap. C_1, C_2, C_3, C_4 are the positions of the crank at admission, cut-off, release, and compression, respectively. *OL*, being the half-travel minus the outside lap, is the greatest opening of the port to steam. *DF*, drawn perpendicular to *AO*, is the displacement of the valve when the crank is on the dead-centre *A*, and *FG*, the intercept on this line between *AO* and a line *GH* drawn parallel to *AO* touching the outside-lap circle, is the lead.

In illustration of the use of this diagram we may take two problems. (1) Given the travel of the valve, the position of the crank at cut-off, and the lead; to find the lap and the angular advance. Taking a base *AO* to represent the direction of motion of the piston, draw OC_2 parallel to the crank at cut-off, and produce it through *O* towards *P*. Draw a line *GH* parallel to *AO* and at a distance from it equal to the given lead. Bisect the angle *GHP*. The centre of the lap circle is on this bisecting line, and its distance from *H* is found by sweeping an arc with centre *O* and radius *OD* equal to the half-travel. When *D* is found the circle is drawn touching *GH* and *PH*, and its radius gives the required lap. By joining *D* with *O* we have the angle *AOD*, which gives the angular advance.

(2) Given the position of the crank at cut-off, the lead, and the greatest opening of the port to steam; to find the lap and the angular advance, and the travel of the valve. Draw OC_2 as before to represent the position of the crank at cut-off and produce it through *O*. Draw *GH* parallel to *AO* at a distance below it equal to the lead: with centre *O* and radius *OL* equal to the given greatest opening of the port draw an arc *L*. Then find a centre *D* such that a circle may be drawn touching this arc and the lines *GH* and *OP*: this is most conveniently done by trial. The radius of the circle gives the lap and *AOD* gives the angular advance as before.

196. Oval diagram. A diagram is sometimes drawn which represents by a single curve the simultaneous displacements of the piston and the valve. When the position of the valve has been determined at various phases of the piston's stroke, whether by Reuleaux's or Zeuner's or any other method, a curve is drawn having for ordinates the displacement of the valve, on a base *AB* (fig. 145) which is the stroke of the piston, the scale of the ordinates being suitably exaggerated to prevent the curve from being inconveniently flat on account of the comparatively small amplitude of the valve's motion. This gives a species of oval figure resembling an ellipse, but somewhat distorted through the influence of the connecting-rod's obliquity. To find the events of the distribution, lines *EE* and *II* are drawn above and below the base at distances from it equal to the outside and inside laps respectively; their points of intersection with the curve at *a*, *b*, *c* and *d* mark the four events for the corresponding end of the cylinder. For the other end the outside-lap line $E'E'$ is to be drawn below the base and the inside-lap line $I'I'$ above it. The distance of the curve beyond

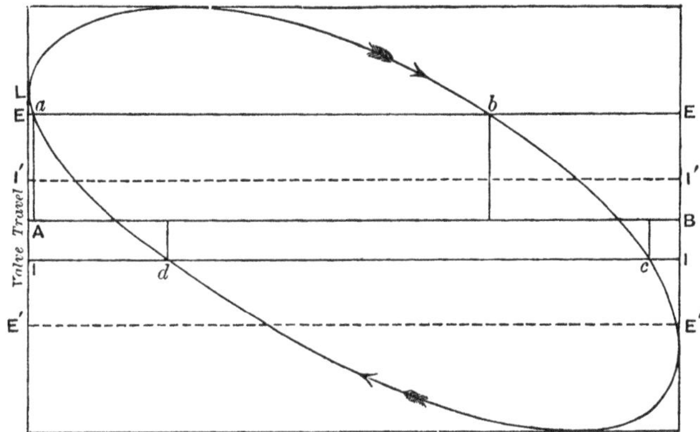

FIG. 145. Oval Diagram for the Slide-Valve.

the outside-lap line shows at any stage in the stroke the extent to which the steam-port is then open. The lead, which is EL, is not well defined in this form of graphic construction.

197. Wave-form diagram. A much more useful diagram is obtained by drawing (preferably on section paper) separate curves to represent the displacements of piston and valve respectively,

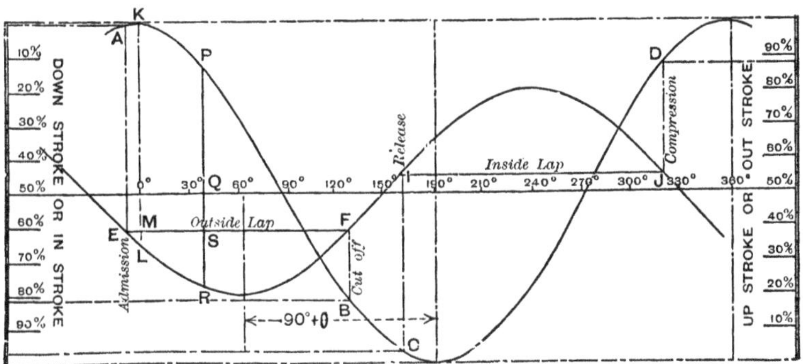

FIG. 146. Wave-form Diagram for the Slide-Valve.

each in relation to the angle turned through by the crank-shaft, rectangular coordinates being used to represent the crank-angle and the displacement. Taking a base (fig. 146) the length of which represents the angle turned through in one revolution, let the curve $ABCD$ be drawn to represent by its ordinates the displacement of the piston from mid-stroke, for all positions of the crank. Similarly let a curve $EFIJ$ be drawn with ordinates which are (on any conveniently exaggerated scale) the displacements of the valve. Owing to the angle between the crank and the eccentric the phase of this curve is $90° + \theta$ in advance of the other: in other words, the valve attains its maximum displacement at a point on the base-line $90° + \theta$ earlier than (or to the left of) the point at which the piston attains its maximum displacement towards the same side. In drawing fig. 146 an angular advance of 30° has been assumed, which makes the total displacement of the valve curve to the left correspond to 120°. When questions have to be considered regarding the effect of varying the angular advance θ, one or other of the curves should be drawn on tracing paper in order that it may readily be slipped over the other into the position that will correspond to any desired angle.

Let any line PQR be drawn perpendicular to the base line to intersect the piston curve in P and the valve curve in R. The displacement of the piston is then PQ and that of the valve is (on another scale) QR. The position of the piston in its stroke is found by projecting P upon the end line of the diagram (to the left) where a scale is marked to show percentages of the stroke. If EF be drawn parallel to the base and at a distance below it equal to the outside lap, SR, which is the excess of the valve's displacement beyond the lap QS, gives the steam opening at the same phase of the stroke. Admission begins at E, and the corresponding position of the piston is found by projecting E upon the piston curve at A and then projecting A upon the scale at the side. The vertical distance from K to A shows the amount of pre-admission. At K, the dead-point of the crank, the valve is open to the extent LM; in other words, LM is the lead. Cut-off occurs at F, and the corresponding position of the piston is found by projecting F upon the piston curve at B, and then projecting B upon the scale at the side. In the same way the positions of the piston at release and compression correspond to the points I and J on the valve curve when the line IJ is drawn at a distance above the base equal to the inside lap. All these events relate to one side of the piston; to obtain the events for the other side the outside-lap line has to be drawn above the base and the inside-lap line below it, and the points found on the piston curve are to be projected upon the scale which is set out on the right-hand side of the diagram in fig. 146. The inequality of lap and lead which is needed to give a symmetrical distribution, and other such problems of design, may be studied by help of this diagram with great ease[1]. Another example of its use will be given below in connexion with separate expansion valves (§ 208).

The ordinates of these curves may be found either by graphic construction or by calculation. As to the valve curve, the length of the eccentric-rod is generally so great that the formula

$$y' = r' \cos \alpha'$$

may be used, r' being the eccentricity or the half-travel of the valve and α' being the angle through which the eccentric has turned from the position that corresponds to the maximum displacement of the valve.

In the piston curve the influence of the rod is usually considerable. Let r be the effective length of the crank AP (fig. 147) and l that

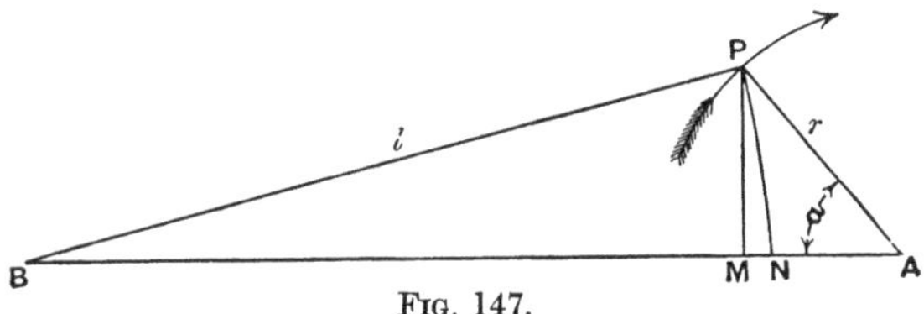

FIG. 147.

of the connecting-rod BP; when the crank has turned through any angle α from the dead-point the displacement of the piston from its middle position is

$$y = AN = AM + MB - l$$

$$= r \cos \alpha + \sqrt{l^2 - r^2 \sin^2 \alpha} - l,$$

or, writing μ for the ratio of the length of the connecting-rod to that of the crank,

$$y = r (\cos \alpha + \sqrt{\mu^2 - \sin^2 \alpha} - \mu).$$

This is always less than $r \cos \alpha$, but approximates closely to that

[1] The writer's attention was drawn by Osborne Reynolds to the merits of this diagram. As a means of solving slide-valve problems it is in several ways superior to the methods more generally used by draughtsmen. The labour of drawing the curves is considerable; but a set of curves drawn once for all, for various ratios of crank to connecting-rod, will allow any particular case of the simple slide-valve to be dealt with readily. Examples of the diagram as applied to link-motions and other valve-gears will be found in Professor Dalby's treatise on *Valves and Valve Gear Mechanisms.*

when μ is very great. An expression of the same form is of course applicable to the displacement of the valve and should be used when the eccentric-rod is so short as to require its length to be taken into account. The angles α (for the crank) and α' (for the valve) are connected by the equation $\alpha' = \alpha + 90° + \theta$.

198. Reversing gear. The link-motion. In locomotives, marine engines, winding engines, traction engines and some other types it is necessary to make provision for reversing the direction in which the engine runs. A primitive way of doing this is to shift the eccentric of the slide-valve round upon the shaft until it takes relatively to the crank the angular position proper to the reversed motion. The eccentric must stand in advance of the crank by an angle equal to $90° + \theta$, and if its position be CE (fig. 148) while the crank is at CK the engine will run in the direction of the arrow A. To set the engine in gear to run in the opposite direction it is only necessary to shift the eccentric into the position CE', when it will still be in advance of the crank by the proper angle, the direction of motion now being that shown by the arrow A'. In some of the older engines this was substantially the actual method of reversal. The valve-rod was temporarily disengaged from the eccentric and the valve was moved by hand in such a way as to make the engine begin to turn backwards. It was allowed to turn until the crank had moved back through an angle equal to ECE', the eccentric meanwhile remaining at rest, and the valve-rod was then re-engaged. To allow the eccentric to remain at rest while the crank turned back through the required angle, the eccentric sheave instead of being keyed to the shaft fitted loosely on it and was driven by means of a spur fixed to the shaft which abutted on one or other of two stops or shoulders projecting from the sheave. Consequently when the engine-shaft began to turn backwards the eccentric sheave did not at once follow it, until it had turned through an angle corresponding to the distance between the two stops. The loose eccentric is a nearly obsolete device; modern engines which require reversing use either the *link-motion* or one of the forms of *radial gear* to be presently described.

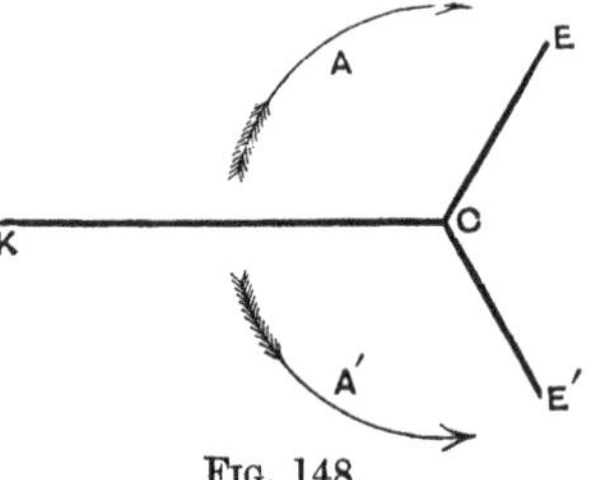

FIG. 148.

In the link-motion there are two eccentrics keyed to the shaft in positions which correspond to CE and CE' in fig. 148, and the ends of their rods are connected to the ends of a link which gives its name to the contrivance. In Stephenson's link-motion—the earliest and still a usual form—the link is a slotted bar or pair of bars curved to a circular arc with radius equal or nearly equal to the length of the eccentric-rods (fig. 149), and capable of being shifted up and down by means of a pendulum-rod to which it is jointed either at one end or at the middle of the link. This suspension by a pendulum-rod also allows the link to move sideways as the eccentrics revolve.

The valve-rod ends in a block which slides within the link, and when the link is placed so that this block is nearly in line with the forward eccentric-rod (R, fig. 149) the valve moves in nearly

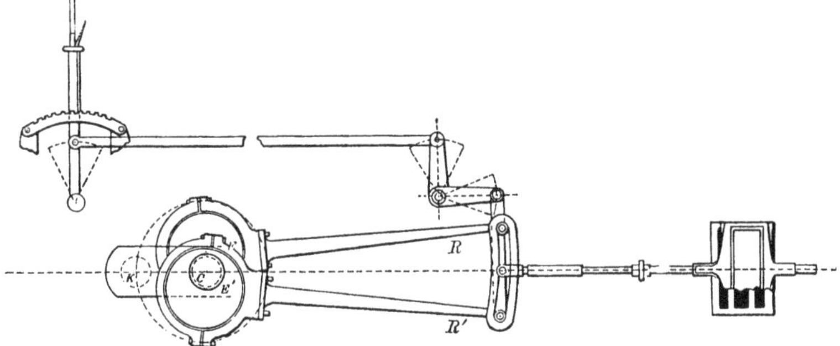

FIG. 149. Stephenson's Link-motion.

the same way as if it were driven directly by a single eccentric. This is the position in "full forward gear." In "full backward gear," on the other hand, the link is pulled up until the block is nearly in line with the backward eccentric-rod R'. The link-motion thus gives a ready means of reversing the engine—but it does more than this. By setting the link in an intermediate position the valve receives a motion nearly the same as that which would be given by an eccentric of shorter throw and of greater angular advance, and the effect is to give a distribution of steam in which the cut-off is earlier than in full gear, and the expansion and compression are greater. Hence the mechanism also serves to adjust the amount of work done by the steam to the demand made on the engine. In mid-gear, which is the position sketched in the diagram, the steam distribution is such that scarcely any work is done in the cylinder. The movement of the link is effected by a hand lever, or by a screw, or (in large engines) by an auxiliary steam-engine. A usual arrangement of hand lever, sketched in fig. 149, has given rise to the phrase "notching up," to describe the setting of the link to give a greater degree of expansion, by bringing it nearer to mid-gear. The eccentric-rods are sometimes crossed instead of being "open" as shown in the sketch.

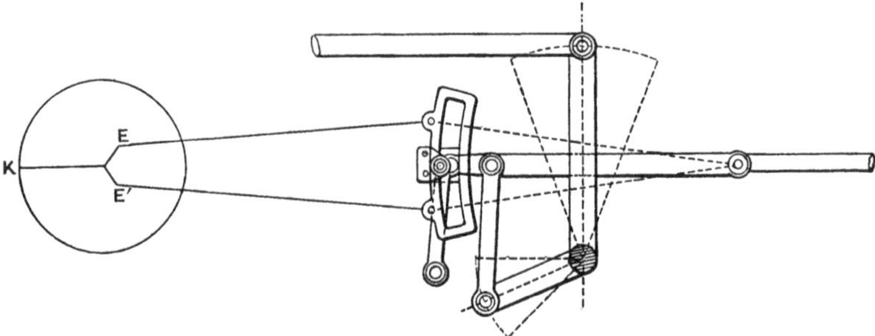

FIG. 150. Gooch's Link-motion.

In Gooch's link-motion (fig. 150) the link is not moved up in shifting from forward to backward gear, but a radius rod between the valve-rod and the link (which is curved to suit the length of this radius rod) is raised or lowered—a plan which has the advantage that the lead is the same in all gears. In Allan's straight link-motion (fig. 151) the change of gear is effected partly by shifting

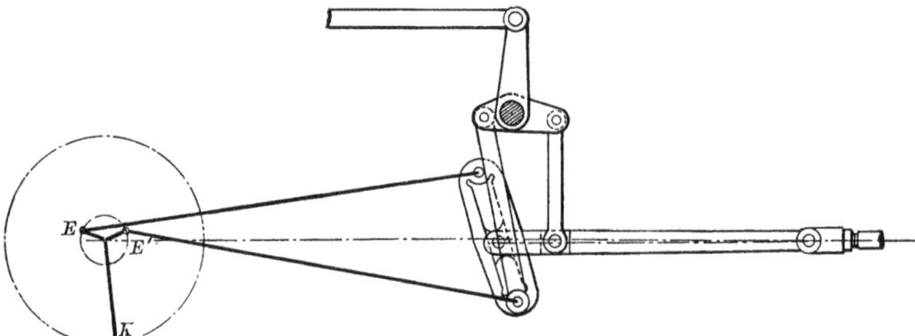

FIG. 151. Allan's Link-motion.

the link and partly by shifting a radius rod. This allows the link to be straight and has also the advantage that the weight of the link with its eccentric-rod on the one hand, and the radius rod on the other, can be arranged to balance one another when a suitable proportion is given to the two arms of the short beam from which the pendulum rods hang. Compared with the Stephenson motion it requires less room for the play of the link between its extreme positions.

199. Graphic solution of the link-motion. The movement of a valve driven by a link-motion may be fully and exactly investigated by drawing with the aid of a template the positions of the centre line of the link corresponding to a number of successive

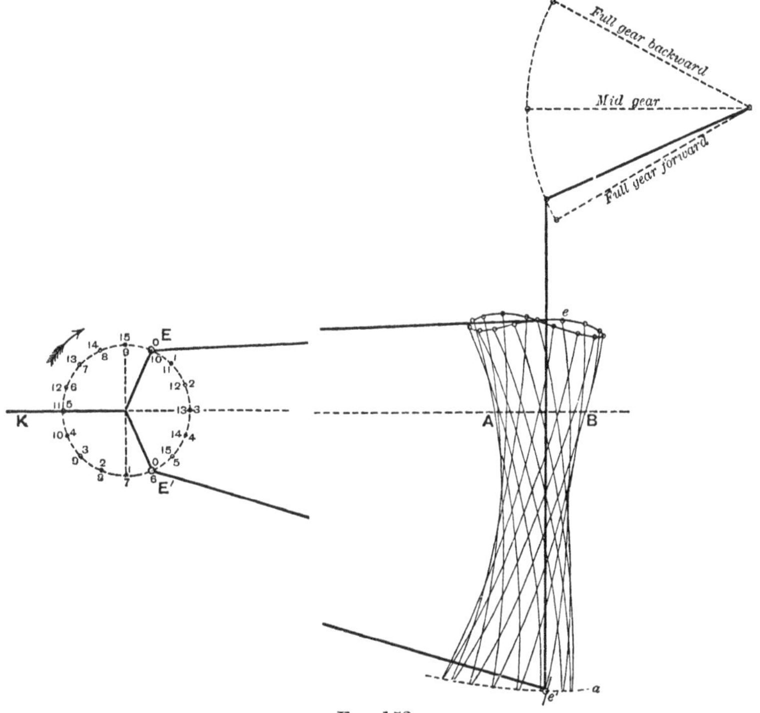

FIG. 152.

positions of the crank. Thus, in fig. 152, two circular arcs passing through e and e' are drawn with E and E' as centres and the eccentric-rods as radii. These arcs are loci of two known points of the link, and a third locus is the circle a in which the point of suspension must lie. By placing on the paper a template of the link, with these three points marked on it, the position of the link is readily found, and by repeating the process for other positions of the eccentrics a diagram of positions (fig. 152) is drawn for the assigned state of the gear. A line AB drawn across this diagram in the path of the valve's travel determines the displacements of the valve, and enables the wave-form diagram to be drawn as in fig. 146, or alternatively the oval diagram as in fig. 145. The example refers to Stephenson's link-motion in nearly full forward gear; with obvious modification the same method may be used in the analysis of Gooch's or Allan's motion. The same diagram serves to determine the amount of sliding motion of the block in the link. In a well-designed gear this sliding is reduced to a minimum for that position of the gear in which the engine runs most usually. In marine engines the suspension-rod is generally connected to the link at that end of the link which is next the forward eccentric, in order to reduce this sliding as much as possible when the engine is running in its normal condition, namely, in forward gear.

200. Equivalent eccentric. A less laborious, but less accurate, solution of link-motion problems is reached by the use of what is called the equivalent eccentric—an imaginary single eccentric, which would give the valve nearly the same motion as it gets from the link under the joint action of the two actual eccentrics. The following rule for finding the equivalent eccentric, in any state of gear, is due to MacFarlane Gray:—

Connect the eccentric centres E and E' (fig. 153) by a circular arc whose radius $= \dfrac{EE' \times \text{length of eccentric-rod}}{2 \times ee'}$, ee' being the length of the link from rod to rod. Then, if the block is at any point B, take EF such that $EF : EE' :: eB : ee'$. CF then represents the equivalent eccentric both in radius and in angular position. If the rods of the link-motion are crossed instead of open—an arrangement seldom used—the arc EFE' is to be drawn convex towards C. Once the equivalent eccentric has been found the movement of the valve may of course be determined by Zeuner's or any of the other methods already described. The method of the equivalent eccentric should not be taken as giving more than a first approximation to the actual motion; for anything like a

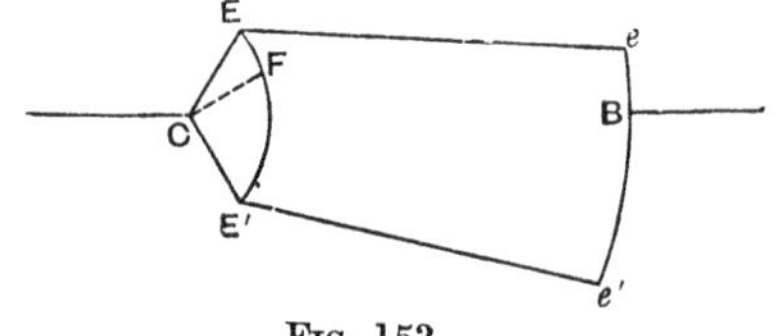

FIG. 153.

complete study of a link-motion the graphic method of § 199 or the use of a model is to be preferred.

201. Resolution of the valve displacement into two components. It is instructive in this connexion to think of the action of a single eccentric in driving a slide-valve as made up of two components, due to two imaginary eccentrics one of which is set opposite to the crank and the other at right angles to it. Thus in fig. 154 the actual eccentric CE gives the valve a motion which is the same as that which would be got by compounding the motions of two eccentrics CM and CN. Hence in the wave-form diagram we might draw separate curves for the two component eccentrics CM and CN, and by adding their ordinates we should get a curve which is the curve given by the actual eccentric CE. The characteristic, then, which must be possessed by any mechanism for driving a slide-valve is that the displacement of the valve must have what we may call for brevity a component M, which differs in phase from the motion of the crank by 180°, and a component N which differs in phase from the motion of the crank by 90°. Further the maximum amount of the M component, namely, the length of the imaginary eccentric CM or the value of CM when the crank is in the initial position sketched, is equal to the outside lap *plus* the lead.

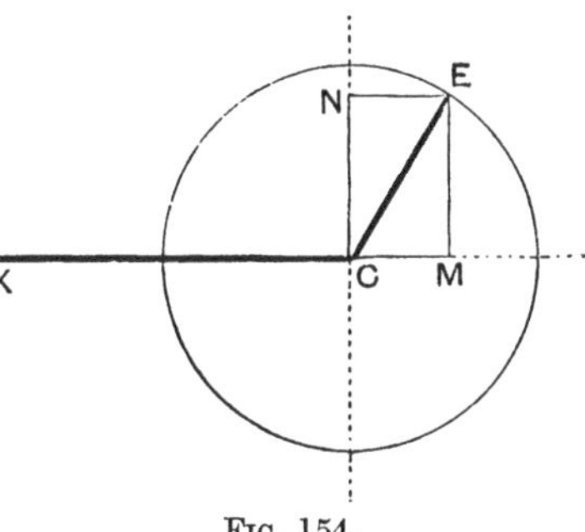

Fig. 154.

Apply this idea to a reversing gear such as a link-motion driven by two eccentrics E and E', fig. 155, and we see that in shifting from forward to backward gear the M component is not altered, but the N component is first reduced, passes through a zero value, and then becomes negative.

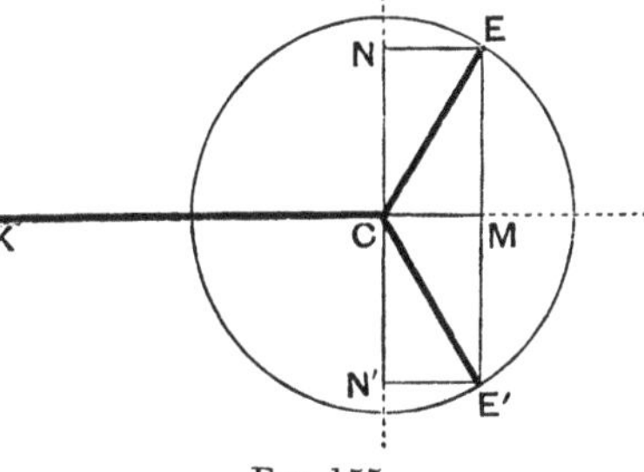

Fig. 155.

In the mid-gear position the distribution of steam is that which would be given by the M component alone, and we approach full gear, in either direction, by introducing more and more of the N component. The amount of steam admission increases with the increase of the N component until full forward gear or full backward gear is reached.

What is wanted therefore in any form of reversing gear is a mechanism which will allow the valve to keep, unchanged, a component of reciprocating motion due to M directly opposite to the motion of the piston, and will superpose on that another component due to N differing 90° from it in phase, which can be varied in magnitude and reversed in sign. The link driven by two eccentrics CE and CE' in any of the three arrangements which have been described offers one means of solving the problem: but there are other means in some of which only one actual eccentric is used and in some none at all. These go by the general name of *radial gears*. Like the link-motion they supply a means of varying the output of power as well as reversing the direction of running. The action of these gears becomes readily intelligible when it is recognized that in every one of them the function of the mechanism is to combine a motion due to a constant component eccentric M, at 180° from the crank, with that due to a component N at 90° from the crank which can change continuously from positive to negative through zero.

202. Radial gears. Hackworth's. Of these radial gears the oldest is Hackworth's, which was patented in 1859, and has been the parent of several others. It dispenses with one of the two eccentrics which are used in an ordinary link-motion, but retains an eccentric which is set directly opposite the crank and gives the valve the M component of the motion. Referring to the skeleton diagram, fig. 156, E is the eccentric, set at 180° from the crank, which drives an eccentric-rod EQ, the mean position of which is perpendicular to the travel of the

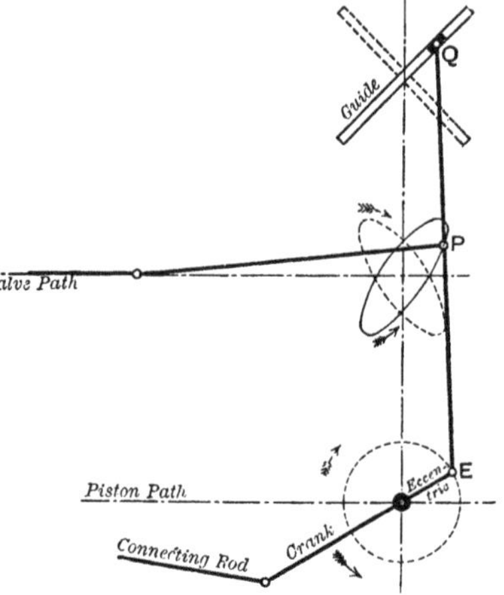

FIG. 156. Hackworth's valve-gear.

valve. This rod ends in a block Q, which slides on a fixed inclined guide-bar or link, and the valve-rod receives its motion through a valve connecting-rod from an intermediate point P of the eccentric-rod, the locus of which is an ellipse. To reverse the gear the guide in which Q moves is tilted over to the position shown by the dotted lines, and intermediate inclinations give various degrees of expansion without altering the lead. When the guide is set at mid gear, pointing directly towards the crank centre, the valve receives a motion which consists of the M component alone. Looking on Q as a fulcrum we see that P is then giving to the valve a motion which is the same, but on a reduced scale, as would be given by the direct action of the eccentric E. Under these conditions there is no N component. But by inclining the guide the N component is introduced. The up and down sliding movement of the block Q, which takes place with a phase differing by 90° from the crank, does not contribute to the motion of the valve so long as the guide is not inclined, but does contribute as soon as the guide begins to be inclined, for the sliding motion of Q then has a lateral component, which is shared by P to an extent depending on the proportion of lengths PE and QE. This is distinct from the lateral motion which P receives in the former case from the eccentric alone, and differs in phase from it by 90°. Thus with the guide inclined we have the two necessary components M and N, the amplitude of the M component being constant while that of the N component ranges from a positive to a negative maximum according to the inclination of the guide.

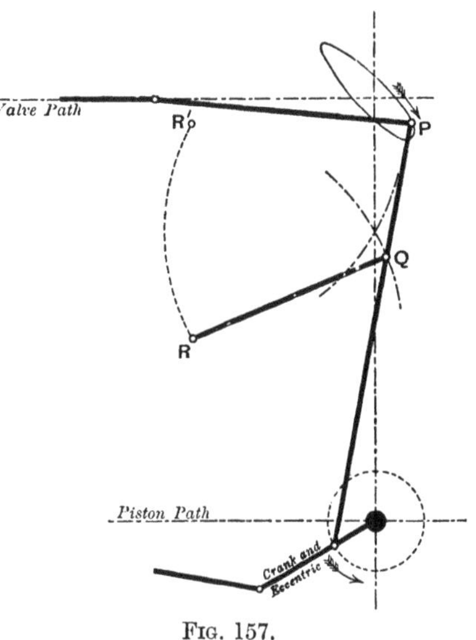

FIG. 157.

An objection to this arrangement of Hackworth's gear is the wear of the sliding block Q and its guide. In a modified form this objection is obviated with some loss of symmetry in the valve's motion by constraining the motion of the point Q, not by a sliding-guide as in fig. 156, but by a suspension-link, which makes the path of Q a circular arc instead of a straight line. To reverse the gear the centre of suspension R of this link is thrown over to the position R' (fig. 157). The same figure also serves to illustrate another modification of the Hackworth gear, namely, the placing of P beyond Q, with no angle between the crank and the eccentric; but P may be between Q and the crank (as in fig. 156), in which case the eccentric is set at 180° from the crank. When P is beyond Q the angle between eccentric and crank is zero, for the eccentric-rod oscillating on a lever about the fulcrum Q itself introduces the 180° difference of phase which is required in the component M. The radius of the eccentric and the ratio of the distances to P and Q are taken such that the maximum value of the M component is equal to the lap plus the lead. By arranging the centre about which the guide-bar (fig. 156) rocks so that it coincides with the position of Q when the crank is at the dead-point we secure the condition that the lead does not alter during the shifting of the gear.

The Hackworth gear has been extensively applied to marine engines, in forms designed by Marshall, Bremme, and others. Fig. 158 illustrates an arrangement by Mr Marshall[1] for a marine engine of the usual vertical type. K is the crank turning about C, V the valve-rod, E the eccentric centre, EPQ the eccentric-rod, and RQ the suspension-rod, which rocks about a centre R. To reverse the gear R is shifted to R' by the screw gear shown at the side.

203. The Walschaerts gear. In this gear, which for long has been so largely used on continental locomotives as to be their normal type of reversing gear, and has now become common on British and American locomotives also, one eccentric is dispensed with. The M component is derived from the cross-head, through a lever which serves to reduce the motion and also to change its phase by 180°. The N component is obtained from an eccentric set at 90° from the crank in a manner which will be understood by reference to the skeleton diagram, fig. 159. There A is a projection fixed to and moving with the cross-head. It is connected to the valve-spindle V by the rod AB and the lever BV, which is pivoted at D on another

[1] *Proc. Inst. Mech. Eng.* 1880.

rod DG. Regarding D for the moment as a fulcrum, it will be seen that this system gives to V a component of motion proportional

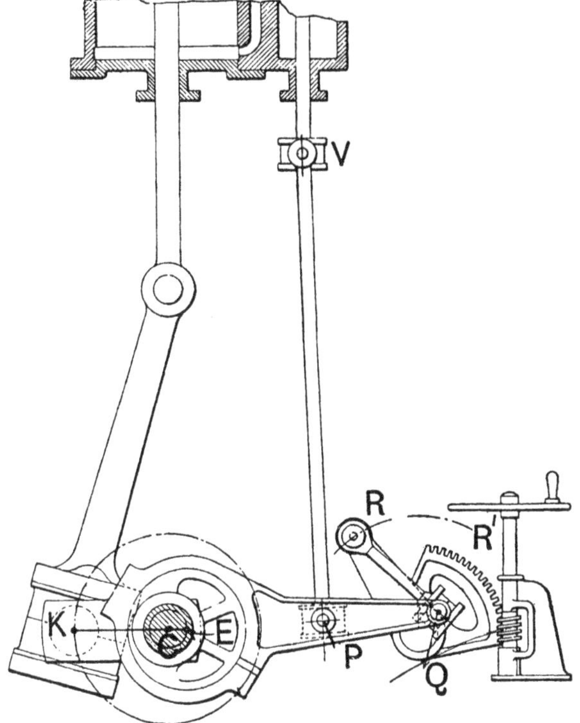

FIG. 158. Marshall's form of Hackworth gear.

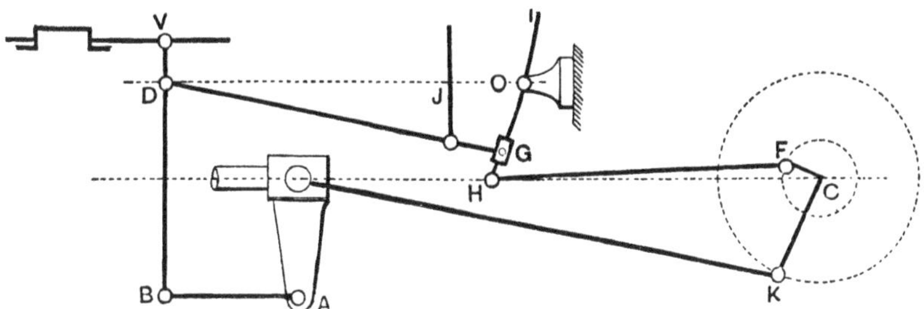

FIG. 159. Skeleton diagram of the Walschaerts gear.

to that of the cross-head, but on a reduced scale, and with its sign reversed. These are the proper conditions for the component M.

To provide the component N there is an eccentric CF, at 90° from the crank CK, the rod of which rocks a link HI about a fixed centre O. D is connected to the link by the rod DG, which has a

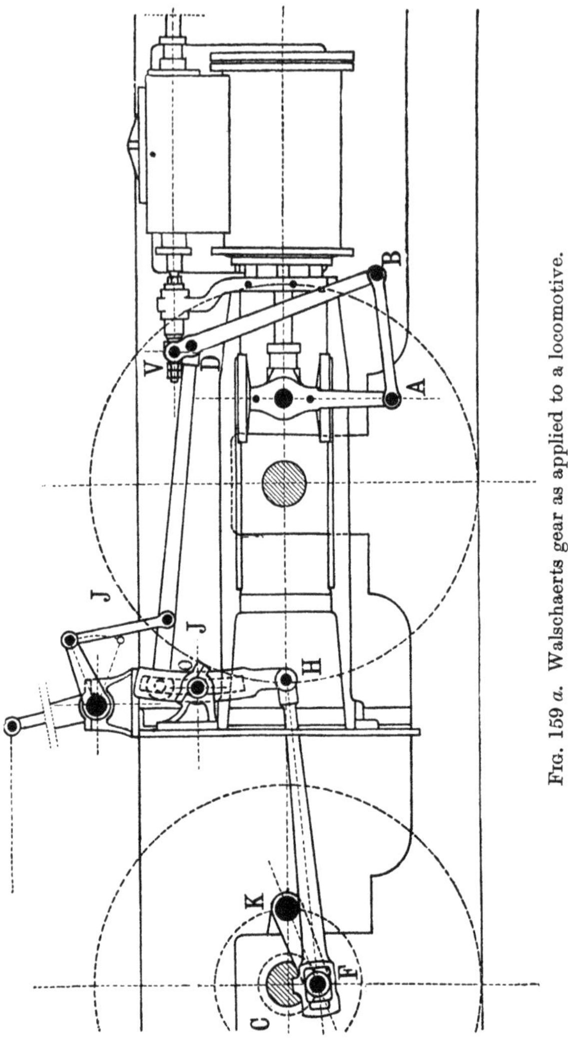

Fig. 159 *a*. Walschaerts gear as applied to a locomotive.

sliding block at G capable of sliding over the whole length of the link, up to I. To reverse the gear this block is shifted to I, on the opposite side of the fixed centre O about which the link oscillates.

The rod DG is shifted by means of the rod J which is connected to the notching up lever.

It will be obvious that the eccentric F provides the N component by giving to D and to V a movement in phase with the eccentric itself. When the block G is shifted to O, which is the position for mid gear, this component disappears, and when the block is at I it is negative: in other words, the component is then such as would be given by an eccentric opposite CF, and the gear is in the position for running backwards. To secure constant lead during change of gear the link is curved to a radius equal to DG, so that, if the crank is kept at its dead-point, D remains at rest while the gear is shifted. An example of the Walschaerts gear applied to a locomotive is given in fig. 159 a[1].

204. Joy's gear. Another type of radial gear is Joy's, which has no eccentric, but presents points of relationship to Hackworth's. The rod SQP corresponding to the eccentric-rod in Hackworth's gear, which supplies the M component of the valve's displacement, receives its motion from a point in the connecting-rod through the linkage OTU shown in fig. 160, and is either

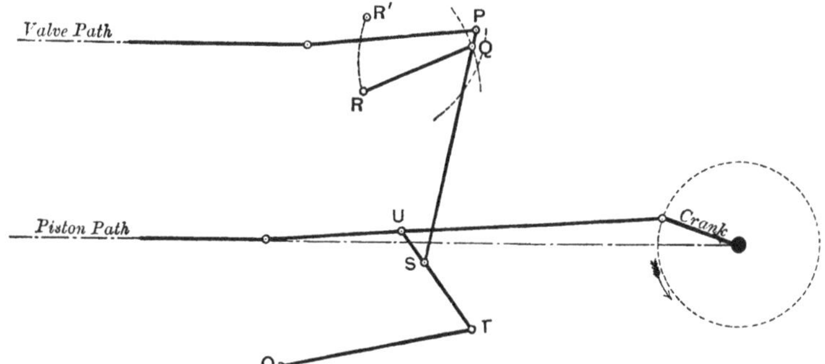

FIG. 160. Skeleton diagram of Joy's gear.

suspended by a rod whose suspension-centre R is thrown over to R' to reverse the motion, or constrained, as in the original form of the Hackworth gear, by a slot-guide whose inclination is reversed. The rod SQP takes its motion from the connecting-rod not directly

[1] In practice the pin H by which the eccentric-rod is connected to the rocking link may be put some little distance behind the centre line of the link, nearer C, to correct for inequalities in the motion arising from the shortness of the rods.

but from a point S in the link UT. O is a fixed centre. Fig. 160 a shows Joy's gear as applied to a locomotive. A slot-guide E is used, and it is curved to allow for the obliquity of the valve connecting-rod AE. C is the crank-pin, B the line of the piston's motion, and D a fixed centre. We may regard the Joy gear as supplying the M component in much the same way as the Walschaerts gear supplies it. If the gear is set in the mid position the M component alone is operative. Though the motion is taken from the connecting-rod the effect (for that position) is the same as if it were taken from the cross-head, as in Walschaerts. But by taking it from the connecting-rod instead Joy gets a vertical sliding of the block in the link E (fig. 160a) resembling the sliding motion in Hackworth, and consequently is able to introduce the N component by inclining the link.

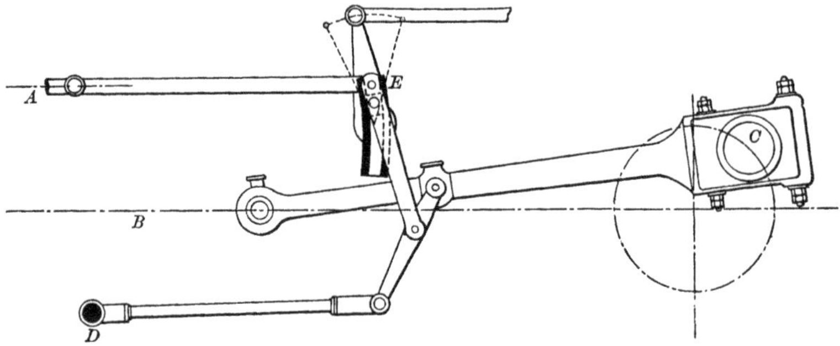

Fig. 160 a. Joy's gear as applied to a locomotive.

It might be thought possible to take the motion of the rod SQP direct from the connecting-rod by putting S at U, but it will be found that this would cause the valve to have a far from symmetrical motion about the mean position, on account of the obliquity of the main connecting-rod.

Reviewing these various radial gears, we see that Hackworth retains one eccentric, getting the M component from it almost directly, and getting also the N component from it indirectly by the device of an inclined sliding-guide or its equivalent. Walschaerts also retains one eccentric, but gets only the N component from it, through the medium of a link which serves to vary the amount and reverse the direction of the motion so taken. He gets the M component almost directly from the cross-head, using a lever only to reduce the motion and also to change its phase by 180°. Joy uses a point on the connecting-rod to supply both components, the M component more or less directly, as in Walschaerts, and the N component indirectly as in Hackworth by the same device of an inclined sliding-guide or its equivalent. He escapes the use of any eccentric, but there are practical drawbacks to taking motion from the connecting-rod which have kept his gear from being widely adopted.

205. Reversing by interchange of steam and exhaust. Steam-steering engines. An engine fitted with the most primitive type of slide-valve, namely, a valve without lap or lead, and therefore with its eccentric set without angular advance, will run backwards if we reverse the usual connexions of the steam and exhaust-pipes, by supplying steam to the central cavity of the valve, and allowing exhaust to take place from the outside of the valve. We have only to open what is usually the exhaust side to steam, and what is usually the steam side to exhaust. This can be done by a separate reversing valve on the steam and exhaust-pipes, which may itself conveniently take the form of a slide-valve. When this valve is moved over by hand the engine has its direction of running reversed. The arrangement does not admit of expansive working, owing to the absence of lap: it gives admission of steam throughout the whole stroke. But its simplicity makes it useful in certain cases where economy of steam is comparatively unimportant.

A notable instance is in steam-steering engines. There the function of the engine is to form a mechanical relay between the helmsman and the rudder, making it possible for the movement of the steering wheel to be effected with scarcely any force, while the rudder, which requires much force, is moved to a corresponding extent. A movement of the wheel in either direction opens a reversing valve of the kind just spoken of, and sets the steering engine in motion in the corresponding direction. This turns the rudder, but the turning has to stop when the rudder has gone over to the proper angle, depending on the extent to which the helmsman has turned his wheel. Accordingly the steering engine is fitted with what is called a *hunting gear*, which is a device for automatically stopping it after the proper amount of movement has been effected. There are many forms of such gear, but all act on the general principle that the reversing valve, which has been shifted by the movement of the steering wheel, is brought back to its central position by the action of the engine itself, though the helmsman continues to hold his wheel over. The result is that when the engine has forced the rudder over to the proper angle it stops, and makes no further movement until the helmsman alters the position of the wheel. When he does so the steering engine at once responds, turning one way or the other according as the helmsman gives the ship more helm or less. Thus the movements of the rudder are kept in complete correspondence with those of the wheel, and the resistance to be overcome by the steersman is no more than the friction of a small valve.

The hunting gear often consists of a screwed sleeve or nut on the spindle of the reversing valve. This nut is geared to the steering engine, so that it revolves so long as the engine is running. It gradually works the spindle back after it has received, from the steering wheel, a movement which has set the engine in motion, and the engine consequently stops after displacing the rudder to a corresponding extent.

The above remarks apply also to the servo-motor or oil relay which was shown in fig. 80 (§ 153) as applied to the governing of steam turbines. There the valve directly actuated by the centrifugal governor was essentially a slide-valve without lap or lead, which admitted oil under pressure to one or other side of a relay piston and thereby caused the main steam throttle-valve to be raised or lowered. The movement of the relay piston also brought into action a hunting device to restore the slide-valve of the servo-motor to its middle position.

206. Operating reversing gears. In small engines fitted with a link-motion or similar reversing gear the gear may be shifted by hand either directly by a lever or with the aid of a screw or worm, but in large engines steam power is employed to do the work.

In Brown's reversing apparatus there are two cylinders arranged in tandem with a single piston-rod common to both, the movement of which pushes over the gear. One of the two is a steam-cylinder to either end of which steam can be admitted, while the other end is put in communication with the exhaust. The other cylinder is a hydraulic brake, the function of which is to prevent the movement of the piston-rod from being jerky. A control valve, operated by hand, admits steam to the steam-cylinder of the apparatus, and is fitted with a hunting device, of the kind mentioned in the preceding section, with the effect that whatever amount of movement is given to the hand control is accurately reproduced in the movement of the steam piston of the apparatus, which is brought to rest in a position corresponding to the position of the hand control, any further supply of steam being then cut off by the action of the hunting device.

Reversing gears are frequently operated by what is called an "all round" power mechanism. The weigh-bar which moves the link is driven from a pin on a worm wheel which may be completely revolved. A half revolution of the wheel shifts the gear from full gear ahead to full gear astern; the next half revolution shifts it back to full gear ahead. The wheel is revolved by a tangent screw which may be turned by a hand wheel or by a small auxiliary steam-engine. By making the wheel revolve continuously the gear may go on being rapidly shifted from full gear ahead to full gear astern and back again, an arrangement which is found convenient by marine engineers when waiting orders from the bridge or when warming the engines before getting under way.

207. Separate expansion valves. When the distribution of steam is effected by the slide-valve alone the arc of the crank's motion during which compression occurs is equal to the arc during which expansion occurs, and for this reason the slide-valve would give an excessive amount of compression if it were made to cut off the supply of steam earlier than about half-stroke. Hence, when an early cut-off is wanted it is necessary either to use an entirely different means of regulating the distribution of steam, or to supplement the slide-valve by another valve—called an expansion valve, and usually driven by a separate eccentric—whose function is to effect the cut-off, the other events being determined as usual by the slide-valve. Such expansion valves may be of two types. In one, which is much the less common, the expansion valve cuts off the supply of steam to the chest in which the main valve works. This may be done by means of a disc or double-beat valve (§ 214), or by means of a slide-valve working on a fixed seat (furnished with one or more ports) which forms the back or side of the main valve-chest. Valves of this last type are usually made in the "gridiron" or many-ported form to combine large steam-opening with small travel. Expansion-valves working on a fixed seat may be arranged so that the ports are either fully open (fig. 161) or closed (fig. 162) when the valve is in its middle position. In the latter case the expansion-valve eccentric is set in line with

FIG. 161. FIG. 162.

or opposite to the crank, if the engine is to run in either direction with the same grade of expansion. Cut-off then occurs when the crank is at P (fig. 163), the expansion eccentric being at P', the shaft having turned through an angle α from the beginning of the stroke. This is because the valve is then within a distance equal to l (fig. 162) of its middle position. The expansion valve reopens when the crank is at Q, and the main slide-valve must therefore have enough lap to cut off earlier than $180° - \alpha$ from the beginning of the stroke, in order to prevent a second admission of steam to the cylinder. In the example shown in fig. 163 the expansion eccentric is set 180° in advance of the crank, which is a usual arrangement when the engine is provided with reversing gear, since it makes the cut-off happen at the same place in the stroke for both directions of running. If this condition need not be fulfilled, the expansion eccentric may have a somewhat different angular position, and in this way a more rapid travel at the instant of cut-off may be secured for one direction of running.

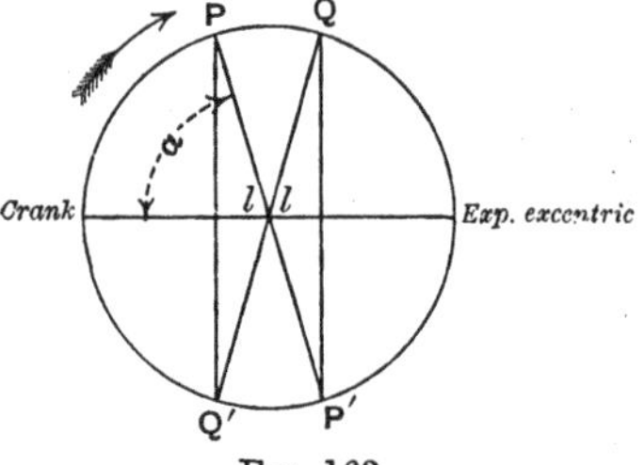

FIG. 163.

Since the separate expansion valve of fig. 161 or 162 acts by cutting off the supply of steam from the steam chest, but not directly from the cylinder, it does not prevent the steam which is stored in the chest from continuing to enter the cylinder until the main slide itself closes the admission port. This effect is avoided in the type of expansion valve next to be described.

208. Meyer's expansion valve. The much commoner type of expansion valve is known as Meyer's. It consists of a pair of plates sliding on the back of the main slide-valve, which is provided with through ports which these plates open and close. Fig. 164 shows one form of this type. Here it is the relative motion of the

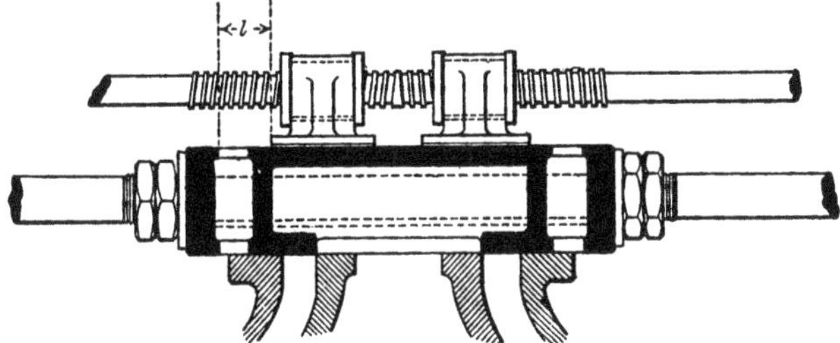

FIG. 164. Meyer Expansion valve.

pair of plates forming the expansion valve with respect to the main valve that has to be considered. If r_a and r_b (fig. 165) are the eccentrics working the main and expansion valves respectively, then *CR* drawn equal and parallel to *ME* is the *resultant* eccentric which determines the motion of the expansion valve relatively to the main valve. Cut-off occurs at *Q*, when the shaft has turned through an angle α, which brings the resultant eccentric into the direction *CQ* and makes the relative displacement of the two valves equal to the distance *l*. Another form of this valve (corresponding to the fixed-seat form shown in fig. 162) cuts off steam at the inside edges of the expansion slides. With the form shown in fig. 164 the expansion eccentric will be set at 180° from the crank if the engine is to run in both directions with the same grade of expansion; otherwise a somewhat different angle may often be chosen with advantage, as giving a sharper cut-off.

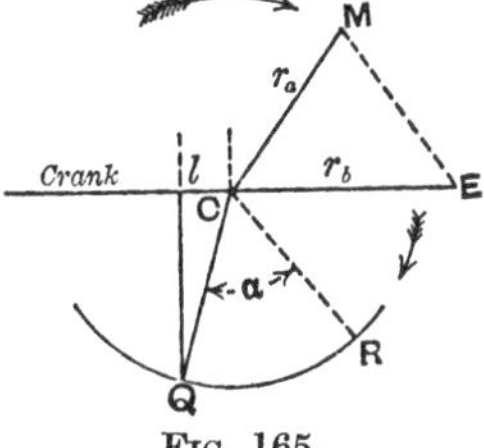

FIG. 165.

The action of Meyer's valve may be examined by the help either of Zeuner's diagram or of the wave-form diagram of § 197. Taking Zeuner's diagram first and assuming, for greater generality, that the expansion eccentric is not set just opposite the crank, let the circles I and II (fig. 166) be drawn to show as in fig. 140, by the lengths of their chords through *C* the amount of absolute displacement of the expansion slide and main slide respectively each from its middle position, when the crank is in the angular position corresponding to the direction of the chord. In drawing these circles the angle XCM is set out to represent the whole angle by which the main valve eccentric is ahead of the crank, as in fig. 140, and the angle XCE (also measured against the direction of the arrow) represents the angle by which the expansion eccentric is ahead of the crank, CM and CE being diameters of the circles II and I respectively. Then if any chord CQP be drawn from C to meet both circles the distance PQ, which is the difference between CP the absolute displacement of the main valve and CQ the

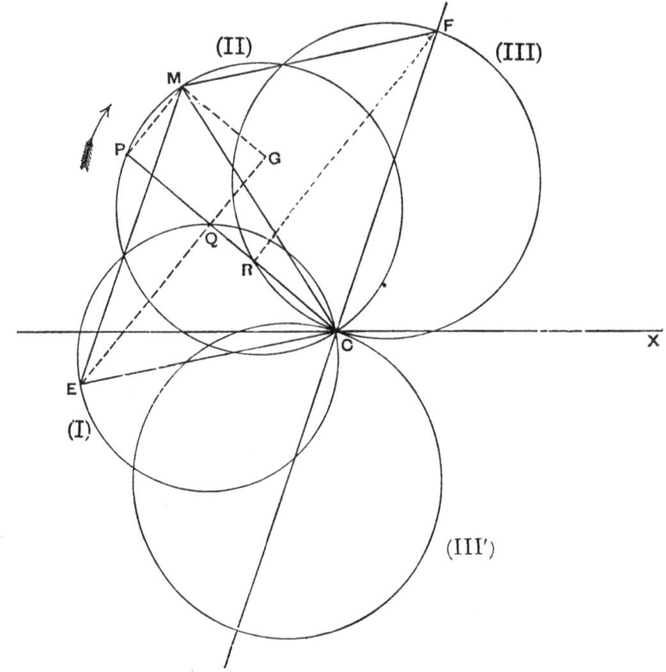

Fig. 166. Application of Zeuner's Diagram to the Meyer Expansion valve.

absolute displacement of the expansion valve, measures the relative displacement of one with respect to the other. This distance PQ is equal to the chord CR of a third circle (III) drawn with CF equal and parallel to EM as its diameter. To prove this make the supplementary construction shown by the dotted lines. Then since the angles at P, Q, R and G are right angles, being angles in semicircles, PQ equals MG in the parallelogram PG, and RC equals MG in the equal triangles EMG and FCR. Hence PQ

equals RC. A similar construction of course applies to the return stroke, for which the circle showing the resultant motion is III′.

Thus by drawing the circles III and III′ we at once determine, by the length of their chords through C, the relative displacement of the two valves for all positions of the crank. Cut-off, on the part of the expansion valve, occurs when the crank is in such a position that the chord CR is equal to l (fig. 164), which is the amount of relative displacement that suffices to close the steam passage through the main slide. The expansion valve reopens when the chord is again diminished to this value, towards the end of the stroke, and care must be taken that the main slide has enough outside lap to close the steam-port leading into the cylinder before this stage in the revolution has been reached.

When the expansion valve is furnished with a means of varying l, as in fig. 164, where right- and left-handed screws enable the blocks which make up the expansion valve to be brought nearer together or separated by turning the spindle, the point of cut-off may be made to take place early or late, the limit of earliness being imposed by the condition that l must not be reduced below the amount which will give a fair steam opening, and the limit of lateness being imposed by the consideration that the main slide itself becomes closed at a position determined by its own outside lap. The events of release, compression and admission, depending as they do on the main slide-valve alone, are found by drawing lap arcs on the main-valve circle I in the same manner as in earlier examples of Zeuner's diagram.

The wave-form diagram of § 197 gives an excellent means of studying the action of Meyer's valve. Three distinct curves having been drawn for the piston, main valve and expansion valve respectively, showing the displacement of each in relation to the angle turned through by the crank-shaft, they are to be superposed as in fig. 167 (using tracing paper as before) with the proper differences of angular position set out by distances measured along the base line between the points at which the maximum displacement towards one side occurs in each. Both valve curves must have the same scale. Then the relative displacement of the valves is everywhere shown by the vertical distance between the main-valve curve and the expansion-valve curve. Cut-off is made to occur at any desired place in the motion by making the quantity l of fig. 164 equal to the distance found by measurement between the two valve curves at the corresponding point of the base line. Thus in fig. 167 if it is wished to cut off steam when the piston has travelled 25 per cent. of its stroke, the corresponding point P is found by projection from the scale at the side, PR is drawn and the intercept TR is measured: this determines the proper length of l. With a smaller value of l the cut-off comes earlier, and in the particular example shown in the figure the admission may be reduced to 10 per cent. of the stroke, or even less, by reducing l. The diagram[1] relates to a case in which the expansion eccentric is set at 180° in advance of the crank and the main-valve eccentric at 130° (making $\theta = 40°$). Both eccentrics have the same throw, giving a travel of 1·55 inches to each valve. The main valve has

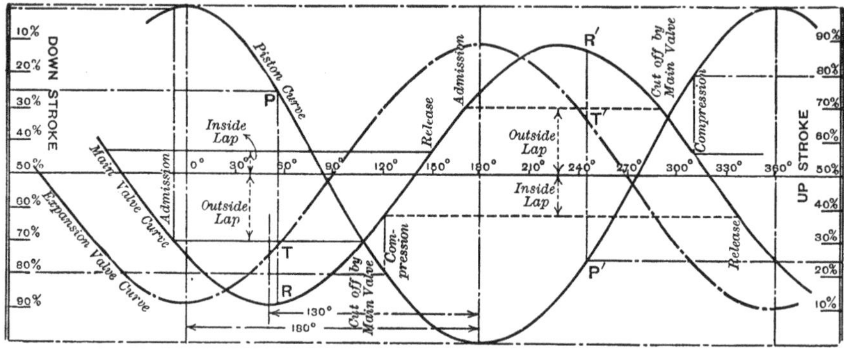

FIG. 167.

an outside lap of 0·4 inch on both sides: this gives equal amounts of lead, namely 0·1 inch, but would make the cut-off unequal on the two sides, namely, at 70 per cent. of the in-stroke or down-stroke and at 62 per cent. of the out-stroke or up-stroke, if the cut-off depended on the main valve. Since the cut-off is accomplished earlier, by means of the expansion valve, this inequality does not matter. The inside laps of the main valve are made unequal so that they give the same compression on both sides, namely, by stopping the exhaust at 80 per cent. in each back stroke; their values, found by projection from points at 80 per cent. on the stroke scale, are 0·24 inch on the front or bottom side and 0·14 inch on the back or top side.

By measuring distances such as TR between the two valve

[1] Drawn by Prof. Dalby for a small vertical experimental engine in the Engineering Laboratory at Cambridge.

curves it will be seen that equal cut-off on the two sides can only be secured by having different values of l at the two ends of the valve. Thus TR is 0·33 inch and $T'R'$, which also corresponds to a 25 per cent. cut-off, is 0·42 inch, or nearly one-tenth of an inch more. A constant difference between the values of l at the two ends is in fact preserved in Meyer's gear while the values of l are varied, and in this case the diagram shows that a constant difference of about one-tenth of an inch suffices to keep the points of cut-off practically symmetrical from say 10 per cent. to 35 per cent. of the stroke. When the cut-off is to be later than 35 per cent. equality can only be preserved by reducing slightly the difference between the values of l. The difference between them can be varied in practice by shifting the expansion valve bodily towards or from its eccentric, provided the valve-spindle in the eccentric-rod be furnished with a screw coupling or other device which permits its length to be altered.

The cut-off is altered in the ordinary form of Meyer's valve by turning the spindle, as already mentioned, to make the two blocks approach or recede from one another. Matters are often so arranged that this adjustment can be made while the engine is running, by means of a sleeve and hand wheel fitted on a prolongation of the valve-rod through the back end of the steam chest. The cut-off may also be varied by altering the travel of the expansion valve, instead of l. In some examples the expansion is varied automatically to suit the varying load upon the engine, the governor being connected to the expansion valve in such a way that either the quantity l or more commonly the travel is varied in response to variation in the speed. When the travel is to be altered a link, oscillating about a fixed centre, is interposed between the valve-rod and the eccentric-rod, and by sliding the end of the eccentric-rod up or down in the link, the link is made to act as a lever of variable length.

In a modified form of this valve, known as Rider's, the expansion valve is a species of piston working in a cylindrical hole bored out of the main valve. The steam passages terminate in a pair of oblique slots within this hole, and the front and back edges of the piston-shaped expansion valve are also cut obliquely, with the result that when the valve is turned about its axis its edges approach or recede from the oblique slots which form the steam-ports. This turning can be effected by the governor.

209. Forms of slide-valves. Double-ported valve. Trick valve. In designing a slide-valve the breadth of the steam-ports in the direction of the valve's motion is determined with reference to the volume of the exhaust steam to be discharged in a given time, the area of the ports being generally such that the mean velocity of the steam during discharge is less than 100 feet per second. The travel is made great enough to keep the cylinder port fully open during the greater part of the exhaust; for this purpose it is $2\frac{1}{2}$ or 3 times the breadth of the port. To facilitate the exit of steam the inside lap is always small, and is often wanting or even *negative*, especially in engines which are designed to run at a high speed. During admission the steam-port is rarely quite uncovered when the valve reaches the end of its travel, particularly if the outside lap is large and the travel moderate. Large travel has the advantage of giving freer ingress and egress of steam, with more sharply-defined cut-off, compression, and release, but this advantage is secured at the cost of more work spent in moving the valve and more wear of the faces. To lessen the necessary travel without reducing the area of steam-ports, double-ported valves are often used, and occasionally there are even three ports at each end. An example of a double-ported valve is shown in fig. 169. Fig. 168 shows the Trick valve, a device which accomplishes the same purpose by giving simultaneous admission in two ways; steam enters directly past the outer edge, as in an ordinary slide-valve, and at the same instant an opening at the other end of the valve is uncovered by passing beyond the edge of the raised seat on which the valve works. This gives a supplementary admission, to the same cylinder port, through a passage cast in the back of the valve itself.

210. Balance-piston. Incidentally, fig. 169 illustrates an arrangement that is usual in heavy slide-valves whose travel is vertical—the *balance-piston*, seen at the top of the figure, which is pressed up by steam on its lower side and so supports the weight of the valve, valve-rod, and connected parts of the mechanism. The space above the piston in the balance cylinder is kept in communication with the condenser.

211. Relief-frames. To relieve the pressure of the valve on the seat, large slide-valves are often fitted with what is called a *relief-frame*, which excludes steam from the greater part of the back of the valve. In a common form of relief-frame a ring fits steam-tight into a recess in the cover of the steam chest, and is

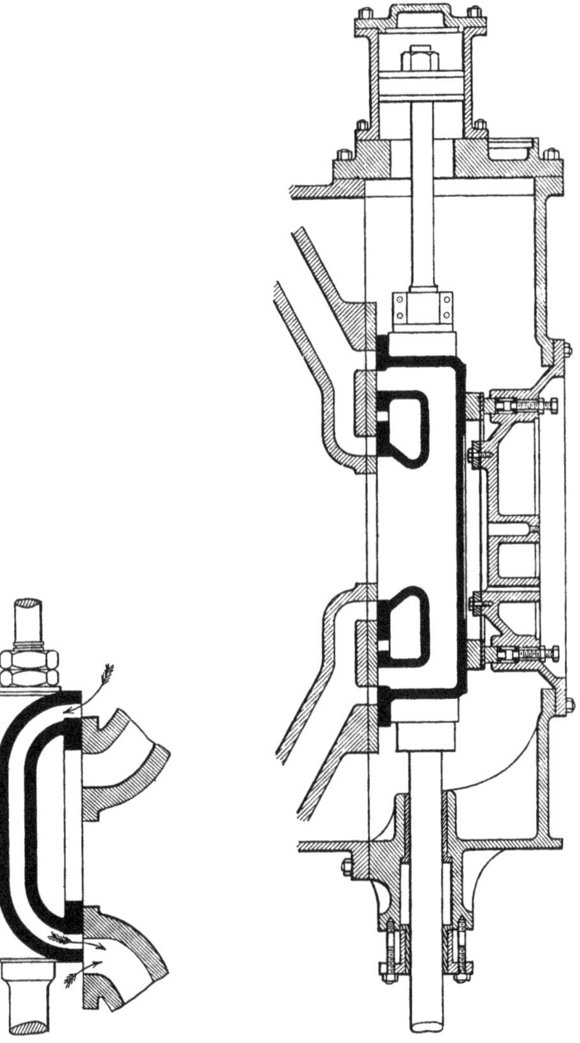

Fig. 168. Trick valve.

Fig. 169. Double-ported valve with balance-piston and relief-frame.

pressed by springs against the back of the valve, which is planed smooth to slide under the ring. Another plan is to fit the ring into a recess on the back of the valve, and let it slide on the inside of the steam-chest cover. Steam is in either case excluded from the space within the ring, any steam that leaks in being allowed to escape to the condenser (or to the intermediate receiver when the arrangement is fitted to the high-pressure cylinder of a compound engine). Fig. 169 gives an example of a relief-ring fitted on the back of a large double-ported slide-valve for a marine engine.

212. Piston-valves. The pressure of valves on cylinder faces is still more completely obviated by substituting a *piston slide-valve* for the flat form. The piston slide-valve may be described as a slide-valve in which the valve face is curved to form a complete cylinder, round whose whole circumference the ports extend. The pistons are often packed like ordinary cylinder pistons by split metallic rings, and the ports are crossed here and there by diagonal bars to keep the rings from springing out as the valve moves over them. In some cases the packing rings are omitted, and in other cases uncut floating rings are used instead of split rings. Fig. 170 shows one form of piston-valve. *PP* are the cylinder ports, and the supply of steam reaches the valve through two distinct inlets at the top and bottom. In locomotives, to which this form is much applied, the valve-spindle is sometimes continued as a tail-rod which projects through the outer end of the valve-chest. When the movement is horizontal this allows the weight of the valve to be taken by the spindle, and the pressure against the working face is only that of the piston-rings. In another form of piston-valve the rod connecting the two pistons is hollow and forms a communication between the steam chambers above and below the valve, thus making one steam inlet suffice. An example of a hollow piston-valve for a marine engine is shown in fig. 171. Steam is admitted to the outer casing and passes freely through the interior of the valve. The exhaust-pipe (which does not appear in the diagram) opens on the inner casing, in the space between the two pistons of the valve.

The increased use of superheating has tended to bring about a substitution of the piston type of slide-valve for the earlier flat type, owing to the difficulty which was experienced in maintaining lubrication between flat surfaces at a high temperature when sliding over one another under pressure. Besides being very generally employed in marine engines, the piston-valve has become a normal form for locomotives using superheated steam. In marine triple and quadruple expansion engines it is usual to fit

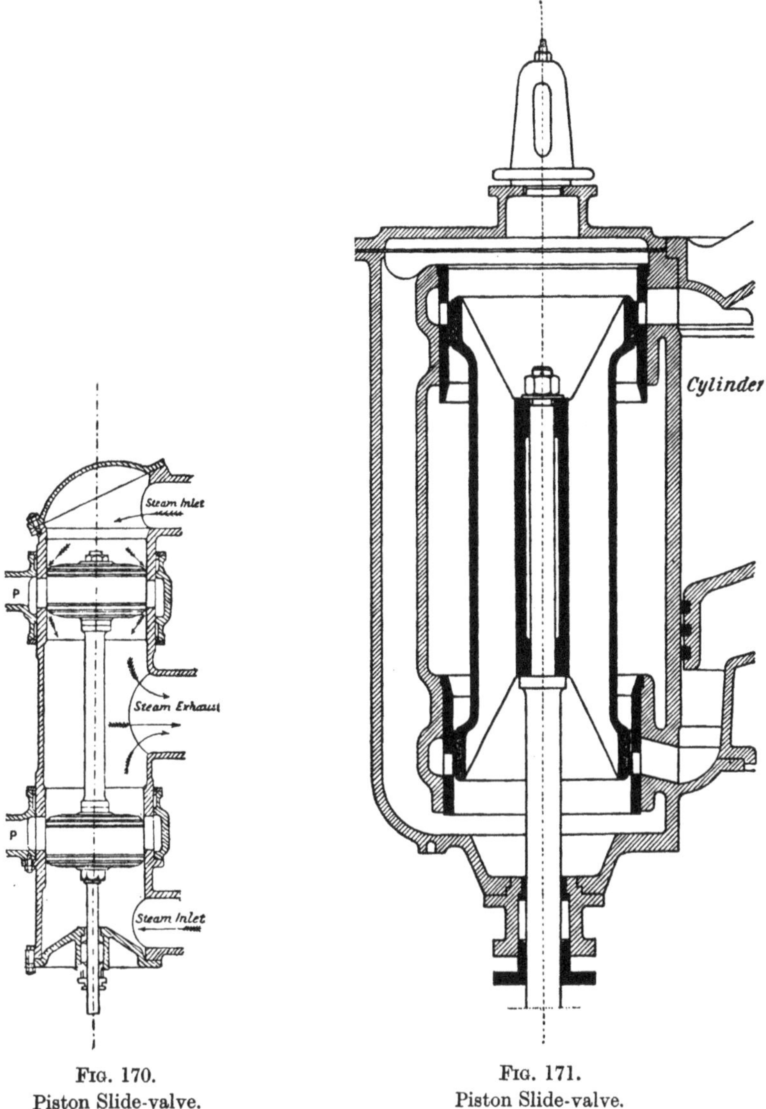

FIG. 170.
Piston Slide-valve.

FIG. 171.
Piston Slide-valve.

piston-valves to the high-pressure and intermediate-pressure cylinders, and flat balanced slide-valves to the low-pressure.

213. Rocking valves. The slide-valve may take the form of a rocking cylinder (fig. 172), the motion of which is produced by the turning of a spindle through the required arc, the valve being a block loosely set on a squared part of the spindle so that it is constrained to turn but at the same time is free to take a close bearing against the cylindrical port-face.

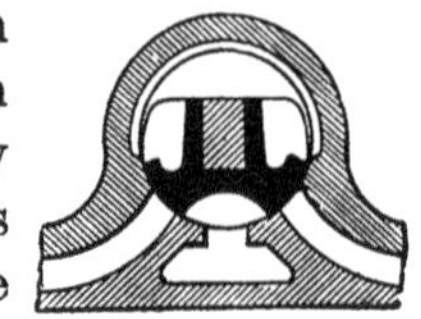

Fig. 172.
Rocking Slide-valve.

In some engines the distribution of steam, for each cylinder, is effected by two rocking valves, as in fig. 173, one for each end of the cylinder. Each valve opens the corresponding cylinder port alternately to steam and to exhaust, and has a lap which closes the port during expansion and compression.

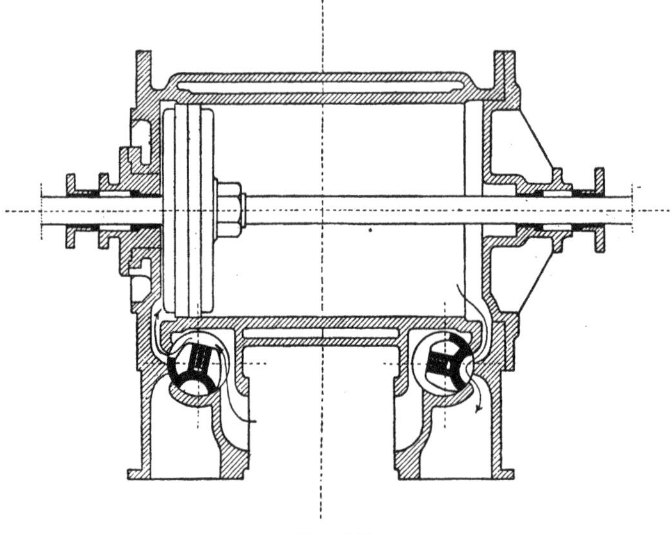

Fig. 173.

In another arrangement four rocking valves are used, namely, a separate steam and exhaust-valve for each end of the cylinder, as in fig. 174, where one end of the cylinder is shown with the steam-valve above and the exhaust-valve below. Valves of this type are called Corliss valves and are frequently found in connexion with the Corliss and other forms of trip gear, which will be mentioned in the next chapter. The characteristic of such gear is that after the steam-admission valve has been opened to the extent required for admission it is disconnected from the eccentric by a trip device which allows it to close suddenly under the action of a spring and thus produces a sharp cut-off.

214. Double-beat lift or drop-valves. In many engines *lift*-valves are used, also called *drop*-valves, worked by tappets, cams, or eccentrics. They open by being raised from their seats, and close by dropping back. They are generally of the double-beat type (fig. 175), in which equilibrium is secured or rather approximated to by the use of two conical faces of nearly the same size,

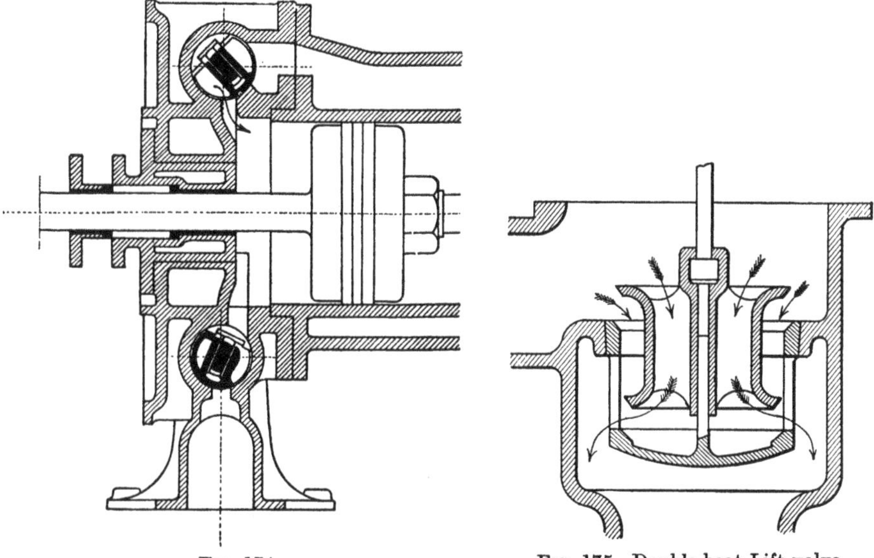

FIG. 174.

FIG. 175. Double-beat Lift-valve.

which open or close together. Thus a valve of large area may be opened against a heavy steam pressure with but little exertion of force.

In many large horizontal stationary engines four double-beat valves control the action; two on top of the cylinder providing for the admission and two underneath providing for the exhaust. The steam-valves in such cases are usually operated by a form of trip gear. Fig. 176 shows one end of a cylinder with steam and exhaust-valves of this type. The arrangement, like that of fig. 174, has the advantage of making it possible to have very small clearance spaces, also of perfectly draining the cylinder, and of separating the channels which are exposed to contact with the hot steam from those which are in contact with the comparatively cold exhaust. Fig. 177 shows an admission valve of this class with the spring for closing it after it has been opened and released by a trip gear. The spring, which is at the top of the valve-rod, is compressed by lifting the valve. This also lifts a piston in a dash-pot beneath the spring, admitting air to a space below. When the valve is released this air has to be expelled through a small orifice at the side, the size of which is regulated by a small screw plug. This prevents the spring from operating too violently, and by adjusting the orifice the valve may be made to come softly to its seat.

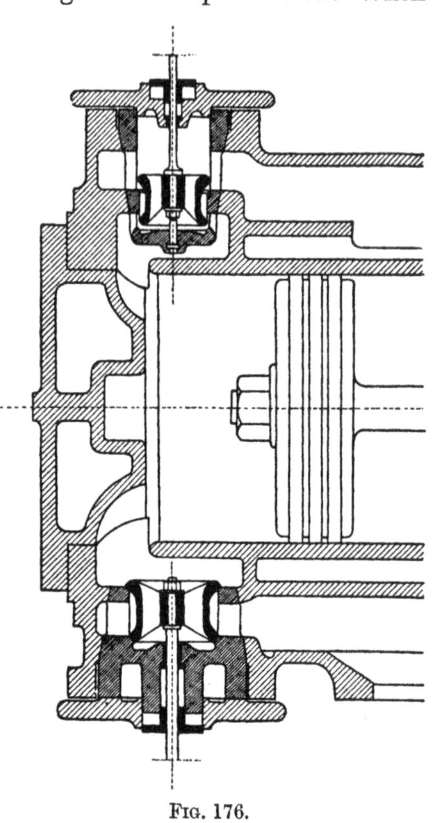

FIG. 176.

The Sulzer engines, which have a high reputation for thermal economy, give one out of many examples that might be cited, in which the admission and exhaust are controlled by lift-valves of the double-beat type. In some Sulzer designs the valves are four-seated, with the effect that a large aggregate opening is given by a small lift. A four-seated valve is shown in fig. 178.

The uniflow engine already referred to (§ 127) has a drop-valve at each end of the cylinder to control the admission, giving a sharp early cut-off and a small clearance. In that engine the advantage of keeping the exhaust and admission passages apart is very completely secured. The main piston, as was there explained, is itself the exhaust-valve; it releases the steam by uncovering exhaust-ports which form a ring round the cylinder at the middle of its length, the piston being so long that they are uncovered only near the end of each stroke. Hence the cylinder is kept hot at both ends, by the presence there of live steam, while the middle of its length, where the exhaust occurs, is relatively cool.

The device of making the piston itself serve as the exhaust-valve, by uncovering ports in the cylinder wall as it approaches the end of its stroke, is used, as will be seen later, in many internal-combustion engines.

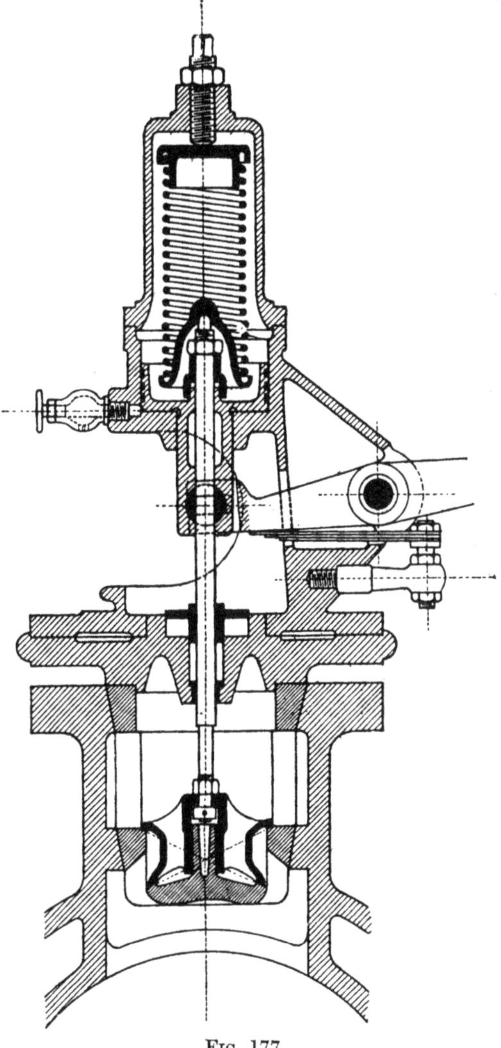

Fig. 177.

215. The Cornish cataract. In the Cornish pumping engine, which for long retained the single action of Watt's early engine, and of which examples may still be seen, three double-beat valves are used, as steam-valve, equilibrium-valve, and exhaust-valve respectively. These are closed by tappets on a rod moving with the beam, but are opened by means of a device called a cataract, which acts as follows. The cataract is a small pump with a weighted plunger, discharging fluid through a stop-cock which can be adjusted by hand when it is desired to alter the speed of the engine. The weighted plunger is raised by a rod which hangs from the beam, but is free in its descent, so that it comes down at a rate depending on the extent to which the stop-cock is opened. When it comes down a certain way it opens the steam and exhaust-valves by liberating catches which hold them closed; the "out-door" stroke then begins and admission continues until the steam-valve is closed; this is done directly by the motion of the beam, which also, at a later point in the stroke, closes the exhaust. Then the equilibrium-valve is opened, and the "in-door" stroke takes place, during which the plunger of the cataract is raised. When it is completed, the piston pauses until the cataract allows the steam-valve to open and the next "out-door" stroke then begins. By applying a cataract to the equilibrium-valve also, a pause is introduced at the end of the "out-door" stroke. Pauses have the advantage of giving the pump time to fill and of allowing the pump-valves to settle in their seats without shock.

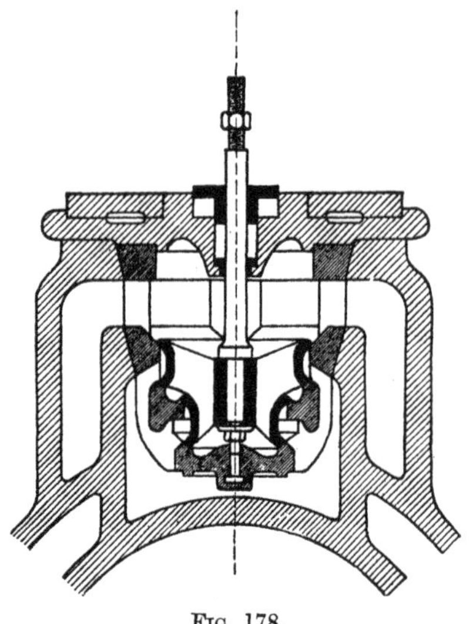

Fig. 178.

CHAPTER XII

GOVERNING

216. Methods of regulating the work done in an engine. To make an engine run steadily an almost continuous process of adjustment must go on, by which the amount of work done by the working substance is adapted to the amount of external work demanded of the engine. Generally the process of governing aims at regularity of speed; occasionally, however, it is some other condition of running that is to be maintained constant, as when an engine driving a dynamo is governed by an electric regulator to give a constant difference of potential on the mains—a condition which may require the engine to run rather faster when it is giving a greater output.

The ordinary methods of regulating a steam-engine of the piston and cylinder type, whether by hand or automatically through a governor, are either (*a*) to alter the pressure at which steam is admitted by opening or closing more or less a throttle-valve between the boiler and the engine, or (*b*) to alter the volume of steam admitted to the cylinder by varying the point of cut-off. The former plan was introduced by Watt, and is still common, especially in small engines. The second plan of regulating is in general to be preferred, especially when the engine is subject to large variations of load, and it is usually followed in large engines of that class. With steam turbines it is of course not possible, and the normal method of governing is by means of a throttle-valve. A method used by Parsons in some of his early turbines, of admitting the steam in a series of regulated blasts, is now obsolete. In impulse turbines with partial admission (§ 153) more or fewer nozzles may be opened to admit steam in the first stage.

217. Automatic regulation by centrifugal speed governors. Watt's conical pendulum governor. Within certain limits regulation of an engine can be effected by hand, but for the finer adjustment of speed some form of automatic governor is necessary. Speed governors are commonly of the *centrifugal* type: a pair of masses revolving about a shaft or spindle which is driven by the engine are kept from flying out by a certain controlling force.

When an increase of speed occurs this controlling force is no longer able to keep the masses revolving in their former path; they move

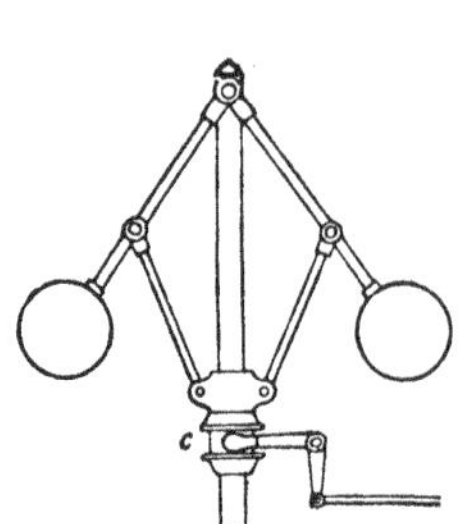

FIG. 179. Watt's Governor.

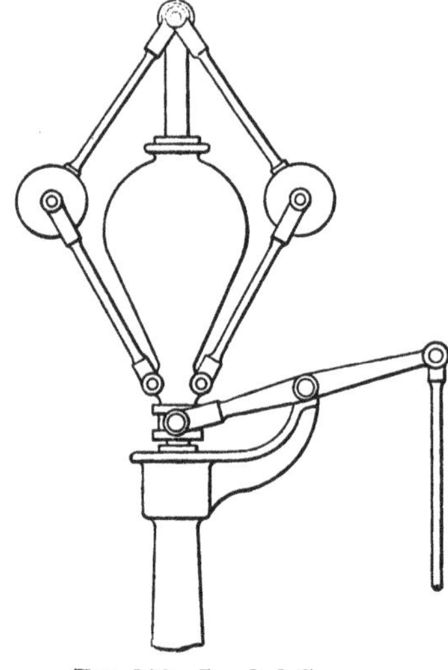

FIG. 180. Loaded Governor.

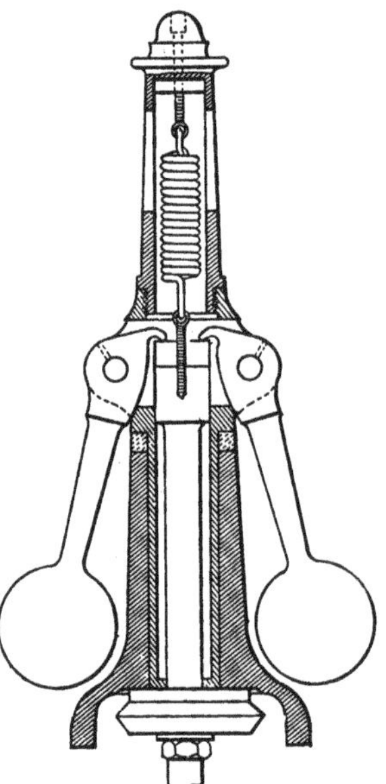

FIG. 181.
Spring Governor (Tangye).

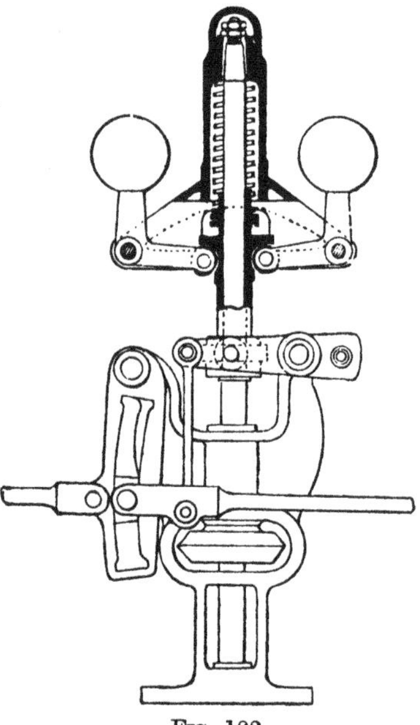

FIG. 182.
Spring Governor (W. Hartnell).

out until the controlling force is sufficiently increased, and in moving out they act on the regulator of the engine, which may be a throttle-valve or some form of automatic gear by which the cut-off is varied. In the conical pendulum governor of Watt (fig. 179) the revolving masses are balls attached to a vertical spindle by rods, and the controlling force is furnished by the weight of the balls, which, in receding from the spindle, are obliged to rise. When the speed exceeds or falls short of its normal value they move out or in, and so raise or lower a collar C which is in connexion with the throttle-valve through a lever. The suspension-rods may be hung from the ends of a T-piece attached to the revolving spindle instead of being pivoted in the axis as in fig. 179, and in some cases they cross each other and the spindle itself (see figs. 184–186).

In many high-speed engines the masses which form the governor revolve about the shaft of the engine itself, in a vertical plane, and the controlling force is furnished by means of springs which oppose their movement away from the axis of rotation.

218. Loaded and spring governors. In a modified form of Watt's governor, known as Porter's, or the *loaded* governor, the balls revolve, as in Watt's form, about a vertical axis, but the tendency which they have to fly out is resisted not only by their own weight but also by a supplementary controlling force which is furnished by a weight resting on the sliding collar (fig. 180). This device is equivalent to increasing the *weight* of the balls without altering their *mass*. In other governors the controlling force instead of being due to gravity only is wholly or partly produced by springs. Fig. 181 shows a governor by Messrs Tangye in which the balls are controlled partly by their own weight and partly by a spring, the tension of which is regulated by turning the cap at the top. Another form of a governor with spring control is Hartnell's, shown in fig. 182. There the balls move in a sensibly horizontal line and consequently their weight contributes nothing towards resisting the tendency to fly out. A certain amount of additional control, however, is supplied by the weight of the sliding collar and the parts which rise with it, unless these are counterbalanced. In this example the governor regulates the volume of steam admitted per stroke by altering the position of a sliding block in a rocking link, which determines the travel of an expansion valve.

219. Shaft governors. This name is given to a type, common in high-speed engines, where the masses revolve in a vertical plane about the engine-shaft, the control being furnished by springs. Shaft governors are sometimes placed within the fly-wheel of the engine. They may be arranged to act on a throttle-valve or to alter the point of cut-off. An example is shown in fig. 183 of a shaft governor designed to produce automatic variations of the cut-off by acting on the slide-valve of the engine. The displacement of the revolving masses M, M changes both the throw and the angular advance of the eccentric, thereby effecting a change in the steam-supply similar to that produced by "notching up" a link-motion. The eccentricity B is altered by the relative displacement of two parts C, D into which the eccentric sheave is divided. This relative displacement not only changes the length of B but gives it more or less of angular advance.

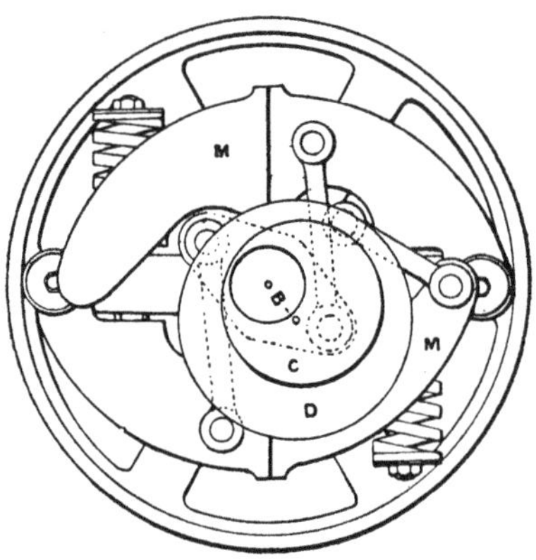

FIG. 183. Shaft Governor.

220. Controlling force. In whatever way the tendency of the balls to fly out be resisted, whether through their own weight or through a supplementary load or through springs, it is convenient to treat the control as equivalent to a certain force F acting on each ball in the direction of the radius towards the axis of revolution. We shall call this the *controlling force*. The value of F varies, in a given governor, when the position of the balls changes. If it were not for friction, the controlling force for any position of the balls could be found experimentally by applying a spring balance to each ball, with the governor at rest, and noting the force required to hold the ball in the assigned position when this force is applied directly away from the axis of the governor. Owing to friction such an experiment would give two extreme values of the force, for if the pull on the spring balance were increased the ball would not move further out until the pull became equal to $F + f$, where f denotes the force due to friction. And if the effect of friction in resisting the return of the ball were the same as its effect in resisting the displacement outwards, the pull of the spring balance might be reduced to $F - f$ before the ball would begin to move in. A mean of these extremes would give the true controlling force in cases where the influence of friction remained unchanged.

When the governor is running the influence of friction is in general less than when it is at rest; but the effect still is to make the actual force which the ball experiences, pulling it towards the spindle, greater or less than F according as the ball is on the point of moving out or moving in.

221. Condition of equilibrium. Once the controlling force F is known for each position of the balls, it is easy to calculate the speed at which the governor must revolve in order that an assigned position shall be taken up. If M be the mass of the ball (in lb.), n the number of revolutions per second and r the radius (in feet) of the path in which the balls revolve, equilibrium will be maintained when the speed is such that the controlling force and the "centrifugal force" are equal, that is to say when

$$F = 4\pi^2 n^2 r M,$$

F, the controlling force, being expressed in poundals. Hence the speed corresponding to the assigned configuration of the governor is defined by the equation

$$n = \frac{1}{2\pi}\sqrt{\frac{F}{Mr}}.$$

For the present, friction is left out of account; its influence on the speed will be considered immediately.

222. Condition of stability. When a governor is running at a steady speed, imagine the balls to be slightly displaced from the position proper to that speed, by temporarily applying some displacing force. Then, assuming the speed to remain constant, let the balls be left to themselves. If the governor is *stable* they will tend to recover their original position. The temporary displacement caused the controlling force to be changed by an amount δF, while the radius of the balls' path was changed by δr. The "centrifugal force" was temporarily changed from $4\pi^2 n^2 r M$ to $4\pi^2 n^2 (r + \delta r) M$ while the controlling force was changed from F to $F + \delta F$. The condition of stability requires that the change of controlling force should be greater than the change of centrifugal force, so that there may be a resultant force tending to bring the balls back.

Thus in order that the governor should be stable δF must be greater than $4\pi^2 n^2 M\delta r$, and this will be the case only if

$$\frac{\delta F}{F} \text{ is greater than } \frac{\delta r}{r}.$$

That is to say, the controlling force must increase more rapidly than in simple proportion to r as the balls move outwards.

When a stable governor is running in a given configuration the effect of any small increase or decrease of speed is (friction apart) to make the balls go out or come in by a finite amount so that they reach a new position of equilibrium corresponding to the new speed. As r increases a higher and higher speed is required to preserve equilibrium in each new position of the balls.

If F varied just proportionally to r, the speed would be constant for all values of r. This state of things would correspond to neutral equilibrium on the part of the governor: its consequences are considered more particularly in § 225 below.

223. Equilibrium of the conical pendulum governor. Height of the governor. When the governor is a simple conical pendulum controlled by gravity only and without load other than the weight of the balls themselves—a condition never quite realized in practice, since the weight of the sliding collar and its attached parts always applies some extra load which adds to the controlling force—F, the controlling force, is the resultant of F_2 the tension in the suspending rod and F_1 or Mg the weight of the ball. The triangle of forces is sketched in fig. 187. This applies whichever of the three forms shown in figs. 184, 185 and 186 is given to the governor. Let the *height* of the pendulum governor, that is the vertical distance from the plane of rotation of the balls to the point where the axis of the suspending rod (produced if necessary) cuts the axis of the spindle, be called h. Then, using absolute units for the forces,

$$F : Mg :: r : h;$$

from which

$$F = \frac{Mgr}{h}, \text{ and } n = \frac{1}{2\pi}\sqrt{\frac{g}{h}}.$$

The condition of stability requires that the speed n should increase as the balls move out, and hence h must diminish when

r increases if a governor of this class is to be stable. This will obviously happen if the form be that of fig. 184 or fig. 185. With

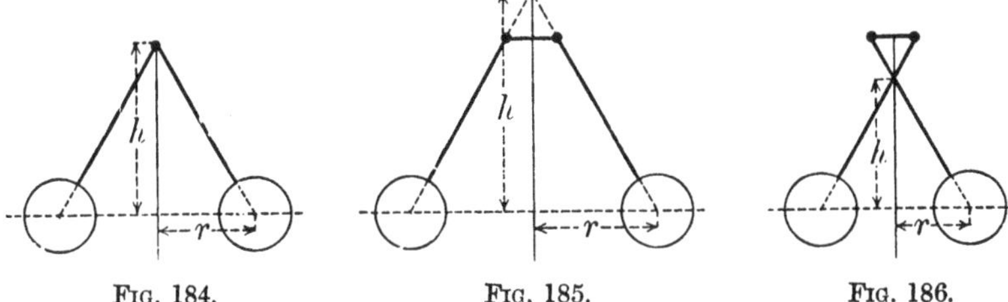

FIG. 184. FIG. 185. FIG. 186.

the crossed-rod form of fig. 186 the height h will increase when the balls rise only if the points of suspension are not far from the axis. By placing them at a particular distance from the axis, h may be kept very nearly constant: in other words, this governor may be arranged to have nearly neutral equilibrium, so that a very small change in the speed n may be associated with a large change in the position of the balls. When the centres of suspension are put further from the axis the mechanism sketched in fig. 186 becomes unstable, and is then unfit to serve as a governor.

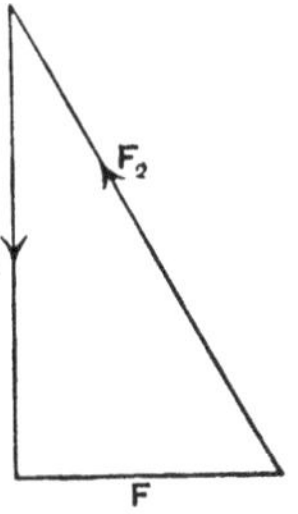

FIG. 187.

224. Equilibrium of loaded governor. The results obtained in the last section are readily adapted to the case of a governor of the Porter type (fig. 180). Let M' be the amount of the extra load, per ball (in general M' is one-half the total extra load), and let q be the velocity ratio of the vertical movement of the load to the vertical movement of the ball—a quantity which is easily found by calculation or graphically when the form of the governor is given. Then each ball, in being displaced outwards, has not merely to raise its own weight but has to raise what is equivalent to an additional weight equal to q times the weight of M'. The effect of the load is therefore to increase the controlling force F from $\frac{Mgr}{h}$, in poundals, to $\frac{(M+qM')\,gr}{h}$. But the condition of equilibrium still is that F should be equal to $4\pi^2 n^2 rM$. Hence the speed n at which the governor must now turn to maintain any assigned height h is

$$n = \frac{1}{2\pi}\sqrt{\frac{(M+qM')\,g}{Mh}}.$$

Compared with the simple or unloaded form, this governor requires a higher speed in the proportion of $\sqrt{M + qM'}$ to $\sqrt{M}$.

In the ordinary construction of the Porter governor the four links form a parallelogram, and consequently the vertical movement of the load borne by the sliding collar is twice that of the balls, or $q = 2$. And as the whole load is divided between two balls, each ball virtually has its weight, but not its mass, increased by an amount equal to the whole weight of the central load.

Another way of considering the equilibrium of the Porter governor may be mentioned. Let the mass of each ball be M and let that of the load be $2M'$ as before. The load, the weight of which is $2M'g$ in poundals, is borne by the tensions in the two lower rods (fig. 188). By drawing the triangle abc (fig. 189) which is the diagram of forces for the load, and in which ab or F_3 is the weight of the load, we find the value of F_4 and F_4' which are the tensions in the lower rods. Then draw bd or F_1 to represent the weight of one ball (namely Mg), and draw the horizontal line de to meet a line ce drawn from c parallel to the direction of the upper rod. The figure $ecbd$ is the polygon of forces acting on the ball, ed being the resultant controlling force F. The speed at which the governor will run is determined by the condition that $4\pi^2 n^2 rM$ is to be equal to this force ed. In the usual case of parallel rods, ace is one straight line and then

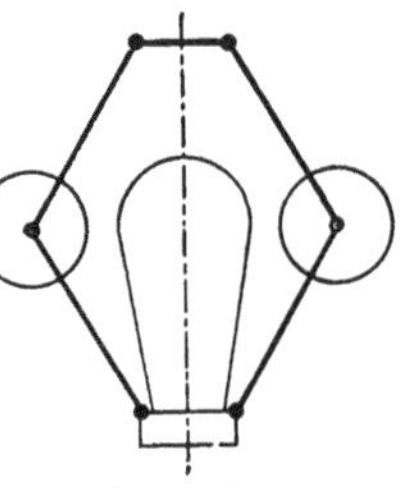

FIG. 188.

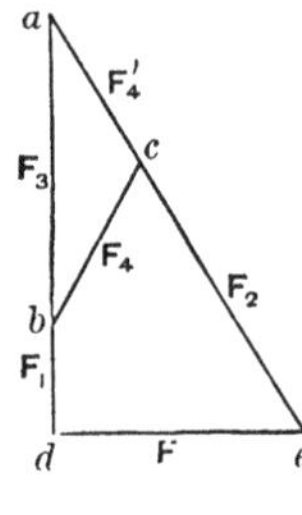

FIG. 189.

$$ed = ad \tan a, \quad \text{or} \quad F = (M + 2M')\, g \tan a,$$

where a is the inclination of the rods to the vertical. Since $\tan a = \frac{r}{h}$ this expression agrees with the one given above.

225. Sensitiveness in a governor. Isochronism. Any change of speed in a governor tends to produce a change in the position of the balls, and if the governor itself and the regulating mechanism connected with it were free from friction, only one position of the governor would be possible for any one speed, provided the condition of stability were complied with. If therefore the supply of steam depends on the position taken up by the governor balls a stable governor does not maintain a strictly constant speed in the engine it controls. Whenever the boiler pressure or the demand for work changes a certain amount of displacement of the balls is necessary to increase or reduce the steam-supply, and the balls can retain their new position only by virtue of continuing to turn slower or faster than before. The maximum change of speed which can occur under the control of the governor is that which will make the balls move from one to the other extremity of their range—namely, from the position which allows the full supply of steam to the position which completely checks the supply. Of course if the engine is overloaded by giving it too much external resistance to overcome, the speed may be further reduced after the governor has done all that it can do to let steam in freely, but the variation of speed for which the governor is responsible is only that which makes the change from no steam to full steam. When a small variation of speed suffices to do this the governor is said to be sensitive, its sensitiveness being measured by the reciprocal of the ratio which this variation of speed bears to the mean speed.

The more stable a governor is the less sensitive is it; on the other hand, when the equilibrium is neutral the sensitiveness is indefinitely great. The controlling force F then varies as r, and hence n is constant (§ 222) at whatever distance from the axis the balls revolve. In other words, the balls are in equilibrium at one speed and only at one (except for friction), and the least variation from this speed suffices to send them to one extremity or the other of their range. A governor having this quality is said to be isochronous. Friction makes the condition of strict isochronism impossible, but many governors are made nearly isochronous by arranging them so that, as the balls are displaced, the controlling force increases only a little more rapidly than r.

226. Isochronism in the gravity governor. Parabolic governor. An ideal frictionless governor, in which the controlling force is furnished by gravity, can be made isochronous if the balls instead of being hung by rods from fixed points are constrained to move in a parabolic path, as in fig. 190, where the cup or channel which holds the ball is so shaped that the locus of the centre of the ball, shown by a dotted curve, is a parabola. The pressure of the ball against the cup is equivalent to the tension of an imaginary suspension-rod PQ; and it is a property of the parabola that the sub-normal QM, which represents h, is constant wherever P be taken along the curve. Hence a ball supported in this way would remain in equilibrium at one particular speed of rotation on the part of the cup, but would fly up to the rim of the cup if the speed were ever so little increased, and would sink to the foot if the speed were ever so little reduced.

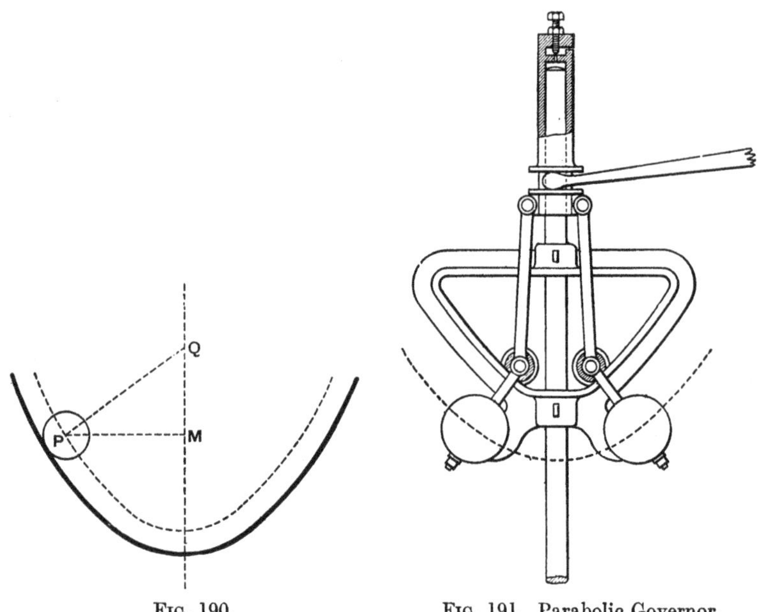

FIG. 190. FIG. 191. Parabolic Governor.

Fig. 191 shows a governor in which the balls are constrained to follow a nearly parabolic path[1]. An important feature is the air-cylinder at the top, forming a dash-pot, which is furnished with a small adjustable orifice through which air is driven out or in as the balls rise or fall. The function of this is to check the tendency which the balls have to fly violently in or out when the speed drops below or rises above the normal value: it limits the rate at which the governor responds to a change of speed and may prevent the balls from over-running the position proper to the new speed.

[1] From J. Head's paper on "A Steam-engine Governor," *Proc. Inst. Mech. Eng.* 1871.

227. Approximate isochronism in pendulum governors. A useful approximation to the condition of isochronism can be reached in the conical pendulum governor by using crossed rods with the centres of suspension at a suitable distance from the axis. If each centre of suspension were so placed as to be at the centre of curvature of a parabolic arc which coincided, at the position corresponding to the normal speed, with the actual circular curve along which the balls rise and fall, the governor would be sensibly isochronous at that speed. By taking points a little nearer the axis for the two centres of suspension a margin of stability, always necessary in practice, is secured, but the governor is left nearly

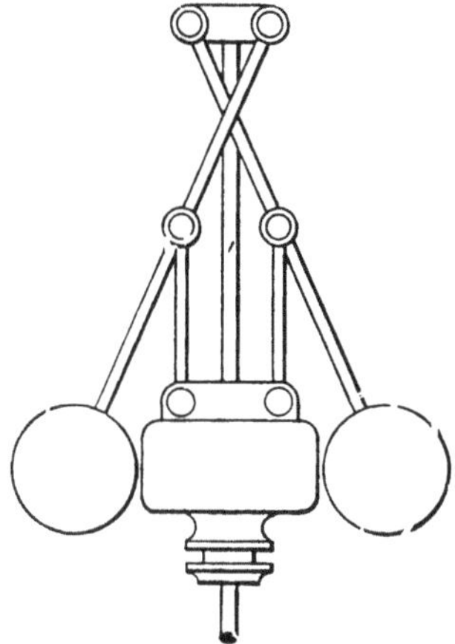

FIG. 192. Loaded Governor with Crossed Rods.

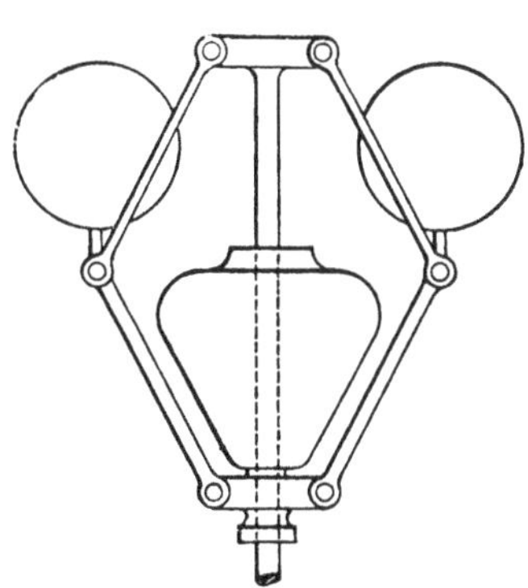

FIG. 193. Pröll's Governor.

enough isochronous to be very sensitive. This crossed-rod type of governor, which is due to Farcot, is often met with in a loaded form. An example is given in fig. 192. Loading a governor (whether the rods are crossed or open) need not affect the sensitiveness; it makes a higher speed necessary, but the proportion of the fluctuation of speed to the mean speed is not changed, provided the links are arranged in such a way that the vertical velocity ratio of the load and the balls does not alter as the balls rise.

Another approximately isochronous form of gravity governor is Pröll's (fig. 193), which exemplifies a different method of reducing the stability of the pendulum type. Let the ball be supported not at the joint between the links as in the ordinary Porter governor but at the end of an arm projecting upwards and rigidly connected to the lower link. By a proper choice of the length of this arm the controlling force may be made as nearly proportional to the radius as may be desired.

Pendulum governors of the stable class are occasionally loaded indirectly, the weight which forms the load being applied at some point in the lever by which the governor is connected with the valve. This allows the load and therefore the speed to be adjusted: further, by applying the load at the end of a cranked arm in the lever in such a way that it becomes less effective when the balls go out, the system can be made approximately isochronous.

228. Governors with spring control. Adjustment of sensitiveness. When springs furnish the controlling force, in whole or part, as in the governors shown in figs. 181 and 182, their tension is generally adjustable. This gives a convenient means of altering the speed; at the same time it affects the sensitiveness of the governor. In spring governors which are constructed so that the radial displacement of the balls produces a proportional change in the tension of the spring, the condition of isochronism can be approached, as nearly as may be wished, by giving the spring a suitable amount of initial tension. Thus in Hartnell's apparatus, fig. 182, where the balls move in a nearly horizontal direction and gravity has almost nothing to do with the control, the governor can be made isochronous by screwing down the spring so that the initial force exerted by the spring (before the balls are displaced) is to the increase of this force by the displacement of the balls, as the initial radius of the ball's path is to the increase of that radius by the displacement. This makes F vary proportionally to r, and therefore (§ 122) requires no change in n as the balls move out. Any greater initial tension would make the governor unstable, and a less tension is in fact necessary, in order that the sensitiveness may not be impracticably great.

229. Determination of the controlling force. Whatever be the method of control, by weights or springs or both, the controlling force F may generally be calculated for any assumed position of the balls. The simple pendulum governor both unloaded and loaded as in fig. 188 has already been considered. A case such as that of fig. 181 or fig. 182 presents no difficulty when the stiffness and initial tension of the spring are given. Slightly less simple cases of the loaded governor present themselves when the balls are not placed at the joints between the upper pair of links and the lower pair which carry the load. Let the ball M (fig. 194) be fixed on the upper link AB, at any place either beyond B or between A and B. First find F_1, the stress in BC, from the consideration that BC and its twin link on the other side are in simple tension and support between them the load. Then the forces acting on ABM, namely F_1, the tension in BC, F_2 which is the weight of the ball, and F which is the force to be determined, are in equilibrium, and hence F is readily found by taking moments about A. We here treat F as the equilibrant instead of the resultant of the forces which are actually applied, for the sake of bringing the system into static equilibrium.

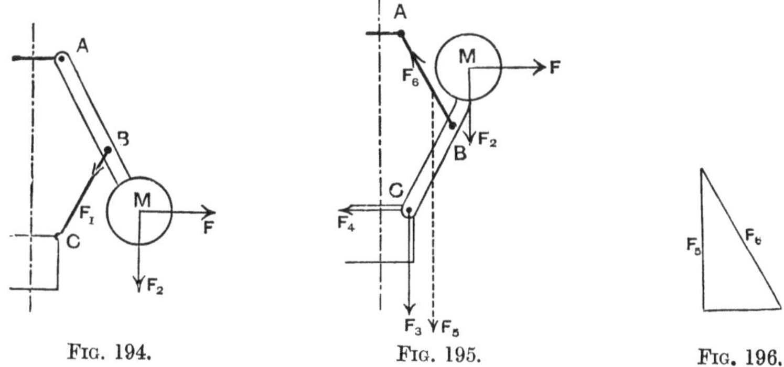

FIG. 194. FIG. 195. FIG. 196.

When the ball is carried by the link BC or by a piece rigidly connected to it as in Pröll's governor we may proceed thus (fig. 195):—The forces concerned in the equilibrium of the rigid piece CBM are (1) F_3 the half weight of the load acting at C, which is the vertical component of the pull at the joint C, (2) the horizontal component F_4 of the pull at the joint C, (3) the tension F_6 in the link AB, (4) the weight of M, or F_2, and finally (5) the force F which is to be determined. The resultant of F_3 and F_4 no longer acts along BC for there is a bending moment on the piece CBM. Compound F_2 and F_3 into a single force F_5. Since F_4 and F are horizontal, this vertical force F_5 must be wholly balanced by the vertical component of the stress in AB. Hence F_6 is found by drawing a right-angled triangle (fig. 196) with a line parallel to AB as hypotenuse and with a vertical side equal to F_5. Having found F_6 we are in a position to take moments about C in order to find F, which is now the only unknown force not acting through C.

230. Influence of friction. Power of the governor. We may express the influence of friction on the behaviour of a governor by treating it as equivalent to a force with some limiting value f, acting radially on each ball, in the same direction with the controlling force F when the balls are moving out and in the opposite direction when they are moving in. This makes the whole controlling force $F+f$ in the former case and $F-f$ in the latter. Let n be the speed proper to the force F alone, then if there were no friction any increase of speed above n would begin to alter the configuration, making the balls move out; but in consequence of friction this does not happen until the speed has increased by some finite amount Δn such that

$$n + \Delta n = \frac{1}{2\pi}\sqrt{\frac{F+f}{Mr}}.$$

Similarly, should the speed fall below the normal speed n proper to any configuration, friction prevents the balls from beginning to move in until the reduction of speed $\Delta' n$ is such that

$$n - \Delta' n = \frac{1}{2\pi}\sqrt{\frac{F-f}{Mr}}.$$

Hence in consequence of friction the speed may alter as much as Δn above and $\Delta' n$ below the normal speed n, while the position of the balls remains unchanged. From the above equations, if f be small relatively to F, as it always should be in practice, so that Δn may be small in comparison with n, Δn and $\Delta' n$ are nearly equal, and we have, approximately,

$$\frac{\Delta n}{n} = \frac{f}{2F}.$$

This variation of speed due to friction is independent of whatever further variation of speed the governor may allow in consequence of its equilibrium being stable (§ 225), and would of course be experienced even with a governor which except for friction was isochronous.

To keep the effects of friction within moderate limits it is essential that F should be great in comparison with f. The frictional resistance f proceeds partly from the joints of the governor itself but mainly from the throttle-valve spindle or other gear whose position the governor has to regulate. A *powerful* governor, namely a governor with a large amount of controlling force F, is therefore required when any considerable amount of frictional resistance in the valve or gearing is to be overcome. With simple pendulum governors, the only way to secure power in this sense is to make the balls large. Loaded governors have the advantage that great power may be secured with comparatively small revolving masses. The quality of powerfulness in a governor is increased whenever the controlling force is increased, whether by gravity loading or by the use of springs. From another point of view, the loaded governor (with the same revolving masses) is more powerful because it runs at a higher speed; but this is just because its controlling force F is greater. A governor in which the control is given by springs, such as a shaft governor, may be made very powerful without the use of large masses, by using stiff springs and a high speed.

231. Curves of controlling force. The consideration of sensitiveness and powerfulness in governors generally is greatly elucidated by using a graphic method, suggested by W. Hartnell[1], of exhibiting the controlling force. Having found the controlling force F for various positions of the balls, let a curve P_1P_2 (fig. 197) be drawn in which abscissæ represent r the radius of the balls' path and ordinates represent F. To find the configuration proper to any assigned speed n draw a line OS at such an inclination that $\tan SOX = 4\pi^2n^2M$, due regard being had to the scales of F and r. When the base OX is taken equal to unity on the scale used in plotting r, the value of SX is equal to $4\pi^2n^2M$ on the scale used in plotting F. Let P be the point in which this line cuts the curve of F. Then since

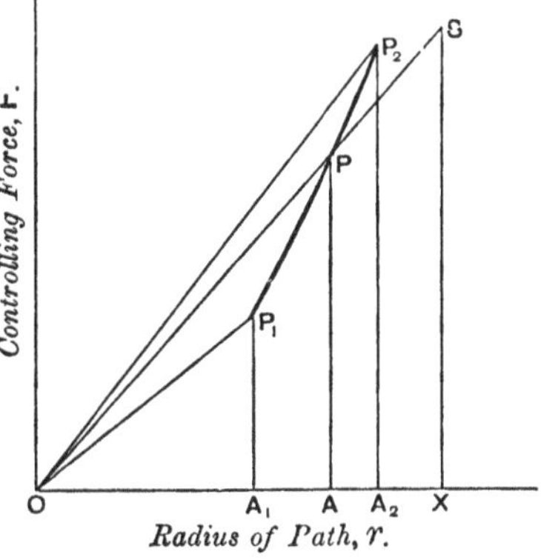

FIG. 197. Curve of Controlling Force.

$$F = PA = OA \tan POA = 4\pi^2n^2rM,$$

it follows that the point of intersection P determines the radius OA at which the governor will run when the speed is n. Similarly the tangent of the angle which is made with the base by any other

[1] *Proc. Inst. Mech. Eng.* 1882.

line drawn from O to meet the curve, such as OP_1 or OP_2, is proportional to the square of the speed at the corresponding path radius OA_1 or OA_2. Thus if OA_1 be set off to represent the least and OA_2 the greatest radius, corresponding to the positions giving full steam and no steam respectively, the inclinations of the lines OP_1 and OP_2 determine the whole range through which the speed will alter in consequence of the stability of the governor (apart from any effect of friction).

Further, if a pair of additional curves Q_1Q_2 and R_1R_2 be drawn as in fig. 198 to represent the values of $F+f$ and $F-f$ respectively, in relation to r, the diagram shows the additional changes of speed that are due to friction. The lowest possible speed is then determined by the inclination of the line OR_1, the highest by that of the line OQ_2. Thus the lower limit of speed, when there is full admission of steam, is

$$\frac{1}{2\pi}\sqrt{\frac{A_1R_1}{M.OA_1}},$$

and the higher limit of speed, when the admission is reduced to nil, is

$$\frac{1}{2\pi}\sqrt{\frac{A_2Q_2}{M.OA_2}}.$$

Again, the whole work done in altering the configuration of the governor, while the balls move out from A_1 to A_2, would be (for each ball) equal to the area $A_1P_1P_2A_2$ if there were no friction to be overcome: actually it is the area $A_1Q_1Q_2A_2$. And as the ball comes in from A_2 to A_1 the part of the stored energy which is recovered is measured by the area $A_2R_2R_1A_1$, the rest having been spent on friction. The area $P_1Q_1Q_2P_2$ is the work spent against friction while the governor is closing the throttle-valve or shifting the expansion gear from full steam to no steam.

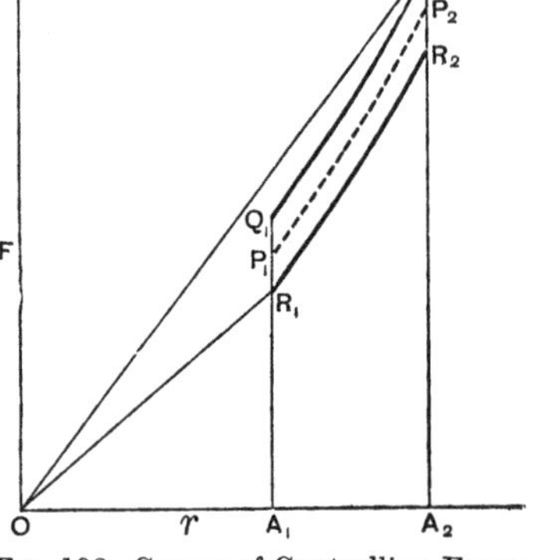

FIG. 198. Curves of Controlling Force, taking friction into account.

The *powerfulness* of the governor is measured in a definite manner by the area $A_1P_1P_2A_2$, namely, the work stored and restored (save for friction) as the governor balls open or close throughout their range. In order that friction should cause no very serious irregularity in speed this area must be many times greater than the area $P_1Q_1Q_2P_2$ or $P_1P_2R_2R_1$. These last areas are equal if the friction f has the same value in closing as in opening the valve (as we have assumed above), but the construction shown in fig. 198 is evidently applicable whether f has or has not the same value during the rise and fall of the balls.

Again, the governor is stable provided the inclination of the curve to the axis OX be greater than the inclination of a line drawn from O to meet the curve at any point within the range of possible positions. Thus in fig. 197 the curve shows the governor to be stable because any line OP is less steep than the inclination of the curve itself at P. This is the condition of stability stated in § 222, namely, that the controlling force must increase more rapidly than the radius. A strictly isochronous governor would have for its curve of F and r a straight line passing, when produced, through O. If this condition were fulfilled by the line without friction P_1P_2, the line with friction Q_1Q_2, which lies above P_1P_2 at a more or less constant distance from it, would in general be less steep than a line from O drawn to meet it, which would mean that friction would make the otherwise neutral governor *unstable*. This is one reason why the isochronous governor is impracticable. The governor of fig. 198 is stable notwithstanding friction.

232. Hunting. Apart from the reason just stated it is indispensable to give a governor some margin of stability, especially when any change of speed takes some time to affect the supply of steam. An over-sensitive governor is liable to produce in the engine which it governs a state of forced oscillation called *hunting*. Several reasons contribute to produce this effect. When an alteration of speed begins to be felt, however readily the governor alters its form the engine's response is more or less delayed. The action of the regulator does not immediately take full effect upon the speed in conseqence of the energy that is stored within the engine itself, not only in its moving parts but also in the steam that has passed the regulator and is still doing work in the engine. If the governor acts by closing a throttle-valve, the engine has still a capacious valve-chest on which to draw for steam. If it acts by changing the cut-off, its opportunity has passed if the cut-off has already occurred, and the control only begins in the next stroke. Such a lagging of effect occurs specially in compound engines, where that portion of the steam which is already in the engine continues to do its work for nearly a whole revolution after passing beyond the governor's control. The result of this storage of energy in an engine whose governor is too nearly isochronous is that whenever the demand for power suddenly falls the speed rises so much as to force the governor into a position of over-control, such that the supply of steam is no longer adequate to meet even the reduced demand for power. Then the speed slackens, and the same kind of excessive regulation is repeated in the opposite direction. A state of forced oscillation is consequently set up. The tendency to hunt depends upon the fact that the rate at which steam does work is not immediately controlled when the load on the engine varies, but that there is a time-lag between any variation in the load and the proper corresponding variation in the action of the steam. A similar time-lag with a consequent tendency on the part of the engine to hunt may proceed from another cause, which is independent of the storage of steam between the regulating valve and the engine piston. A sensitive governor, especially when it is of the relay type described below (§ 237), may take some time to come to its new position when the load is suddenly reduced. The governor begins to close the throttle-valve or to hasten the cut-off. But this takes time, and meanwhile the supply of steam is excessive and spends itself on the fly-wheel of the engine, giving it increased speed. By the time the supply is adjusted the speed has risen beyond its normal value, and a stage is reached when the regulating mechanism is carried too far and the supply is too much reduced. Thus a condition of forced oscillation may be set up even in cases where there is no storage of steam. The tendency is especially noticeable in engines with heavy fly-wheels running under light loads: under a heavier load the same engine may govern well without hunting. Again, hunting may be caused by the friction of the governor and of the regulating mechanism. Friction prevents the governor and regulator from beginning to change its position until the speed has changed by a finite amount, and when once the movement begins it goes beyond the point proper for steady control. The effect is aggravated by the momentum which the governor balls acquire in being displaced. Oscillations of the governor due to its own inertia are often prevented by introducing a *viscous* resistance to the displacement of the governor, which prevents the displacement from occurring too suddenly, without affecting the ultimate position of equilibrium. For this purpose many governors are furnished with a *dash-pot*, which is a hydraulic or pneumatic brake, consisting of a piston connected to the governor, working loosely in a cylinder which is filled with oil or with air. An instance of the use of a dash-pot has already been mentioned in speaking of the parabolic governor of fig. 191.

233. Throttle-valve and automatic expansion gear. The throttle-valve, as introduced by Watt, was originally a disc turning on a transverse axis across the centre of the steam-pipe. It is now usually a double-beat valve (§ 214) or a piston-valve. When regulation is effected by varying the cut-off, and an expansion valve of the slide-valve type is used, the governor generally acts by changing the travel of that valve. Fig. 182 illustrates one usual mode of doing this, by giving the expansion valve its motion from an eccentric-rod through a link the throw of which is varied by the displacement of the governor balls. In some forms of automatic expansion gear the governor acts upon the lap of the expansion valve. In others it acts by shifting the expansion eccentric round upon the shaft and so changing its angular advance. In others, again, it acts on an ordinary slide-valve through some form of link-motion or in such a way as has just been described.

234. Trip gear for governing by varying the expansion. In large stationary engines the most usual plan of automatically regulating the expansion is to employ some form of trip gear, the earliest type of which was introduced in 1849 by G. H. Corliss of Providence, U.S. In this system the valves which admit steam are distinct from the exhaust-valves. The latter are opened and closed by a reciprocating piece which takes its motion from an eccentric. The former are opened by a reciprocating piece, but are closed by springing back when released by a trip- or trigger-action. The trip occurs earlier or later in the piston's stroke according to the position of the governor. The admission valve is opened by the reciprocating piece with equal rapidity whether the cut-off is going to be early or late. It remains wide open during the admission, and then, when the trip-action comes into play, it closes suddenly. The indicator diagram of an engine fitted with trip gear consequently has a nearly horizontal admission line and a sharply defined cut-off. In the original trip-gear engines of Corliss the valves were cylindrical plates turning in hollow cylindrical seats extending across the width of the cylinder, of the rocking type mentioned in § 213 as Corliss valves, and this type is

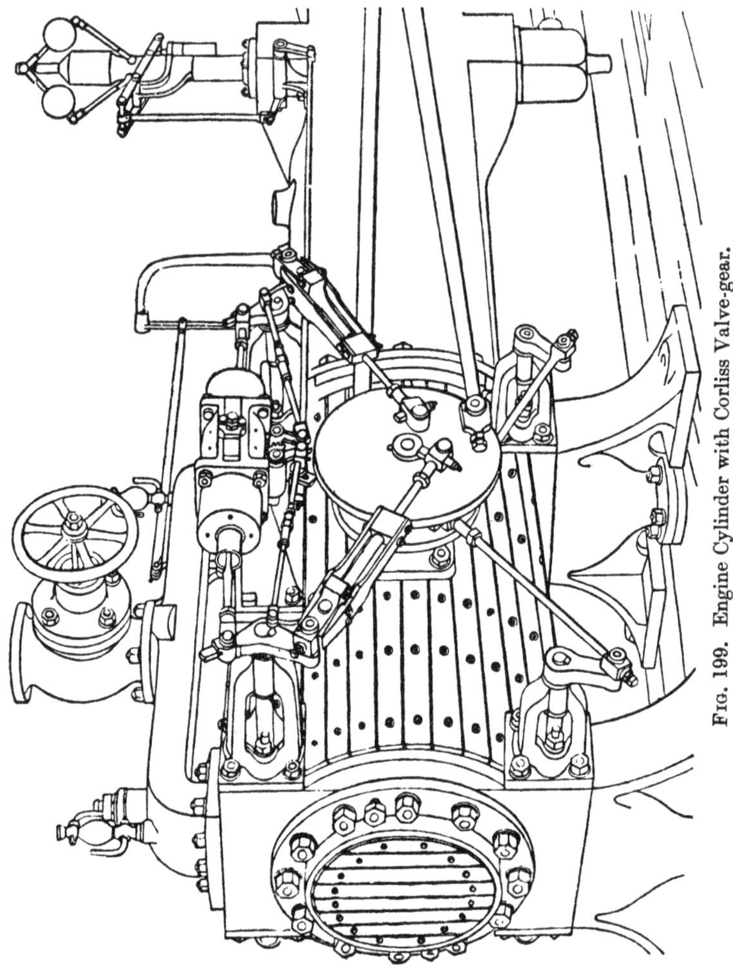

Fig. 199. Engine Cylinder with Corliss Valve-gear.

still found in some trip-gear engines. Often, however, the admission valves are of the double-beat type, and spring into their seats when the trip gear acts. Valves of this type were brought into use by Sulzer Brothers, who took up the manufacture of trip-gear engines in 1867 and have brought them to great perfection. Many forms of trip gear have been devised by Corliss himself, and by others. One of these, the Spencer Inglis[1] trip gear, by Messrs Hick, Hargreaves and Co., is shown in figs. 199 and 200. A wrist-plate A, which turns on a pin on the outside of the cylinder, receives a motion of oscillation from an eccentric. It opens the cylindrical rocking valve B of the Corliss type, by pulling the link C, which consists of two parts, connected to each other by a pair of spring clips a, a. Between the clips there is a rocking cam b, and as the link is pulled down this cam places itself more and more athwart the link, until at a certain point it forces the clips open. Then the upper part of the link springs back and allows the valve B to close

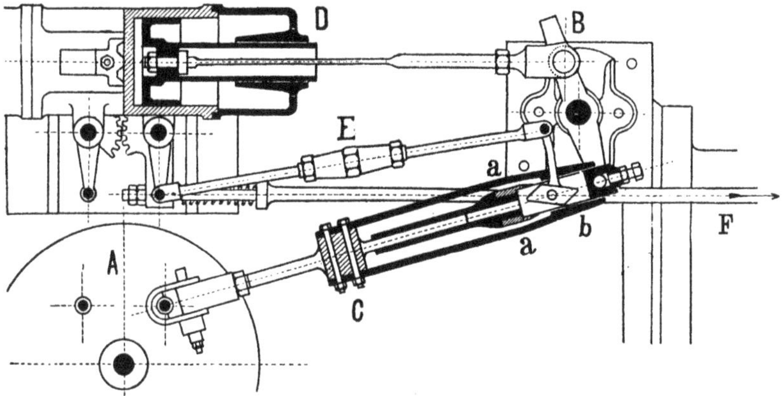

FIG. 200. Corliss Valve-gear, Spencer Inglis form.

by the action of a spring in the dash-pot D. When the wrist-plate makes its return stroke the clips re-engage the upper portion of the link C, and things are ready for the next stroke. The rocking cam b has its position controlled by the governor through the rods F and E in such a way that when the speed of the engine increases it stands more athwart the link C, and therefore causes the clips to be released at an earlier point in the stroke. A precisely similar arrangement governs the admission of steam to the other end of the cylinder. The exhaust-valves, which are also rocking valves of the Corliss type, are situated on the bottom of the cylinder, at the ends, and take their motion from a separate wrist-plate which oscillates on the same pin with the plate A.

[1] *Proc. Inst. Mech. Eng.* 1868.

Besides securing a sharp cut-off, without wire-drawing of the steam, trip gears have the advantage that the force to be exerted by the governor in changing the position of the trip or trigger is very slight in comparison with the forces that would be involved in moving the valve.

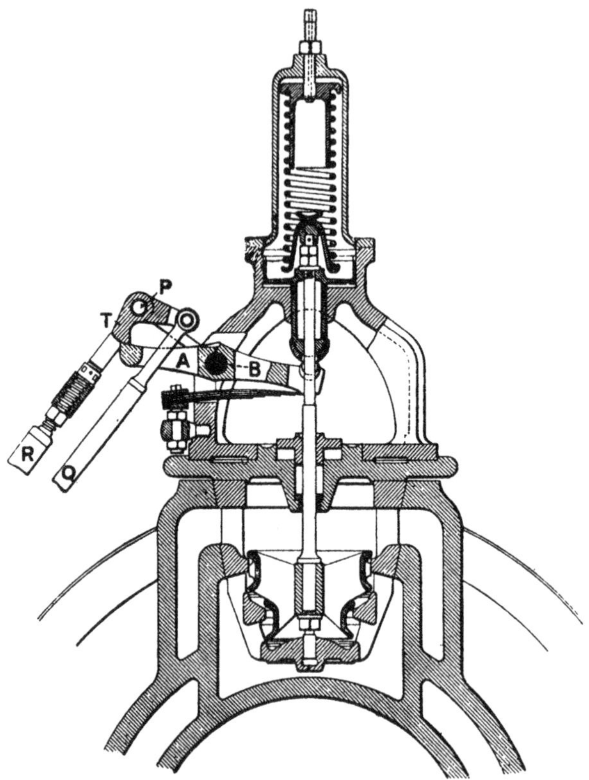

FIG. 201. Trip gear (Sulzer).

An example of Sulzer's gear is shown in fig. 201. A shaft, not shown in the figure, revolves alongside of the cylinder, giving motion to an eccentric which causes the rods R and Q to oscillate in the direction of their length. The top of R is pinned at P to a short radius rod which oscillates about the fixed centre B. On the same fixed centre there is a lever A by which the valve is opened. A is caused to open the valve by the downward movement of the rod R, through the action of the bell-crank trigger-piece T.

The angular position of T is controlled by Q, which in its turn is controlled by the governor. In each downward movement of R there is a rocking of the bell-crank T, and at the appropriate moment, determined by the governor, A escapes from the edge of

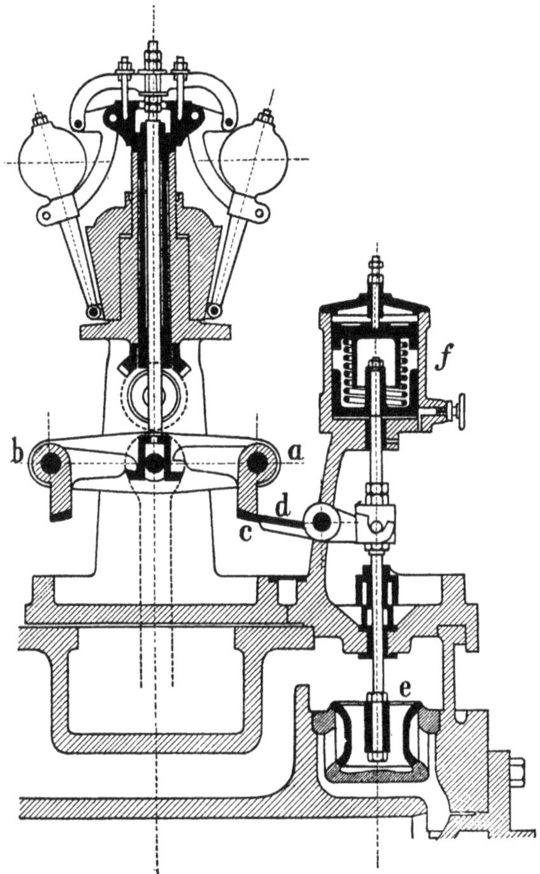

Fig. 202. Pröll's Trip gear.

the trigger and the valve is closed by the spring, at the bottom of which there is a dash-pot to make the valve settle gently on its seat.

Fig. 202 shows a compact form of trip gear by Pröll. A rocking lever ab is made to oscillate on a fixed pin through its centre by a connexion to the cross-head of the engine. When the end a

rises, the bell-crank lever *c* engages the lever *d*, and when *a* is depressed the lever *d* is forced down and the valve *e* is opened to admit steam to one end of the cylinder. As *a* continues moving down a point is reached at which the edge of *c* slips past the edge of *d*, and the valve is then forced to its seat by a spring in the dash-pot *f*. The disengagement occurs early or late according to the position of a central fulcrum piece, on which the heel of the bell-crank *c* rests during the opening of the valve. The position of the fulcrum piece is determined by the governor, which is of the kind already mentioned in § 227. A similar action, occurring at the other end of the rocking bar *ab*, supplies steam to the other end of the cylinder.

235. Hit and miss governors. A method of controlling the speed of gas-engines and oil-engines is to prevent any of the combustible from entering the cylinder when the speed is in excess of a certain limit. This may be done by means of a hit and miss arrangement, in which the fuel-admission valve is operated by a reciprocating piece through an intermediate piece which is lifted out of the way by the governor when the speed-limit is exceeded, so that no opening of the valve takes place and no charge is taken in. The governor itself may be of any ordinary centrifugal type. In small engines the hit and miss is sometimes effected without the use of a centrifugal governor by giving the intermediate piece the form of a bell-crank, pivoted to the reciprocating piece, with a heavy end projecting at right angles to the direction in which reciprocation takes place. When the speed of reciprocation exceeds the desired limit the inertia of this heavy end causes the bell-crank to tilt into such a position that the end of the valve-spindle is missed and the valve remains closed.

236. Disengagement governors. With the ordinary form of centrifugal governor the position of the throttle-valve, or the expansion link, or the Corliss trigger, depends on the configuration of the governor, and is definite for each position of the balls. In disengagement governors, of which the governor *A* shown on the right-hand side in fig. 203 is an example, any reduction of speed below a certain value sets the regulating mechanism in motion, and the adjustment continues until the speed has been restored. This is done by means of the wheel *c* which comes into gear with a wheel on the end of the spindle *a* when the speed falls below a certain limit. Similarly a rise of speed above a certain limit sets the regulating mechanism in motion in the other direction by putting *b* in gear with *a*. If the spindle *a* is connected to the regulator so as to give more steam when it turns one way and less when it turns the other, the speed at which the engine will run in equilibrium must lie between narrow limits, since at any speed high enough to keep *b* in gear with *a* the supply of steam will go on being reduced, and at any speed low enough to bring *c* into gear with *a* the supply will go on being increased. This mode of governing, besides being nearly isochronous, has the important advantage that the power of the governor is not limited by the controlling force on the balls, since the governor acts by applying a portion of the power that is being developed by the engine to the work of moving the regulator. It is rarely applied to steam-engines, mainly because its action is too slow. This defect was remedied in the supplementary governor of W. Knowles, who combined a disengagement governor with one of the ordinary type in the manner shown in fig. 203.[1] Here the spindle *a*, driven by the supplementary or disengagement governor *A*, acts by lengthening the rod *d* which connects the ordinary governor *B* with the regulator. It does this by turning a coupling nut *e* which unites two parts of *d*, on which right- and left-handed screws are cut. Any sudden fluctuation in speed is immediately responded to by the ordinary governor. Any more or less permanent change of load or of steam pressure gives the supplementary governor time to act. It goes on adjusting the supply until the normal speed is restored, thereby converting the control of the ordinary governor, which is stable, and therefore not isochronous, into a control which is isochronous as regards all fluctuations of long period.

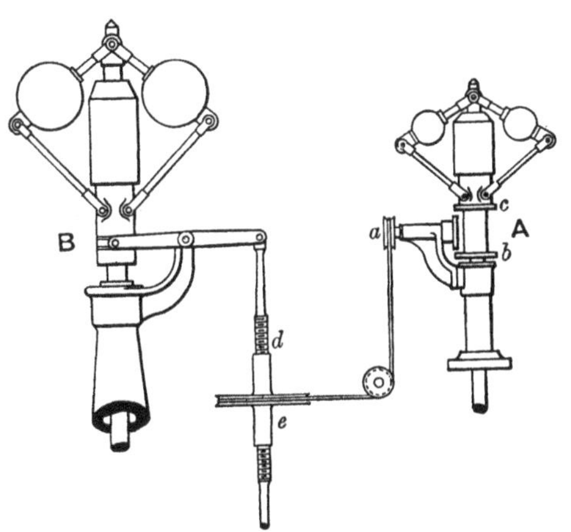

FIG. 203. Knowles' Supplementary Governor.

[1] *Proc. Inst. Mech. Eng.* 1884.

The power of the combination, however, is limited to that of the ordinary governor *B*.

237. Relay governors. Other governors which deserve to be classed as disengagement governors are those in which the displacement of the governor affects the regulator, not directly by a mechanical connexion, but by admitting steam or other fluid into what may be called a relay cylinder, whose piston acts on the regulator. This, as was pointed out in Chapter VIII, is a usual method of governing steam turbines, the piston of the relay cylinder or "servo-motor" being caused to move by oil under pressure, which is admitted to one or other side of it by a slide-valve without lap, under the control of the governor. In order that a governor of this class should work without causing the engine to hunt, the piston and valve of the relay cylinder should be connected by what is termed differential gear, the effect of which is that for each displacement of the valve by the governor the piston moves through a distance proportional to the displacement of the valve. An example of differential gear is shown in fig. 204. Suppose that the rod *a* is connected with the governor so that it is raised by an acceleration of the engine's speed. The rod *c* which leads from the relay piston *b* to the regulator serves as a fulcrum, and the valve-rod *d* is consequently raised. This admits steam (or oil under pressure) to the upper side of the piston and depresses the piston, which pulls down *d* with it, since the end of *a* now serves as a fulcrum. Thus by the downward movement of the piston the valve is again restored to its middle position and the movement of the regulator then ceases until a new change of speed occurs. The use has already been mentioned (§ 205) of a similar contrivance in steam-steering engines to make the position of the rudder follow, step by step, every movement of the hand wheel[1]; also in the steam reversing gear which is applied to large marine engines, to make the position of the drag-link follow that of the hand lever. The effect of adding a differential gear such as this to a relay governor

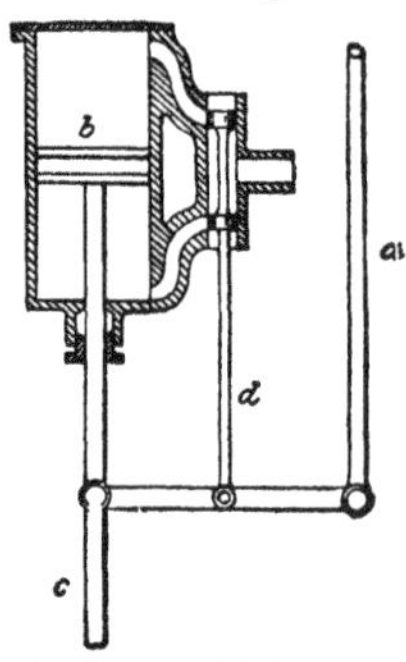

FIG. 204. Differential gear for Relay Governor.

[1] See a paper by J. MacFarlane Gray, *Proc. Inst. Mech. Eng.* 1867.

or other disengagement governor is to convert it from the isochronous to the stable type.

238. Differential or dynamometric governors. Another group of governors is exemplified by the "differential" governor of Sir William Siemens[1] (fig. 205). A spindle *a* driven by the engine drives a piece *b* (whose rotation is resisted by a friction brake) through the dynamometer coupling *c*, consisting of a nest of bevel-wheels and a lever *d* which is loaded, the weight of the load acting at right angles to the plane of the paper. So long as the speed remains constant the rate at which work is done on the brake is constant and the lever *d* is steady. If the speed increases, more power has to be communicated to *b*, partly to overcome the inertia and partly to meet the increased resistance of the brake, and the lever *d* is displaced. The lever *d* works the throttle-valve or other regulator, either directly or by a steam relay. The governor is isochronous when the force employed to hold *d* in position does not vary; if the control of *d* is arranged so that the force tending to hold it in position increases when *d* is displaced, the governor is stable. A governor of this class may properly be called a dynamometric governor, since it regulates by endeavouring to keep constant the rate at which energy is transmitted to the piece *b*.

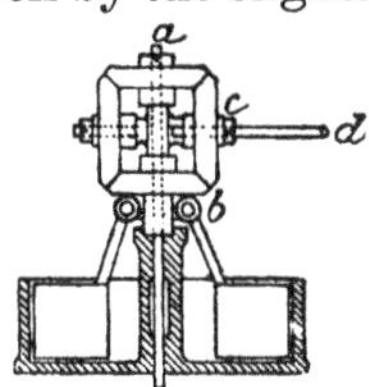

Fig. 205. Siemens' Governor.

In one form of Siemens' governor[2] the friction brake is replaced by a sort of centrifugal pump, consisting of a paraboloidal cup, open at the top and bottom, whose rotation causes a fluid to rise in it and escape over the rim when the speed is sufficiently great. Any increase in the cup's speed augments largely the power required to turn it, and consequently affects the position of the piece which corresponds to *d*. Siemens' governor is not itself used to any important extent, but the principle it embodies has found application in several other forms.

One of these is a marine-engine regulator by Messrs Durham and Churchill[3], in which the rotation of a piece corresponding to *b* is resisted by means of a fan revolving in a case containing a fluid, and the coupling piece which is the mechanical equivalent

[1] *Proc. Inst. Mech. Eng.* 1853.
[2] *Proc. Inst. Mech. Eng.* 1866; or *Phil. Trans.* 1866.
[3] *Proc. Inst. Mech. Eng.* 1879.

of *d* in fig. 205 acts on the throttle-valve, not directly but through a steam relay. In Silver's marine governor[1] the only friction brake that is provided to resist the rotation of the piece which corresponds to *b* is a set of air-vanes. The inertia is, however, very great, and any acceleration of the engine's speed consequently displaces the dynamometer coupling, and so acts on the regulator in its effort to increase the speed of *b*.

Another example of the differential type is a governor by Allen[2], which has a fan directly geared to the engine, revolving in a case containing a fluid. The case is also free to turn, except that it is held back by a weight or spring and is connected to the regulator. So long as the speed of the fan is constant, the moment required to keep the case from turning does not vary, and consequently the position of the regulator remains unchanged. When the fan turns faster the moment increases, and the case has to follow it (acting on the regulator) until the spring which holds the case from turning is sufficiently extended, or the weight raised. The term "dynamometric governor" is equally applicable to this form; the power required to drive the fan is regulated by an absorption-dynamometer in the case instead of by a transmission-dynamometer between the engine and the fan.

239. Pump governors. Pump governors form another group closely related to the differential or dynamometric type. An engine may have its speed regulated by working a small pump which supplies a chamber from which water or other fluid is allowed to escape by an orifice of constant size. When the engine quickens its speed the fluid is pumped in faster than it can escape, and the accumulation of the fluid in the chamber may be made to act on the regulator through a piston controlled by a spring or in other ways. This device has an obvious analogy to the cataract of the Cornish pumping engine (§ 215), which has, however, the somewhat different purpose of introducing a regulated pause at the end of each stroke, or rather serves this purpose in addition to regulating the number of strokes per minute. The "differential valve-gear" applied by H. Davey to pumping engines, combines the functions of the Cornish cataract with that of a hydraulic governor for regulating the expansion[3]. In this gear, which is shown diagram-

[1] *Brit. Ass. Rep.* 1859, p. 123. [2] *Proc. Inst. Mech. Eng.* 1893.
[3] *Proc. Inst. Mech. Eng.* 1874.

matically in fig. 206, the valve-rod of the engine (*a*) receives its motion from a lever *b*, one end of which (*c*) copies, on a reduced scale, the motion of the engine piston, while the other end (*d*), which forms the fulcrum, has its position regulated by attachment to a subsidiary piston-rod, which is driven by steam in a cylinder *e*, and is forced to travel at a nearly uniform rate by a cataract *f*. The point of cut-off is determined by the rate at which the main piston overtakes the cataract piston, and consequently comes early with light loads and late with heavy loads.

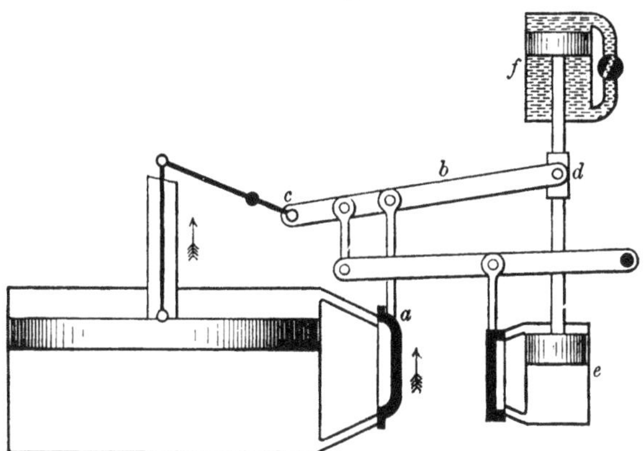

FIG. 206. Davey's Differential Valve-gear.

240. Governing marine engines. The governing of marine engines is peculiarly difficult on account of the sudden and violent fluctuations of load to which they are liable through the alternate uncovering and submersion of the screw in a heavy sea. However rapidly the governor responds to increase of speed by closing the throttle-valve, an excess of work is still done by the steam in the valve-chest and in the high-pressure cylinder. To check the racing which results from this, it has been proposed to supplement the control which the throttle-valve in the steam-pipe exercises by throttling the exhaust or by spoiling the vacuum. With the same object Messrs Jenkins and Lee gave supplementary regulation by causing the governor to open a shunt-valve connecting the top with the bottom of the low-pressure cylinder, thus allowing a portion of the steam in it to pass the piston without doing work. In Dunlop's pneumatic governor[1] an attempt was made to anticipate

[1] *Proc. Inst. Mech. Eng.* 1879.

the racing of the screw by causing the regulator to be acted on by the changes of pressure on a diaphragm connected by an air-pipe to an open vessel fixed under the stern of the ship. None of these devices has met with general adoption, but many marine steam-engines use a contrivance called Aspinall's governor, which acts on the hit and miss principle to close a throttle-valve when the speed exceeds the normal by more than a certain margin. A small mass hinged on one of the reciprocating parts of the engine becomes tilted, by virtue of its inertia, out of its usual position when the margin of speed is passed, and thereby causes the reciprocating part to engage with a rod by which the throttle-valve is closed. The valve is automatically reopened when the speed falls to the normal limit. Aspinall's governor is also applied to marine engines of the internal-combustion (Diesel) type to prevent delivery of oil-fuel to the cylinder when the engine begins to race. The rod which engages with the hinged mass, when that becomes tilted by undue acceleration, lifts the suction-valve of the fuel-pump and holds it open, with the result that so long as the speed exceeds a set limit the pump returns its contents to the suction-chamber instead of causing them to be injected into the cylinder.

CHAPTER XIII

DYNAMICS OF THE RECIPROCATING ENGINE

241. Fluctuations of speed during any single revolution: function of the fly-wheel. Besides the variations of speed which it is the function of the governor to check, there are variations within each revolution of a piston and cylinder engine over which the governor has no control. These are due to the varying rate at which work is done on the crank-shaft in different parts of a single revolution. To keep them within reasonable limits is the function of the fly-wheel. It acts by forming a reservoir of energy to be drawn upon during those parts of the revolution in which the work done on the shaft is less than the work done by the shaft, and to take up the surplus in those parts of the revolution in which the work done on the shaft is greater than the work done by it. To accomplish this alternate storing and restoring of energy the fly-wheel must undergo slight fluctuations of speed; the range of these depends on the ratio which the alternate excess and defect of energy bears to the whole stock of energy the fly-wheel holds in virtue of its motion. The duty of the fly-wheel may be studied by drawing a *diagram of crank-effort*, which shows the work done on the crank in the same way that the indicator diagram shows the work done on the piston. A diagram of crank-effort serves another useful purpose in determining the twisting and bending stress in the crank.

242. Diagram of crank-effort. The diagram of crank-effort is best drawn by representing, in a curve drawn with rectangular coordinates, the relation between the torque or moment which the connecting-rod exerts to turn the crank and the angle turned through by the crank. When the angle is expressed in circular measure, the area of the diagram is the work done on the crank. Or instead of selecting the turning moment and the angle turned through as the two coordinates, we may take the tangential effort on the crank-pin as one coordinate, namely the force which is found when the thrust against the crank-pin is resolved along the tangent to the crank-pin's path, the other component being directed towards the centre of the crank-shaft and consequently exerting no turning moment. The linear motion of the crank-pin in its circular path is then taken as the other coordinate of the crank-effort diagram; and the area still represents the work done upon the crank.

Neglecting friction for the present, and supposing in the first place that the engine runs so slowly that the forces required for the acceleration of the moving masses are negligibly small, the moment of crank-effort is found by resolving the thrust P of the

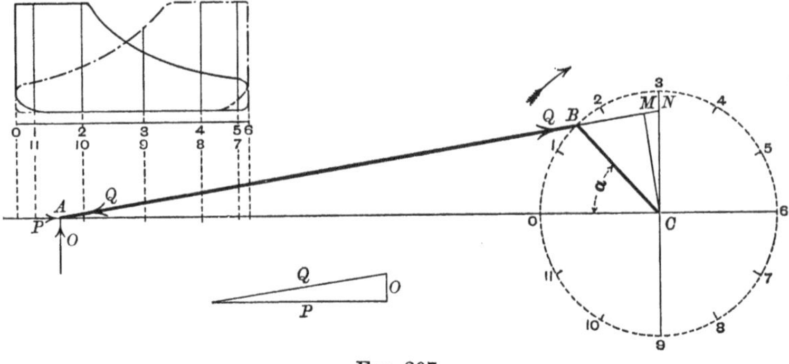

FIG. 207.

piston-rod into a component Q along the connecting-rod and a component O normal to the surface of the guide (fig. 207). The moment of crank-effort is

$$Q \,.\, CM = P \,.\, CN = Pr \sin \alpha \left(1 + \frac{r \cos \alpha}{\sqrt{l^2 - r^2 \sin^2 \alpha}}\right),$$

where CN is drawn perpendicular to the centre line or travel of the piston, r is the crank, l the connecting-rod, and α the angle ACB which the crank makes with the centre line. A graphic determination of CN is the most convenient in practice, unless the connecting-rod is so long that its obliquity is negligible, when the second term in the above expression vanishes. Fig. 208 shows the diagram of crank-effort determined in this way for an engine whose connecting-rod is $3\frac{1}{2}$ times the length of its crank, and

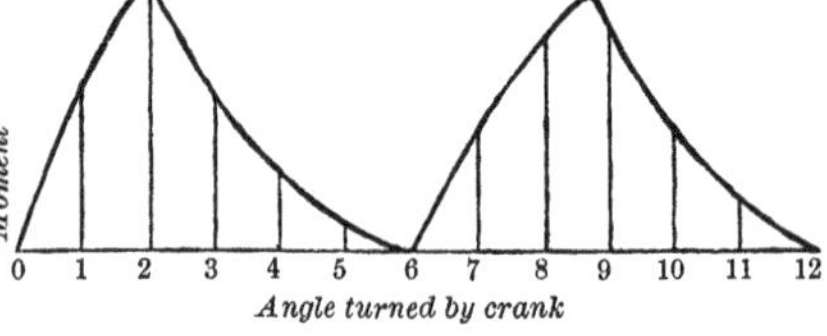

FIG. 208. Diagram of Crank-Effort.

in which steam is cut off at about one-third of the stroke. The thrust P is determined from the indicator diagrams of fig. 207 by taking the excess of the forward pressure on one side of the piston over the back-pressure on the other side, and multiplying this effective pressure by the area of the piston. The diagram is drawn for a complete revolution, each of the twelve numbered steps corresponding to 30°. The area of the diagram of crank-effort is the work done per revolution.

In the example for which this diagram is drawn it happens that there is very little compression of steam at the end of each back stroke, and consequently the forward pressure is greater than the back-pressure throughout the whole of the stroke. In many cases, however, the back-pressure rises so much toward the end of the stroke that the resultant thrust on the piston opposes its motion, the diagram of resultant steam pressure taking a form such as that sketched in fig. 209, and consequently the ordinates in the corresponding part of the crank-effort diagram become negative.

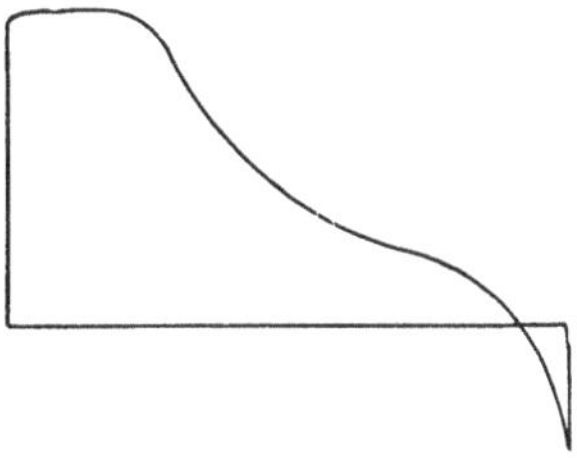

FIG. 209.

Another way of expressing the relation of the moment of crank-effort to the thrust P on the piston is to resolve the thrust Q along the rod into a component T in the direction of the tangent at B, and a component along BC. The former alone exerts a moment on the shaft, and its moment is $T \,.\, CB$. To find T we have, by the principle of work, $T \,.\, V_B = P \,.\, V_O$, where V_B and V_O are the velocities of the crank-pin and piston respectively. Hence

$$T = \frac{P \,.\, V_O}{V_B} = \frac{P \,.\, IO}{IB},$$

where I is the instantaneous centre for the movement of the connecting-rod. I is found graphically by producing CB to meet a perpendicular to OC from O. Since $\frac{IO}{IB} = \frac{CN}{CB}$, the expression $T \,.\, CB$ for the moment of crank-effort has the same value as that found before, namely $P \,.\, CN$.

243. Effect of friction. The friction of the piston in the cylinder and the piston-rod in the stuffing-box is easily allowed for, when its amount is known, by making a suitable deduction from P. If we treat the friction at the guides, the cross-head and the crank-pin as being proportional to the pressure, then the stress at each of these places will be inclined to the rubbing surfaces at an angle ϕ, the angle of repose, whose tangent is the coefficient of friction[1]. Hence, on that basis, the thrust O of the guide upon the cross-head, instead of being normal to the surface of the guide, is inclined at the angle ϕ in the direction which resists the piston's motion (fig. 210); and the thrust along the connecting-rod, instead of passing through the centre of each pin, is displaced far enough to make an angle ϕ with the radius at the point where it meets the pin's surface. To determine this displacement of the line of thrust let a "friction-circle" be drawn about the centre of each pin, namely a circle with radius equal to $p \sin \phi$, where p is the actual radius of the pin. Any line drawn tangent to this circle will make

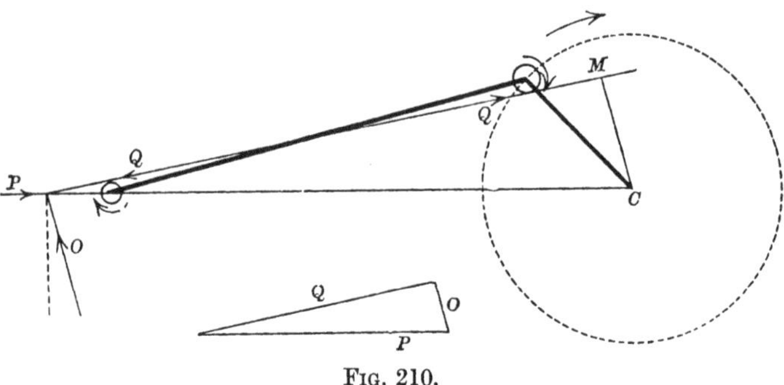

FIG. 210.

the angle ϕ with the radius of the pin at the surface of the pin and will therefore satisfy the required condition as regards friction. The thrust of the connecting-rod must be tangent to both circles; it must therefore be drawn as in fig. 210, so that it resists the rotation of the pins relatively to the rod. The direction of rotation of the pins is shown by curved arrows in the figure, where the friction-circles are drawn to a greatly exaggerated scale. Finally, P (after allowing for the friction of piston-packing and stuffing-box) is resolved into O and Q, and then $Q \cdot CM$, the moment of Q on the shaft, is determined. This gives a diagram of crank-effort whose area is no longer equal to that of the indicator diagram. The difference, however, does not represent the whole work lost through

[1] This is what would take place if the friction at these surfaces were of the same kind as between dry solid bodies. In fact however the presence of the oil, in a properly lubricated engine, makes the action different, and the construction given in the text does not apply quantitatively in practical cases. It serves however to illustrate the general effect of friction.

friction in the mechanism, since the friction of the shaft itself, and of the valves and other parts of the engine which it drives, has still to be allowed for if the frictional efficiency of the engine as a whole is in question.

244. Effect of the inertia of the reciprocating pieces. The diagram of crank-effort is further modified when we take account of the inertia of the piston and connecting-rod, and the influence of inertia is generally much more important than that of friction. For the purpose of investigating the effects of the inertia of the reciprocating pieces, we may assume that the crank is revolving at a sensibly uniform rate of n turns per second. Let M be the mass of the piston, piston-rod, and cross-head in pounds, and a its acceleration at any instant in feet per second per second; the force required to accelerate it is Ma/g, in pounds-weight, and this is to be deducted in estimating the effective value of P. The effect is to reduce P during the first part of the stroke and to increase it towards the end, thereby compensating to some extent for the variation which P undergoes in consequence of an early cut-off. If the connecting-rod is so long that its obliquity may be neglected the piston has simple harmonic motion, and

$$a = -4\pi^2 n^2 r \cos \alpha,$$

when the crank has turned through any angle α from its dead-point. More generally, whatever ratio the length l of the connecting-rod bears to that of the crank r,

$$a = -4\pi^2 n^2 r \left(\cos \alpha + \frac{rl^2 \cos 2\alpha + r^3 \sin^4 \alpha}{(l^2 - r^2 \sin^2 \alpha)^{\frac{3}{2}}}\right)^*.$$

* To prove this, let θ be the angle BAC of fig. 207; then

$$\theta = \sin^{-1}\left(\frac{r \sin \alpha}{l}\right).$$

Hence,
$$\frac{d\theta}{dt} = \frac{r \cos \alpha}{\sqrt{l^2 - r^2 \sin^2 \alpha}} \frac{d\alpha}{dt} = \frac{2\pi nr \cos \alpha}{\sqrt{l^2 - r^2 \sin^2 \alpha}}.$$

Differentiating again, and remembering that $\frac{d^2\alpha}{dt^2} = 0$ since the rotation is assumed to be sensibly uniform, we obtain

$$\frac{d^2\theta}{dt^2} = \frac{-r(l^2 - r^2) \sin \alpha}{(l^2 - r^2 \sin^2 \alpha)^{\frac{3}{2}}} \left(\frac{d\alpha}{dt}\right)^2 = \frac{-4\pi^2 n^2 r (l^2 - r^2) \sin \alpha}{(l^2 - r^2 \sin^2 \alpha)^{\frac{3}{2}}}.$$

Again, writing x for AC (fig. 207),

$$x = r \cos \alpha + l \cos \theta,$$

$$\frac{dx}{dt} = -r \sin \alpha \frac{d\alpha}{dt} - l \cos \theta \frac{d\theta}{dt},$$

The effect is to make, on the diagram of P, a correction of the character shown in fig. 211 where the straight line cd refers to the case of an indefinitely long connecting-rod and the full line aeb to the case of a connecting-rod $3\frac{1}{2}$ times the length of the crank. In a vertical engine the weight of the piston and piston-rod is to be added to or subtracted from P.

The form of the inertia line aeb of fig. 211 may be determined much more shortly and with sufficient accuracy for any graphic application by finding the points a and e and b as follows, and then sketching a smooth curve through these three points. The position of the point e in the stroke is found from the fact that since the acceleration is then zero the velocity of the piston is a maximum: this happens when the crank and connecting-rod are nearly at right angles (see next two sections). The acceleration at a is the centrifugal acceleration due to the sum of the curvatures of the path of the crank-pin and of the arc AA' struck with l for radius (fig. 212). Similarly the acceleration at b (fig. 211) is due to the difference of these curvatures. Hence at a the acceleration is $\frac{v^2}{r} + \frac{v^2}{l}$, where v is the velocity of the crank-pin, and at b it is $\frac{v^2}{r} - \frac{v^2}{l}$. Substituting $2\pi nr$ for v in these expressions the acceleration of the piston is found to be

$$4\pi^2 n^2 r\left(1 + \frac{r}{l}\right) \text{ and } 4\pi^2 n^2 r\left(1 - \frac{r}{l}\right)$$

at a and b respectively.

245. Graphic construction for finding the acceleration of the piston. A graphic method of finding the acceleration of the piston at any point in the stroke is shown in fig. 213. Produce AP to N and describe a circle with centre P and radius PN. Bisect the connecting-rod in E, and with E as centre and EP as radius draw a circular arc cutting the first circle in F and G. Join FG

and $$a = \frac{d^2x}{dt^2} = -r\cos\alpha\left(\frac{d\alpha}{dt}\right)^2 - l\cos\theta\left(\frac{d\theta}{dt}\right)^2 - l\sin\theta\,\frac{d^2\theta}{dt^2}.$$

Substituting the values found above for $\frac{d\alpha}{dt}$, $\frac{d\theta}{dt}$ and $\frac{d^2\theta}{dt^2}$, and putting $r\sin\alpha$ for $l\cos\theta$ and $\sqrt{l^2 - r^2\sin^2\alpha}$ for $l\cos\theta$, this gives the expression in the text,

$$a = -4\pi^2 n^2 r\left(\cos\alpha + \frac{rl^2\cos 2\alpha + r^3\sin^4\alpha}{(l^2 - r^2\sin^2\alpha)^{\frac{3}{2}}}\right).$$

and produce it when necessary to cut the line AC in H. Then the length HC, when multiplied by the square of the angular velocity of the crank, gives the acceleration of the piston. In other words, if the length CP represents the radial acceleration of the crank-pin, HC represents the acceleration of the piston[1].

To prove this, draw CM perpendicular to PN and HK parallel to LM. Find also I the instantaneous centre of the connecting-

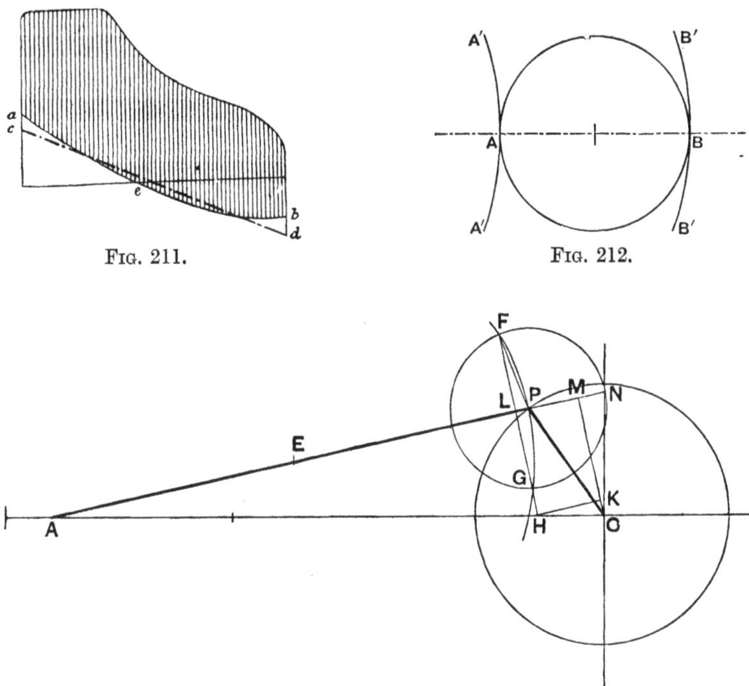

FIG. 211.

FIG. 212.

FIG. 213.

rod by producing CP through P to meet a line drawn at A perpendicular to AC. The triangle AIP (not shown in the diagram) is similar to NPC and $\frac{IP}{CP} = \frac{AP}{PN}$. For brevity we shall write ω for the angular velocity of the crank and ω' for the corresponding angular velocity of the connecting-rod. If v is the velocity of the crank-pin, $\omega = v/CP$ and $\omega' = v/IP$. Hence $\omega' = \omega . \frac{CP}{IP} = \omega . \frac{PN}{AP}$.

[1] J. F. Klein, *Journal of the Franklin Institute*, vol. CXXXII, 1891. See also Kirsch, *Zeitschrift des Vereines deutscher Ingenieure*, 1890, p. 1320.

The motion of the rod may be regarded as made up of (1) a translation with velocity v in the direction of the tangent to the crank-pin circle at P, and (2) an angular movement about P with angular velocity ω'. The acceleration of P along PC is $\omega^2 CP$. Resolve this into components along the rod and perpendicular to it. The component along the rod is $\omega^2 PM$, and this is also the acceleration of A in the direction AP, so far as A receives acceleration in consequence of the rod's movement of translation. The acceleration of A due to rotation of the rod about P is

$$\omega'^2 AP = \frac{\omega^2 . PN^2}{AP} = \frac{\omega^2 . PF^2}{AP} = \omega^2 LP, \text{ since } \frac{LP}{PF} = \frac{PF}{AP}.$$

This acceleration is in the direction AP.

The total acceleration of A in the direction AP is therefore $\omega^2 (LP + PM) = \omega^2 LM$, the other component of acceleration being perpendicular to AP.

Hence a, the acceleration of A along AC, is

$$\omega^2 LM . \frac{HC}{HK} = \omega^2 HC,$$

which was to be proved.

Further, the component acceleration of A in the direction perpendicular to AP is $\omega^2 CK$. But this is made up of (1) a component of the general acceleration of the rod $\omega^2 CP$ due to its translation, namely $\omega^2 CM$, and (2) the acceleration due to rotation about P with angular velocity ω', namely $AP . d^2\theta/dt^2$.

Hence $$\omega^2 CK = \omega^2 CM - AP . \frac{d^2\theta}{dt^2}.$$

Or $$AP . \frac{d^2\theta}{dt^2} = \omega^2 LH.$$

Collecting the results we have:—

The angular velocity of the connecting-rod varies as PN, being equal to $\omega . PN/AP$.

The angular acceleration of the connecting-rod varies as LH, being equal to $\omega^2 . LH/AP$.

The acceleration of the piston varies as CH, being equal to $\omega^2 CH$.

246. Position of the crank for which the piston has no acceleration. By means of this construction, or otherwise, the position of the crank may be found for which the acceleration of the piston is zero and its velocity a maximum. This happens in the diagram (fig. 213) when H coincides with C. The corresponding crank-angle α is given by the cubic equation in $\sin^2 \alpha$,

$$\sin^6 \alpha - n^2 \sin^4 \alpha - n^4 \sin^2 \alpha + n^4 = 0,$$

where n is the ratio of the length of the connecting-rod to that of the crank[1]. The following table gives values of the angle for various values of n, and also values of the angle at which the connecting-rod is tangent to the crank-pin circle.

Ratio of connecting-rod to crank	Angle from the dead-point at which the velocity of the piston is a maximum	Angle at which the connecting-rod is perpendicular to the crank
2	67° 42′	63° 26′
3	73° 11′	71° 34′
4	76° 43′	75° 58′
5	79° 7′	78° 41′
6	80° 48′	80° 32′
7	82° 2′	81° 52′
8	82° 59′	82° 52′
9	83° 44′	83° 40′
10	84° 20′	84° 17′

On comparing the two it will be seen that the rough approximation which is obtained by taking as the position of no acceleration the place where the rod is tangent to the crank-pin circle introduces no serious error in ordinary cases.

247. Inertia of the connecting-rod. The inertia of the connecting-rod presents a rather more complex problem. A fair approximation to the real effect is often arrived at by supposing part of the whole mass of the rod to be gathered at the cross-head, forming an addition to the mass which has simply a reciprocating motion, and the remainder to be gathered at the crank-pin, forming an addition to the rotating mass of the fly-wheel.

To obtain an exact solution, the motion of the rod may be analysed as consisting of translation with the velocity of the cross-head, combined with rotation about the cross-head as centre. By means of this analysis, the force required for the acceleration of the rod is determined as the resultant of three components, namely, F_1, the force required for the linear acceleration a (which is the same as that of the piston); F_2, the force required to cause angular acceleration about the cross-head; and F_3, the force towards the centre of rotation, which depends on the angular

[1] See Minchin's *Uniplanar Kinematics*, 1882, p. 48.

velocity, and is equal and opposite to the so-called centrifugal force. Let θ as before be the angle BAC (fig. 214), so that $d\theta/dt$ is the angular velocity of the rod about A, and $d^2\theta/dt^2$ is its angular acceleration, and let M' be the mass of the rod. Then, using gravitational units,

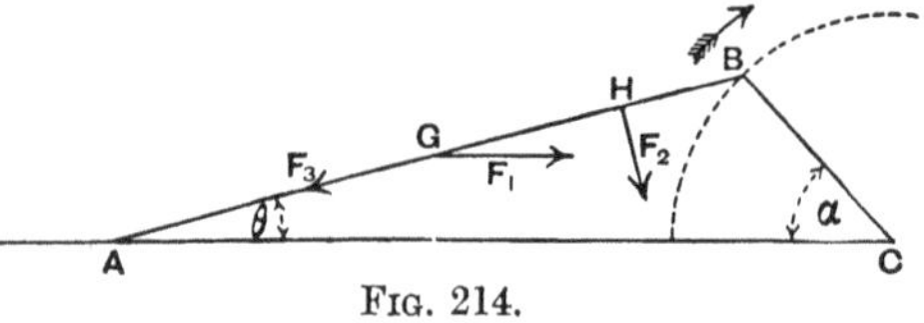

FIG. 214.

$$F_1 = \frac{M'a}{g},$$

and acts through the centre of gravity G, parallel to AC;

$$F_2 = \frac{M'\,(AG)}{g}\,\frac{d^2\theta}{dt^2},$$

and acts at right angles to the rod through the centre of percussion H;

$$F_3 = \frac{M'\,(AG)}{g}\left(\frac{d\theta}{dt}\right)^2,$$

and acts along the rod towards A.

The values of a and of $d\theta/dt$ and $d^2\theta/dt^2$ in relation to the crank-angle α have already been given, in the footnote to § 244.

If now we imagine the directions of the forces F_1, F_2, F_3 to be reversed, these reversed forces will, when taken along with the weight of the rod, equilibrate the external forces applied to the rod at A and B. To draw the diagram of forces, refer to the joints A and B each of these reversed forces and also the weight. Then treat the rod as if it were a member in a frame, loaded at the joints and exerting simple thrust along its length. At A all the forces are known in direction, but two of them are unknown in magnitude. These are found by drawing the polygon of forces for A, and then the polygon of forces for B gives the magnitude and direction of the force on the crank-pin.

248. Treatment of inertia and friction together. When in addition to the inertia of the rod, the friction at the cross-head and crank-pin is to be taken account of, the whole group of forces acting on the rod may be considered as follows. Compound forces equal and opposite to F_1, F_2, and F_3 into a single force R (fig. 215), which may be called the resultant resistance to acceleration of the connecting-rod. If the weight of the rod is to be considered, let it also be taken as a component in reckoning R. Then the rod may in any position be regarded as in equilibrium under the action of the forces Q, R and S, where Q and S are the forces exerted on it by the cross-head and crank-pin respectively. These three forces meet in a point p in the line of action of R, which point is to be

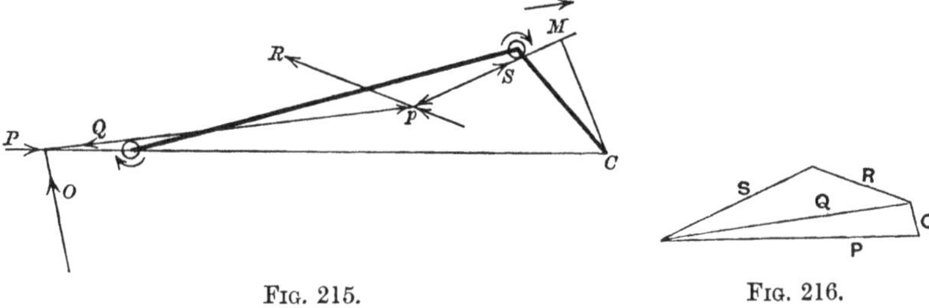

FIG. 215. FIG. 216.

found by trial, the condition being that in the diagram of forces, fig. 216, after the triangle POQ has been drawn, and the force R set out, the force-line S shall be parallel to a line drawn from p tangent to the friction-circle of the crank-pin, as shown in fig. 215. When this condition has been satisfied by trial, the value of S, which is the thrust on the crank-pin, is determined, and then

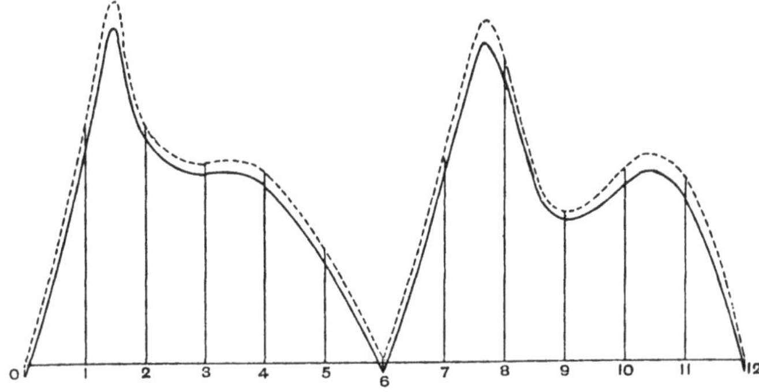

FIG. 217. Crank-Effort Diagram.

$S.\overline{CM}$ is the moment of crank-effort. This method is due to Fleeming Jenkin, who applied it with great generality to the investigation of mechanism in two important papers[1], the second of which deals in detail with the dynamics of the steam-engine. Fig. 217, taken from that paper, is the diagram of crank-effort in

[1] *Transactions of the Royal Society of Edinburgh*, vol. XXVIII, p. 1 and p. 703.

a horizontal direct-acting engine—the full line with friction, and the dotted line without friction—the inertia of the piston and connecting-rod being taken account of, as well as the weight of the latter. It shows well how the inertia of the reciprocating parts tends to equalize the crank-effort when there is an early cut-off. The cut-off is supposed to occur pretty sharply at about one-sixth of the stroke. The engine considered is of practical proportions, making four turns per second; the initial steam pressure is 50 lb. per sq. inch. It appears from the diagram that by making the speed slightly higher, or the rods more massive, a better approach to uniformity in the crank-effort might be secured, especially as regards the stroke towards the crank, which comes first in the diagram; on the other hand, by unduly increasing the mass of the reciprocating pieces or their speed the inequality due to expansion would be over-corrected and a new inequality would come in.

In drawing crank-effort diagrams it is not necessary in practice to take account of the friction of the guide and of the pins, indeed the treatment of friction in the way here indicated is, for the reason already given, to be regarded as little more than an academic exercise. But the inertia of the piston, piston-rod, and connecting-rod is of the utmost importance, especially in high-speed engines. The graphic method which is exhibited in figs. 215 and 216 of finding S, the thrust on the crank-pin, after R, the resistance to acceleration of the connecting-rod, has been determined, may of course be as readily applied when friction is neglected as when it is taken into account. Though for the sake of greater generality the influence of friction at the joints has been included, it must not be supposed to be a practically important factor in the problem. Neglecting friction, the lines of action of the forces P, O, and Q meet in the centre of the cross-head pin, and O is perpendicular to P. The line of action of S passes through the centre of the crank-pin. The forces P and R are known. By trial a point p is found, in the line of action of R, such that the forces Q and S in the force diagram (corresponding to fig. 216) shall be parallel to lines joining p with the cross-head centre and the crank-pin centre respectively.

249. Forms of crank-effort diagrams. When two or more cranks act on the same shaft a complete diagram, showing the resulting turning moment, may be drawn by combining the separate diagrams of crank-effort for the several cranks. An example is shown in fig. 218, where the dotted lines are the separate diagrams for two cranks set at right angles to each other and the full line is the combined diagram. It is obvious that the inequalities of crank-effort are vastly reduced by using two cranks instead of one, and with three cranks the effort becomes still more uniform. An illustration of this is given in fig. 219, which also exemplifies the circular form in which the diagram of crank-effort is sometimes

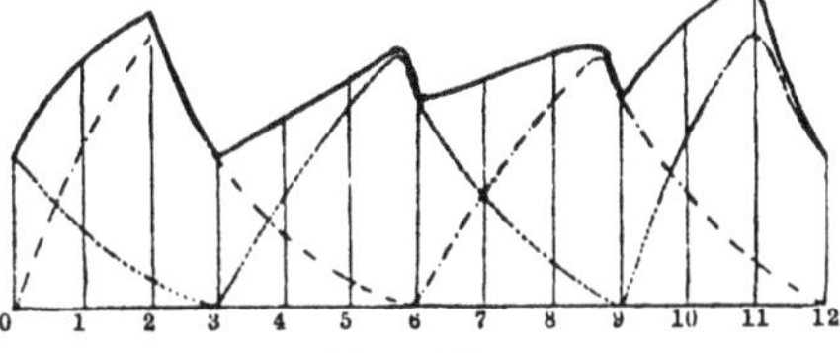

FIG. 218.
Crank-Effort Diagram for Two Cranks.

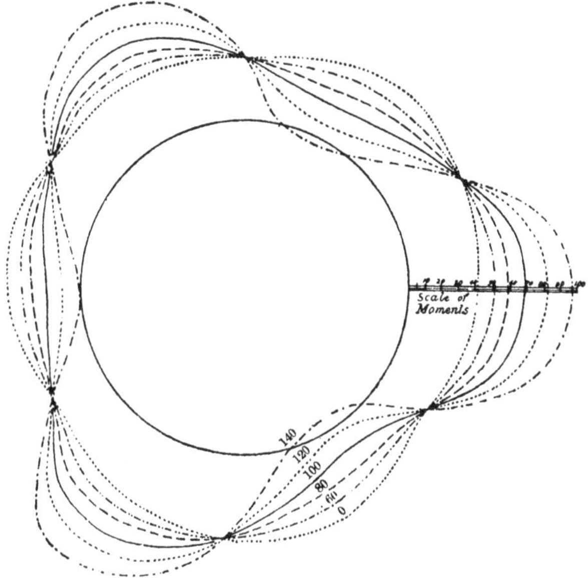

FIG. 219. Circular Diagram of Crank-Effort for a Three-Cylinder Engine.

drawn. In this construction lines proportional to the moment are set off radially from a circular line which represents the zero of moment. The figure is one drawn by Kirk for a triple-expansion marine engine with three cranks at 120° from each other. The curves show the resulting crank-effort, as determined from actual indicator diagrams and as affected by the inertia of the reciprocating parts. They are drawn for various numbers of revolutions per minute, which are indicated by the distinguishing numbers, the line marked 0 referring to an indefinitely slow motion.

An opposite extreme to the nearly uniform crank-effort that may be obtained in a steam-engine by the use of three or more cranks is found when we have a single-cylinder single-acting internal-combustion engine using the Otto or four-stroke cycle, in which an impulse is given to the crank only in one single stroke out of two revolutions, two of the other three strokes being idle, and the third being that in which the contents of the cylinder are compressed before ignition (see Chapter XVII). It is an interesting exercise to draw a crank-effort diagram for such an engine, extending the diagram over two revolutions to get a complete cyclic period, and then to apply the method described below of determining the size of fly-wheel which is necessary to prevent the speed from fluctuating beyond assigned limits. In this case, however, it is not practically necessary to take into account the inertia of the reciprocating parts in order to find the amount of energy that has to be alternately absorbed and given out by the fly-wheel. That is readily determined, from the indicator diagram, by comparing the work done by the gas on the piston during the single effective stroke, and the work done by the piston on the gas during the compression stroke, with the mean amount of work got from the engine during the four strokes which make up the cycle.

250. Fluctuation of speed in relation to the energy of the fly-wheel. The extent to which the fly-wheel has to act as a reservoir of energy in any engine is found by comparing the diagram of effort exerted on the crank-shaft by the piston or pistons with a similar diagram drawn to show the effort exerted by the crank-shaft throughout the revolution, in overcoming the resistance of the mechanism which it drives as well as the resistance due to its own friction. Like the driving effort, this resistance may be expressed as a torque or moment, or (dividing the moment by the radius of the crank) we may state the equivalent resistance referred to the crank-pin as a force acting always tangent to the crank-pin's path. In general, except in dealing with direct-acting pumping and blowing engines, or engines which are compressing air or other gases, the resistance may be taken as having a constant moment on the shaft, and the diagram of effort exerted by the crank-shaft is then a straight line, as $EFGHIJKL$ in fig. 220. At F, G, H, I, J, and K the rate at which work is being done on and by the shaft is the same; hence at these points the fly-wheel is neither gaining nor losing speed. The shaded area above FG is an excess of work done on the crank, and raises the speed of the fly-wheel from a minimum at F to a maximum at G. From G to H the fly-wheel supplies the defect of energy shown by the shaded area below GH, which represents the amount by which the demand for work exceeds the supply; the speed of the wheel again reaches a minimum at H, and again a maximum at I. The excesses and defects balance in each revolution if the engine is making a constant number of turns per second. In what follows it is assumed that the greatest excess or defect is only a small fraction of the whole energy stored up by the fly-wheel in virtue of its revolution, and consequently that the variations in speed are small in comparison with the mean speed. In practice the dimensions and speed of the fly-wheel are chosen so that this is the case: indeed the chief object of the investigation is to find what amount of energy must be given to the wheel in order that the variations in speed may not exceed a narrow prescribed range.

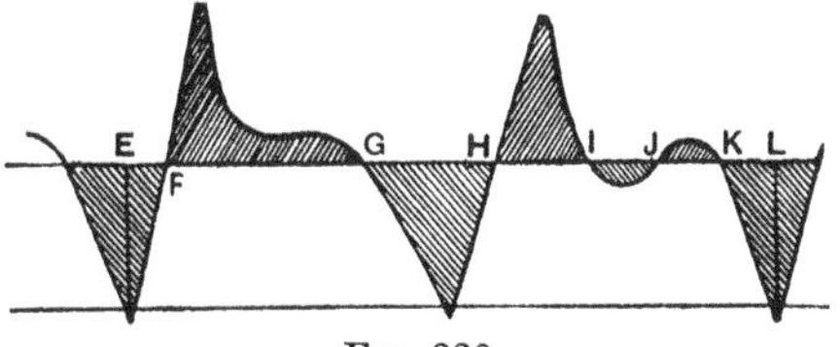

FIG. 220.

Let ΔE be the greatest single amount of energy that the fly-wheel has to give out or absorb, which is determined by measuring the shaded areas of the diagram and selecting the greatest of these areas; and let ω_1 and ω_2 be the maximum and minimum values of the wheel's angular velocity, which occur at the extremes of the period during which it is storing or supplying the energy ΔE. The mean angular velocity of the wheel ω_0 will be sensibly equal to $\frac{1}{2}(\omega_1 + \omega_2)$ if the range through which the speed varies is moderate. Let E_0 be the energy of the fly-wheel at this mean speed. Then

$$E_0 = \tfrac{1}{2} I \omega_0^2,$$

where I is the moment of inertia of the fly-wheel. Also

$$\Delta E = \frac{I(\omega_1^2 - \omega_2^2)}{2} = I\omega_0(\omega_1 - \omega_2) = 2E_0 \frac{(\omega_1 - \omega_2)}{\omega_0}.$$

The quantity $\dfrac{\omega_1 - \omega_2}{\omega_0}$, which we may write q, is the ratio of the extreme range of speed to the mean speed, and measures the degree of unsteadiness which the fly-wheel leaves uncorrected. If the problem be to design a fly-wheel which will keep q down to an assigned limit, the energy of the wheel must be such that

$$E_0 = \frac{\Delta E}{2q}.$$

The periodic fluctuations of speed which are due to the limited capacity the fly-wheel has for storing energy may be examined experimentally by causing a vibrator, such as a tuning-fork electrically maintained in vibration, to scribe its oscillations on a surface which moves with the fly-wheel shaft. A sheet of smoked paper clasping the shaft forms a convenient surface, on which the fork draws an undulating line by means of a bristle or light pointed spring attached to one of its prongs. The fork should be mounted on a carrier such as the slide-rest of a lathe so that it may be kept moving slowly in a direction parallel to the axis of the shaft, in order that the records of successive revolutions may be traced on fresh portions of the smoked surface.

251. Influence of the inertia of the reciprocating parts on the working stresses. In the design of high-speed engines the stresses which arise from the movements of the reciprocating parts have to be taken account of in determining the proper form and dimensions, from the point of view of strength. Thus in designing, say, the forked end of the connecting-rod, which embraces the cross-head, we have to estimate the greatest thrust which will be exerted on it through the cross-head pin. This thrust will in general be divided equally between the two limbs of the fork and will produce a bending moment tending to open the fork. The thrust exerted on the connecting-rod through the cross-head pin is found by compounding the force of the steam on the piston with the force required to accelerate the mass of the piston, piston-rod, and cross-head. In a vertical engine the weight of these parts also enters into the account. The mass whose inertia comes into question here is the whole moving mass between the joint under consideration and the steam. Similarly in considering the thrust at the crank-pin we have to take account of all the reciprocating mass between that joint and the steam: this will include, in addition to the mass of the piston, piston-rod, and cross-head, a proportion of the mass of the connecting-rod. For the purpose of estimating the effects of the inertia of the connecting-rod at the two ends of the stroke, where the forces in question have their greatest values, we may regard the rod as dynamically equivalent to two masses, one concentrated at the crank-pin, and therefore contributing nothing to the forces due to reciprocation, and the other concentrated at the cross-head, and therefore forming part of the reciprocating mass. In this reckoning the whole mass of the rod is to be divided into two parts, inversely proportional to the distance of the two ends from the centre of gravity of the rod, and these are to be regarded as concentrated at the ends.

This treatment applies to the external effects of the rod's inertia. In addition to such effects, there are within the rod itself stresses due to the acceleration of each portion of its own mass, the effect of which is to produce a bending moment on the rod. Similarly the coupling-rod of a locomotive, every part of which is subject to the same acceleration, is equivalent to a beam carrying a load which is uniformly distributed if the rod has the same cross-section throughout its length, and therefore the same mass per inch run.

252. Reversal of thrust at the joints. Let the diagram of resultant steam thrust upon the piston be represented by the line *SS* as in fig. 221 for the two successive strokes of a revolution,

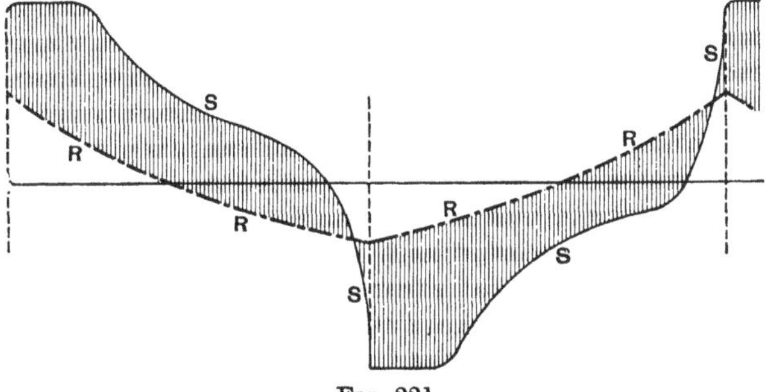

FIG. 221.

the line being drawn in such a way as to show that the steam is pushing the piston towards the crank when it lies above the base, and is pulling the piston away from the crank when it lies below the base. Let the line *RR* represent in the same way the forces that are used up in producing the acceleration of the reciprocating pieces. Then the points at which the steam curve *SS* crosses the inertia curve *RR* mark the places at which the direction of thrust at the bearings become reversed. If in drawing *RR* the mass of the piston, piston-rod, and cross-head only is taken account of the intersection of the two curves will show at what places the thrust changes its sign at the cross-head pin. But if a suitable proportion of the mass of the connecting-rod also has been added in calculating the forces represented by this curve, the points where *SS* crosses *RR* will relate to the reversal of thrust on the bearing surfaces of the crank-pin. Two inertia lines may be drawn, one referring to the masses between the steam and the cross-head pin, the other to the whole reciprocating mass, up to the crank-pin. Since the bearings are necessarily somewhat loose a sudden reversal of the thrust from pull to push at either joint may give rise to a knock. To prevent an engine from knocking badly is partly a question of lubrication, but the clearance at the bearings should be kept small, and the form of the thrust diagram (fig. 221) should be such that the steam and inertia curves cross each other as gradually as may be practicable.

253. Preventing reversal of the thrust in single-acting engines. Before the steam turbine became the recognized means of dynamo driving in central stations, engineers gave much attention to the design of single-acting reciprocating steam-engines which should run with a very high frequency of revolution and entirely avoid any reversal from thrust to pull. In the Willans engine, for example, a vertical type at one time much used, steam was admitted on the top of the piston only, and the piston-rod and connecting-rod were kept in compression throughout the whole action. During the up-stroke there was nothing happening in the cylinder, except a little compression towards the end of the stroke, to provide the force that was required to reduce the velocity of the reciprocating pieces after the point of maximum velocity (near mid-stroke) had been passed. Hence unless special provision were made for this force the connecting-rod would be pulling instead of pushing the crank-pin during the later portion of the up-stroke. The special provision consisted in an air-cylinder, the piston of which was set tandem with the steam piston. The air began to be compressed early in the up-stroke and was more and more compressed to the end, the work spent in compressing it being given out again during the down-stroke. The force exerted by the compressed air was arranged to be in excess of that required for the (negative) acceleration of the pistons and rods, and hence the thrust both at the cross-head and at the crank-pin was continuously a push, never a pull.

An example will help to make this clear. Let *dd*, fig. 222, be the indicator diagram of a single-acting vertical engine to which steam is admitted on the top of the piston only. It is required to find what amount of air compression on the part of an air-buffer piston will serve to keep the thrust on the crank-pin from changing its sign at any point in the revolution. The line *aa* is the atmospheric datum line, representing the constant pressure which acts below the piston of the steam-cylinder. The line *ii* represents the forces

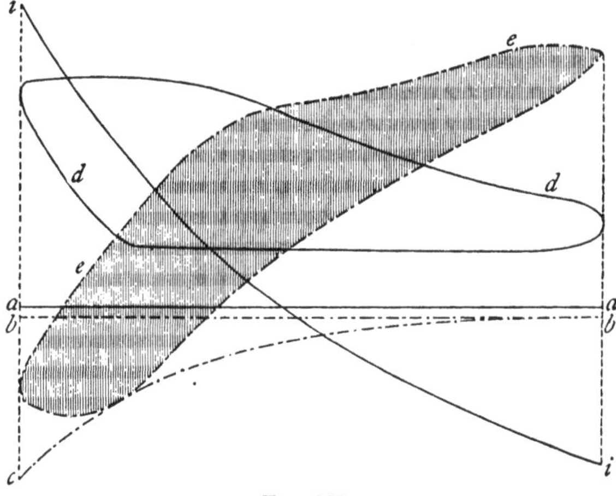

FIG. 222.

due to the inertia of the whole reciprocating mass which is carried by the crank-pin—namely, the piston and piston-rod of the steam-cylinder and of the air-buffer and also the connecting-rod. The forces due to inertia are represented per sq. inch of piston area, to the same scale as the steam-pressures. Let the diagram *ee* be drawn to compound the forces due to inertia with those due to steam pressure. In other words, let its ordinates above or below the datum line *aa* be the excess of the ordinates of *dd* above those of *ii*. So long as the figure *ee* lies above the datum line *aa*, the steam pressure pushing the piston down exceeds the force necessary for acceleration, and consequently there is push, not pull, at the crank-pin. But when the figure *ee* comes below the line *aa* the force required for acceleration exceeds the force exerted by the steam. We must however take account of the weight of the reciprocating pieces, which assists the steam pressure. This is readily done by shifting the datum line down to *bb*, the distance *ab* representing the weight, expressed in lb. per sq. inch of piston area. The forces which are left to be balanced by the compression of air in the air-buffer are represented by the projection of *ee* below *bb*. Any such compression line for the air-cylinder as *cc*, touching or lying wholly below the projecting part of *ee*, will therefore serve to prevent the force at the crank-pin from ever changing from a push into a pull. In the Willans engine there were often two, sometimes three, steam pistons arranged tandem on each piston-rod, and the steam diagram to be used in the foregoing construction was one which represented the sum of the pressures on them.

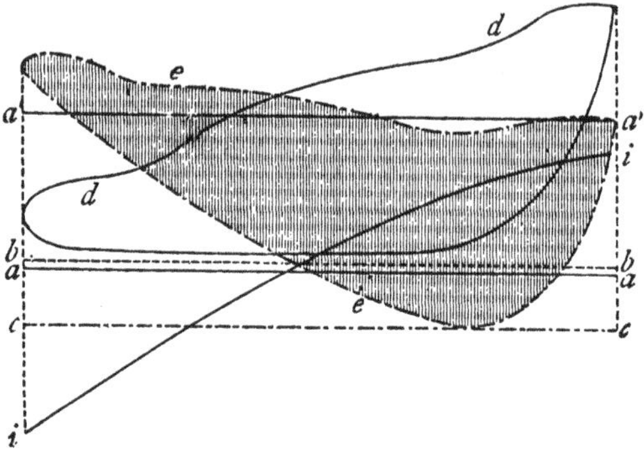

FIG. 223.

A similar problem arises when steam is admitted to the underside of the piston in a single-acting engine, and the rods are to be kept always in tension. This has been done by adding a balance-piston at the top of the cylinder which is continuously exposed to the full pressure of the steam on its lower side. The case is illustrated in fig. 223. There *dd* is the indicator diagram and *ii* the inertia line as before. The datum line *aa* is not the atmospheric line *a'a'* but a line showing the pressure in the space above the piston, which space is connected with the condenser through the exhaust-pipe. As before, the indicator diagram *dd* is compounded with the inertia line to give the figure *ee*: the datum line is shifted, in this case, *up* to *bb*, to allow for the weight of the reciprocating parts, which now opposes instead of helping. Then the projection of *ee* below *bb* shows what has to be provided by the steam-balance piston; the pressure which the steam exerts on it must be at least equal to the height *bc* in order to prevent a reversal of force at the crank-pin.

254. Balancing. An important matter in the kinetics of the steam-engine is the balance of its working parts. A machine is said to be perfectly balanced when the relative movements of its parts have no tendency to make it vibrate as a whole. In other words, perfect balance implies that the reactions of those forces that are required for the acceleration of the parts should neutralize each other in every phase of the motion, so that no resultant reaction is ever felt by the bed-plate of the machine. A perfectly balanced machine would be self-contained as regards the stresses between the parts and would run steadily without foundations. Actual machines rarely do more than approximate to this condition. The question of balance is especially important in marine engines and locomotives, and in the internal-combustion motors of road vehicles and aircraft. Even in stationary engines, however, it often requires attention, for a want of balance may cause serious nuisance through the vibrations it sets up in the neighbourhood.

The inertia effects in an engine are due in part to masses which simply revolve, such as cranks or eccentric sheaves, and in part to masses which reciprocate, such as pistons or valves. For the complete consideration of balance these have to be dealt with separately.

255. Balance of revolving masses. Consider first the conditions which will give balance in a system of masses *revolving* about an axis, such as a number of cranks on the same shaft. It is convenient for this purpose to take as a plane of reference one situated anywhere along the axis and normal to it. Let the masses be M_1, M_2, M_3, etc.; let r_1, r_2, r_3, etc. be the respective radii at which they revolve, and let x_1, x_2, x_3, etc. be their respective distances from the assumed plane of reference. Calling the angular velocity of rotation ω, the centrifugal forces exerted by the masses, radially away from the axis and, in each case, in the plane in which the mass revolves are $M_1\omega^2r_1$, $M_2\omega^2r_2$, $M_3\omega^2r_3$, and so on. Each of these forces is equivalent to an equal and parallel force acting in the plane of reference, together with a couple the moment of which is equal to the value of the force multiplied by the distance

x through which it has been shifted to bring it into that plane. Thus the system as a whole is equivalent to a group of forces in the plane of reference, namely,

$$M_1\omega^2 r_1,\ M_2\omega^2 r_2,\ M_3\omega^2 r_3,\ \text{etc.},$$

together with a group of couples

$$M_1\omega^2 r_1 x_1,\ M_2\omega^2 r_2 x_2,\ M_3\omega^2 r_3 x_3,\ \text{etc.}$$

To secure balance the resultant of each of these two groups must vanish. In dealing with the proportions between the forces or couples we may omit ω^2 which enters as a constant factor into all. Hence if we draw, in the plane of reference, a system of vectors $M_1 r_1$, $M_2 r_2$, $M_3 r_3$, etc. in the directions of the several forces at any instant, lines equal and parallel to these vectors must, if the system is in balance, form a closed polygon; and further, if we draw, to represent the couples, a system of vectors $M_1 r_1 x_1$, $M_2 r_2 x_2$, $M_3 r_3 x_3$, etc., lines equal and parallel to them must also form a closed polygon. These vectors, which represent the moments of the couples, are to be drawn in the direction of the axes of the several couples: or, what will do equally well, they may be drawn respectively parallel to the radii r_1, r_2, r_3 at which the masses for the moment are revolving[1].

We therefore have, as the conditions of balance, the requirement that two polygons shall close, one for the forces and one for the moments of the couples which are introduced in transferring the several forces to the plane of reference. The balance is imperfect if there is either a resultant force or a resultant couple. The effect on the bed-plate which carries the revolving shaft may be described as hammering, if there is a resultant force, and as tilting, if there is a resultant couple.

It follows from these considerations that if two masses are to balance they must revolve in the same plane, and that if three masses are to balance they either must lie in one plane of revolution or, if they are in different planes, one (the middle one) must be 180° in advance of the other two: in other words, they must all three lie either in a plane of revolution, or in a plane which contains the axis of the shaft.

[1] This graphic method of investigating the conditions of balance appears to have been first given by Mr D. W. Taylor (*Journal of the American Society of Naval Engineers*, 1891). It was developed by Professor Dalby, whose book on *The Balancing of Engines* should be referred to for a fuller account of the subject.

Four masses can be arranged to balance in an indefinite number of ways, by suitable selection of the relative values of Mr, of the crank-angles, and of the distances apart. This, as will be seen presently, forms a very important practical problem, especially with reference to the design of marine engines. With four masses to be considered, revolving at defined distances from the axis, there are in all nine independent variables, namely, the ratio of Mr for the second, third and fourth masses to the first, the crank-angles made by the second, third and fourth cranks respectively with the first, and the three distances apart of the cranks. It should be noted that the balance depends on the ratio of the masses, or rather of Mr, and not on the absolute values. Generally, if we have n masses the total number of independent variables is

$$3(n-1),$$

made up of $n-1$ values of Mr, $n-1$ crank-angles, and $n-1$ distances from crank to crank along the shaft.

The two conditions for balance are that the Mr polygon shall close and the Mrx polygon shall close. The closing line in each polygon is determined by the lines already drawn. Each closing line fixes two independent variables. Hence the number of the independent variables which may be assigned beforehand, as data of the problem, is short of the whole number by four. In other words,

$$3(n-1)-4, \text{ or } 3n-7,$$

is the largest number of variables which may be fixed beforehand without making the problem of attaining a balance insoluble.

256. Four-crank system. Thus in a four-crank system we may fix five of the quantities to begin with, and then by drawing the two polygons the other four quantities are found. We may, for example, have as data two masses at given radii, with the angle between their cranks, and the axial distances between the planes of all four masses. This means one ratio of Mr, one angle, and three distances: five quantities in all. By drawing the polygons we find the other two values of Mr and the other two crank-angles which are required to complete the determination of the system. The problem presents itself in this form in the balancing of locomotives when masses treated as revolving at the two cranks are to be balanced by two other masses in the planes of the wheels. Or, to take a case arising in the design of a four-cylinder marine engine, we may fix to begin with the four planes and three of the masses (namely, three distances and two ratios of Mr), leaving the fourth mass and all the crank-angles to be found.

In the solution of any such problem it is a convenient artifice to take as the plane of reference the plane in which one of the unknown masses revolves. This eliminates one value of Mrx from the polygon of moments. That polygon is then drawn, and with the information afforded by the closing line the force polygon may next be drawn, which completes the solution.

As an example, say that given masses M_2 and M_3 (fig. 224) are revolving on given cranks, and are to be balanced by masses at

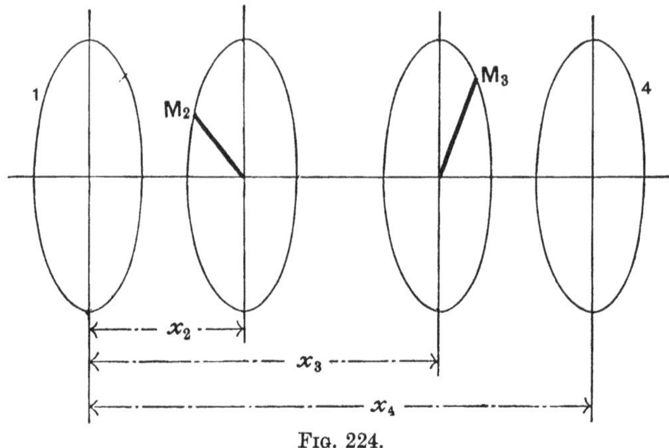

FIG. 224.

planes 1 and 4, the distances x_2, x_3, x_4 being given. Take (1) as plane of reference. Calculate $M_2r_2x_2$ and $M_3r_3x_3$. Since x_1 is zero, $M_1r_1x_1$ vanishes. Draw for the polygon of moments a triangle with $M_2r_2x_2$ and $M_3r_3x_3$ as sides and complete it by a third line. This line represents $M_4r_4x_4$, and since x_4 is known it fixes M_4r_4 both in magnitude and in angular position. Next take the known values of M_2r_2, M_3r_3 and M_4r_4, and with them draw the force polygon, which is a four-sided figure, the closing line of which is M_1r_1. Hence M_1r_1 is also fixed in magnitude and direction. No difficulty will be found in applying a generally similar method to other cases or in extending it to a larger number of cranks.

257. Balance of reciprocating masses. It is easy to extend these considerations to the important practical case of reciprocating masses such as the pistons of a steam-engine, and so to determine conditions which will avoid hammering, or unbalanced resultant force, and tilting, or unbalanced resultant couple. To simplify the problem we shall as a first approximation treat the pistons as having sensibly simple-harmonic motion—a condition which is more and more closely approximated to the longer the connecting-rods are relatively to the cranks. A reciprocating mass such as a piston cannot be completely balanced by a revolving mass, or by any combination of masses revolving on one shaft, for the force due to acceleration of each reciprocating mass acts only in the line of its motion and therefore has an invariable direction, while the forces due to revolving masses change their direction continually. But a system of reciprocating masses may be arranged to secure a complete state of balance amongst themselves, by an application of the same method as has been applied in the foregoing section. Any reciprocating mass M exerts on the frame of the engine a force Ma, where a is the acceleration, and since the motion is treated as simply harmonic, a is $\omega^2 r \cos\theta$, where θ is the angle between the crank and the centre line. Thus the disturbing force, which acts along the line of motion, is $Mr\omega^2 \cos\theta$. This is equal to one component of the force $Mr\omega^2$ which would be exerted by the mass at radius r if it were revolving instead of reciprocating, namely, the component in the direction of the stroke, the other component being taken at right angles to the stroke. In other words, if we imagine an equal revolving mass, concentrated at the crank-pin, to be substituted for the actual reciprocating mass, and resolve its centrifugal force along and across the line of stroke, the component along the line of stroke represents the disturbing force which the reciprocating mass in fact produces. Hence when we have a number of masses reciprocating in parallel lines, as for instance in the ordinary vertical form of marine engine, if we substitute for each an imaginary revolving mass the conditions which would secure balance among the imaginary revolving masses would also secure balance among the reciprocating masses of the actual machine. These conditions, as we have seen, are that the two polygons should close which are drawn with their sides equal to Mr and to Mrx respectively. All that has been said regarding the balance of revolving masses, the number of independent variables, and the number of these variables which may be assigned as data of the problem, applies equally to reciprocating masses, when these are considered as a system by themselves. It must however be understood that in an engine where there are real revolving masses to be considered as well as reciprocating masses, a state of complete balance can be secured only by making the revolving masses form a balanced system among themselves and the reciprocating masses also form a balanced system among themselves, the two systems being dealt with separately.

In such calculations the connecting-rod is to be regarded as belonging in part to the revolving masses and in part to the reciprocating masses, a portion of it being treated as concentrated at the crank-pin and the remainder at the cross-head. These parts are to be taken (as in § 251) in the inverse ratio of the distances of crank-pin and cross-head respectively from the mass-centre of the rod.

258. Four-crank marine engine. Yarrow-Schlick-Tweedy system. These principles have received particular application in marine engines, where a want of balance, especially in high-speed engines, may cause much trouble by setting up vibrations of the hull. The consequences are specially serious when the speed of the engines is such that the impulses happen to agree in frequency with the free vibrations of which some part of the elastic structure of the ship is susceptible. Under these conditions the vibrations may become very large, and if the synchronism is approximate they may increase to a maximum, subside and increase again periodically.

In the usual form of marine engine the cylinders are arranged vertically over the crank-shaft, with their centre lines in a fore-and-aft plane. The three-cylinder type which became common on the introduction of triple expansion does not admit of balance except by setting the cranks at 180° from one another, an arrangement which would be very objectionable on account of difficulty in starting and non-uniformity in crank-effort. Taking the three-cylinder engine, Sir Alfred Yarrow showed in 1892 that its reciprocating masses could be effectively balanced by adding two others, which he called "bob-weights[1]," making five masses in all. In 1894 Mr Schlick[2] pointed out that when the engine has four cranks the condition of balance may be secured in a practical manner by a suitable choice of crank-angles and relative masses.

[1] A. F. Yarrow, *Trans. Inst. Nav. Arch.* 1892.

[2] Otto Schlick, *ibid.* 1894.

This arrangement has found extensive application under the name of the Yarrow-Schlick-Tweedy system. The solution in any particular case is readily arrived at by the graphic method already described, in which a plane of reference is taken through one of the cranks, and the two polygons, one of moments and the other of forces, are successively drawn. In general the distances along the shaft are assumed, and the masses for three of the pistons, which may in the first instance be calculated from the dimensions which, for other reasons, the pistons should have: with these data the fourth mass and the crank-angles are found. If necessary the masses of one or more of the pistons can be modified in the design by adding material for the purpose, until the solution gives suitable values both for them and for the angles. The designer may start by assuming values for the angles between three of the cranks, instead of three masses.

For a complete treatment of the practical problem the valves as well as the pistons have to be taken into account. For each valve there are in general two cranks, namely, the two eccentric sheaves of the link-motion. The engine is treated as running in full forward gear, that being the state for which balancing is most important, and the mass of the valve and valve-spindle accordingly reciprocates with the ahead eccentric. In a four-crank engine the consideration of the valve-gear brings the number of reciprocating masses up to twelve, but a good approximation may be obtained by substituting for each pair of eccentrics a single imaginary resultant eccentric[1].

259. Three-crank engine. Balance of forces but not of couples. If the conditions are such that the force polygon closes but the moment polygon does not, there is no resultant force tending to move the engine as a whole in any one direction, but there is a resultant couple tending to make it pitch on its bed; in other words, there is no hammering, but there is tilting. Take for example a three-cylinder vertical marine engine with cranks 120° apart, equal reciprocating masses M_1, M_2, M_3, equal strokes, and equal distances along the shaft (fig. 225). Here the force polygon is an equilateral triangle, and there is no resultant force on the

[1] For a numerical example of the solution taking valve-gears into account see Professor Dalby's book already cited. See also his paper on "A comparison of Five Types of Engines, etc." *Trans. Inst. Nav. Arch.* 1902.

engine as a whole. There is left however an unbalanced couple which tends to make the engine pitch in the fore-and-aft line AB. The closer the spacing of the cranks the smaller this couple will be.

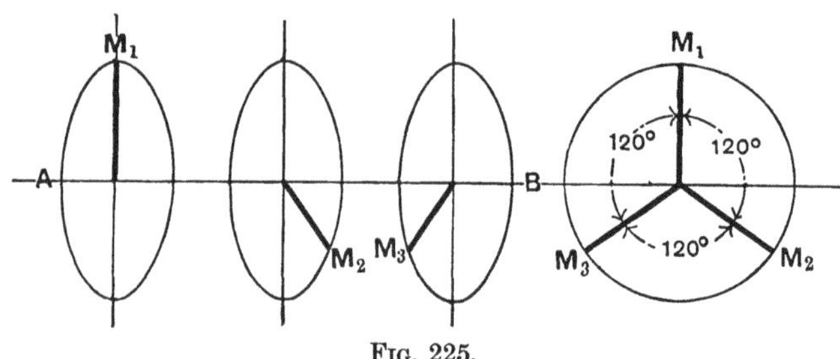

Fig. 225.

260. Balanced six-crank and five-crank engines. Taking the engine of fig. 225 let it be duplicated by the addition of three more cylinders but with cranks in the reversed order, namely, M_4, M_5, and M_6 in fig. 226, the second set being like the first but forming a mirror image of them about a transverse plane through the centre C. The second set, considered by themselves, will have no resultant force, but will have a resultant couple equal and opposite to that of the first set. Hence the couples neutralize each other, and this six-crank engine is completely balanced as regards both forces and couples.

From this it is an easy step to derive a completely balanced five-crank engine. We have only to suppose the two sets brought close together so that the crank of M_4 coincides with that of M_1. The reciprocating mass for that crank must then be equal to $M_1 + M_4$: the other masses remain unaltered. This gives the five-crank engine of fig. 227, which is also in a state of complete balance, having neither resultant force nor resultant couple.

It is not essential that the masses should be equal and the pitches of the cranks equal, provided the following relations hold:—

$$M \text{ or } M_1 + M_4 = M_2 + M_5 = M_3 + M_6,$$

$$M_2 x_2 = M_5 x_5, \quad M_3 x_3 = M_6 x_6.$$

Hence it is sufficient that the masses M_2, M_3, M_5 and M_6 should bear the following ratios to the central mass M,

$$M_2 = \frac{M x_5}{x_2 + x_5}, \quad M_3 = \frac{M x_6}{x_3 + x_6}, \quad M_5 = \frac{M x_2}{x_2 + x_5}, \quad M_6 = \frac{M x_3}{x_3 + x_6}.$$

The distance between the cranks can therefore be chosen first, and masses determined from them which will form a balanced system, as regards both forces and couples.

It is assumed throughout these statements about balancing that the engine bed is sufficiently rigid to admit of no distortion, otherwise we could not regard the geometrical balance of forces and moments as resulting in no disturbance of the ground on which the engine stands, or of the hull of the ship.

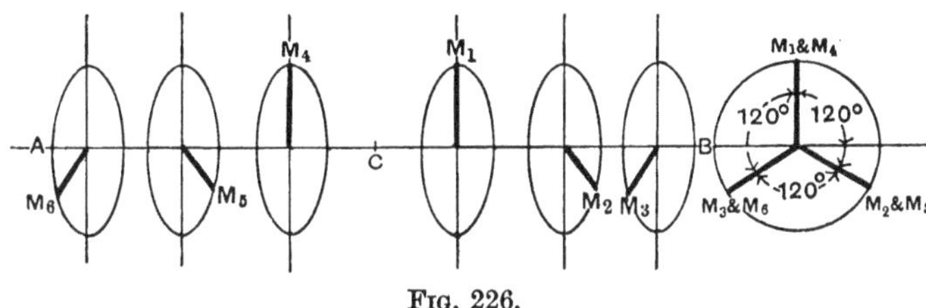

Fig. 226.

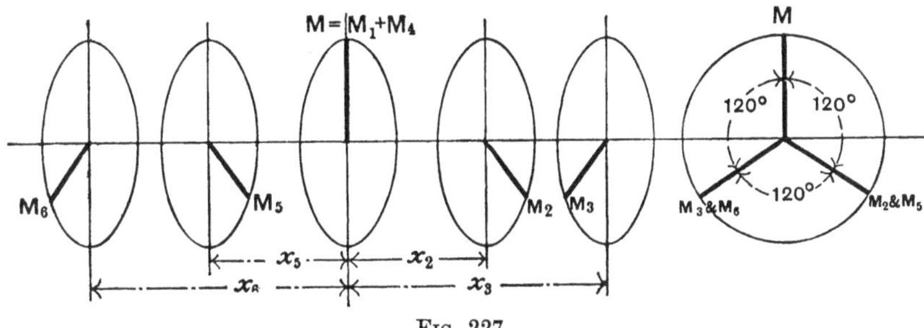

Fig. 227.

261. Use of revolving masses to produce partial balance of reciprocating masses. Although a reciprocating mass cannot be correctly balanced by revolving masses it is often useful to neutralize, wholly or partially, the longitudinal forces on the engine-frame due to a reciprocating mass by applying revolving balance masses, though a necessary effect of this is to introduce a new disturbance by giving rise to transverse forces.

The simplest case is that of a single-cylinder engine. Here we have a reciprocating mass, consisting of the piston, piston-rod, cross-head and a portion of the connecting-rod: we have also a revolving mass, consisting of the crank and the remainder of the connecting-rod. The revolving mass is readily balanced by prolonging the two crank-cheeks towards the opposite side from the crank and adding sufficient mass there. This leaves unbalanced a longitudinal force only, due to the reciprocating mass. If now we increase the balance masses on the crank-cheeks, on the opposite side from the crank, we may reduce the longitudinal force to any desired extent. But the extra masses so added exert transverse forces when they are out of the line of stroke. Thus, taking for example a horizontal engine, we may neutralize, in whole or in part, the forces which make the frame tend to move in a horizontal plane, but we do this only by introducing forces which tend alternately to raise it up from and push it down upon its foundations. Similarly in an engine with two or more cranks we may neutralize to any degree the longitudinal forces and couples by adding revolving masses, but only at the expense of introducing transverse forces and couples. A partial balance of longitudinal forces in this way may give the best compromise in practical cases.

262. Balancing locomotives. Such a compromise is in fact carried out in the locomotive, where we have, in general, two horizontal cylinders, coupled by cranks at right angles. The practical problem is to add such revolving masses as will, on the whole, make the resulting disturbance as little objectionable as possible. If the reciprocating masses were left wholly unbalanced the forces due to their acceleration would not only cause violent variations in the tractive force but, what is still more serious, would apply a couple tending to make the leading wheels of the engine sway from side to side, and this effect might in extreme cases cause the engine to leave the rails. On the other hand, if the longitudinal forces were wholly balanced by means of revolving masses, large unbalanced vertical forces would arise causing variation of pressure between each wheel and the rail, to which the name of "hammer-blow" is given. The name is scarcely appropriate, for the pressure varies not suddenly but gradually from a minimum to a maximum and back in each revolution of the wheel. But in extreme cases (not permissible in practice) the centrifugal force might exceed the total weight borne by the wheel, and in that case the wheel would actually lift when the force is acting upwards and would return to the rail with a blow. Large variations of vertical pressure are objectionable in any event, not only because they might cause slipping between wheel and rail by reducing the frictional adhesion, but because of their effect on railway bridges. When a locomotive passes over a bridge the so-called hammer-blow or pulsating force on the rails may add substantially to the load which the bridge has to bear. On a long bridge its regular recurrence tends to set the structure into a state of vertical oscillation, which may become violent when the speed is such that the frequency of the pulsating force agrees with the natural frequency of oscillation of the loaded structure.

A usual practical compromise is to make the revolving masses balance about two-thirds of the forces due to reciprocating masses. The revolving masses introduced for this purpose are placed in the wheels. In determining their amounts the masses of the coupling-rods, which also revolve, have to be taken into consideration. The problem is readily solved on the lines already explained, treating the whole as a system of revolving masses, two-thirds of each reciprocating mass being supposed concentrated at the corresponding crank.

263. Balancing an engine with two cylinders at right-angles to one another. An interesting special case is met with in a form of engine where two cylinders are combined as in fig. 228, with their pistons both driving on one crank P. The cylinders may be horizontal and vertical, as in the sketch; or both inclined at 45° to the vertical as in examples for internal-combustion use on motor cycles and aircraft. Treating the motion of each piston as simple harmonic, and making the pistons have equal mass, a complete balance as regards primary forces (see § 264) may be secured by adding revolving balance masses opposite the crank, at Q. For the force due to a mass at Q, in any position, may be resolved into a horizontal and a vertical component, and these balance the respective pistons. The possibilities of balance which this type of coupled engine offers were at first

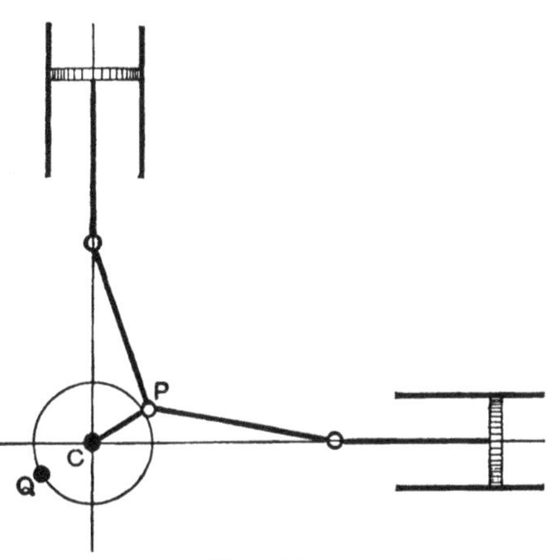

FIG. 228.

overlooked[1]. To prevent vibration it is only necessary to make the masses of the two reciprocating systems equal, and add a revolving mass at Q equivalent to the mass of either.

264. Secondary balancing. In what has been said about the balance of reciprocating parts we have treated them as having simple harmonic motion, ignoring the influence of shortness in the connecting-rod. In this way the conditions of what may be called *primary* balance have been arrived at. It remains to refer briefly to the forces that arise in consequence of the shortness of the rod, and to the conditions under which a further balance may be attained, among these forces. To this refinement of the problem the name *secondary balancing* is applied.

On the assumption of simple harmonic motion the force along the line of the stroke due to a reciprocating mass M is $M\omega^2 r \cos\theta$, where θ is the angle between the crank and the direction of the stroke. This would apply to a real piston if the connecting-rod were indefinitely long. At the two ends of the stroke, when the force is at its maximum, the value is $M\omega^2 r$.

With a rod of finite length l the force is (by § 244)

$$M\omega^2 r\left(1+\frac{r}{l}\right)$$

at the end A of the stroke furthest from the crank-shaft, and

$$M\omega^2 r\left(1-\frac{r}{l}\right)$$

at the end B of the stroke nearest the crank-shaft. Comparing this with the force $M\omega^2 r$ which would be found if the motion were simple harmonic we see that the actual force is obtained by adding $\left(\frac{rM}{l}\right)\omega^2 r$ at one end and subtracting the same quantity at the other.

Hence, so far as the ends of the stroke are concerned, the actual effect is the same as if a mass equal to rM/l were introduced, reciprocating with a simple harmonic motion due to a crank of the same radius but of twice the angular velocity, in addition to the mass M reciprocating with simple harmonic motion. This imaginary crank is to be thought of as coinciding with the real

[1] Engines of this class, but not balanced in the manner here explained, working at the East Greenwich Electric Power Station were found to disturb instruments at the Royal Observatory, half a mile away.

crank on the near dead-point, that is to say, when $\theta = 0°$. The imaginary mass rM/l is then exerting a force in the same direction as the real mass. When the real crank has made half a revolution the imaginary crank has made a complete revolution and come round again to the near dead-point, and the force due to the reciprocation of the imaginary mass rM/l is then opposed to the direction of the force due to M. Thus when $\theta = 0°$ the term $rM\omega^2 r/l$ is to be added, and when $\theta = 180°$ it is to be subtracted, which is the proper correction for the shortness of the rod.

Now suppose there is an engine comprising a system of reciprocating masses in "primary" balance, that is to say, properly balanced for simple harmonic reciprocation, but with connecting-rods which make their motion differ sensibly from true simple

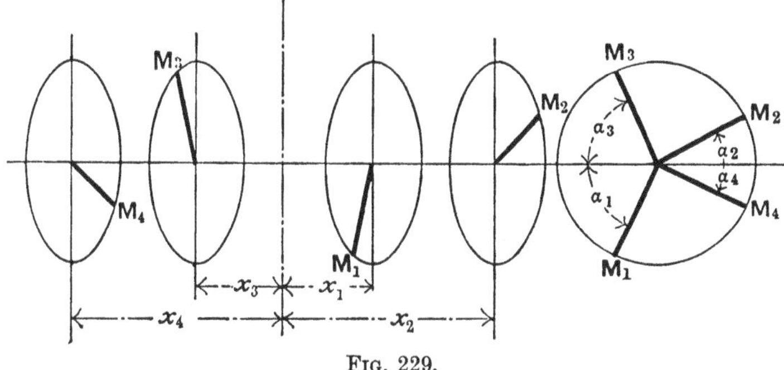

FIG. 229.

harmonic form. Think of each as having associated with it an imaginary crank, agreeing in phase when at the near dead-point, revolving with double the angular velocity, and driving (also with simple harmonic motion) an imaginary reciprocating mass equal to rM/l, where l is the actual length of the connecting-rod. Let these imaginary masses be balanced amongst themselves: this secures the condition of "secondary" balance.

When it is attempted to apply these principles to the design of a four-crank engine, with the object of securing secondary as well as primary balance, it will be found impossible to satisfy the conditions completely. But the four-crank engine can be arranged in such a way that besides being balanced as regards primary forces and primary couples it is also balanced as regards secondary forces though not as regards secondary couples. To do this the engine is generally made symmetrical, as in fig. 229, so that $x_1 = x_3$, $x_2 = x_4$,

$M_3 = M_1$, $M_4 = M_2$, $\alpha_3 = \alpha_1$, $\alpha_2 = \alpha_4$, and under these conditions balance of secondary forces (but not of secondary couples), along with complete primary balance, is secured by a suitable choice of the ratios of mass M_1/M_2, of distance x_1/x_2, and of angles[1].

In designing a four-cylinder engine on this basis it is usual to place the two relatively heavy pistons over the central cranks (1) and (3), with the lighter pistons over (2) and (4), and to make the distances between (1) and (2) and between (3) and (4) smaller than that between (1) and (3).

In the practical solution of this problem the valve masses have to be taken into account, which leads to some modification of the crank-angles. An example given by Dalby[2] of a balanced four-crank engine is shown in figs. 230–232, along with the resulting

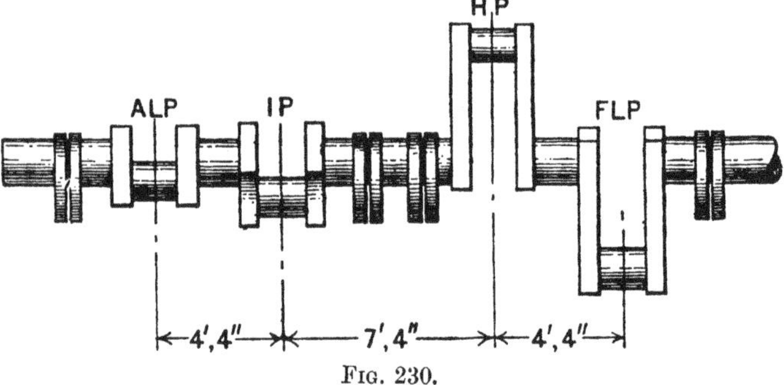

FIG. 230.

unbalanced forces and moments. Here a first approximation is reached by designing a symmetrical arrangement for the four main masses only, and the design is then adjusted so as to include the valves. As the revolving masses are not separately balanced they are included with the reciprocating masses in the reckoning, and this has the effect of leaving a small unbalanced horizontal force. The vertical couple is the only serious part of the system that is left unbalanced, and as the diagram shows it is a secondary couple, for its period is half that of the revolution. It is instructive to compare, as

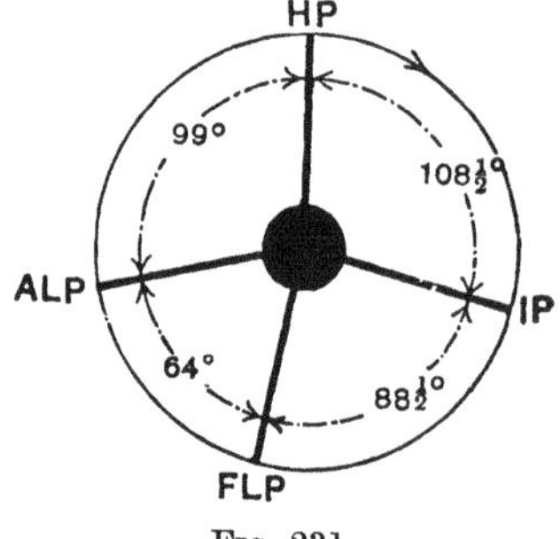

FIG. 231.

[1] O. Schlick, "On Balancing Steam Engines," *Trans. Inst. Nav. Arch.* 1900.
[2] W. E. Dalby, *Trans. Inst. Nav. Arch.* 1902.

Dalby does, the force and couple diagrams of a well-balanced engine such as this with those of other types in which balancing considerations have been more or less overlooked.

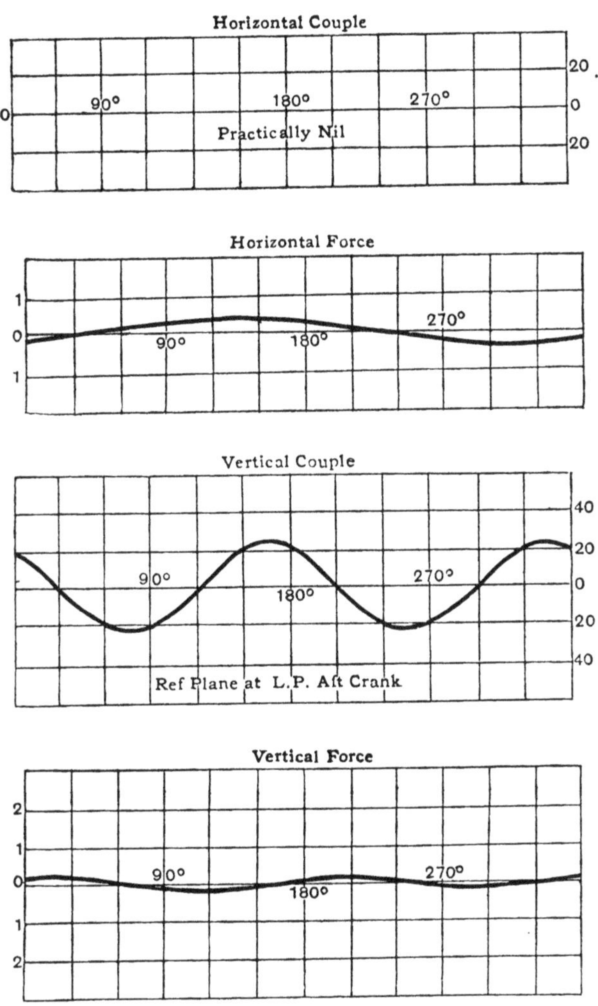

FIG. 232.

265. Application of Fourier's series. The disturbances due to the shortness of the connecting-rod may usefully be treated in another manner, by analysing them into a series of simple harmonic terms[1]. The acceleration of the piston at any instant in its motion may be expressed in the form

$$\omega^2 r\,(\cos\theta + A\cos 2\theta - B\cos 4\theta + C\cos 6\theta \ldots),$$

where A, B, C, etc. are coefficients whose magnitude depends on the ratio of the length of the crank to that of the connecting-rod[2].

Thus a complete dynamical equivalent for the real reciprocation of the mass M is arrived at by superposing the effects of, first, an equal mass moving with simple harmonic motion with the actual period of the piston, then a smaller mass equal to A times M reciprocating with twice the frequency, then a still smaller mass equal to B times M with four times the frequency and so on, all these imaginary masses having simple harmonic motion with a stroke equal to that of the real piston. From this point of view we may state the condition for perfect balance in any system of actual pistons by saying that when for each piston this series of imaginary pistons is substituted, every one of the subordinate imaginary systems must be in balance within itself. That is to say, not only must the primary system balance but the system of imaginary pistons whose masses are A times M_1, M_2, and so on, all of which reciprocate with double frequency, must balance in itself, and so must each of the systems with higher frequency. Speaking generally, it is the primary and secondary systems only that are important in practice.

With certain arrangements of engines it is possible to satisfy the condition of obtaining secondary as well as primary balance.

Take for instance the six-crank engine of fig. 226, where the crank-angles are 120°. We have already seen that this is in complete primary balance, as regards both forces and couples. Now think of the imaginary masses, forming the second group, which move with double frequency. When each real crank is at the near

[1] See papers by J. H. Macalpine, *Engineering*, Oct. 22, 1897, and by Prof. C. E. Inglis, *Trans. Inst. Nav. Arch.* 1911 (1), p. 248.

[2] Their values are:—

$$A = \frac{r}{l} + \frac{r^3}{4l^3} + \frac{15r^5}{128l^5} + \ldots,$$

$$B = \frac{r^3}{4l^3} + \frac{3r^5}{16l^5} + \ldots,$$

$$C = \frac{9r^5}{128l^5} + \ldots.$$

dead-point the corresponding imaginary crank is also at that dead-point, and when the real crank has moved through 120° the imaginary one has moved through 240° and so on. Hence the imaginary masses, forming the second system, also form a system with crank-angles of 120° and are in balance amongst themselves, both as regards forces and couples. This conclusion also applies to the five-crank engine of fig. 227. In both of these cases the conditions which secure primary balance secure secondary balance as well. The forces and couples due to the $A \cos 2\theta$ term are in balance, but those due to the $B \cos 4\theta$ term are not.

Applying this method to a four-crank engine, Inglis (*loc. cit.*) shows that in addition to primary hammering and primary tilting, either secondary hammering or secondary tilting can be eliminated, but not both. If the designer chooses to balance secondary tilting (as well as both primary effects) he must put the cranks 180° apart. This arrangement, though well adapted for a single-acting internal-combustion engine working on the four-stroke cycle, as in a motor car, is unsuitable for a marine steam-engine. In a steamship the best result is got by balancing primary and secondary hammering as well as primary tilting. This can be done not only in symmetrical designs such as are usually adopted in applying the Yarrow-Schlick-Tweedy system, but by various unsymmetrical groupings of the four cranks, provided the mass-ratios, distances, and angles satisfy certain equations which are discussed by Inglis in his paper.

CHAPTER XIV

THE FORMATION OF STEAM

266. Heating surface. After the potential energy of the fuel has been converted by combustion into actual heat, which shows itself in the incandescent solid and gaseous contents of the furnace, the heat has to reach the *heating surface* of the boiler before it can pass, by conduction through the metal, to the water and steam within. It reaches the heating surface partly by *convection* on the part of the hot gases, which are brought into intimate contact with the heating surface on their way from the combustion chamber to the chimney: they carry heat from the place where it is formed to the various parts of the heating surface to which they give it up. But besides this transfer of heat by convection and contact, there is another method by which the heating surface receives heat, namely by *radiation*, a process which does not depend on movement of the furnace gases, and does not even require that there should be any material medium between the hot body that is radiating heat and the colder body on which the radiated heat falls. The action is the same, in kind, as that by which the sun warms the earth. It is by absorbing radiation from the burning fuel on the grate and from the glowing gas that any parts of the boiler which are, so to speak, in sight of the fire receive most of their heat. Those portions of the heating surface which enclose the combustion chamber take up much heat in this way, and thus may contribute largely to the whole heating effect. When the hot gases pass from the combustion chamber into tubes or flues, they radiate comparatively little, and the amount radiated lessens very rapidly as the gases become cooler. Consequently the more distant parts of the heating surface receive heat almost wholly by direct contact with the moving gases.

Once the heat has entered, at any point, the outer surface of the metal plate or tube, it passes by conduction to the inner surface, with no more than a few degrees of difference in temperature between one side and the other, even when the flow of heat is as rapid as ever occurs in a boiler. It next passes into the adjacent water by conduction, and this is followed by a movement of the heated fluid. Provided the water side of the plate or tube is clear of scale and has water in contact with it, which requires that the bubbles of steam should escape freely as soon as they are formed, the passage of heat from the metal to the water is easy and the metal will nowhere (at any point of its thickness) be much hotter than the water, however fierce the fire to which it is exposed. Here the process of convection comes again into play. It is by the formation and movement of the bubbles and by circulation of the heated but still unvaporized water that the heat which has passed through the plate is carried away.

Hence, except in so far as the metal receives heat by absorbing radiation, the transfer from the furnace gas to the water requires a continuous circulation on both sides of the plate. On the water side, the steam generated at the metal surface would oppose conduction if it were not continually giving place to water; on the other side the film of gas which has parted with heat to the metal must continually be scrubbed away and be replaced by hot gas. Even in the earliest stages of the process of getting up steam an automatic circulation of water is set up, to which the transfer of heat from the plate is mainly due. Differences in the specific conductivity of the metal, or in its thickness, are almost without effect on the quantity of heat that passes across a given area of plate in a given time, for the chief resistance to the flow of heat is in its passage from the gas to the metal: after that a small gradient of temperature suffices to do the rest. There is a big drop of temperature between the metal and the hot gas near it, across the chilled film, but no considerable drop at any other part of the flow. Effects of radiation will be discussed later; for the present we are concerned with the process of transfer by conduction, associated with movement of the fluid past the surface of the metal.

Under ordinary conditions of conduction of heat through the plates or tubes of a boiler, by far the greater part of the whole temperature difference between the furnace gases and the water is used up in overcoming the resistance of the chilled gaseous layer. Nicolson has shown as a simple deduction from the known thermal conductivity of steel and the known rate of evaporation in a boiler that in ordinary cases the side of the plate next the gases is only about 5° C. hotter than the side next the water, although the temperature of the gases, a little way from the plate, is many hundreds of degrees higher[1].

[1] J. T. Nicolson, *Proc. Inst. Junior Engineers*, Jan. 1909. See also Dalby's Report on Heat Transmission, *Proc. Inst. Mech. Eng.* Oct. 1909, where an important synopsis is given of papers relating to the subject.

267. Influence of the velocity of the gases on the effectiveness of the heating surface. It was pointed out by Osborne Reynolds[1] in 1874 that the rate at which heat is given off by a hot gas to any colder surface over which it flows depends primarily on the velocity of flow. If the gas were stagnant or nearly stagnant it would part with heat very slowly, for the portion of it in immediate contact with the metal surface would become chilled and would form a protective layer of very feeble conductivity. The passage of heat across the layer would be determined by gaseous diffusion between the coldest parts, close to the metal, and the hotter parts at greater distances from the metal, and diffusion would be competent to produce only a very gradual transfer of heat. But if the gas is moving past the surface at a sufficient speed to give rise to turbulence in the motion (and a very moderate speed will do that) we may regard this layer as being continuously formed and continuously broken by eddies which stir it up, making a mixture between its contents and the hot gases of the warm stream. Such is the normal action at the heating surface of a boiler. The faster the stream moves the more effective will this mixing or scrubbing action be, with the result that the rate at which the gases part with heat to the metal surface will be correspondingly increased.

Consequently the effectiveness of any part of the heating surface in a boiler depends, most materially, on the rapidity of movement of the hot gases over it. The faster the gases move the more heat they give up, and within wide limits the heat transmitted through a given surface is increased *pari passu* with the velocity of the gases, provided their temperature is kept up by burning fuel at a correspondingly greater rate.

The same considerations apply in the transfer of heat to, or from, a liquid[2]. It is important to have a scrubbing action on the water side, as well as on the gas side, of the heating surface, not only to facilitate the escape of the steam as it forms but also to promote rapid conduction from the metal to the water. This is supplied,

[1] O. Reynolds, "On the Extent and Action of the Heating Surface for Steam Boilers," *Proc. Lit. and Phil. Soc. of Manchester*, 1874; *Collected Papers*, vol. I, p. 81.

[2] See experiments and theoretical discussion by Dr T. E. Stanton, "On the passage of heat between metal surfaces and liquids in contact with them," *Phil. Trans. Roy. Soc.* 1897.

more or less, by convection currents in the water, and by the circulation which the escaping steam itself sets up, but there are wide differences in different boilers in this respect. A mere statement of the number of square feet of heating surface conveys little indication of a boiler's evaporative capability, for a given area may be ten-fold more effective in one situation or in one design than in another, through rapid circulation of the hot gases and of the water, as well as through higher temperature in the gases themselves.

An illustration of these principles is afforded when a boiler is *forced*, that is to say when, by artificially increasing the draught, we compel a larger amount of fuel to be burnt in a given time, with the result that a larger amount of steam is generated, the area of the heating surface of course remaining unchanged. The rate at which heat passes across the surface is augmented by the higher velocity of the gases, and, in a less degree, by the higher velocity of circulation of the water. Each square foot of heating surface becomes much more effective, and the efficiency of the boiler, which is the proportion of heat transferred to heat generated, is not in general much reduced. The gases, at the end of their course over the heating surface, have given up nearly the same proportion of the heat they have received.

The formula given by Reynolds for the amount of heat Q taken up per second by each square foot of heating surface in a boiler flue or fire-tube traversed by hot gases of density ρ, in pounds per cub. foot, moving with a velocity v, in feet per second, is

$$Q = (A + Bv\rho)\,(T_g - T_m),$$

where T_m is the temperature of the metal surface, and T_g is the temperature of the gas, taken at a point a little way from the surface so as to be clear of the chilled film. Instead of $v\rho$ we may write n/a, where a is the area of section of the flue in square feet, and n is the number of pounds of gas passing through it per second. A and B are (to a first approximation) constants.

From experiments by H. P. Jordan[1] it appears that when hot air is passed through a clean water-cooled tube (of something like one inch bore) the expression becomes approximately

$$Q = (0{\cdot}0015 + 0{\cdot}0008n/a)\,(T_g - T_m),$$

for velocities such as may occur in boilers. In a locomotive, for

[1] *Proc. Inst. Mech. Eng.* Dec. 1909.

example, where under heavy steaming n/a may be 4 lb. per second, per square foot of fire-tube section, this corresponds to 0·0047 thermal unit per second, per square foot of heating surface, for each degree by which the gas is hotter than the metal. The heat which passes in this way through the heating surface is directly proportional to the difference of temperature. As we shall see in the next section, the heat which is taken up by radiation follows a very different law.

268. Heating by radiation. In many boilers, especially in some of modern design, much of the heat reaches the metal by radiation from the glowing hearth or from flame. By a well-established principle of physics, known as Stefan's Law, the greatest amount of heat which can be radiated from a body varies as the fourth power of its absolute temperature. Hence when two bodies at very different temperatures are in juxtaposition, as for example the hearth and the crown of a locomotive fire-box, the hearth can radiate enormously more heat to the crown than the crown can radiate to the hearth. By Stefan's Law an upper limit for the amount of heat which can pass by radiation in such a case from the hot body to the colder one is determined by the formula

$$\sigma (T_a^4 - T_b^4),$$

where T_a and T_b are the absolute temperatures of the two bodies and σ is a constant called the coefficient of "full" or "black-body" radiation. When the rate of transfer of heat is expressed in ergs per second per square centimetre of the opposed surfaces, the value of σ, according to the best measurements, is about $5{\cdot}7 \times 10^{-5}$.

A "black-body" or "full radiator" in this connexion means a body whose surface absorbs all the radiation, of whatever wave-length, that falls upon it, neither reflecting nor transmitting any of the incident radiation. Such a body itself radiates, at any temperature, the greatest amount which can be radiated from any surface at that temperature. A perfect absorber is also a full radiator.

The "black-body" may be raised to a white heat and at any stage it is more actively radiant than any other body would be at the same temperature. No real body confirms strictly to this definition; carbon in the form of coke or soot comes very near it. The surface of a flame in which particles of carbon are being consumed, or the glowing bed of a furnace on which coal or coke is burning approximates more or less closely to being a full radiator. Such surfaces are sending out radiant energy at a rate which is not far short of the greatest rate which their temperature allows. Again, the oxidized surface of furnace plates or tubes behaves not very differently from a black body in respect of radiation and absorption. Stefan's Law indicates a limit which, though it must be in excess of the heat actually transferred, serves to show how large, in favourable circumstances, the part played by radiation may become.

Taking the value of σ given above, the formula for full radiation may be written

$$Q_R = 10^{-8}\,(T_a^4 - T_b^4),$$

where Q_R is the quantity of heat in pound-calories per hour per square foot of radiating surface, T_a and T_b being the absolute temperatures in centigrade degrees. With a fire-box full of flame, this sets an upper limit to the rate at which heat can be received by radiation per square foot of the exposed heating surface.

The absolute temperature of a hearth or flame may easily be as high as 1800°; that of a boiler plate in action is generally not quite 500°. When these are taken as representative figures it will be seen that the T_b^4 term is relatively insignificant, and Q_R becomes 104,000 lb.-calories, or enough to convert about 200 lb. of water into steam. This may be compared with the average evaporation over the whole heating surface of a boiler, which is usually of the order of 5 or 10 lb. per square foot per hour. The comparison shows how comparatively active the parts exposed to radiant heat from the fire may be, even when allowance is made for the fact that the figure for radiation is a limit not realized in actual boilers, because the whole fire-box is not full of flame, and the flame surface radiates with less than the full or "black-body" intensity.

Modern practice tends, in large power plants, to the use of pulverized coal which is injected in such a manner as to keep a considerable part of the combustion chamber full of flame. This is also true of the oil-burning furnaces now used with many marine and other steam-engines. In such cases the combustion chamber may with advantage be completely or almost completely surrounded by water-cooled walls or groups of water-tubes which are part of the effective heating surface, so that all the available radiant energy may be received by water-cooled metal, care being taken that the chamber is big enough, in relation to the volume of the flame, to let combustion be complete before the gas touches the metal, otherwise imperfectly burnt products may be carried into the flues.

269. Additional heating surface obtained by the use of a feed-water heater. As the gases traverse the flues or fire-tubes their temperature falls, until they finally escape at a temperature which is necessarily somewhat higher than that of the water to which they have been yielding up their heat. The temperature of the gases, however, need not be higher than that of the steam, or even as high, for after ceasing to be in contact with the boiler proper the gases may continue to give up heat to an *economizer* or *feed-water heater*, that is a set of tubes which are exposed to the hot gases on their way to the chimney, and through which the comparatively cold feed-water passes on its way to the boiler. The feed-water heater virtually forms an extension of the heating surface, with the advantage that it may be more effective for the transfer of heat than an equivalent extension of the boiler surface proper would be, on account of the lower temperature of the contents; and it allows the initial temperature of the feed-water, instead of the temperature of the steam, to form the lower limit which the temperature of the gases may approach. Conduction, however, would become so slow if the temperature of the gases were near this limit that in practice they are always considerably hotter. When the draught through the fire is maintained simply by means of a chimney there is an independent reason for allowing the gases to escape at a relatively high temperature: the draught then depends on the contents of the chimney being lighter than the air outside, and this lightness is secured by their being considerably hotter than the atmospheric air.

270. Draught. The furnace gases are made up of the products of combustion, along with a quantity of air of dilution which passes through the furnace without undergoing chemical change. For the complete combustion of each pound of coal about 12 lb. of air are required to furnish the necessary oxygen, and usually about 12 lb. more enter as air of dilution. The greater part of this air comes in through the grate, between the fire-bars on which the burning fuel rests, but some air has to be admitted above the fire to complete the burning of the combustible gases. This is specially necessary when fresh coal has been thrown on the fire and volatile hydrocarbons are being given off. The furnace door has apertures to allow a small part of the air to pass through it, and these are often made adjustable in area.

A *natural* or *chimney* draught is one which is produced wholly by the lightness of the contents of the chimney. A *forced* draught is one in which other means are taken to produce a difference between the pressure of the air inside and outside of the furnace. A fan, for instance, may be used to force the draught, either by extracting the gases from the flues or by blowing air into a closed room or channel from which the furnace takes its supply. Or a jet of steam may be allowed to escape up the chimney, producing a partial vacuum there on the principle of the jet pump.

With a forced draught it is easy to produce much more difference in air pressure above and below the grate than can readily be produced by means of a chimney, and consequently to compel the entrance of a larger quantity of air through the fuel, with the result that more coal can be burned per square foot of grate. A furnace using chimney draught does not as a rule burn more than 20 lb. of coal per hour per square foot of grate, but with forced draught the combustion may go on at four or five times this rate and still be fairly perfect.

Further, when the draught is forced the combustion is intensified and localized, and it is found that a smaller proportion of air will suffice for dilution. Instead of the 24 lb. or so of air which chimney draught requires per lb. of coal, 18 lb. or less will serve. Hence with a forced draught the temperature of the furnace gases is higher. The effectiveness of the heating surface is also increased through the more rapid movement of the gases. Again, since the proportion of air passing through the furnace is reduced by forcing the draught, the proportion of heat lost in the hot gases is also reduced, provided the action of the heating surface is such that they leave the flues at no higher temperature than before.

But the theoretical advantage of forced draught in respect of efficiency does not stop here. When the draught does not depend on the action of a chimney there is no need to let the escaping gases have any higher temperature than is imposed by the condition, already indicated, that they must be reasonably hotter than the temperature of the feed. With a chimney, on the other hand, as much heat is necessarily wasted as will keep the temperature of the escaping gases up to the comparatively high value necessary to maintain the draught. A chimney being an exceedingly inefficient form of heat-engine, the heat which is expended in maintaining its draught is much greater than the equivalent of the work that a fan would do in producing the same draught, or even than the heat that would have to be supplied to an engine employed in driving the fan.

In many instances in which the draught is forced, as in locomotives and in some marine and land engines, the theoretical advantages of forced draught, in respect of efficiency, are imperfectly realized. The draught is forced with the object of increasing the power of a given boiler rather than of securing a high efficiency. The motive is to burn more coal per square foot of grate surface, and thus to evaporate more water in a boiler of given weight. It is clear, however, that the most efficient boiler would be one using a strong mechanically forced draught, with a relatively small area of grate and with a heating surface adapted both by its area and by the speed of the gases over it to extract as much as possible of the heat, supplemented by the use of a feed-water heater, or, alternatively, of an air pre-heater by which the residual heat in the furnace gases, after they leave the boiler, is transferred to the incoming air. Under these conditions not only would the gases be cooled as far as possible before escaping, but the proportion of air to coal would be as small as is consistent with thorough combustion.

271. Sources of loss of heat. Ordinarily about seven-tenths and rarely more than eight-tenths of the potential energy of the fuel are conveyed to the steam. The remaining two or three-tenths are accounted for as follows:—(1) waste of fuel in the solid state by dropping through the grate; (2) waste of fuel in the gaseous and smoky state by imperfect combustion; (3) waste of heat by external radiation and conduction; and (4) waste of heat in the escaping gases due mainly to their high temperature, but partly also to their containing as one of the products of combustion a certain amount of water-vapour which passes off uncondensed. Of these sources of waste the first is generally trifling and the fourth is the most important. If we assume the air of dilution to be 12 lb. the whole quantity of gas escaping from the chimney is 25 lb. per lb. of coal burnt. The specific heat of this gas is nearly the same as that of air, say 0·24 thermal unit. Hence about 6 thermal units are lost, per lb. of fuel burnt, for every degree by which the temperature of the escaping gas is allowed to exceed the lowest attainable limit. It is not unusual to have a chimney temperature as much as 250° or even 300° C. higher than the limit which is imposed by the temperature of the feed-water. This represents a more or less preventable loss of 1500 or 1800 thermal units (pound-calories) per lb. of coal, or in round numbers fully one-fifth of the whole energy of the coal.

With forced draught a great part of this loss might be avoided, by extracting more heat from the gases before they escape. At the best, however, some loss results from their high temperature, and it is important to keep this loss down by supplying no more air than is really necessary. Any unnecessary dilution of the gases means additional loss by increasing the quantity of the discharge relatively to the quantity of fuel that is consumed. It is obvious that to minimize the loss two conditions should be aimed at: the quantity of air admitted should be no more than is required for satisfactory combustion, and the temperature of the gases should be brought down to the lowest practicable figure by conduction of heat to the water or to the incoming air before they are discharged. In large modern power plants using pulverized fuel with forced draught a careful attention to these conditions brings the boiler efficiency to a figure approaching nine-tenths[1].

272. Chimney draught. In a chimney draught the "head" (usually stated in inches of water pressure) under which the current of air is kept up is equal to the amount by which the weight of a column of air in the chimney falls short of the weight of a corresponding column of outside air. Except for their excess of temperature the contents of the chimney would be heavier than the air outside, namely, in the ratio of $n+1$ to n, where n is the number of pounds of air which have taken up 1 lb. of fuel in passing through the furnace. The actual density of the gases is less than that of the air outside in the proportion $\frac{T}{T_0}\left(\frac{n+1}{n}\right)$ to 1, where T and T_0 are the absolute temperatures inside and outside respectively. The difference in actual density multiplied by the height of the chimney gives the effective head. This head is used up partly in setting the column of air in motion and partly in overcoming the

[1] An efficiency of 0·92 is claimed in some test results. See *Trans. of the First World Power Conference*, vol. II, p. 1317.

resistance to its passage which is offered by the flue, by the chimney itself, and by the grate. With a forced draught and a short chimney the resistance of the grate is the most important of these items; with a tall chimney on the other hand the resistance of the chimney itself comes to be so considerable that an increase of height produces almost no increase of draught, and may even diminish the draught if the sectional area is at all reduced in the added part. Under such conditions also there is a limit in the extent to which the draught will be assisted by letting the chimney temperature remain high. In raising the temperature of the chimney gases a stage is reached at which the gain in head and consequently in velocity of current is more than counterbalanced by the diminution of density, and if the gases are hotter than this the amount of gas passing through the chimney in a given time is actually reduced. No advantage whatever is gained by making the temperature higher than corresponds to maximum draught, and on the score of thermal efficiency a lower temperature is of course to be preferred, as diminishing the heat lost in the escaping gases.

273. Coal and its combustion. The chief constituents of coal are (1) carbon, (2) certain volatile hydrocarbons which may be distilled out by the application of heat, along with a little free hydrogen, and (3) incombustible ash. In bituminous coal the hydrocarbons are present in considerable quantity, and when the coal begins to burn some of them exude in liquid form, tending to make it cake. Anthracite is a variety of coal in which they are almost entirely absent; coals which are classed as anthracitic are free from any tendency to cake. In the complete combustion of coal all the carbon, including that which is present in the hydrocarbons, should be converted into carbon dioxide (CO_2), and the hydrogen of the hydrocarbons is also completely oxidized, becoming water-vapour. Any appearance in the chimney gases either of carbon, in the form of smoke, or of carbon monoxide (CO) is evidence of incomplete combustion. Incomplete combustion may be due to an insufficient supply of air, but it often occurs, although the air-supply is sufficient or even excessive, as a result of the gases having their temperature prematurely lowered, which generally occurs through their being brought into contact with surfaces which cool them below the point at which combustion will continue, before the combustion is complete. When fresh coal is thrown on the fire the volatile constituents are quickly distilled, and would escape unburnt or very imperfectly burnt if air for their combustion were not admitted above the fire, and if care were not taken in the design of the furnace to provide a combustion chamber of sufficient size. Incompleteness of combustion in the distillates from fresh coal is readily detected by the presence of smoke. But a smokeless fire is no proof that the air-supply is rightly regulated. The supply may be excessive, which, as we have seen, results in serious waste of heat, or it may be insufficient, for the gases may contain carbon monoxide, which is an invisible gas. Important information may be obtained by testing the furnace gases for CO and also for CO_2. If CO is found the air-supply is defective. The greater the quantity of CO_2 the more satisfactory is the combustion, provided there is neither CO nor smoke, for this means that the combustion is perfect with a small supply of air. Having measured the proportion of CO_2 it is easy to calculate what is the proportion of carbon in the gas escaping to the chimney, and hence to infer what excess of air is being supplied. With suitable apparatus the test is readily applied[1], and instruments have been devised for giving a continuous indication of the proportion of CO_2 as a check on the efficiency of the firing. In very favourable examples the proportion of CO_2 by volume may be 15 per cent. or more[2].

The calorific value of coal varies somewhat widely. In some of the best kinds used for steam raising it is about 8300 thermal units, in others of high quality it is 8000 units, but in coals containing much moisture or ash or both it may be only 7000 units or less. It is usually stated as a quantity which does not include the latent heat of any H_2O that may be present in the fuel or be formed as one of the products of combustion, since that passes away from a furnace in the gaseous state. This is called the "lower" calorific value of a fuel; the "higher" calorific value does include the latent heat of the water-vapour. The calorific value is best determined

[1] Such tests are usually carried out by the Orsat apparatus, in which measured volumes of the furnace gas are passed through three vessels containing substances which absorb CO_2, CO, and oxygen respectively.

[2] Air contains approximately 21 per cent. by volume of oxygen. CO_2 occupies the same volume as the oxygen from which it is produced. Hence a pure carbon fuel, if it could be completely burnt without excess of air, would yield a furnace gas containing 21 per cent. by volume of CO_2. With 50 per cent. air of dilution the volume percentage of CO_2 would be 21/1·5 or 14 per cent.; with 30 per cent. air of dilution it would be 21/1·3 or fully 16 per cent.

by a laboratory experiment on a small scale, using a combustion calorimeter in which a sample of the coal previously ground to powder is electrically ignited and burnt in a closed vessel with a supply of oxygen to maintain the combustion, the vessel being enclosed in another containing a known weight of water, through which the burnt gases are allowed to escape. The rise in temperature of the containing water serves to measure the heat produced, after the calorimeter has been calibrated by observing the rise in temperature which is produced by the development within it of a known quantity of heat generated electrically.

In the process of burning which goes on in a furnace much the same considerations apply as have been mentioned in regard to the action of gases in giving up heat to a metal surface. On the surface of every fragment of ignited coal there must, if combustion is to go on reasonably fast, be a scrubbing action which removes the inert layer consisting of the gaseous products of combustion and the nitrogen of the burnt air, and brings up fresh supplies of oxygen. This inert layer is continually being formed as combustion proceeds, and continually being removed through the action of the draught. Under a strong forced draught its removal takes place very rapidly: that is to say, the average amount of inert layer becomes very small. Hence not only is combustion much stimulated but a smaller proportion of the whole supply of oxygen passes the fuel without being consumed. In other words, with a strong forced draught less air of dilution is required over and above the air whose oxygen actually enters into union with the fuel. Experience shows that with a natural draught, which is necessarily not very strong, the total quantity of air supplied has generally to be 24 or 25 lb. per lb. of coal, but under favourable conditions with forced draught this may be reduced to 18 or even 16 lb., without sacrificing completeness of combustion.

274. Cornish and Lancashire boilers. Large stationary boilers of the forms known as the "Cornish" and "Lancashire" are very common. They are internally fired, that is to say, the furnaces are enclosed within the water space of the boiler. The shell of these boilers is a long horizontal cylinder with flat ends, and inside this, stretching from end to end within the water space, is a single large tube in the Cornish form and two parallel tubes in the Lancashire form each tube containing a furnace at one end and communicating

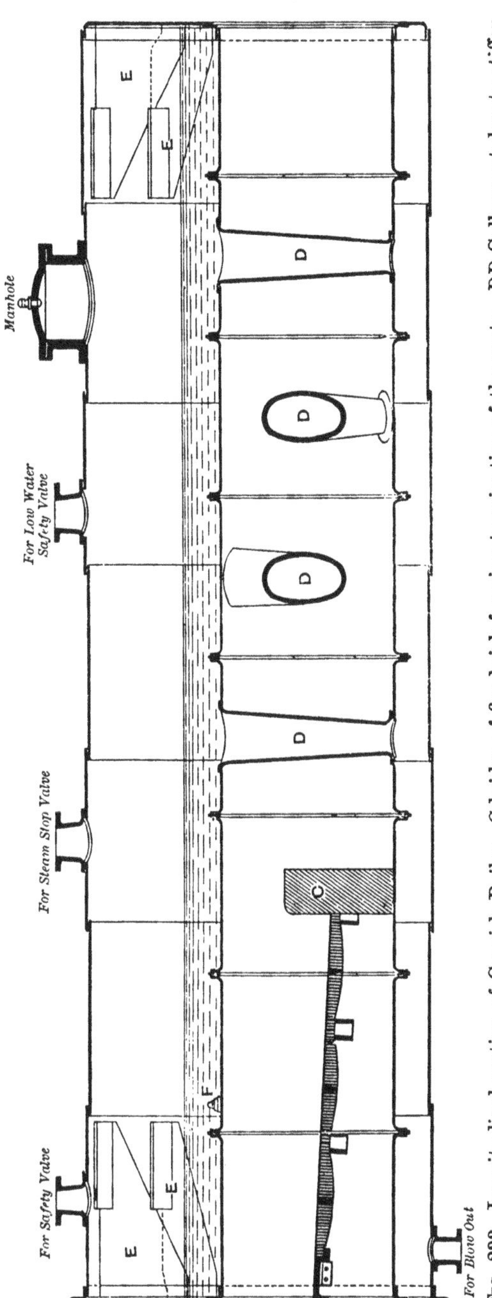

FIG. 233. Longitudinal section of Cornish Boiler. *C* bridge of fire-brick forming termination of the grate. *DD* Galloway tubes to stiffen the flue tube, to promote circulation of the water and turbulence of the gases, and to increase the heating surface. *EE* gusset stays to support the flat ends. *F* fusible plug.

at the other end with external flues which are so arranged as to make nearly all the external surface of the shell below the water line act as part of the heating surface. The remainder of the heating surface is given by the large tube or tubes which contain the furnace, often with the addition of several short cross tubes containing water, called Galloway tubes, which traverse the main furnace tube at right angles to its length and not only serve the purpose of enlarging the heating surface, but promote circulation in the water, help to make the motion of the gases turbulent, and strengthen the main tube. Fig. 233 shows a Cornish boiler in

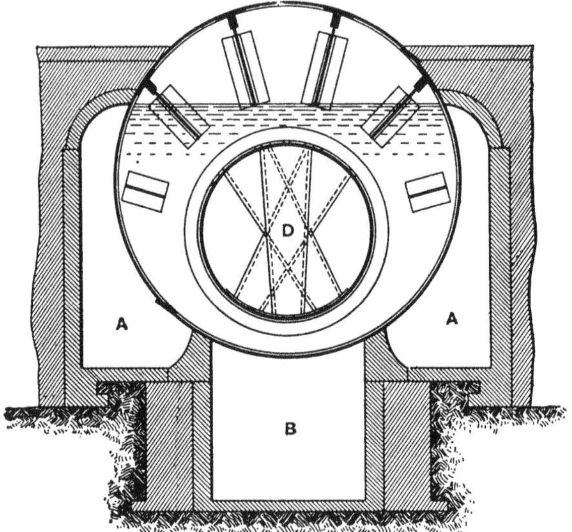

FIG. 234. Transverse section of Cornish Boiler.

longitudinal section, and fig. 234 is a cross-section which shows the arrangement of the external flues. The furnace extends from the front up to the bridge of fire-brick *C*. In continuing their passage beyond this through the main tube or flue the hot gases come in contact with the Galloway tubes, *DD*, which have a somewhat conical form so that they may allow the steam formed in them to rise readily. At the end of the internal flue the gases are diverted downwards into the external flue *B*, and having traversed it towards the front of the boiler they are made to rise into the two side flues *AA*, by which they again pass to the back end and thence to the chimney. The form of the Lancashire boiler is essentially the same, except that there are two furnace tubes placed side by side, the diameter of the shell being larger. Fig. 235 is the cross-section of a Lancashire boiler. In a modified form of this boiler, introduced by Galloway, the two furnace tubes unite beyond the bridge into one with a flat section, which is prevented from collapsing by having a number of Galloway tubes in it to act as stays. In ordinary Cornish and Lancashire boilers Galloway tubes are often omitted, and in many instances the flues are corrugated

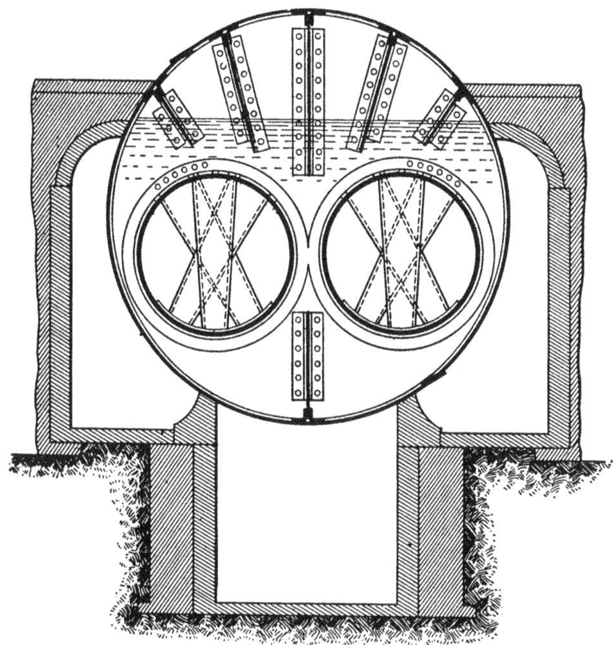

FIG. 235. Transverse section of Lancashire Boiler.

along part, or the whole, of their length, which makes them more stable in resistance to external pressure and more elastic in meeting effects of unequal temperature.

The shell of a Lancashire boiler is commonly from 28 to 30 feet long, with a diameter of 8 or 9 feet, which allows each of the two flues to be about 3 feet wide. A boiler of this size has a heating surface of about 1000 sq. ft.; burning 20 lb. or so of coal per hour per square foot of grate, it will evaporate 6000 to 7000 lb. of water per hour. Its working pressure commonly ranges from 160 to 200 lb. per sq. inch.

In boilers of this type the curvature of the cylindrical shell and furnace tubes enables them to resist the pressure of the steam: when the ends are flat they require to be stayed. This is usually done by means of gusset stays *EE* (fig. 233), which tie the end plate to the circumference of the shell. Often, however, the ends are dished, that is to say, they are curved to form part of a spherical surface the radius of which should be equal to the diameter of the shell, as in fig. 236. The flues are made up of a series of short internal lengths united by joints which give the whole tube stiffness to resist collapse, but leave it some freedom to bend when the top expands more than the bottom through unequal action of the fire, especially in lighting up. To provide for unequal expansion is one of the most important points in the design of a boiler: when it is neglected a racking action occurs which induces leakage at the joints and tends to tear the plates. For this reason the furnace flues are attached only to the end plates and not to the cylindrical part of the shell, the stays of flat end plates are arranged to leave these plates some freedom to bulge out and in when the flues lengthen and contract, and the flues often have a corrugated portion as in this example. Fig. 236 illustrates the use of a flue tube which is corrugated, in this case along part of the length.

275. Boiler mountings. The steam-dome, which used to be an ordinary feature in boilers of this type, is now omitted, and steam is taken direct from the steam space within the shell through a perforated "antipriming" pipe, seen to the right in fig. 236. The other openings on the top of the shell are the manhole, on which a cover is bolted, and openings for two safety-valves. One of these valves is frequently of the dead-weight type, as here, in which the force by which the valve is held closed is furnished by the direct action of a pile of weights: many safety-valves however use springs or weighted levers. In one form of spring-loaded safety-valve known as a "pop" valve the shape of the valve is such that when steam begins to escape its pressure is felt over a greater area of the valve: this produces instability and the valve opens widely, making a characteristic noise. The second safety-valve is often arranged to form what is called a low-water safety-valve, being connected to a float in such a way that the valve will open if the water is allowed to sink below a safe level. At the bottom of the shell there is a nozzle for the blow-out cock, and in the front plate, below the furnace tubes, there is another manhole. Feed-water is supplied by a pipe which enters through the front plate on one side near the top of the water and extends a good way in, distributing the water by holes throughout its length. A pipe at the same level on the other side serves to collect scum. On the front plate are a pair of glass gauge-tubes showing the level of the water within and a pressure-gauge of the Bourdon type. This important fitting consists of a metal tube, oval in section, which is bent into a nearly circular form. One end is closed and is free to move: the other is held fixed and is open to the steam. The pressure of the steam produces an elastic strain tending to make the oval section rounder and to straighten the tube. The free end

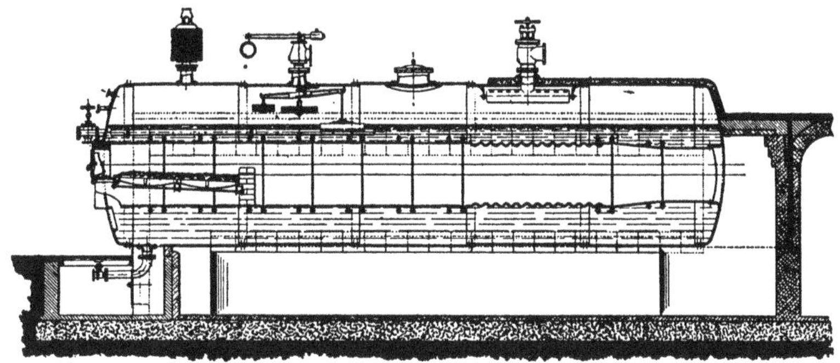

FIG. 236. Lancashire Boiler with dished ends and partly corrugated flues (Galloways).

accordingly moves through a small distance which is proportional to the excess of pressure within the tube above the atmospheric pressure to which its outer surface is exposed, and this movement is magnified by an index turning on a dial. Some of the fittings which have been mentioned are illustrated in the figure, and most of them are common to boilers of other types.

276. Multitubular boilers. In several forms of boiler an extensive heating surface is obtained by the use of a large number of small tubes (called fire-tubes) through which the hot gases pass. This construction is followed in locomotive and marine boilers, and boilers of the typical locomotive and marine forms (to be presently described) are, especially the former, frequently used with stationary engines. The multitubular construction is also applied in some instances to boilers of the ordinary cylindrical form by making a host of small tubes take the place of that part of the flue or flues which lies behind the bridge, or by using small tubes as channels through which the gases return from the back to the front after they have passed through the main flue. A form known as the *dry-back* boiler consists of a short cylinder with flat ends, containing one or more cylindrical furnace tubes through which the gases pass at once into an external chamber behind the boiler. From this they return to the front through a group of many small horizontal return tubes placed at the side of or above the furnace tubes. Still another form of tubular boiler is an externally fired horizontal cylinder filled with return tubes extending from back to front: in this case also the boiler is a plain circular cylinder with flat ends. In all these forms the tubes are placed within the water space of the boiler. Except in some locomotives the tubes are commonly of iron, and a usual diameter is about 3 inches. They are secured in the tube-plates in which they terminate by expanding the ends so as to make a steam-tight joint. They give so much heating surface that the outside surface of the shell need not be used, and hence in a tubular boiler the external flues are dispensed with which are a necessary feature of the Cornish or Lancashire type.

277. Locomotive type. This is a multitubular boiler which, in its normal form, has a nearly rectangular fire-box at one end, surrounded by narrow water spaces, and a long cylindrical barrel extending horizontally to the other end, containing numerous horizontal tubes through which the gases pass from the fire-box to the smoke-box under the funnel. The heating surface is made up of the tubes and of the crown and sides of the fire-box. In most English locomotives the inner part of the fire-box, in which is the grate, is sufficiently narrow to be contained between the frames and wheels of the engine, but in America the need of locomotives of great power, together with the absence of restrictions as to size which are imposed by existing tunnels, etc., has led to the almost universal use of a wide fire-box extending over the frames and trailing wheels.

The plate (figs. 237 and 237 *a*) reproduces a drawing[1] of a British

[1] Kindly prepared for this book (March 1926) by Sir Henry Fowler, Chief Mechanical Engineer of the L.M.S. Railway.

locomotive of moderate power. Fig. 237 is a longitudinal section through the middle, and fig. 237 *a* is a half end-view (from the cab) and a half-section through the fire-box. In this example the length of the barrel is nearly 12 ft. and that of the fire-box is 9 ft. Between the tube-plates there are 146 small tubes, with an outside diameter of $1\frac{3}{4}$ in., and 21 large tubes with an outside diameter of $5\frac{1}{8}$ in. which contain the superheater elements. The fire-box is of copper and is nearly rectangular, with a grate that slopes downwards towards the barrel end. Round its sides, and at the back and front (except where the fire-door comes), is a narrow water space. The flat sides of the fire-box are tied, across this space, to the flat sides of the shell by numerous stay-bolts, the ends of which are seen in fig. 237. Above the crown of the fire-box the shell has a flat horizontal surface, and the two are tied together by stays which are shown on the right-hand side in fig. 237 *a*. Across the same space, above the fire-box, there are transverse stays holding together the two flat sides of the fire-box in its upper portion. Longitudinal stays run from end to end of the steam space, tying the front plate of the boiler to the upper part of the smoke-box tube-plate: in the lower part the tubes themselves serve as stays. A sloping bridge of fire-brick partially separates the upper part of the fire-box from the lower and prevents the flame from striking the tubes too directly. Underneath the grate is an ashpan which serves to control the supply of air under the fire by means of a damper at the lower end. On top of the barrel is a steam-dome from which the steam is taken, through the regulator *H* and main steam-pipe *I*, to the smoke-box end, where it is admitted to the superheater before passing on to the valve-chest. The superheater elements are long tubes with an external diameter of $1\frac{1}{2}$ inches placed within the large fire-tubes, extending nearly their whole length; they are doubled and redoubled on themselves so that their exit and entrance ends are close together. This allows them to be connected in parallel between two steam headers situated in the smoke-box, so that together they form a multiple channel through which all the steam passes before entering the pipe *K* by which it passes on to the valve-chest. The exhaust steam from the cylinder is discharged through the blast-pipe *N*, and forces the furnace draught by producing a partial vacuum in the smoke-box on the principle of the jet pump. Above the fire-box end of the boiler are seen a pair of safety-valves F. In this boiler the pressure is 200 lb. by gauge: the grate area is 28·4 sq. ft.; the heating surface is 1328 sq. ft., of which the fire-box contributes 127 sq. ft. and the tubes the rest. In addition, the superheater elements have a heating surface of 291 sq. ft. Working at its highest power it would produce about 23,000 lb. of steam per hour, superheated to 330° C., with a combustion of about 135 lb. of coal per hour per square foot of grate.

In more powerful locomotives these dimensions may be much exceeded. Thus the boilers of the three-cylinder "Pacific" engines on the L.N.E. Railway, designed by Mr Gresley in 1922, have a grate area of 41¼ sq. ft., an evaporative heating surface of 2930 sq. ft., and a superheating surface, in 32 elements, of 525 sq. ft. The length between the tube-plates is 19 ft. and barrel has a diameter of 6 ft. 5 ins.[1]

American locomotives, designed for railroads where restrictions such as those imposed by the British loading gauge do not apply, often develop much higher powers. Some of the engines built by the American Locomotive Company and by the Baldwin works for the New York Central and Pennsylvania lines have boilers with more than 5000 sq. ft. of evaporating surface; their superheating surface ranges from 1200 to over 2000 sq. ft., and their grates from 84 to 88 sq. ft. Still larger heating surfaces are provided in "articulated" locomotives of the Mallet type: in some of the 2–10–10–2 engines of this class, probably the largest steam locomotives that have been built, the evaporating surface is over 8600 sq. ft. and the area of the grate is as much as 108 sq. ft.

The British practice of using copper for the fire-box is not followed in America: there the fire-box, like the rest of the shell, is of mild steel.

The locomotive type of boiler is used for portable and semi-portable engines, and to a considerable extent for stationary engines of small power. It has also been used in marine practice in cases where lightness is of special advantage, as in torpedo craft, but it has now given place on board ship to boilers of the water-tube type.

278. Cylindrical marine or Scotch boiler. The boiler that is most commonly used in the mercantile marine is a short circular

[1] *Proc. Inst. Mech. Eng.* May 1925.

horizontal cylinder of steel, closed by flat plates at the ends, with internal furnaces in cylindrical flues, internal-combustion chambers, and return fire-tubes above the flues. This type is called the Scotch boiler. Very often it is double-ended with furnaces at both ends of the shell, each pair leading to a combustion chamber in the centre that is common to both, or to separate central-combustion chambers with a water space between them.

Fig. 238 shows a double-ended marine boiler of this class. At each end there are three furnaces in corrugated flues. The use of corrugated plates for flues, due originally to Mr Fox, makes thin flues able to resist collapse, and allows the flues to accommodate themselves easily to changes of temperature. In this example one combustion chamber is common to each pair of furnaces. It is strengthened on the top by girder stays and on the sides by stay-bolts to the neighbouring chamber and to the shell. A certain number of the tubes (called stay-tubes) have screwed ends to make them serve effectively as stays between the tube-plate of the combustion chamber and the front of the boiler. The upper part of the front plate is tied to the opposite end of the boiler by long stays extending from end to end above the combustion chambers. The uptakes from both ends converge to the funnel base above the centre of the boiler's length. The boiler shown is one of a pair, which lie side by side in the vessel, the uptake at each end being common to both. Each boiler in this example has a steam-drum, but that feature is now generally omitted.

The following particulars of a double-ended Scotch boiler, by Messrs Beardmore, for a working pressure of 220 lb. per sq. inch, will serve to illustrate modern practice. The diameter of the shell is $16\frac{1}{2}$ ft. and its length 22 ft. At each end there are four furnaces in corrugated furnace flues, each with a mean diameter of $3\frac{1}{2}$ ft. and a length of 7 ft. 8 ins. Each of the eight furnaces opens into a separate combustion chamber. The tubes have an external diameter of 3 ins.: there are 800 of them in all, of which 244 are stay-tubes with screwed ends. They contribute 4860 sq. ft. to a total heating surface of 6280 sq. ft. The boiler is oil-fired; if grates were used their aggregate surface would be 147 sq. ft., or rather less than one-fortieth of the heating surface. The boiler is entirely of steel, except the tubes, which are of lap-welded wrought iron. The shell plates are $1\frac{5}{8}$ ins. thick; the end plates (in the steam space) $1\frac{1}{4}$ ins.; the tube-plates $\frac{15}{16}$ ins., and the corrugated flues $\frac{5}{8}$ ins. The

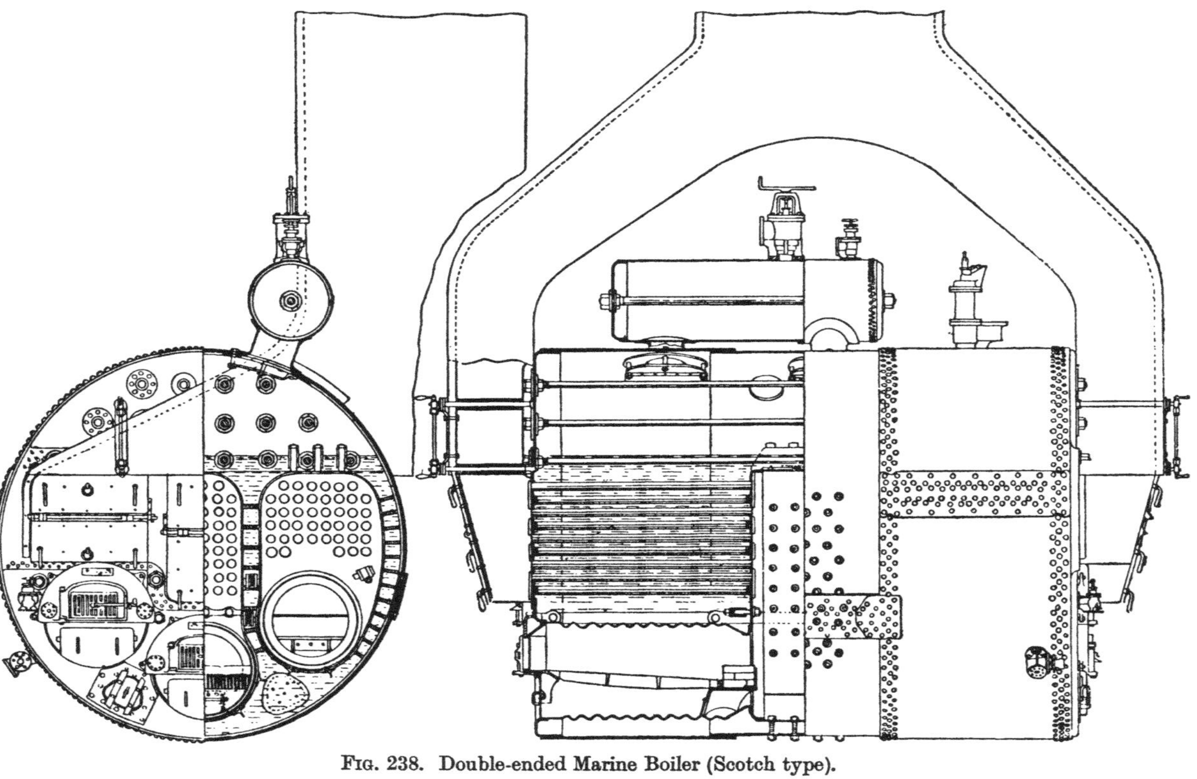

Fig. 238. Double-ended Marine Boiler (Scotch type).

crowns of the combustion chambers are stiffened by girder stays. In the space above the chambers there are 18 long solid steel stay-rods $3\frac{3}{8}$ ins. in diameter tying the end plates together: they have screwed ends and hold each plate between a pair of nuts.

The single-ended marine boiler is practically half a double-ended boiler. The furnace doors are at one end only, and the boiler terminates in a flat end plate which leaves only a few inches of water space between it and the back of the combustion chambers, the end plate and the back plate of each chamber being tied together across this space by short stay-bolts.

With but few exceptions the boilers used on shipboard are either of the Scotch type or of one of the water-tube types which have still to be described. In merchant vessels the Scotch boiler is still the most usual form, but in war ships, where saving in weight and rapidity in steam raising are specially important, the use of water-tube boilers has become universal.

279. Vertical boilers. In the boilers which have been referred to the axis of the cylindrical shell is horizontal. But the cylinder

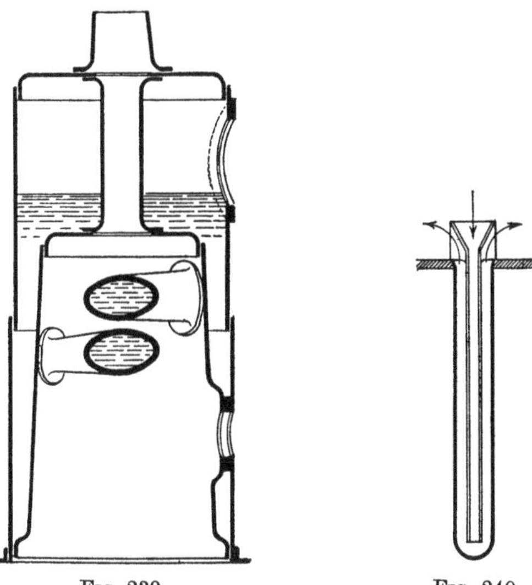

Fig. 239. Fig. 240.

may be turned up on end and the boiler take a vertical form, the grate of course remaining horizontal and forming the floor of a fire-box to which access is given by a door in the side of the cylindrical shell. This type is a very usual one for boilers of small power. It has the advantage of needing no brick setting and of occupying little floor space, with the drawback that the free surface of the water from which the steam rises is comparatively small and consequently the steam rises with a higher velocity, which increases the risk of priming. There are many designs of vertical boilers, some with multitubular flues for the hot gases, and others with

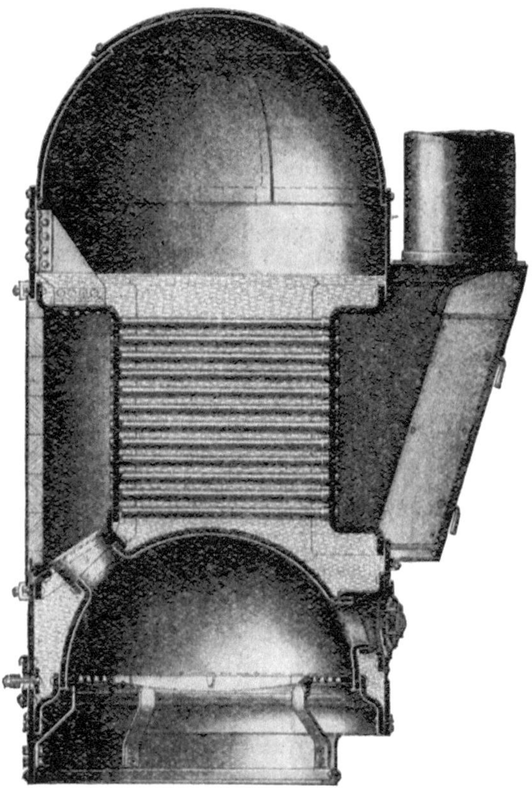

FIG. 241. Cochran Boiler.

water-tubes. Fig. 239 shows a small vertical boiler with Galloway tubes across the upper part of the fire-box, but for efficient action a larger heating surface is required. In some forms the heating surface is increased by water-tubes which hang from the crown of the fire-box and are closed at the lower end, circulation of water being maintained in them by means of a partition in the form of an inner tube inside of which water flows down to allow an upward movement of water and steam to be maintained between the inner tube and the outer. Tubes of this kind, closed at one end and provided with an inner pipe to secure circulation of the water, are called Field tubes. A section of a Field tube is given in fig. 240, with arrows to indicate the manner in which the water circulates. A simple water-tube with a closed end forming a pocket would be quickly burnt out through being left dry if no provision were made for systematic circulation, but the inner tube of the Field combination makes the provision that is required.

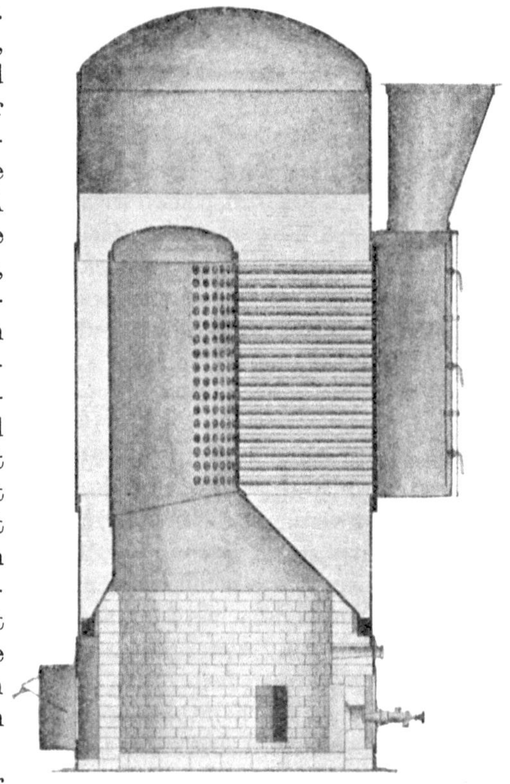

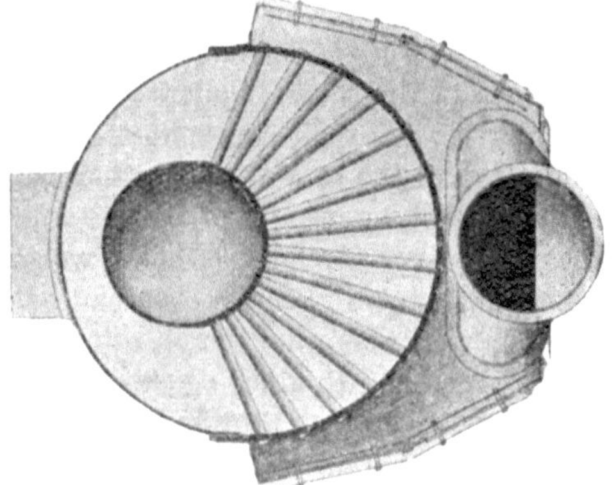

FIG. 242. Blake-Beardmore Boiler.

A vertical boiler much used on land and as an auxiliary at sea is Cochran's, in which most of the heating surface is provided by horizontal fire-tubes (fig. 241). The tube-plates are flat and extend from back to front in the cylindrical shell: the whole nest of tubes occupies a nearly cubic space below which is the hemispherical dome of the fire-box. The top of the shell is dished or hemispherical. The tubes serve as stays between the tube-plates and scarcely any other staying is required. The construction is simple and gives easy access for inspection. A boiler of this class 15 ft. high and 7 ft. in diameter has a heating surface of 600 sq. ft.

The Blake-Beardmore vertical boiler is another simple design with horizontal fire-tubes. Its arrangement will be obvious from fig. 242. All staying is dispensed with; the combustion chamber as well as the shell have curved surfaces throughout. There are no flats and the tubes do not have to act as stays. The lofty wet-back combustion chamber makes this boiler specially suitable for oil firing (as in the example illustrated). It furnishes a capacious volume for the flame and a large surface for absorption of the radiant heat.

280. Water-tube boilers. Many forms of boiler have been designed in which the water is inside the tubes, and its circulation is kept up by virtue of differences in average density which arise mainly because steam is continuously forming in the tubes. In ordinary boilers the circulation is more or less casual; when a bubble of steam is detached from any part of the heating surface its place is taken by water which may come in from any side. In a properly designed water-tube boiler the circulation is systematic; water enters each of the tubes at one end and passes through in a continuous stream, becoming partly converted into steam as it goes. The tubes generally deliver into a separating vessel, from the upper part of which the steam-pipe takes its supply, while water collects in the lower part to be returned by gravity to the lower end of the tubes. Boilers of this type can be constructed so as to have, with their contents, a relatively small total weight in proportion to the rate at which they can make steam, which is a distinct merit in respect of marine and especially of naval use. They also carry a relatively small weight of water, which makes them quick steam raisers. They can be readily forced to give on emergency an abnormally high rate of evaporation. Their design is easily modified to allow of effective superheating. For erection in confined situations they have the advantage that they can be brought together in small pieces. Further they are easily made strong enough to resist very high pressure, owing to the absence of any large shell: and there is scarcely any limit to the practicable size. For these reasons, especially the last two, water-tube boilers have now become the ordinary means of raising steam in power stations on the largest scale.

281. Babcock and Wilcox boiler. A successful example of this type, largely used both on land and on board ship, is the Babcock and Wilcox boiler, of which fig. 243 shows a form for use on land. The heating surface is almost wholly composed of a series of straight inclined tubes up which water circulates in parallel streams. These are joined at their ends by wrought-steel con-

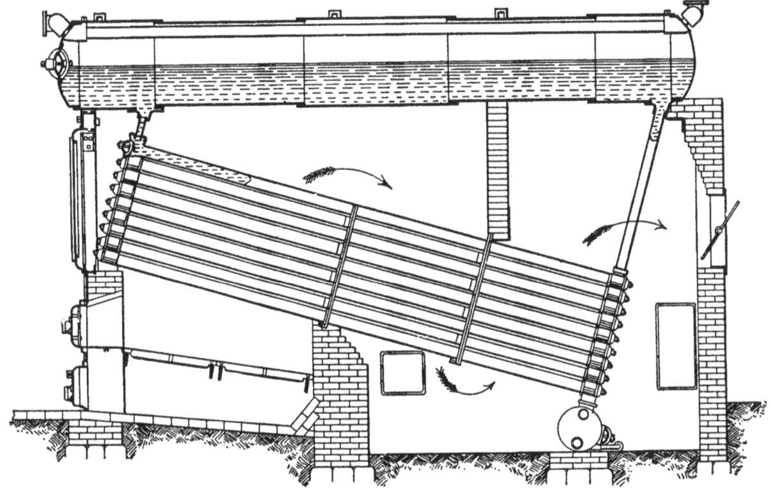

FIG. 243. Babcock and Wilcox Boiler.

necting boxes or "headers" to one another and also to a horizontal drum on the top, in which the mixture of steam and water which rises from the tubes undergoes separation. At the lowest point of the boiler is another drum for the collection of sediment. The route taken by the hot gas is indicated by arrows in the figure: it passes three times across the tubes. When the boiler is in action a regular circulation of water is established down from the drum to the rear header, up the sloping tubes in which evaporation takes place, and back by the front header to the drum. A superheater, consisting of a set of small **U**-shaped or **W**-shaped tubes, is often added, in the triangular space between the main tubes and the top drum. There the hot gases, having lost some of their heat by one passage across the nest of water-tubes, are at a suitable temperature for acting on the superheater tubes before they make their second and third passages across the main nest.

For use on land this boiler generally has tubes about $4\frac{1}{2}$ inches in diameter. In marine forms smaller tubes are often used, a group of four tubes each about $1\frac{3}{4}$ inches in diameter taking the place of a single large tube. The tubes in the marine type (which is also much used on land) slope up towards the back instead of down;

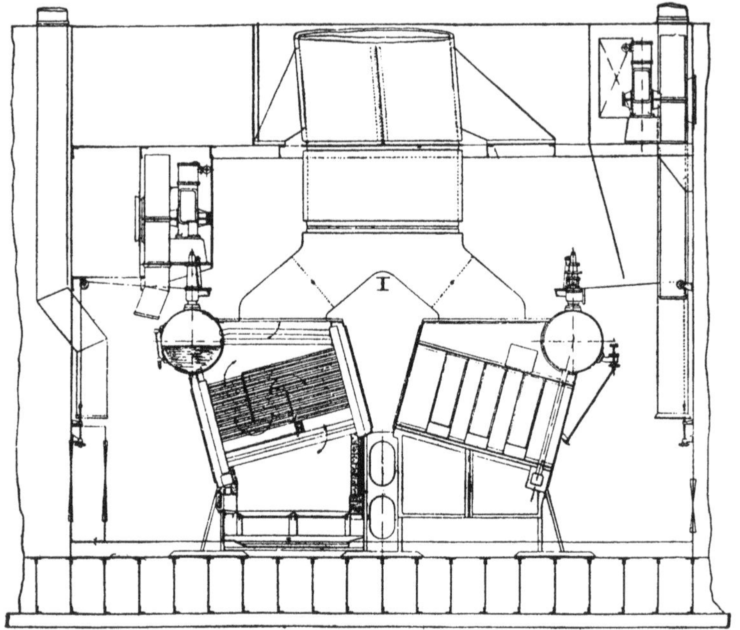

FIG. 244. Babcock and Wilcox Marine Boilers in place on board ship.

the steam-drum extends transversely across the top of the boiler, in front, and a group of nearly horizontal tubes serves to connect it with the top ends of the back headers. Fig. 244 is a longitudinal section through part of a steamship showing a pair of Babcock and Wilcox boilers of the marine type in place in the stokehold. The fire-doors are under the lowest part of the sloping tubes.

Another design for marine use is shown in fig. 245. There the superheater is placed in a gap made for the purpose in the main bank of tubes, so that the gases may reach it while they are still sufficiently hot to give a high degree of superheating. The U-tubes of the

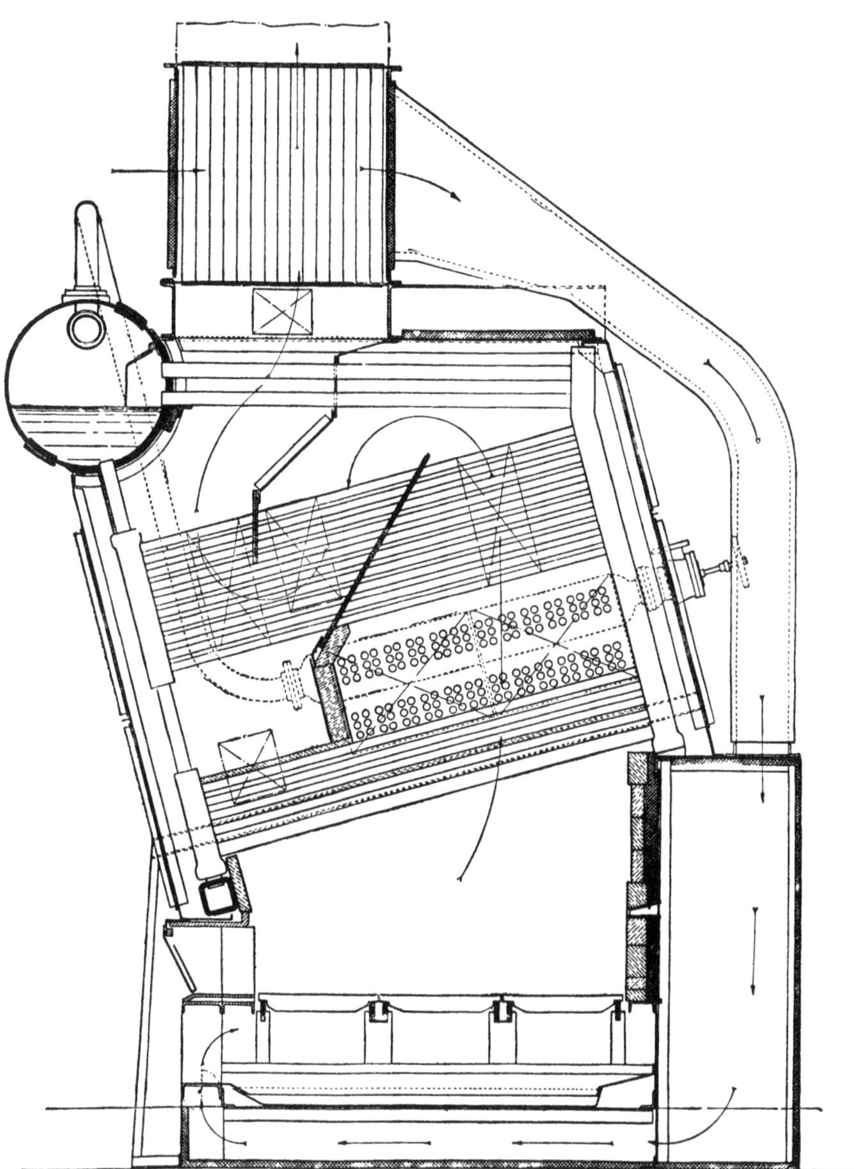

FIG. 245. Babcock and Wilcox Marine Boiler, with superheater and air heater.

superheater are seen in section; they lie across the main tubes. On top is an air heater consisting of a group of vertical tubes through which the gases escape to the chimney; the air entering the furnace crosses these, taking heat from them, goes down a channel behind the boiler and the combustion chamber, and passes on beneath the ashpan to the space below the grate. In this example coal is burnt on an ordinary fixed grate. Often, however, such boilers are fitted for burning oil, and also, in land installations, for use with mechanical stokers or with pulverized fuel.

The Babcock and Wilcox boiler finds very wide application both ashore and afloat. It can readily be made in large sizes, and for the highest pressures. It is the type of steam generator most used in great power stations, where pressures of 600 lb. or more are not uncommon. The Weymouth Station of the Edison Company of Boston, referred to in § 154, has a boiler of this type, working at a pressure of 1200 lb. per sq. inch, in which the steam-drum is of seamless forged steel, 4 ft. diameter inside and 4 ins. thick.

The Vickers-Spearing boiler closely resembles the Babcock and Wilcox except that the header from which water enters the sloping tubes takes its supply from a water or "mud"-drum at the bottom, protected from the action of the fire. The water descends from the under side of the steam-drum to the mud-drum through one or more external downcomers. One of these boilers appears in fig. 257 of § 291, below[1].

282. Stirling boiler. Another water-tube generator in very extensive use is the Stirling boiler, which has taken somewhat varied forms in the hands of several makers. In this boiler the heating surface is supplied almost wholly by water-tubes, the general direction of which is not far from vertical. One usual arrangement is illustrated in fig. 246. There the main tubes connect three steam-drums above with a water-drum below. At the back end the tubes stand in a nearly vertical position: at the front they slope steeply over the fire, providing a lofty combustion chamber. They are grouped in three banks, with fire-brick baffles between, which make the hot gases take a zigzag course over the tubes. In addition to the main tubes there are short cross tubes, lying more

[1] For particulars of steam generation on a large scale see papers by Jacobus, Monroe, Kemnal, Junkersfeld and Orrock, and others in *Trans. of the First World Power Conference* 1924, vol. II. Also Kemnal, *Inst. of Engineers and Shipbuilders in Scotland*, Feb. 1926; Robson, *Min. Proc. Inst. C.E.* Vol. 220, p. 129.

or less horizontally, which connect the three steam-drums with one another. The feed-water is delivered to the backmost steam-drum and has first to go down the backmost bank of tubes, for there is no other exit below the water line. The tubes are straight throughout the greater part of their length, but curve with easy bends towards each end so that they may enter the drum in a

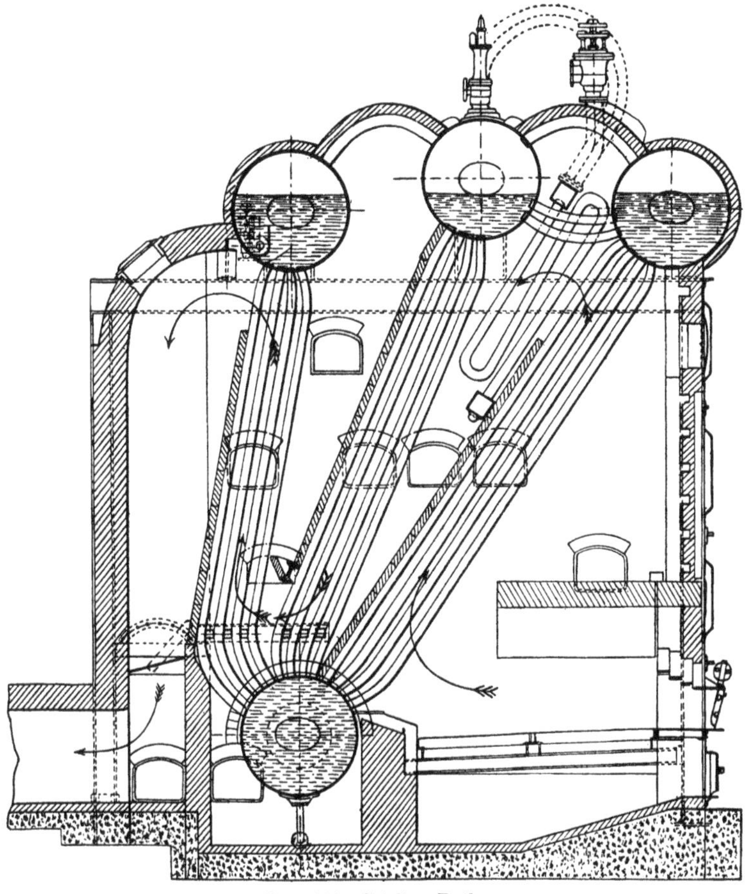

Fig. 246. Stirling Boiler.

radial direction. The ends are expanded into the drum. The design is well adapted to save the structure from racking through changes of temperature. The tubes have a diameter of about $3\frac{1}{4}$ inches. Circulation takes place up the front bank of tubes and also up the forward tubes of the second bank, and down the remainder. The backmost bank are in effect feed-water heaters rather than steam generators. Steam is taken from the middle drum and passes, in this example, through a superheater between the first and second bank of tubes.

In many Stirling boilers there are five drums, three above and two below, with four banks of tubes between them forming the legs of a **W**. The water then circulates down the fourth bank of tubes, up the third, down the second, and up the front bank, where the formation of steam is of course most active. There are cross-connecting tubes between the lower as well as between the upper drums. In another form there are only three drums, two above and one below, but the single water-drum has a partition making it in effect serve as two.

Water-tube boilers of a more or less similar type but with straight tubes are made by John Thompson, Ltd., and others. In these the drums, at the places where the tubes enter, are generally pressed out of the cylindrical shape to provide flat surfaces to receive the tubes, perpendicular to their direction. In Thompson's boiler these flats take the form of small steps or wart-like bulges. In such boilers there are as many drums above as below, usually two or three of each, and the same number of banks of tubes. Each lower drum is simply connected by a bank of straight tubes with one upper drum. The lower drums, as well as the upper, communicate among themselves by short curved connecting-tubes.

283. Belleville boiler. This very different form of water-tube boiler was at one time extensively used in British as well as French war ships. Since 1902, when it was unfavourably reported on by an Admiralty Committee, it has not been put into ships of the Royal Navy. In the Belleville boiler the tubes are grouped in sets, each set forming a flattened helix through the whole length of which the water rises from the sediment chamber to the separating drum. The tubes, about $4\frac{1}{2}$ inches in diameter, slope up with a gradient of about 1 in 25, alternately to left and right, forming a zigzag of straight lengths, which are made continuous by junction boxes. Each set or section has in all a length of about 150 feet through which the current of steam and water passes from end to end. Eight or more such sections stand side by side to make up the complete boiler. A non-return check-valve at the bottom assists in preventing the flow from taking place in the wrong direction. When the boiler begins to make steam the circulation occurs in a series of gusts, the check-valve closing while each gust makes its way up through the zigzag of tubes.

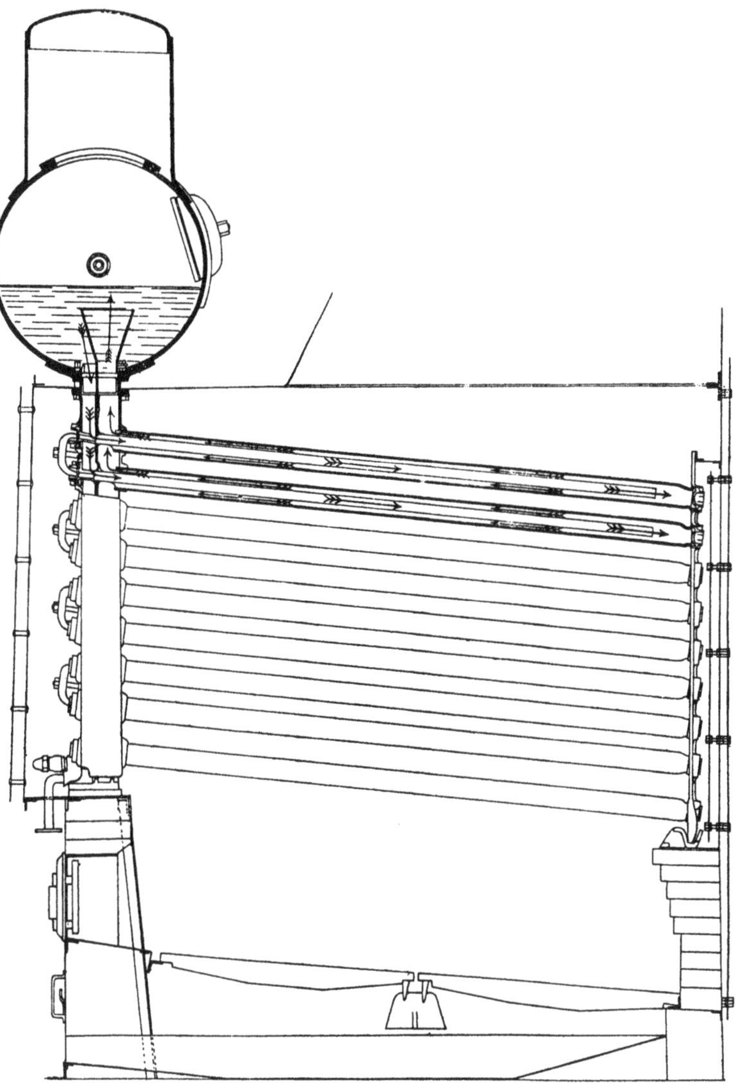

Fig. 247. Niclausse Boiler.

284. Niclausse and Dürr boilers. These are two forms much resembling one another, in both of which the heating surface is furnished by inclined "Field" tubes. The Niclausse boiler is illustrated in fig. 247, and fig. 247 *a* is a section on a larger scale through two tubes and headers. The water comes down from the drum at the top through the front header and is fed to each tube through the inner supply pipe. After evaporation the stream of mixed steam and water rises to the drum through the somewhat wider header nearer the tubes, which is separated by a partition from the down-coming header. The Niclausse boiler has been extensively used in the French Navy as well as on land[1]. The Dürr, a German naval boiler, is generally similar but differs in header details.

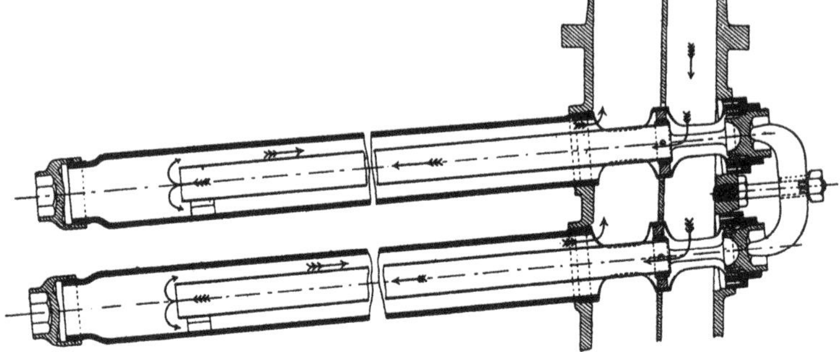

FIG. 247 *a*. Tubes of Niclausse Boiler.

285. Thornycroft, Yarrow, and White-Forster boilers. These boilers, all originally designed for marine use in high-speed naval craft where lightness and quick steam-raising are essential, but now much more widely applied, may suitably be grouped together, for they have this in common that the heating surface is provided essentially by banks of small water-tubes placed, like the legs of an inverted **V**, between a steam-drum which forms the apex and two parallel water-drums at the feet. There is no cross-connexion between the water-drums. The whole makes a sort of arch or short tunnel within which is the fire. It is a design well adapted for heating by radiation. The hot gases pass out sideways and upward through both banks into a casing which converges on a chimney above. A boiler of the same general type was developed in France by Normand. In the earliest examples, introduced by

[1] Niclausse, *Proc. Inst. Mech. Eng.* July 1914.

Sir John Thornycroft about 1885, the tubes were much curved, and entered the upper or steam-drum above the water level: they now always enter it below the water level, and in most modern designs the tendency is towards straight or nearly straight tubes. The use of straight tubes, a feature which makes for simplicity and facilitates convection, was from the first characteristic of Sir Alfred Yarrow's designs. In the White-Forster boiler the tubes are moderately curved to a constant radius, which is the same for all.

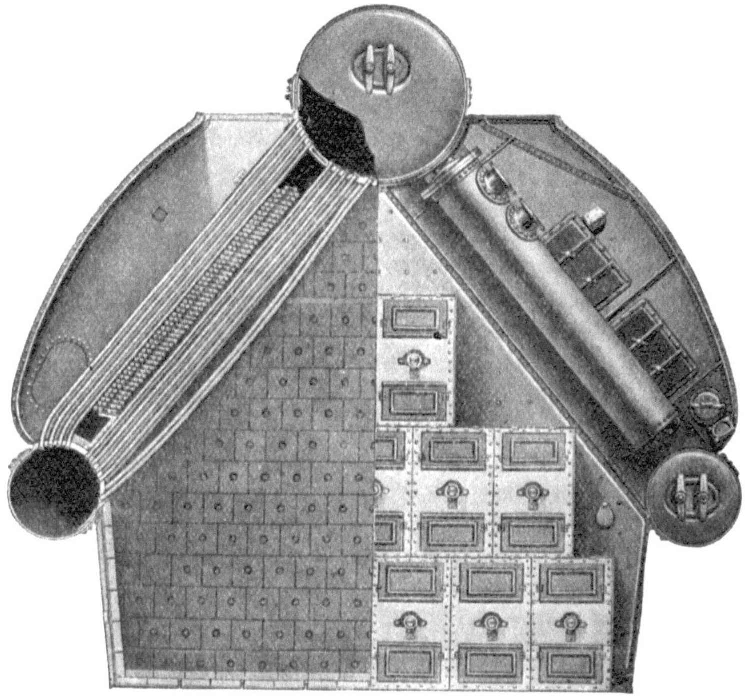

FIG. 248. Thornycroft Boiler.

A recent example of the Thornycroft boiler is shown in fig. 248. The tubes nearest the fire are slightly curved: the others are straight except for a little bending near the lower end, where they enter the comparatively small water-drums. In early boilers of this type large tubes were provided at the ends of the boiler, outside the action of the fire, to serve as downcomers, but these are now dispensed with, for it is found that the circulation establishes itself readily enough without them, the water passing down some of the small tubes and up the others. Superheaters are fitted on both sides, in gaps formed for that purpose, among the water-tubes: each of the two is made up of U-shaped tubes lying horizontally, grouped in parallel, with their ends connected to headers which are formed by having a partition in a cylindrical chamber in front of the boiler. The boiler shown uses oil-fuel, which is sprayed into the combustion chamber through a number of nozzles at three different levels. Boilers of this class are made with heating surfaces that range from 300 to 9000 sq. ft.

Naval types of Yarrow boiler, also fitted for burning oil-fuel, are illustrated in figs. 249 and 249 *a*. If made without a superheater, the boiler would have a symmetrical form; but when a superheater is fitted it is usually placed on one side only, as in these figures, and fewer water-tubes are put on that side than on the other. The chimney is divided, and a damper in it controls the flow of the gases on the superheater side. The headers of the superheater are formed by having a longitudinal partition in a separate superheater drum. The water-tubes are completely straight, except (fig. 249) for the two rows nearest the fire. Some of them consequently enter the water-drums rather obliquely, but this is not found to cause inconvenience when, in the drilling of the holes, "jigs" are used which guide the drills to penetrate the tube-plates in the right direction. Formerly the water-drums of Yarrow boilers had a D-shaped section to provide a more or less flat surface for the reception of the tubes, but now a circular section is preferred. In all such boilers the drums have a comparatively thick plate on the side from which the tubes come. When much superheat is desired, the bank of water-tubes on one side of the boiler is divided into two parts, each with a separate water-drum, and the superheater occupies the space between, as in fig. 250.

That figure illustrates one of the forms taken by the Yarrow boiler when designed for use on land. The grate is low and its operating front is at the side of the boiler tunnel, instead of at the end. Incidentally the figure illustrates the use of a chain-grate, one of the kinds of mechanical stoker which will be referred to later (§ 296). There is also in this example a preheater for the air: on each side, above the boiler, there is a stack of tubes through which the gases flow on their way to the chimney, and the cold air, which is blown in through an orifice between these, passes across them and down through flues which admit it to the fire.

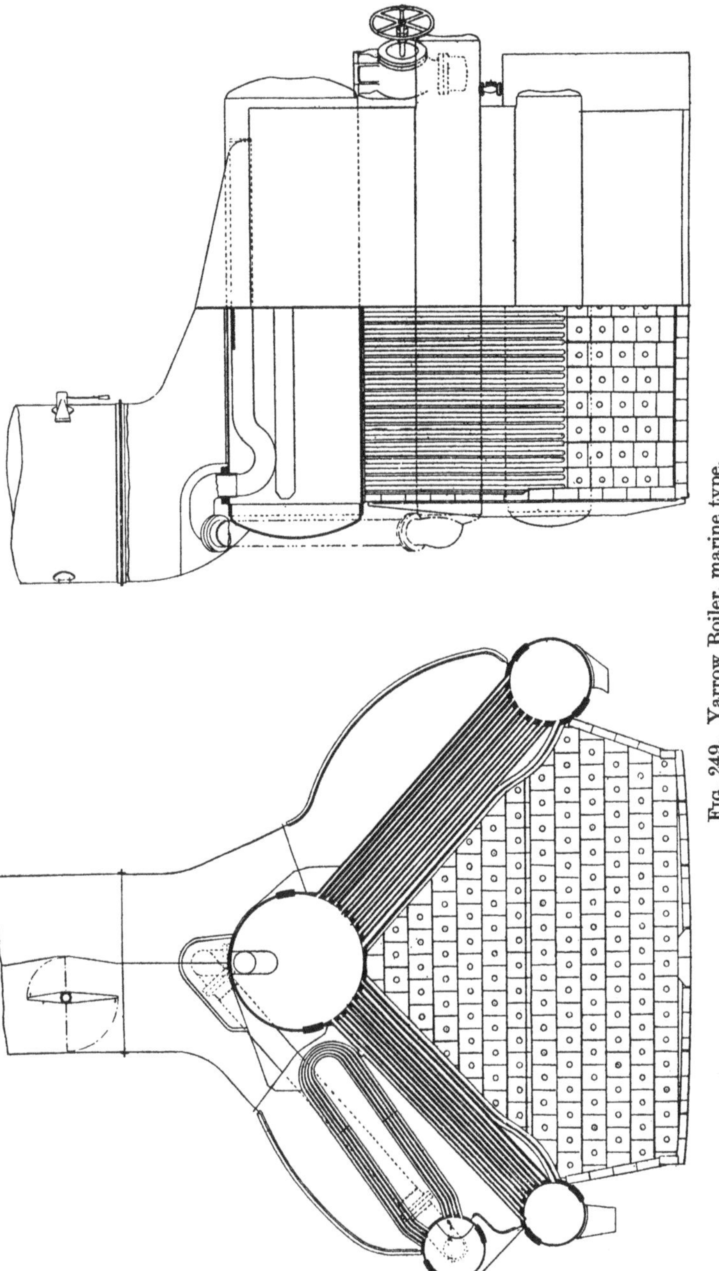

Fig. 249. Yarrow Boiler, marine type.

Boilers of this class lend themselves well to forcing. A Yarrow boiler in a torpedo-boat destroyer may be made to generate steam enough for 12,000 horse-power, equivalent to the evaporation of about 25 lb. per hour per square foot of heating surface. The form is also well adapted for very high pressures. Yarrow boilers in the Barking power station work at 400 lb. and superheat the steam to 725° F. In a paper communicated to the Institution of Naval Architects in 1926, Mr Harold Yarrow describes boilers under construction for a turbine-driven ship to work at 575 lb. with

Fig. 249*a*. Yarrow Boiler, marine type.

superheat to 750° F. In these the steam-drum is a single forging $3\frac{1}{2}$ feet in diameter inside, with rounded ends which are forged solid with the cylindrical part, so that riveted joints are entirely avoided. Apart from boiler mountings the only bolts used are those that secure two internal manhole doors, one at each end of the drum. The water-drums and superheater drums are cylindrical forgings with one solid end; the other end, which contains the manhole, is riveted on. To suit special conditions in the ship the boiler is unsymmetrical; on one side there is only a comparatively

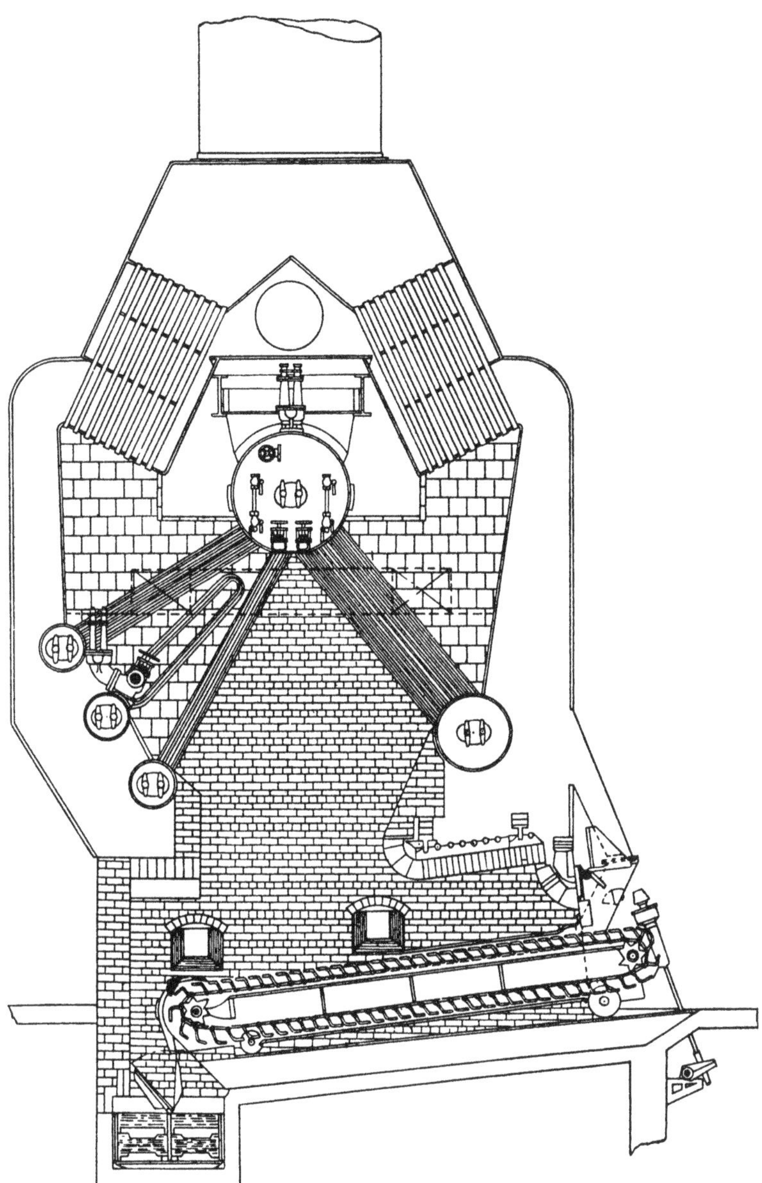

Fig. 250. Yarrow Boiler, land type.

thin bank of tubes, receiving heat by radiation; on the other side there are two banks with the superheater between them, and all the products of combustion go up on that side, passing on their way an air preheater which has a heating surface nearly two-thirds as great as that of the boiler. The design will be clear from fig. 251, which shows the boiler in course of erection, when only a few of the superheater tubes are in place and before a partition is inserted

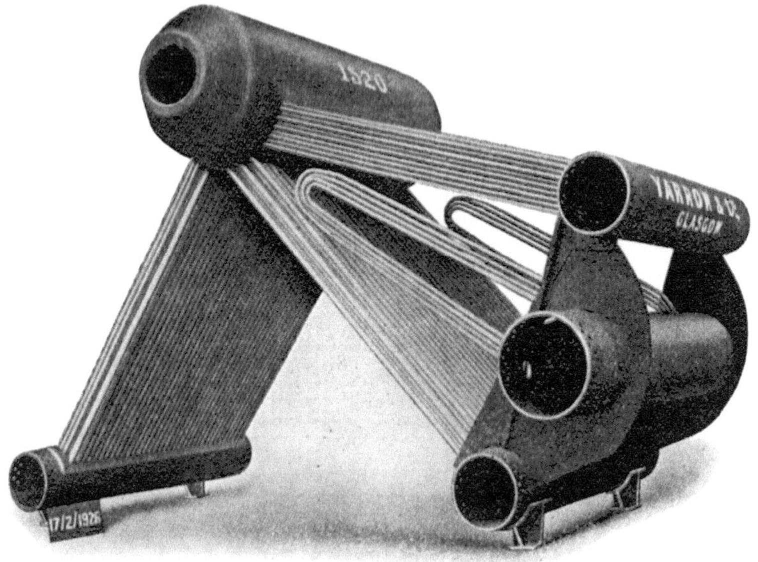

Fig. 251. Yarrow Marine Boiler for high pressure.

in the superheater drum to convert it into two headers. In the same paper Mr Yarrow gives a design for a boiler to work at 1000 lb. per sq. inch and develop 4000 shaft horse-power under moderate conditions of steaming[1].

The White-Forster boiler is illustrated in fig. 252; the longitudinal and transverse sections there show how the tubes, which are all curved to the same radius, are arranged; the figure also shows how a tube or a cleaning-brush may be inserted through the manhole.

[1] H. E. Yarrow, *Trans. Inst. Nav. Arch.* March 1926. His paper in the same *Transactions* for 1912 should also be referred to for particulars of tests of a Yarrow boiler.

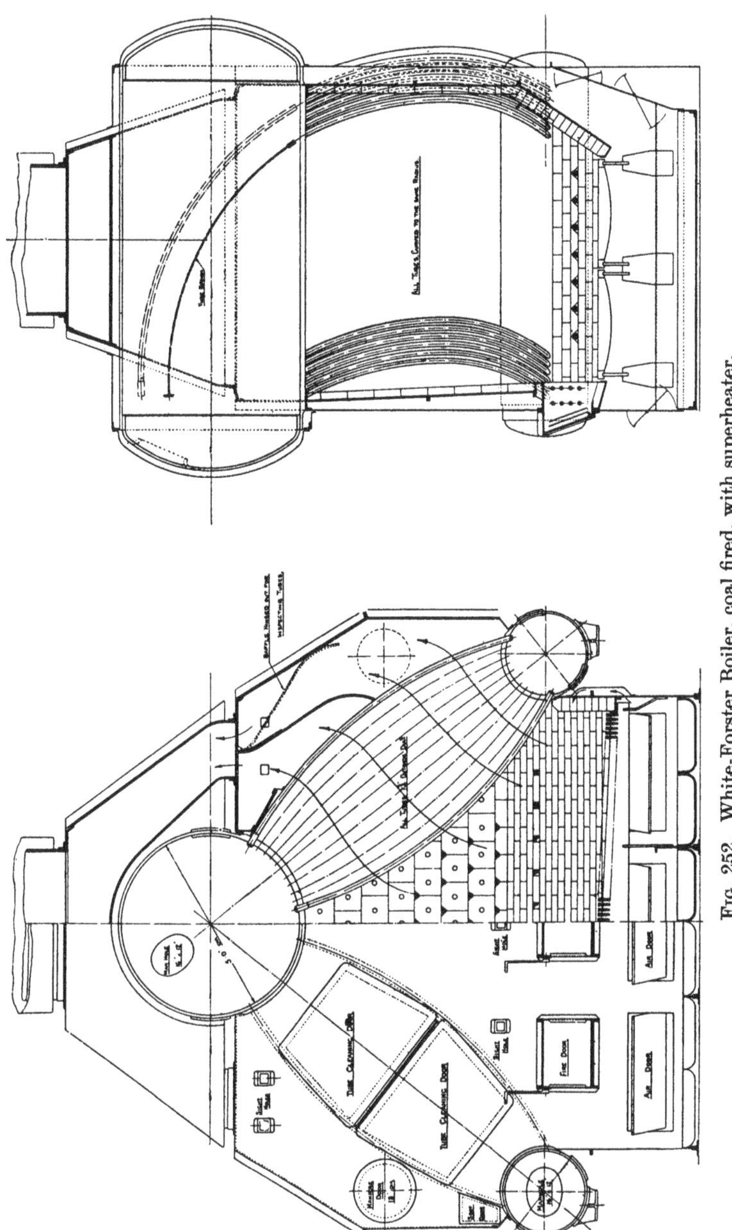

FIG. 252. White-Forster Boiler, coal fired, with superheater.

In this example a grate is provided for coal burning. There may be a superheater in two parts, one on either side, above the tubes; but for higher degrees of superheat it would be placed amongst them. Boilers of this kind are often fitted for oil firing; they are made in large sizes for both marine and land use.

286. Superheaters. From the examples of superheaters which have been mentioned in describing boilers, it will have been seen that they usually consist of a large number of tubes of small diameter, grouped in parallel and bent on themselves in U or S or W fashion, with their ends brought into two headers, one of which is supplied with steam from the boiler, while from the other the steam passes on to the engine, having become superheated in passing from one header to the other through the tubes. The superheater may be independently fired, but much more commonly it forms an integral part of the boiler, in the sense that it uses the same furnace, taking heat from the gases after they have traversed part of the boiler's heating surface. If only a moderate amount of superheat is to be given, the superheater may be so placed that the gases complete their passage of the heating surface before they reach it; in that case, however, a feed-water heater or air preheater is particularly necessary, on grounds of economy, to reduce the temperature of the gases further before they are discharged. Usually the superheater is arranged to meet the gases at an earlier stage, but after they have already parted with some heat, and in this way any amount of superheat that is desired may be given.

In Cornish or Lancashire boilers the superheater is generally placed at the after end of the boiler, where the gases after traversing the internal furnace flue or flues have still to give up more heat to the bottom and the sides. Provision may be made for protecting the superheater tubes from becoming overheated while little or no steam is passing through them, by diverting the course of the gases, but this is not in general necessary, for the bent form of the superheater tubes enables them to suffer wide variations of temperature without racking. In locomotives they are, as fig. 237 illustrates, placed in some of the fire-tubes of the boiler, made larger than the rest in order to accommodate them. Similarly in the Scotch marine boiler the superheater tubes may be placed within the fire-tubes, and each superheater tube may thread several fire-tubes backwards and forwards in its course from header to header, the two headers being close together in front of the boiler. The various positions that superheaters may occupy in water-tube boilers have been sufficiently exemplified. To a great extent superheaters are protected from over-heating by the fact that violent firing occurs only under conditions of rapid delivery of steam. Sometimes, as in a locomotive boiler or an unsymmetrical boiler of the Yarrow type, it is only part of the furnace gas that acts on the superheater, the remainder acting only on evaporating surfaces.

287. Supply of water to boilers. Feed-pumps. Feed-water is forced into boilers either by a feed-pump or an injector. In some steam plants the feed-pump is driven by the engine; this is often done, for example, in reciprocating marine engines, but in many cases an auxiliary engine is used for the purpose. The most common forms of steam-pump for feeding boilers are direct acting; the steam-cylinder and pump-cylinder are in line with one another, a single piston-rod serving for both. Steam is admitted for the whole, or nearly the whole, of the stroke. Such pumps are often arranged to form a "duplex" combination by having two placed side by side, connected so that the piston of one actuates the valve of the other. In the Worthington duplex pump, for example, each of the two piston-rods is connected to an oscillating lever which gives motion to the slide-valve of the other cylinder. By this means steam is admitted to make one piston begin its stroke just as the stroke of the other is about to be completed, with the result that a smooth and continuous action is secured. In several other steam-pumps the arrangement is "simplex" and the valve receives its motion from tappets on the piston-rod. Generally this is done through a steam relay, the direct effect of the tappets being to move an auxiliary steam-valve which then causes the main steam-valve to be thrown over by admitting steam to one or other of its ends which are shaped so that they serve as pistons to enable the action to take place. In Weir's feed-pump—a much used example of this class—there is a regulating device by which the steam may either be cut off some time before the stroke is completed or admitted until the end. The adjustment of this secures silent running. The auxiliary valve slides on the back of the main valve, parallel to the piston, and the main valve moves transversely to this when it is thrown over by the action of the steam.

The widespread adoption of water-tube boilers in modern

practice has been assisted by the development of trustworthy automatic feed regulators. Without such devices the keeping of a constant water level would need incessant attention, owing to the small volume of water in the boiler and the relatively high and often variable rate of evaporation. The principle on which they usually act is that a float, placed in a separate chamber connected with the boiler so that it may rise or fall with the water line, controls a small valve which opens or closes a leakage hole for water delivered under pressure by the feed-pump. The check-valve, through which feed-water is forced into the boiler against the pressure of the steam, has a piston attached to it which prevents it from opening when this leakage occurs. The effect is that the feed automatically ceases when the water level rises above the normal, and is resumed when it falls.

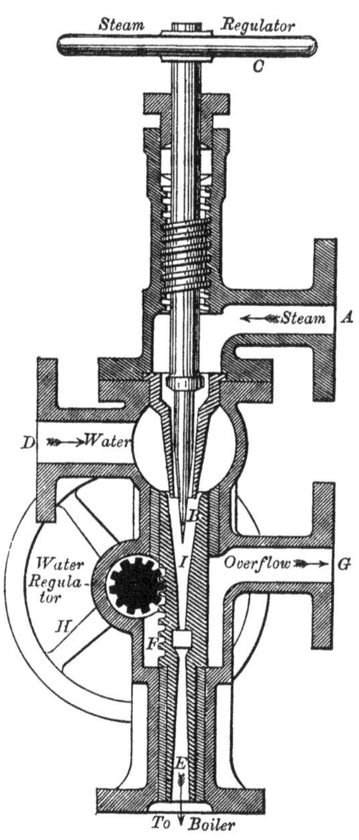

FIG. 253. Giffard's Injector.

288. Injectors. The injector, invented about 1858 by Giffard, is a form of jet pump in which steam is employed to set a stream of water into rapid motion and the momentum of the stream is then utilized to overcome the pressure which opposes its admission to the boiler. It is used almost universally on locomotives and largely on land boilers, where it often serves as an auxiliary to a feed-pump.

To explain the action of the injector we may refer to fig. 253, which illustrates an early form. Steam enters at A and blows through a nozzle at B, the amount of opening for the steam being regulated in this example by means of a conical spindle which can be drawn back by turning the hand wheel C. When B is opened slightly the first effect is that the steam jet emerging from B drags the surrounding air along with it, producing a partial vacuum in the space round the nozzle. The feed-water is allowed to enter this space, through D. It mixes with the steam in the "combining" nozzle I, and condensation of the steam takes place. The momentum of the condensing jet is given up to the water, which consequently streams through I at a high velocity. Beyond the combining nozzle I is the delivery nozzle E, which is divergent. In this, owing to the enlargement of sectional area, the stream loses velocity, and consequently, by a well-known hydrodynamical principle, gains pressure, until at the end its pressure is higher than the pressure in the boiler. It can therefore pass into the boiler, and it does so through a non-return valve which opens to admit it. The orifice F, which in this example is placed in the neck between the combining nozzle and the delivery nozzle, leads to the overflow pipe G. Its function is to allow the injector to start into action by providing a means of escape for the steam and water until the stream has acquired enough momentum to force its way into the boiler. The action in this form of injector depends on a rather nice adjustment of the supply both of steam and of water. The supply of water is regulated by the handle H, which by moving the tube containing the nozzles I and E changes the annular opening through which the water enters, round the point of the steam nozzle.

In modern forms of the injector devices are used which to a great extent do away with the necessity of regulating by hand. In some forms the injectors are classified as *non-lifting*: that is to say, they are designed for use in cases where the water does not have to be drawn up from a lower level and consequently the steam jet does not have to produce a partial vacuum before the water arrives to condense it. This allows the arrangements for steam admission to be simplified, for in a *lifting* injector the initial vacuum is generally obtained by using a finer jet of steam than is afterwards required when the injector is in full action. It is possible to make an injector lift water through as much as 20 feet, but usually there is a much shorter lift, or none at all. Again, the action of an injector is liable to be interrupted by mechanical shocks, air getting in, and so forth, and in such cases the older forms required that the steam should be shut off and the action then restarted by hand. Many modern injectors are arranged as *restarting*, that is to say, they will restart automatically by means of a device which causes the overflow to open freely when the injector fails to act and closes it more or less completely when the action is resumed. Provision is also made in some for an automatic adjustment of the opening through which the feed-water enters, to suit variations in the pressure of the steam.

In a compound injector such as that shown in fig. 254 the work is done in two stages. There is first a "lifter," on the left, which brings in the feed-water and delivers it at moderate pressure into a chamber from which it passes into the "forcer" on the right, which sends it into the boiler. Here, in each case, the combining nozzle and delivery nozzle are in one piece, and provision for overflow comes after the delivery nozzle. Each steam nozzle is divergent—which, as we have seen in Chapter VIII, is the right form. The steam nozzles are closed by valves at the top, which are lifted by turning a handle which raises the rod R and at the same time turns the cock at the bottom which allows overflow to take place first from S_1 and afterwards from S_2 as the injector is starting. The rod R carries a loose cross-piece which opens the valve on the left hand first, admitting steam to nozzle No. 1. The valve on the right hand opens later, by a further movement of the handle, and admits steam to nozzle No. 2. In the first movement, which sets No. 1 in action, there is free overflow for the left-hand injector. The jet of steam flows through freely, creating a sufficient vacuum to bring in the water,

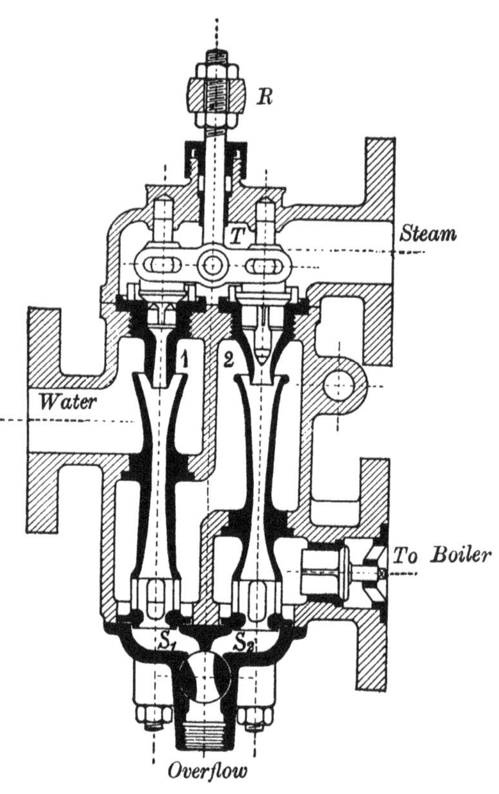

Fig. 254. Compound Injector.

and when that appears at the overflow the handle is turned further, which opens the steam-valve of No. 2 and also closes the overflow S_1 of the left-hand injector or lifter. Finally, when the forcer (No. 2) has come into action a further movement of the handle in the same direction closes the overflow S_2 as well, and delivery to the boiler then goes on through a check-valve. This combination of two injectors in series makes it possible not only to have a high lift, but also to deliver water to the boiler at a high temperature, owing to the relatively high pressure at which water is supplied to the "forcer."

A good example of an automatic restarting injector is Holden and Brooke's, shown in fig. 255, which also illustrates a device for adapting the instrument to work at any one of a wide range of possible boiler pressures. For this purpose an adjusting handle is provided at the top, with a pointer which reads on a graduated circular rim on which various steam pressures are marked. By turning this handle the amount of steam opening at A is adjusted simultaneously with the amount of water opening at B, for the steam nozzle between the two is raised or lowered through the action of a screw, in a manner which will be clear from a study of the figure. When the steam nozzle is raised the steam opening is contracted, between the top of the steam nozzle and the conical point of the screw itself, which preserves a fixed level. At the same time the water opening at B is increased. This adapts the injector to work with steam at a higher pressure. On the other hand, to set it for a lower pressure the handle is turned so as to increase the steam opening and by the same movement to decrease the water opening. The handle having been suitably set, steam

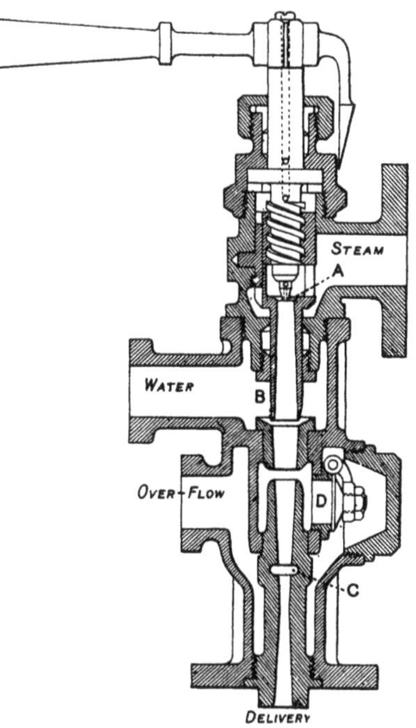

FIG. 255. Restarting Injector.

may be turned on by a separate valve. The free blow-through which is required to start the action (or to restart should it for any reason fail) is provided by having a gap in the combining nozzle with a hinged flap at D which opens to the overflow when steam is passing. When the water arrives and condensation begins the partial vacuum produced in the combining nozzle causes this flap to close, leaving only the small permanent overflow opening C. Another substantially equivalent device is to make one side of the combining nozzle itself in the form of a hinged flap, which opens to allow free escape of steam in the initial stage and then closes, contracting the nozzle to its proper form, when water begins to pass. Another device is to have part of the combining nozzle slide axially in such a manner as to open a free way to the overflow until the stream of water is established, after which it returns to its normal place.

Considered as a heat-engine the injector is far from efficient, for the work done in the delivery of the water is something like one per cent. of the energy contained in the steam. But from another point of view its efficiency as a boiler feeder may be described as perfect, for (if there is no overflow) it returns to the boiler all the heat it takes from the boiler. It is in fact a feed-heater, as well as a feed-pump, and though it converts into work very little of the heat which it takes in, none of the rejected heat is lost, for all goes to heat the feed.

The *exhaust-steam injector* works by steam from the exhaust of non-condensing engines, instead of taking live steam from the boiler. The steam orifice is then larger in proportion to the other parts, the volume of the steam-supply being greater. It affords a useful means of restoring to the boiler heat which would otherwise be wasted. Where delivery has to take place against a high boiler pressure a small supply of live steam may be required to aid the exhaust steam in working the injector, and in some cases there is a compound arrangement in which the exhaust steam drives an injector which delivers water to a second injector using live steam.

Reference was made in Chapter VIII to the use of the injector as a jet pump for maintaining the vacuum in a condenser, and to the high vacuum obtainable by means of a combination of two such pumps in series.

289. Feed-water heaters. The first motive in heating feed-water before it is delivered to a boiler is to save what would otherwise be waste heat. Thus with a non-condensing engine, the feed-water may be heated by letting it take up part of the heat left in the exhaust steam. If there are non-condensing auxiliaries associated with a condensing engine plant their exhaust steam can be utilized in a similar way to warm the main boiler feed. Or again the chimney gases, after giving up as much heat to the boiler as it will take from them, may have more heat extracted when they are brought into contact with pipes through which the colder feed-water flows on its way to the boiler.

Apart from the question of thermal economy there are incidental advantages in heating the feed-water. By bringing it to a temperature nearly as high as that of the steam some risk of racking stresses is escaped. Further, heating the feed-water removes air, and in cases where the water contains solid matter in solution of a kind less soluble in hot water than in cold it is advantageous that these should be deposited in an outside vessel by heating the water there, rather than on the surface of the boiler itself where they might form a scale difficult of removal. Accordingly, where other means of heating the feed are not available, it is sometimes useful to heat it even by taking live steam from the boiler for that purpose.

Green's "economizer" is a feed-water heater largely employed in stationary plants, where the water is heated by cooling the chimney gases to a temperature considerably lower than that at which they leave the heating surface of the boiler. It consists of a stack of vertical pipes through which the feed-water is pumped on its way to the boilers. These are placed in the flue between the boilers and the chimney shaft, and a by-pass is provided through which the hot gases can be diverted if necessary if for any reason the feed-heater is not in action. There is a tendency for soot to deposit on the outside of the pipes, and to prevent this scrapers are provided consisting of loose collars which are kept slowly moving up and down to clean the surfaces.

Any heater of this type is best arranged on the "contraflow" principle: that is to say, the general direction in which the water passes through the heater should be opposite to the direction of flow of the gases which are giving up heat. Thus the water enters where the gases are coldest, and is drawn off where the gases are hottest, and the gases finally pass into the chimney after being in thermal contact with the water in its coldest state. This makes the transfer of heat more nearly reversible, in the thermodynamic sense, than it would be if the direction of flow of one of the two fluids were reversed.

In Weir's feed-heater, applicable to compound engines, a portion of the steam was taken from the last receiver and mixed with the feed-water. This may be regarded as a first step towards the modern regenerative process of heating the feed in a series of steps by "bleeding" steam for the purpose at various stages in its expansion. The use of this method in large turbine plants has been referred to in § 161, where its thermodynamic advantage in bringing the cycle of operations nearer to that of Carnot was explained. As was pointed out there, it makes preheating of the air essential to economy of working, for the cold feed-water is no longer available as a means of usefully extracting from the flue gases the heat that remains in them when they leave the heating surface of the boiler.

290. Separation of oil from the condensate. The condensate from a reciprocating engine generally contains some oil or grease in suspension, from mixture with the lubricant applied to the cylinder, piston-rod and valve-rod; and in cases where this occurs to such an extent as might cause trouble in the boiler when the condensate is returned, grease-filters are employed through which the contaminated water is passed on its way back to the boiler. These generally consist of a screen or bed of canvas, saw-dust, cocoa-nut fibre, or such material adapted to catch the grease and exposing a large area of filtering surface. They are usually arranged in pairs, so that one may be in action while the other is being cleaned, and a by-pass is provided in case of the filter becoming choked. It is a conspicuous advantage of the steam turbine, over engines of the piston and cylinder type, that no lubricant need mix with the steam; hence in turbine installations no separating device is required.

291. Evaporators. Prevention of scale. In marine boilers the fresh water which is required to make up the waste of working substance in the circulating system of boiler, engine and condenser, is often obtained by an apparatus called an evaporator in which sea water is distilled. The heat is generally supplied by a coil of pipe into which steam is passed, and the steam used for this purpose is in many cases furnished by collecting the exhaust steam from auxiliary engines into a system of pipes forming what is called a closed exhaust. The evaporator also serves to supply fresh water for other purposes in the ship. Two evaporators are sometimes connected to form a compound system, the steam produced by evaporation in the first being used in turn to form the distilling agent of the second, in the tubes of which it is condensed.

If salt water be admitted as make-up, or if it gain access through leakage from condenser tubes, or if certain other mineral substances are present in the feed, a hard deposit called scale is apt to form on the boiler surfaces. Among methods of preventing this, one is to introduce a colloidal substance which, by attaching itself to the particles of mineral as they come out of solution, will prevent their caking into a solid crust. In a device called a "filtrator" this is done by placing a closed vessel containing linseed with a small quantity of soda some feet above the boiler and admitting live steam to it. A mucilage is extracted from the linseed which slowly drains back to the boiler and causes the particles to be deposited in a non-adherent state[1].

292. Circulation in boilers. Hydrokineter. The time within which steam may be safely raised from cold water when the fires are lighted depends on the circulation of the water, which is much more rapid in boilers of the water-tube type than in boilers that contain a relatively large volume of water. In the Scotch type of marine boiler especially there is much water under the furnaces which tends to remain cold and to escape circulation, owing to the absence of external flues. If steam were raised too quickly this would give rise to serious differences of temperature and consequent straining of the shell. The process may be accelerated by using, within the boiler, a sort of jet pump called a hydrokineter. It consists of a fixed nozzle under the water, to which steam is supplied from an auxiliary boiler. Concentrically around the nozzle is a screen formed of two cones, spaced a little way apart and so arranged that the jet of steam which issues from the central nozzle, and is immediately condensed, induces a strong current of water which enters the cones at the wider end and is discharged through the narrow end. A lively circulation is thereby set up in what would otherwise be nearly still water. The action resembles that of the injector. It can be continued till the pressure within the boiler becomes equal to that of the auxiliary steam.

[1] *Engineering*, May 9, 1924.

293. Use of zinc to prevent corrosion. Electrolytic protection. To prevent corrosion in boilers it is usual to introduce blocks of zinc in metallic connexion with the shell. These are set in the water space, preferably at places where corrosion has been found specially liable to occur. Their function is to set up a galvanic action, in which zinc plays the part of the active element, and is dissolved while the metal of the shell is kept electro-negative. Otherwise there would be a tendency for differences of electric quality between different parts of the shell to set up galvanic actions between the parts themselves, by which some parts, being positive to others, would be attacked. The zinc raises the potential of the whole shell enough to bring it in all parts above that of the water. Each block of zinc sets up a local current, which passes from the block through the water to portions of the boiler surface in its neighbourhood. The zinc blocks have to be renewed from time to time, as they dissolve. To make them act effectively care must be taken to keep them in good metallic contact with the shell.

The same principle is sometimes carried out by using, instead of blocks of zinc, an external battery to maintain a small current from anodes, or positive poles immersed in the water, to the plates and tubes of the boiler. Current is led to each anode from a storage battery or generator outside, through a rod projecting inwards which serves to support the anode and is carefully insulated from the shell. The anode may be of iron; it is slowly dissolved away by the electrolytic action of the current which passes from it into the water and thence to the shell, but the substance of the boiler is protected. The boiler surface acts as cathode in the electrolytic circuit.

294. Methods of forcing draught. The simplest but by no means the most economical way of forcing the draught is to let a jet of steam from the boiler discharge itself up the chimney. It tends to carry the furnace gases with it and so to reduce the air pressure in the space where the jet escapes. Allusion has already been made to the system which is universal in locomotive boilers of utilizing the exhaust steam from the engine as a means of forcing the draught. As an alternative to a steam jet, a fan may be fitted in the uptake of a boiler flue to assist mechanically in exhausting the gases. Two other methods of mechanically forcing the draught have come into extensive use. One plan is to box in the stokehold and keep the air in it at a pressure which may correspond to as much as two or three inches of water, by the use of blowing fans. This is known as the *closed stokehold* system and is common in war ships. In the other system, known as Howden's, the furnace room or stokehold is open to the atmosphere, and air for the furnace is sent by a blowing fan, under moderate pressure, into a duct which conveys it to the fire. The duct is so arranged as to deliver the air partly into a closed ashpit under the fire, and partly to openings above, when a coal-burning grate is used. The Howden system is very generally associated with preheating of the air.

By either of these means the power of the boiler, that is to say, the rapidity of steaming, is increased in the ratio of 3 to 2, or even more, as compared with its power under chimney draught. The efficiency of the boiler is, in general, slightly but not very materially reduced by severe forcing. Arrangements are sometimes made which allow the chimney draught to serve in ordinary steaming and the fan to be resorted to when an exceptional demand for power has to be met; in other cases the pressure at which air is supplied and consequently the rate of combustion of fuel on the grate is regulated by varying the speed of the fan. Forced draught enables a boiler to adapt itself to a wide range of power. An ordinary marine boiler burns 15 to 20 lb. of coal per hour per square foot of grate with natural draught, and this is easily raised to 30 lb. or more under forced draught. A locomotive using the steam-blast will ordinarily burn something like 70 or 80 lb. per hour per square foot of grate, but the consumption may be forced up to 140 lb. or so. In all cases of extreme forcing the efficiency is low, for the combustion is not very perfect and the temperature of the escaping gases is apt to be high.

295. Air preheaters. The ordinary way of heating the air before its admission to a boiler furnace is to place, somewhere in the air duct, a stack of tubes through which the outgoing flue gases pass on their way to the chimney; heat is conducted through the substance of the tubes from the hot gases inside to the air moving over them outside. With the advent of steam "bleeding," as a means of heating the feed-water, increased importance attaches to the preheating of air, as a means towards efficiency in large steam generating plants. This has led to the development of a less simple but more effective method of transferring the heat—the Howden-Ljungström air preheater—in which the flue gases give up their heat to a group of moving metal plates which carry it into the path of the incoming air. The heat consequently does not have to pass from one side to the other of conducting partitions: it enters and leaves the same metallic surfaces, which are alternately in contact with flue gases and with air. The plates are carried in a slowly revolving rotor which stands (fig. 256) above two large duct pipes: the flue gas passes up one of these and across one half of the rotor to an extracting fan above. Cold air is blown in by another fan, passing down through the other half of the rotor into a duct leading to the furnace. In this example the rotation is about a vertical axis: in others, the channels and the axis are horizontal. The heat-storing and restoring elements of the rotor are alternate corrugated and plain sheets of steel, about one-sixtieth of an inch thick, arranged so that the corrugations supply a multitude of small channels for the incoming air and the outgoing gases. There are partitions in the rotor and in the casing in which it revolves to obviate any considerable mixture of the products of combustion with the fresh air. Provision is made for the occasional use of a steam jet to blow soot off the plates. The device has the advantage of "contraflow" action, so that the transfers of heat approximate to being reversible.

The Perry air heater is a similar device, also with heat-absorbing elements consisting of stacks of plain and corrugated plates. Instead of revolving, the elements oscillate vertically up and down in straight guides. They are grouped in two pairs and are suspended in cages from a chain passing over a pulley on the top of the casing, so that as one goes down it balances another coming up. They make three or four oscillations a minute, and the movement brings each element alternately into the hot flue and the fresh air channel[1].

296. Mechanical stoking. Many appliances have been devised for the mechanical supply of coal to boiler furnaces, to escape the labour of hand-stoking, and, incidentally, to secure a more uniform condition of the fire and to avoid the inrush of cold air which occurs when the fire-door is opened for the admission of fuel. Several

[1] *Engineering*, April 2, 1926.

methods of mechanical stoking are effective for this purpose, and have found much application in large stationary boiler plants. They are not used in marine work, nor, except very rarely, in locomotives.

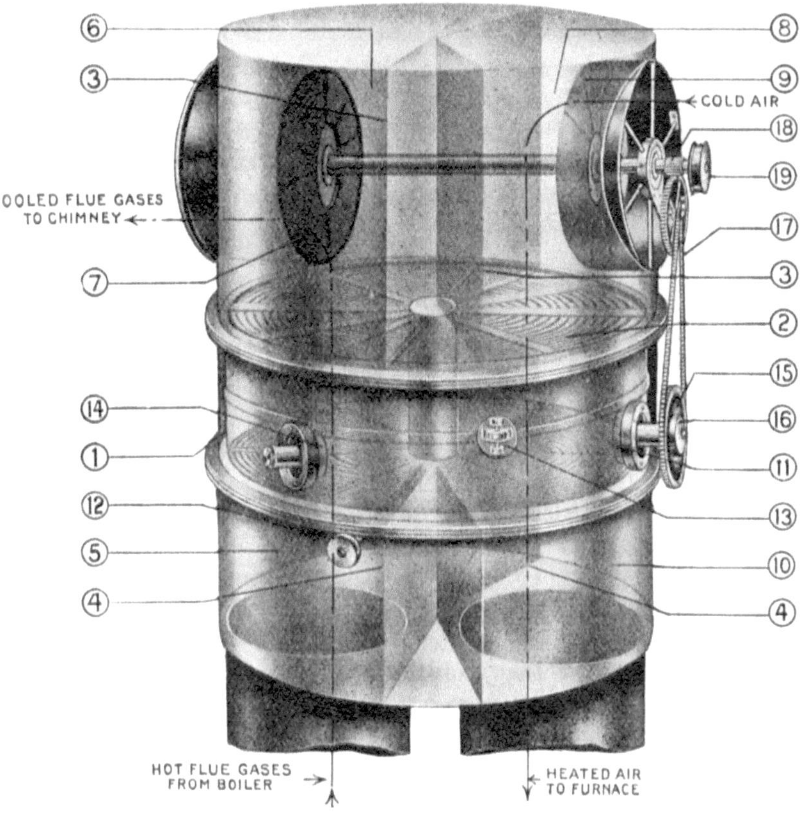

FIG. 256. Howden-Ljungström air heater.

1. Rotor.
2. Heating elements.
3. Upper partition wall.
4. Lower partition wall.
5. Lower flue gas chamber.
6. Upper flue gas chamber.
7. Flue gas fan.
8. Upper air chamber.
9. Air fan.
10. Lower air chamber.
11. Driving roller.
12. Soot blowing device.
13. Guide roller.
14. Carrying roller.
15. Sprocket wheel.
16. Clutch.
17. Chain.
18. Speed reducing gears.
19. Pulley.

The most suitable cases for their application are those in which there is an extensive battery of boilers, and a mechanical conveyer can be used to carry the coal from the bunkers to the hopper of each furnace. By the combination of mechanical conveyance with mechanical stoking in a large installation, such as a power-supply station, much labour may be saved.

Mechanical stokers have the further advantage that they remove the limitation which hand stoking imposes on the size of the grate. With them it becomes practicable to have a very large grate, and consequently to give to the combustion chambers of water-tube boilers the immense size which is a conspicuous feature of modern power-station practice.

In some mechanical stokers the coal is deposited, from a hopper, on the front of the grate and is then slowly carried towards the back: in others it is thrown in at intervals by a sprinkling shovel which distributes it at once more or less uniformly over the fire.

To the first of these types belongs the Juckes chain-grate, which is one of the earliest forms of mechanical stoker. The furnace bars are in short lengths, pinned to one another to form a chain or web which is made endless by returning over pulleys at the front and rear. The upper part of the endless web constitutes the grate, the lower part hangs in the ashpit, and the whole is kept in slow continuous motion, so that coal dropped from a hopper in front moves with the travelling grate towards a bridge piece at the rear. This catches the fuel remaining unburnt, and the excess drops over it into a chamber at the back of the ashpit. Owing to the burning of the coal the fire becomes thinner towards the rear, and this tends to let an excess of air enter through the rear portion of the grate. To remedy this the Coxe stoker has a series of air chambers below the bars supplied with air at graduated pressures, the chamber with highest pressure being in a region a little way from the front where the coal layer is completely ignited but still comparatively thick.

In other forms, such as Vicars' or Meldrum's coking stoker, the coal on dropping from the hopper is pushed by plungers on to what is called a coking plate, which is a dead plate forming the front part of the furnace. Here the volatile constituents are driven off, and the coked fuel is then pushed in further to the fire-bars which have a reciprocating motion through a few inches in the direction of their length. In this motion the bars first move together towards the rear, and then they return towards the front singly or by the successive movement of two sets made up of alternate bars. This gradually works the charge on the grate inwards.

In the sprinkling stoker of Proctor charges of fuel from the hopper drop into a small chamber from which they are thrown on to the fire by the sudden release of a spring shovel. The shovel is drawn outwards against the tension of a stiff spring, by means of a revolving wheel which carries three tappets. As each of these passes a lever attached to the shovel the shovel jerks forward, scattering its charge on the fire; and the three tappets are so shaped that the three successive jerks are graduated to throw the coal to various distances, and so ensure a fairly uniform thickness of fire. To keep the fire dressed and to move the clinkers towards the bridge there is also an alternating motion of the fire-bars through a small distance, sets of alternate bars being periodically lifted and dropped a little as well as moved backward and forward.

Such stokers can be used with internally fired as well as externally fired boilers. With externally fired boilers there is more space for a mechanical grate, and other forms have been successfully developed, especially in America, where the boilers of big power stations give much opportunity for mechanical stoking, and also for the use, to be described immediately, of pulverized fuel. Both of these methods of firing allow boilers and furnaces to be designed on a gigantic scale. In some of the American stations the grates measure as much as twenty feet each way, and the combustion chambers have about that height, below the tubes.

An interesting type is represented by the Rooney stoker, in which the grate is made up of a series of T-shaped bars, set transversely and stepped so as to form an incline sloping downwards away from the fire-door at an angle of about 35°. The tops of the T's are nearly horizontal, but form a series of steps, each T being a little lower than the one in front. Coal is fed from a hopper on to the front of the grate, and to make it pass gradually down the slope the T-bars are all tilted periodically through a small angle by movement about an axis in the direction of the length of each bar. The ashes and clinkers which reach the bottom of the slope are collected on a small horizontal grate there which is hinged at the back, and this is dropped from time to time to discharge them into the ashpit. The Wilkinson stoker is another example of the inclined grate type. Its bars extend down the slope; they are notched to provide a stepped bed for the fire, and have a slight reciprocating motion in alternate groups to make the burning fuel travel downwards.

A much used type is the "underfeed" stoker, an American development of an old device. The coal is fed to a trough which forms a depression in the grate, and is pushed by a ram or screwed by a screw conveyer along this trough towards the rear. It becomes heaped up over the whole length of the trough and spreads laterally over the grate, which extends on either side.

291. Use of pulverized coal. Another old device, which has been revived and is now widely adopted, is to pulverize the fuel before introducing it into the furnace. In Crampton's dust fuel furnace the coal was ground to powder and fed by rollers into a pipe from which it was blown into the furnace by an air-blast. This gave so intimate a mixture of fuel and air that the excess of air required for dilution was only one-fifth of the amount required for combustion[1].

The process of pulverizing is expensive: on the other hand labour is saved; there is a clean boiler house; with no opening of fire-doors the combustion can be kept uniform and regulated at will. The air-supply can be nicely adjusted to burn the fuel completely with a comparatively small excess of air, a high furnace temperature, and a consequent gain both in power and boiler efficiency. Since 1922 powdered fuel has come to be applied in many power stations and other steam plants, where the individual boilers usually have a heating surface ranging from 6000 to 30,000 square feet.

In a large number of these the Lopulco system is used, the general features of which are shown in fig. 257. The coal is first dried, preferably by means of the waste heat in flue gases, then powdered in a "Raymond" mill and conveyed by fan-blast to a "cyclone" separator which removes any particles that are too big to burn in the flame and returns them to the mill for further grinding. The coal powder then passes into a bin from which it is fed to the furnace by a screw feeding device and an air-blast from another fan. The combustion chamber is made specially large, as the illustration shows; it projects beyond the boiler so that the burners may deliver the coal mixed with air in vertical streams. The flame takes the form of a U, and combustion is complete before it touches the tubes of the boiler. Part of the air of combustion is blown in with the coal; the rest enters the furnace through openings in the wall in which it becomes heated before admission.

[1] *Proc. Inst. Mech. Eng.* 1869.

The other walls of the furnace are "water-lined"; that is to say, they are screened by tubes through which the boiler water circulates, and at the bottom there is also a very open screen of water-tubes

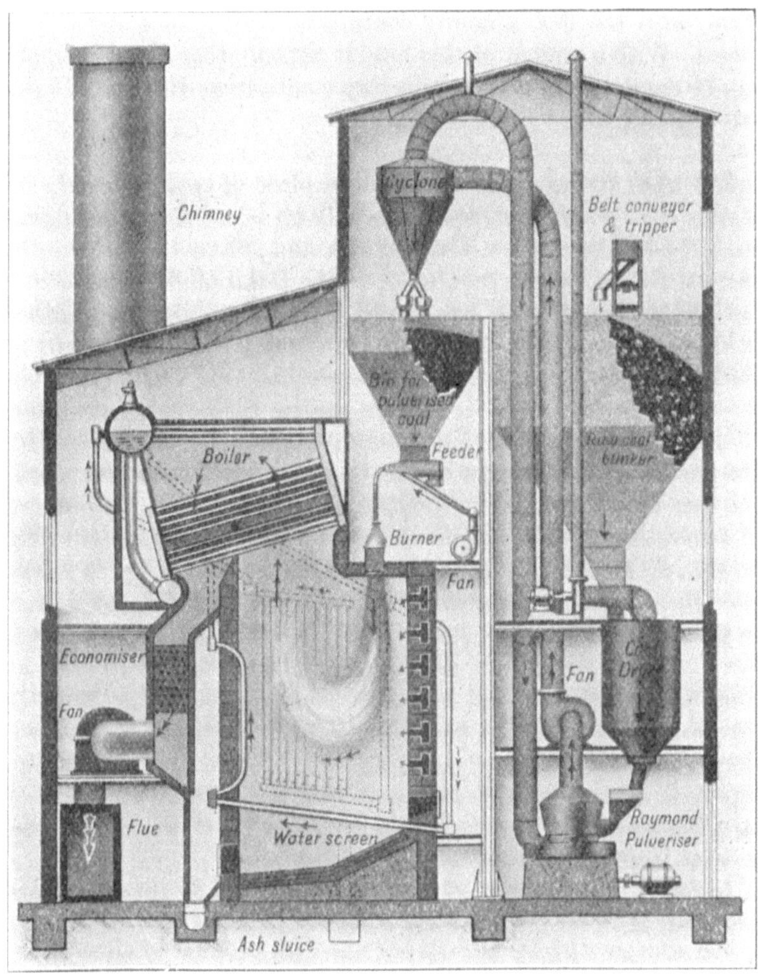

Fig. 257. Use of pulverized coal.

through the interstices of which the ash drops to a space below. The side screens and bottom screen add materially to the effective heating surface, especially by absorbing radiant heat, and the water screen at the bottom also causes the molten particles of ash to solidify as they fall and so keeps them from clinkering. The water-tubes which line the vertical walls of the furnace have flat fins extending on either side, and are set close enough to form, with the fins, a nearly continuous surface of water-cooled metal. With a system of this kind it appears that the air-supply can be regulated to give satisfactory combustion with fully 15 per cent. of CO_2 in the flue gases.

298. Oil firing. The use of oil in place of coal as boiler fuel is now exceedingly common, especially on board ship. Beginning with steamships on the Caspian Sea and locomotives in south-eastern Russia, where proximity to the Baku oil-fields gave it a particular advantage, it has spread, with the opening up of other fields and the development of oil transport in bulk (by means of tank steamers) until it has become world-wide. The cost of oil, per unit of available heat, remains greater than that of coal, but that is held to be more than counterbalanced, in many cases, by the facility with which the oil can be shipped and stored on board, the ease and cleanliness with which it can be handled, the absence of ashes, the saving of labour in stoking, and the greater steaming power and efficiency which it gives to the boilers. Its bulk is less than that of coal, for the same energy content, and it can be stowed in parts of the hull where the space could not be otherwise utilized. For naval purposes these advantages are so conspicuous that as soon as the supply of oil as fuel could be considered sufficiently secure, it was adopted in war ships to the exclusion of coal. From them it has extended to a large part of the mercantile marine, especially to passenger ships. Since 1920 the aggregate tonnage of new steamships built with boilers arranged for oil firing has been greater than that of new coal-fired ships.

In round figures, a pound of oil-fuel has about 30 per cent. more potential thermal energy than a pound of coal, but the greater boiler efficiency which attends the use of oil allows it to make about 50 per cent. more steam. The greater efficiency arises in the main from steady action and from there being no need to open the furnace doors to admit the fuel.

In the early days of oil burning it was customary to spray the fuel into the furnace by compressed air or by steam, and also to let the flame impinge on a fire-brick wall or bridge. Now, however, fire-brick is generally omitted from the combustion chamber: the injected and ignited oil completes its combustion without touching anything; and the jet is formed without the aid of steam or of compressed air. The oil itself, after being filtered and moderately warmed to give it greater fluidity and to facilitate spraying, is delivered under high pressure to a nozzle or nozzles projecting into the furnace. From the orifice of each nozzle it issues as a fine cloud-like spray, in the form of an open cone to which the air needed for combustion is supplied from an annular channel surrounding the nozzle. The orifice is so arranged as to give the oil a whirling motion which helps to "atomize" it by centrifugal action. The air also is often given a whirling motion by passing through spiral channels as it enters. Ignition takes place after the jet has travelled a few inches from the nozzle, the burning cone continues to spread, and the effect of the several jets, parallel and suitably spaced for a large boiler, is more or less completely to fill the combustion chamber with flame.

The pressure under which oil is delivered to the nozzles generally ranges from 50 to 150 lb. per sq. inch. The burners or sprayers contain channels for its admission, which lead tangentially into a small "spinning chamber" where the oil acquires a vortex motion just before it passes through the small hole by which it escapes at a high speed into the furnace. The issuing cone usually has an angle of about 60°. When the pressure of delivery and the air-supply are properly adjusted no oil reaches the furnace walls or the fire-tubes in an unconsumed state.

In some steamship boilers oil firing and coal firing are combined, each of the furnaces being so arranged that an oil sprayer can be inserted above the fire door and both kinds of fuel be used, either together or separately; or the door may be removed and a sprayer inserted instead, the oil thus serving as an occasional substitute for coal.

We have here to do only with the application of oil as fuel for the formation of steam. Its direct use as a producer of power in the internal-combustion engine will be considered in Chapter XVII.

CHAPTER XV

FORMS OF THE STEAM-ENGINE

299. Terms used in classification. A broad distinction is to be drawn between engines in which steam jets act by impulse or reaction, or both, and engines in which work is done by the changes in volume of a chamber or chambers containing steam under pressure. Steam turbines, of the various kinds described in Chapter VIII, represent the former type; to it also belong the injector and other forms of steam jet pump. The other general sort may be collectively described as *pressure engines*: it is with them only that we are concerned in this chapter. They are in general of the *reciprocating* or piston-and-cylinder type. Then there is a general distinction of *condensing* from *non-condensing* engines, with a subdivision of the condensing class into those which act by surface condensation and those which use jet condensers. Next there is the division into *simple* or non-compound and *compound* engines, with a further classification of the latter, according to the number of stages in the compounding, as double-, triple-, or quadruple-expansion engines. Again engines are *single-acting* or *double-acting* according as the steam acts on one side only or alternately on both sides of the piston. Some engines, such as steam-hammers and certain kinds of steam-pumps, are *non-rotative*, that is to say, the motion of the piston does work simply on a reciprocating piece; but generally an engine does work on a continuously revolving shaft, and is termed *rotative*. In most cases the crank-pin of the revolving shaft is connected directly with the piston-rod by a connecting-rod, and the engine is then said to be *direct-acting*; in other cases, of which the *beam engine* of Watt is the historical example, a lever resembling the beam of a balance is interposed between the piston and the connecting-rod leading to the crank. The same distinction applies to non-rotative pumping-engines, in some of which the piston acts directly on the pump-rod, while in others it acts through a beam. The position of the cylinder is another element of classification, giving *horizontal*, *vertical*, and *inclined cylinder* engines. In *oscillating cylinder engines* the connecting-rod is dispensed with; the piston-rod works on the crank-pin, and the cylinder oscillates on trunnions to allow the piston-rod to follow the crank-pin round its circular path. In *trunk engines* the piston-rod is dispensed with; the connecting-rod extends as far as the piston, to which it is jointed, and a trunk, or tubular extension of the piston through the near end of the cylinder, gives room for the rod to oscillate. In *rotary* engines there is no piston in the ordinary sense; the steam does work on a revolving piece, and the necessity is thus avoided of afterwards converting reciprocating into rotary motion. Still another mode of classification speaks of engines in reference to the conditions under which they are at work, as *stationary*, *locomotive*, or *marine*. Locomotive, marine and some kinds of stationary engines such as those employed in heavy rolling mills belong to the *reversing* class, having valve mechanism which enables them to run either way. Other descriptive terms will be mentioned in the sections which follow.

300. Beam engines. In the single-acting atmospheric engine of Newcomen the beam was a necessary feature; the use of water-packing for the piston required that the piston should move down in the working stroke, and a beam was needed to let a counterpoise pull the piston up and actuate the plunger of the pump. Watt's improvements made the beam no longer necessary; and in one of the forms he designed it was discarded—namely, in the form of pumping-engine known as the Bull engine, in which a vertical inverted cylinder stands over and acts directly on the pump-rod. But the beam type was generally retained by Watt, and for many years it was preferred by the builders of engines. The early engines were designed to pump water out of mines, and the beam was highly convenient for pumping. The cylinder could stand on a firm foundation near the mine shaft and the beam project over the shaft. Moreover the beam was a convenient driver for the valve-rod, feed-pump rod and air-pump rod of the engine itself, and it lent itself to the method invented by Watt of guiding the end of the piston-rod to move in a straight line, by what is called the "parallel motion." In a modern direct-acting engine the end of the piston-rod is constrained to move in a straight line by means of a block sliding in guides, but in the early days of engine building accurate methods of shaping suitable surfaces had not been developed, and it was easier to secure a good approximation to straight line motion by means of Watt's device. In modern practice the direct-acting type of engine has almost wholly displaced the beam type.

The "parallel motion" has enough historical interest to deserve a brief description. It is illustrated diagrammatically in fig. 258, where MN is the path in which the piston-rod head, or cross-head, as it is often called, is to be guided. ABC is the middle line of half the beam, C being the fixed centre about which the beam oscillates. A link BD connects a point in the beam with a radius link ED, which oscillates about a fixed centre at E. A point P in BD, taken so that $BP : DP :: EN : CM$, moves in a path which coincides very closely with the straight line MPN. Any other point F in the line CP or CP produced is made to copy this motion by means of the links AF and FG, parallel to BD and AC. In the ordinary application of the parallel motion a point such as F is the point of attachment of the piston-rod, and P is used to drive a pump-rod. Other points in the line CP produced are occasionally made use of, by adding other pairs of links parallel to AC and BD.

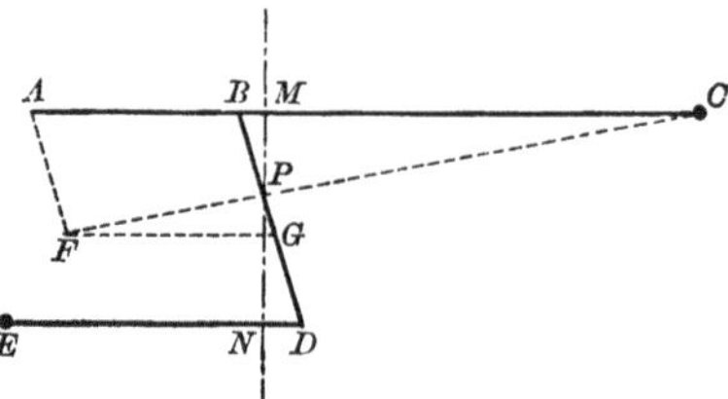

Fig. 258. Watt's Parallel Motion.

Watt's linkage gives no more than an approximation to straight-line motion, but in a well-designed example the amount of deviation need not exceed one four-thousandth of the length of the stroke. It was for long believed that the production of an exact straight-line motion by pure linkage was impossible, until the problem was solved by the invention of the Peaucellier linkage, which however has not been applied to the steam-engine except in isolated cases.

301. Direct-acting horizontal and vertical engines. The use of guiding surfaces and sliding motion for the piston-rod head is practically universal in engines of the direct-acting class: the piston, the connecting-rod, the crank and the frame or bed-plate of the engine constitute a kinematic chain of four elements, of the "slider-crank" type.

No form of steam-engine is so common as the horizontal direct-acting. For small powers the engine is generally self-contained, in other words, a single frame or bed-plate carries all the parts, including the main bearings in which the crank-shaft with its fly-wheel turns. The cylinder either rests on the bed-plate or overhangs at the back, being in the latter case bolted to a vertical part of the frame which forms a cover for the front end of the cylinder. The frame is often given what is called a girder shape, which brings a portion of it more directly into the line of thrust between the cylinder and the crank centre, and allows the upper as well as the lower of the two surfaces which serve as guides for the cross-head to be formed on the frame itself. This construction is usual in large as well as in small horizontal engines. The feed-pump plunger is usually driven from a separate eccentric: in some cases it is directly attached to the cross-head, and in others to the valve-rod. When a condenser is used with a small horizontal engine it is usually placed behind the cylinder, and the air-pump, which is within the condenser, is a horizontal plunger or piston-pump worked by a "tail-rod"—that is, a continuation of the piston-rod past the piston and through the back cover of the cylinder. In large horizontal engines the condenser often stands in a well between the cylinder and the crank-shaft, and the pump, which has a vertical stroke, is worked by means of a bell-crank lever attached by a link to the cross-head of the engine. A tail-rod, however, may be introduced in such engines to assist in supporting the weight of the piston, which would otherwise tend to cause undue wear on the lower side of the cylinder.

When uniformity of driving effort or the absence of dead-points is specially important, two independent cylinders often work on the same shaft by cranks at right angles to each other, an arrangement which allows the engine to be started readily from any position. Such engines are called *coupled.* The ordinary locomotive is an example of this form. Winding engines for mines and collieries, in which ease of starting, stopping, and reversing is essential, are very generally made by coupling a pair of horizontal cylinders, with cranks at right angles to each other, on opposite sides of the winding-drum, with a link-motion as the means of operating the valves.

Direct-acting engines of the larger class are generally compounded either (1) by having a high and a low-pressure cylinder side by side, working on two cranks at exactly or nearly right angles to each other, or (2) by placing one cylinder behind the other, with the axes of both in the same straight line. The former is called the *cross-compound* and the latter is called the *tandem* arrangement.

Tandem engines usually have a piston-rod common to both cylinders. In a few compound engines the large cylinder is horizontal, and the other lies above it in an inclined position, with its connecting-rod working on the same crank. In another form, already referred to in speaking of the balancing of engines, the combination consists of one horizontal and one vertical cylinder, with connecting-rods working on the same crank.

In tandem engines, since the pistons move together, there is no need to provide a receiver between the cylinders. It is practicable to follow the "Woolf" plan (§ 180) of allowing the steam to expand directly from the small into the large cylinder; and in some instances this is done. For the reasons stated in Chapter x it is more usual to work with a moderately early cut-off in the low-pressure cylinder. Unless it is desired to make the cut-off occur before half-stroke, an ordinary slide-valve will serve to distribute steam to the large cylinder. For an earlier cut-off than this a separate expansion valve is required on the low-pressure cylinder, to supplement the slide-valve; and in any case, by providing a separate expansion valve, the point of cut-off is made subject to easy control, and may be adjusted so as to reduce drop or to divide the work as may be desired between the two cylinders. For this reason it is not unusual to find an expansion valve, as well as a common slide-valve, on the low-pressure cylinder even in tandem engines. In many cases, however, the common slide-valve only is used. In the high-pressure cylinder of compound engines the cut-off is often effected either by an expansion slide-valve or by some form of Corliss or other trip gear.

For mill engines the compound tandem and compound coupled types are the most usual, and the high-pressure cylinder is very generally fitted with trip gear. In the compound coupled arrangement the cylinders are on separate bed-plates, and the fly-wheel is between the cranks. In triple expansion horizontal engines it is not unusual to use a tandem arrangement for two of the cylinders but to cross-couple as regards the third, and in some cases the third stage is divided between two cylinders, making four in all, which are grouped as two cross-coupled pairs, one pair consisting of the high-pressure cylinder tandem with one low-pressure cylinder, and the other pair consisting of the intermediate cylinder tandem with the other low-pressure cylinder. Where considerations of floor space admit of it a horizontal arrangement is generally preferred with separate admission and exhaust-valves for each end of each cylinder. The condenser is generally put in a pit at a somewhat lower level, and the air-pump is worked by a bell-crank lever.

The general arrangement of vertical engines differs little from that of horizontal engines. The cylinder is usually supported above the shaft by a cast-iron frame resembling an **A** or inverted **V**, whose inner sides are kept parallel for a part of their length to serve as guides for the cross-head. Sometimes one side of the frame only is used, and the engine is stiffened by one or more wrought-iron columns between the cylinder and the base on the other side. The vertical type is very usual in marine practice and will be referred to further in that connexion. *Wall-engines* are a vertical form with a flat frame or bed-plate, which is fixed by being bolted against a wall; in these the shaft is generally at the top. Vertical engines are compounded, like horizontal engines, either by coupling parallel cylinders to cranks at right angles (or at 120° if triple expansion is to be used with three cylinders) or, tandem fashion, by placing the high-pressure cylinder above the other. In vertical condensing engines the condenser is often situated near the base under the back limbs of the frame, and the air-pump, which has a vertical stroke, is worked by a horizontal lever connected by a short link to the cross-head.

302. Uniflow or central-exhaust engine. This engine, which was remarked on in § 127 as avoiding much of the loss that occurs in other reciprocating engines through periodic exchanges of heat between the metal and the steam, is characterized by having no exhaust-valve other than the piston itself. The advantage of uniflow working is widely recognized, and the engine is built, in horizontal form, by many makers for application to pumping, mill-driving and other purposes. An example by Messrs Sulzer is shown in fig. 259. Admission takes place at each end through a drop-valve in the cylinder cover, which is hollow and serves both as steam-chest and end jacket. The drop-valves are operated from a lay shaft at the side to give an early cut-off at a point determined by the governor. Exhaust occurs when the piston, which (at its circumference) is nearly as long as the stroke, uncovers a ring of ports opening into the exhaust-pipe at the middle of the cylinder's length. There is very little clearance; consequently

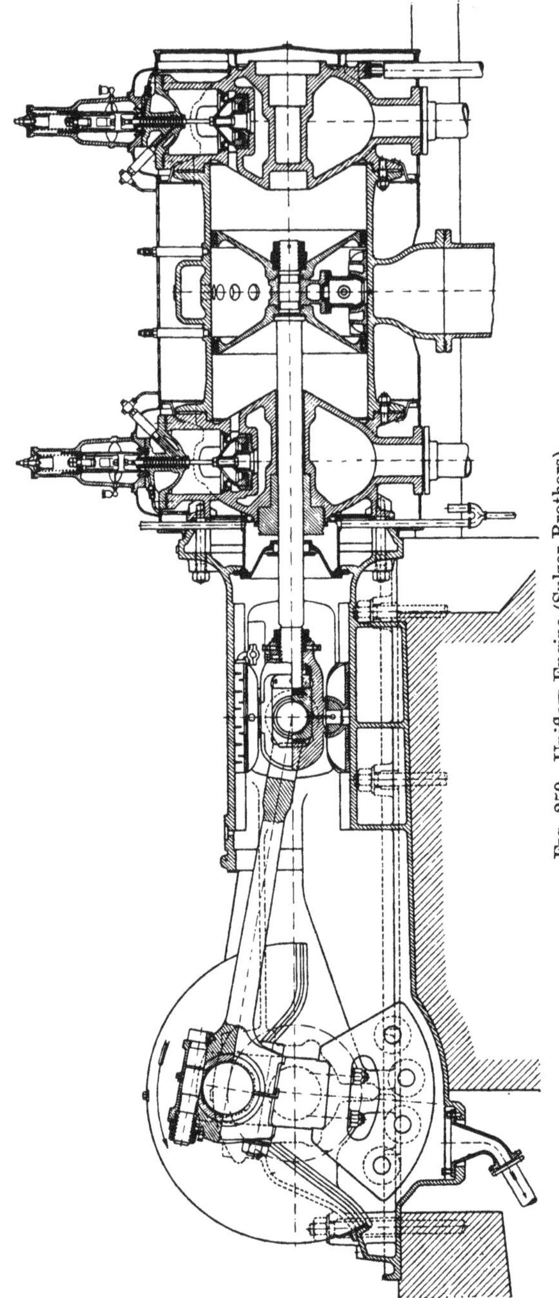

Fig. 259. Uniflow Engine (Sulzer Brothers).

such exhaust steam as remains in the cylinder when the ports close on the return stroke is compressed nearly to admission pressure. A spring relief valve is provided in case through failure of the vacuum the compression pressure should become excessive. With an early cut-off, at say one-tenth of the stroke, steam at 170 lb. or so is economically expanded in this way, especially if superheated. With considerably higher pressures, however, it is desirable to make the engine compound by adding a high-pressure cylinder of the ordinary type and treating the uniflow cylinder as the low-pressure element of a compound system.

303. Back-pressure or extraction engines. The practice, already referred to in connexion with steam turbines, of first utilizing in an engine steam which is afterwards to be employed, at a comparatively low pressure, for process work in a factory or for heating, is becoming increasingly common with reciprocating engines also. In chemical works, breweries, paper-mills and so forth where much low-pressure steam is wanted, as well as some power, it is economical to use a high-pressure supply and treat the engine as a reducing valve in which the steam does work before it passes on to serve another purpose. The engine is accordingly designed to exhaust at a pressure such as the other process requires. The higher pressure of supply somewhat increases the heat of formation, and the steam, by doing work as it expands, becomes wet. But there is a clear thermodynamic gain on the whole, when the comparison is with a system in which the two supplies of steam are independent and the engine steam, after doing its work, rejects heat uselessly to the condensing water or to the atmosphere instead of contributing to carry out the purpose for which low-pressure steam is wanted.

304. Condensers. Many land engines use a simple jet form of condenser, but surface condensation is resorted to when the available supply of water is unsuited for boiler feed, and it is therefore important to return the condensed water to the boiler. For the same reason surface condensation is universal in marine engines. In large steam-turbine plants, as was pointed out in Chapter VIII, surface-condensers are preferred because the steam turbine can take full advantage of the high vacuum which they make practicable. The most usual form of surface-condenser is a group of brass tubes extending horizontally between two tube-plates, with an arrangement by which the cooling water circulates towards one side through the lower half of the group and back towards the other side through the upper half, while the steam to be condensed comes into contact with the outside of the tubes. The aggregate surface of the tubes is roughly about $1\frac{1}{2}$ sq. feet per indicated horse-power, and the quantity of cooling water is some 20 or 30 times the quantity of condensed steam. The amount depends on the vacuum aimed at as well as on the temperature of the available supply. When the supply is limited means may be taken to cool the condensing water, as by letting it trickle down over a high scaffolding constituting a *cooling tower*, before it is used over again. But an interesting alternative in such cases is to employ what is called an *evaporative condenser*, consisting of a stack of pipes into which the exhaust steam is admitted and over which a comparatively small amount of cold water is allowed to drip. Such a condenser is placed in the open air, generally on the roof, sometimes with a fan blowing air over it to stimulate evaporation of the water dripping on the surface, and the amount of water used by it need not exceed the amount that is condensed. It can therefore be applied to what would otherwise be a non-condensing engine, giving the thermodynamic advantage of condensation without any additional expenditure of water, the feed-water being saved by condensation while the quantity of cooling water is no more, and may even be less, than the quantity which would be required for feed. A vacuum of 24 or 25 inches is obtained by this device.

Any of these forms of condenser requires the use of an *air-pump* to keep up the vacuum by continuously removing whatever air or non-condensable gas comes over with the steam. The same pump will serve to remove the condensed water and (in a jet condenser) the water admitted by the jet. But in many cases two pumps are provided, one for the water and another, called a *dry air-pump*, for air and vapour only. The former draws water from the bottom of the condenser; the latter has its intake above the water line. The combination allows a high vacuum to be maintained.

With some land engines it is practicable to remove the water from the condenser by a *gravity drain*, namely a pipe with a vertical drop of about 35 feet, ending in a water seal to prevent the return of air, a dry air-pump doing the rest.

Reference has already been made in Chapter VIII to the *vacuum augmenter* devised by Parsons for the purpose of securing a higher vacuum than is usually obtainable with an air-pump of the ordinary kind, and also to the use of steam-jet pumps for service as air-pumps, while the condensate (from a surface-condenser) is extracted by a separate centrifugal or other water pump.

The *ejector condenser* is a device for obtaining jet condensation without the use of any air-pump. It acts on the principle of the injector, delivering water against the pressure of the atmosphere and at the same time condensing the exhaust steam from the engine. It consists of a series of coaxial cones set in a vertical line with the wider ends above, with annular spaces between them through which the exhaust steam enters, and with a central channel down which the condensing water flows. To secure a good vacuum and steady working the water should be supplied under a head of about 20 feet. The group of cones may be said to form a combining nozzle, in which the exhaust steam, entering by the circumferential channels, meets the stream of water in the centre and is condensed. Then the stream passes on to a divergent delivery-nozzle, in which its velocity is gradually reduced and its pressure consequently increased, and this terminates in the hot-well. As the pressure there is that of the atmosphere it follows that in the combining nozzle, where the velocity of the stream is greater, the pressure is much less than atmospheric, and in fact a vacuum equivalent to about 25 inches of mercury may readily be maintained by this device, with a single jet of water, and one of 26 or 27 inches when, as is often the case, several jets of water are made to act together in what is called a multiple-jet ejector condenser.

305. Single-acting and double-acting high-speed engines. With the development of electric lighting, soon after 1880, a demand arose for a type of engine which would run fast enough to drive a dynamo directly, without the use of belting or other intermediate gearing, the speed of the dynamo being often as much as 500 revolutions per minute or even more. To some extent this demand was met by adapting the direct double-acting engine for high-speed work, by enlarging the bearing surfaces and reducing, as far as possible, the masses of the reciprocating parts. But the single-acting type of engine came into favour as being specially suited for this class of work, and for a good many years had a great vogue. The most successful example of this type (now obsolete) was the "central-valve" engine of P. W. Willans. It owed much of its success to the excellence of its workmanship, and to the scientific manner in which Willans dealt with questions of steam-engine efficiency, in the researches to which reference has been made in Chapter VII. The dynamical problem of preventing a reversal of stress at the joints of single-acting engines has been dealt with in § 253.

When it was realized that a double-acting engine could be made to run smoothly at a high speed by the device of continuously pumping oil into the bearings under a moderately high pressure the special merit of the single-acting engine disappeared. For the same power the double-acting type has of course the advantage of saving weight and space. Messrs Belliss and Morcom introduced in 1890 a double-acting high-speed engine with forced lubrication which has been highly successful. It is made in sizes rated up to 2500 horse-power, and for speeds ranging from 650 revolutions per minute in the small sizes to 200 in the largest. It is of the inverted vertical class and is enclosed. The bottom of the crank case serves as a reservoir for the lubricating oil which is pumped up by a pump worked from the valve-eccentric and delivered through a system of small pipes to all the bearings at a pressure of 10 to 30 lb. per sq. inch. These engines are generally two-crank compound or three-crank triple with piston-valves for all cylinders. In one of the two-crank compound classes an ingenious arrangement of two piston-valves, placed one above the other between the cylinders, enables both valves to be worked by one eccentric-rod. The cranks are set at 180° apart, and the valves have therefore to be opposite in phase. Both valves are of the hollow piston type: they are set in tandem on one rod, and the ports are so arranged that the high-pressure valve admits steam by its inner edges and exhausts at its outer edges, while the other valve admits by its outer edges and exhausts by its inner edges. A shaft governor with external spring control regulates the speed, through an equilibrium throttle-valve, but in the larger engines automatic cut-off gear may be fitted. The tension of the external controlling spring is adjustable by means of a hand wheel while the engine is running. From published tests it appears that the steam consumption of these engines, in sizes of about 1000 horse-power and at speeds of 250 revolutions per minute, using triple expansion, may be as low as 11 lb. per brake horse-power-hour, with steam at 180 lb. per sq. inch and about 125° C. of superheat.

306. Pumping engines. In engines for pumping water and other liquids, or for blowing air, it is not essential to drive a revolving shaft, and in many forms the reciprocating motion of the steam piston is applied to produce the reciprocating motion of the pump piston or plunger without the intervention of any revolving part. On the other hand, pumping and blowing engines are frequently made rotative by introducing a fly-wheel, and even when there is a direct transference of the effort of the piston-rod to the pump-rod, a crank and fly-wheel are often added, in order that such differences as may occur in various phases of the stroke between the work given out by the piston and the work taken by the pump may be alternately stored and restored by the fly-wheel.

The beam, which was for long a very common feature in pumping-engines, is not found in modern designs. Often an inverted vertical triple expansion engine is used, resembling the usual marine form. When water has to be delivered against a considerable head, as in city water-supply, Messrs Hathorn Davey and Co. place the three vertical cylinders over the pumps, with a pump plunger in line with each piston-rod. Between the two is a crank-shaft, but the thrust of the piston-rod is transmitted from the four corners of a square cross-head directly to the pump plunger by a group of four vertical rods, two on each side of the shaft, with room between them for the crank to turn. There is a connecting-rod from the middle of the cross-head to the crank, but only a small part of the power is transmitted through it to and from the fly-wheel, the effect being to equalize the effort on the pump. The steam-cylinders have Corliss valves, and the cranks are set 120° apart. They run at a slow speed, about 30 revolutions per minute. With steam at 180 lb. and moderate superheat a consumption of 10·6 lb. per pump horse-power-hour is realized[1].

In some instances the water has to be lifted from a deep well before being delivered against a pressure-head, and separate

[1] Plant of this type is used, for example, at the Metropolitan Waterboard's pumping stations at Lea Bridge and Walton. A description and analysis of tests of the Lea Bridge engine is given by H. R. Lupton, *Min. Proc. Inst. C. E.* 1926.

pumps, driven from the same engine, perform the two operations. The force-pumps are set immediately below the steam pistons and driven by them direct, in the manner just described, but enough power is transmitted through the connecting-rods to the crank-shaft to let it drive the lift-pumps. For this purpose it is prolonged, over the well, and may carry three cranks 120° apart driving a set of three-throw pumps side by side in the well; or a single well-pump of the "concertina" class may be used, with two buckets driven by cranks which differ in phase by 180°. The pump-rod for one of the buckets passes through the other so that the buckets alternately approach and recede from one another in the same pump cylinder.

Engines for deep-well pumping without any fly-wheel, like the old Cornish pump, are able to work expansively in consequence of the great inertia of the reciprocating pieces, the chief of which are the long and massive pump-rods. Notwithstanding the comparatively low frequency of the stroke, enough energy is stored in the movement of the rods to counterbalance the inequality with which the expanding steam works in different portions of the stroke, and the rate of acceleration of the system adjusts itself to give, at the plunger end, the nearly uniform effort which is required in the pump. In other words, the motion, instead of being almost simply harmonic as it is in a rotative engine, is such that the form of the inertia curve, when drawn as in fig. 211, is nearly the same as that of the steam curve, with the result that the distance between the two, which represents the effective effort on the pump plunger, is nearly constant. The massive pump-rods may be said to form a reciprocating fly-wheel.

It is however only to deep-well pumping that this applies, and a very numerous class of direct-acting non-rotative steam-pumps have too little mass in their reciprocating parts to allow such an adjustment to take place at any ordinary speed. This is the case for example in the direct-acting feed-pumps already referred to in speaking of steam boilers (§ 287). In such engines an auxiliary rotative element is often introduced, partly to secure uniformity of motion and partly for convenience in working the valves; a connecting-rod, for instance, is sometimes taken from a point in the piston-rod to a crank-shaft which carries a fly-wheel, or a slotted cross-head is fixed to the rod and gives motion of rotation to a crank-pin which gears in the slot, the line of the slot being perpendicular to that of the stroke. But many pumps of this class are purely non-rotative, and in such cases the steam is admitted throughout the whole or nearly the whole of the stroke, since the inertia of the parts is not sufficient to give a means of reconciling uniformity of pump-effort with expansive working.

The Worthington duplex engine, referred to as a boiler pump in § 287, has been extensively applied, on a large scale, to deliver water for the supply of towns and to force oil through "pipe-lines" in the United States. In the larger sizes it is made compound or triple, each of the two parallel pump-rods having two, or three, steam-cylinders on it arranged in tandem. To allow of expansive working, a device was introduced to compensate for the inequality of effort on the pump piston that would result from an early cut-off. A cross-head A (fig. 260) fixed to each of the piston-rods is connected to the plungers of a pair of cylinders B, B, which are free to oscillate on fixed trunnions. These cylinders contain water which can pass between them and a reservoir in which the pressure is maintained at 200 or 300 lb. per sq. inch, the water passing to and fro through the trunnions, which are made hollow for the purpose. When the stroke (which takes place in the direction of the arrow) begins these plungers are at first forced in, and hence work is at first done by the main piston-rod, through the compensating cylinders B, B, by forcing water into the reservoir against this pressure. This continues until the cross-head has advanced so that the oscillating cylinders stand at right angles to the line of stroke. Then for the remainder of the stroke their plungers assist in driving the main piston, and the reservoir gives out the energy which it stored in the earlier portion. Any leakage from the compensating cylinders or their reservoir is made good by a small pump. One advantage which this method of equalizing the effort of a steam-engine piston has (as compared with making use of the inertia of the reciprocating masses) is that the effort, when adjusted to be uniform at one speed, remains nearly uniform although the speed be changed, provided the inertia of the reciprocating parts be small. In the Worthington "high-duty" engine this plan has been used to make it possible to have an early cut-off in the steam-cylinders.

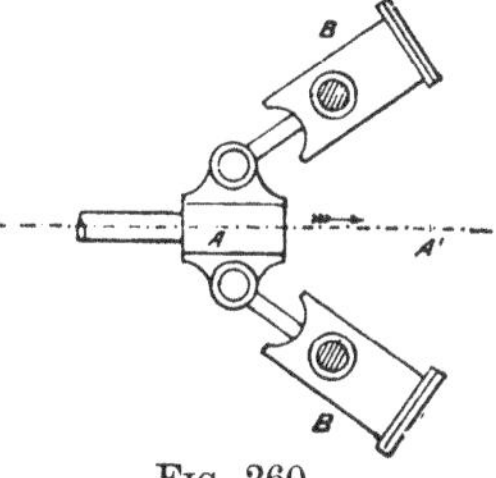

Fig. 260.

307. The Pulsometer. Hall's "pulsometer" is a peculiar pumping-engine without cylinder or piston, which may be regarded as a modern representative of the engine of Savery (§ 6). The sectional view, fig. 261, shows its principal parts. There are two chambers A, A', narrowing towards the top, where the steam-pipe B enters. A ball-valve C allows steam to pass into one of the chambers and closes the other. Steam entering (say) the right-hand chamber forces water out of it past the clack-valve V into a delivery passage D, which is connected with an air-vessel. When the water level in A sinks so far that steam begins to blow through the delivery-passage, the water and steam are disturbed and so brought into intimate contact, the steam in A is condensed and a partial vacuum is formed. This causes the ball-valve C to rock over and close the top of A, while water rises from the suction-pipe E to fill that chamber. At the same time steam begins to enter the other chamber A', discharging water from it, and the same series of actions is repeated in either chamber alternately. While the water is being driven out there is comparatively little condensation of steam, partly because the shape of the vessel does not promote the formation of eddies, and partly because there is a cushion of air between the steam and the water. Near the top of each chamber is a small air-valve opening inwards, which allows a little air to enter each time a vacuum is formed. When any steam is condensed, the air mixed with it remains on the cold surface and forms a non-conducting layer. Further, when the surface of the water has become hot the heat travels very slowly downwards so long as the surface remains undisturbed. The pulsometer of course cannot claim high efficiency as a thermodynamic engine, but its suitability for situations where other steam-pumps cannot be used, and the extreme simplicity of its working parts, make it valuable in certain cases. Trials of its performance have shown that under favourable conditions a pulsometer may use no more than 150 lb. of steam per effective horse-power-hour. This consumption, large as it is when judged by the standard of an efficient large engine using steam expansively, does not compare very unfavourably with that of small non-rotative steam-pumps[1].

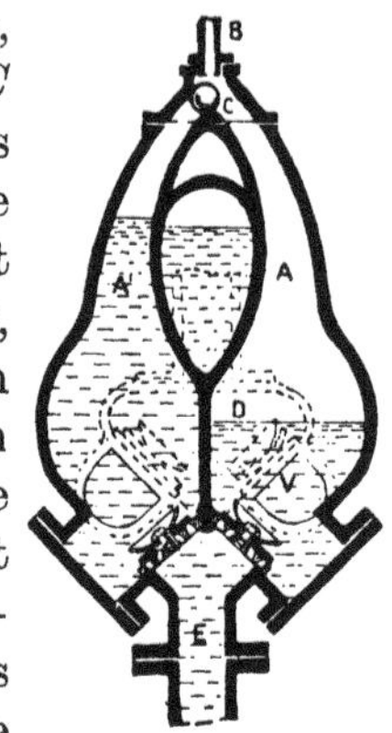

Fig. 261. Pulsometer.

[1] *Proc. Inst. Mech. Eng.* 1893, p. 456.

308. Rotary engines. From the earliest days of the rotative engine attempts have been made to avoid the intermittent reciprocating motion which an ordinary piston engine first produces and then converts into motion of rotation. The design of *rotary* engines, to use the name generally applied to non-reciprocating forms, has exercised the ingenuity of many inventors, with results which have little value or interest except to the student of applied kinematics. Murdoch, the contemporary of Watt, proposed an engine consisting of a pair of spur-wheels gearing with one another in a chamber through which steam passed by being carried round the outer sides of the wheels in the spaces between successive teeth[1]. This device, reversed in function so that it acts as a pump, is often now used for the delivery of oil (§ 153).

In another gear-wheel engine (Dudgeon's) the steam was admitted by ports in side-plates into the clearance space behind teeth in gear with one another, just after they had passed the line of centres. From that point to the end of the arc of contact the clearance space increased in volume; and it was therefore possible, by stopping the admission of steam at an intermediate point, to work expansively. The difficulty of maintaining steam-tight connexion between the teeth and the side-plates on which the faces of the wheels slide is obvious; and the same difficulty has prevented the success of other forms of rotary engine. These have been devised in immense variety, in many cases, it would seem, with the fallacious idea that a distinct mechanical advantage in respect of power was to be secured by avoiding the reciprocating motion of a piston. In point of fact, however, very few forms entirely escape having pieces with reciprocating motion. In all rotary engines, with the exception of steam turbines—where work is done by the kinetic impulse or reaction of steam—there are steam chambers which alternately expand and contract in volume, and this action usually takes place through a more or less veiled reciprocation of working parts[2].

A once interesting example of the rotary type in which reciprocating motion occurred only to a trifling extent was the spherical engine of Beauchamp Tower[3]. This engine was, like

[1] See Farey's *Treatise on the Steam-Engine*, p. 676.

[2] A large number of proposed rotary engines are described, and their kinematic relations to one another are discussed, in Reuleaux's *Kinematics of Machinery*, translated by Sir A. B. W. Kennedy.

[3] *Proc. Inst. Mech. Eng.* March 1885.

several of its predecessors[1], based on the kinematic relations of the moving pieces in a Hooke's joint. Imagine a Hooke's joint, connecting two shafts set obliquely to one another, to be made up of a central disc to which the two shafts are hinged by semicircular plates, each plate working in a hinge which forms a diameter of the central disc, the two hinges being on opposite sides of the disc and at right angles to one another. Further, let the disc and the hinged pieces be enclosed in a spherical chamber through whose walls the shafts project. As the shafts revolve each of the four spaces bounded by the disc, a hinged piece, and the chamber wall will suffer a periodic increase and diminution of volume, between limits which depend on the angle at which the shafts are set. In Tower's engine this arrangement was modified by using spherical sectors, each nearly a quarter sphere, in place of semicircular plates, for the hinged pieces in which the shafts terminate. The shafts were set at 135° to each other. Each of the four enclosed cavities then altered in volume from zero to a quarter sphere, back to zero, again to a quarter sphere, and again back to zero in a complete revolution of the shafts. One shaft was a dummy and ran free, the other was the driving-shaft. Steam was admitted and exhausted by ports in the spherical sectors, whose backs served as revolving slide-valves.

Another rotary engine of the Hooke's-joint family was Fielding's, in which a gimbal-ring and four curved pistons took the place of the disc. Two curved pistons were fixed on each side of the gimbal-ring, and as the shafts revolved these worked in a corresponding pair of cavities fixed to each shaft.

No rotary engine has been permanently successful, and with the advent of the steam turbine, which does all that a rotary engine ever attempted to do, and far more, interest in this class of mechanism may be said to have ceased.

309. Marine engines. The early steamers were fitted with paddle-wheels, and the engines used to drive them were for the most part modified beam engines. Bell's "Comet" (§ 21) was driven by a species of inverted beam engine, and another form of inverted beam, known as the side-lever engine, was for long a favourite with marine engineers. In the side-lever engine the

[1] One of these, the disc-engine of Bishop, was used for a time in the printing-office of *The Times*, but was discarded in 1857.

cylinder was vertical, and the piston-rod projected through the top. From a cross-head on the rod a pair of links, one on each side of the cylinder, led down to the ends of a pair of horizontal beams or levers below, which oscillated about a fixed gudgeon at or near the middle of their length. The two levers were joined at their other ends by a cross-tail, from which a connecting-rod was taken to the crank above. The side-lever engine is now obsolete.

An old form of direct-acting paddle-engine was the steeple engine, in which the cylinder was set vertically below the crank. The piston-rod was divided so that its parts should pass the crank, and they were united by a cross-head sliding in vertical guides, from which a return-connecting-rod led to the crank.

Paddle-wheel engines are sometimes *oscillating cylinder engines* but more generally *diagonal engines*. In the former the cylinders are set under the crank-shaft, and the piston-rods are directly connected to the cranks. The cylinders are supported on trunnions which give them the necessary freedom of oscillation to follow the movement of the crank. Steam is admitted through the trunnions to slide-valves on the sides of the cylinders. In some instances the mean position of the cylinders is inclined instead of vertical; and oscillating engines have been arranged with one cylinder before and another abaft the shaft, both pistons working on one crank. Diagonal engines are direct-acting engines of the ordinary connecting-rod type, with the cylinders fixed on an inclined bed and the guides sloping up towards the shaft.

When the screw-propeller began to take the place of paddle-wheels in ocean-steamers, the increased speed which it required was at first supplied by using spur-wheel gearing in conjunction with one of the forms of engines then usual in paddle-steamers. After a time types of engine better suited to the screw were introduced, and were driven fast enough to be connected directly to the screw-shaft. Several horizontal forms were applied, but later these were abandoned in favour of the inverted vertical engine which is in universal use for those sea-going steamships that have not adopted the steam turbine. In many instances the engine has three cranks, using triple expansion, but the tendency is to prefer four cranks even when triple expansion is retained, especially in ships designed for high speed. With a four-crank triple-engine the steam in the last stage of the expansion is divided between two cylinders usually of the same size. This is done partly to avoid the structural objections to a large cylinder, which would be serious in engines of very high power, and partly for the sake of the better balance which, as was pointed out in § 258, is attainable with four cranks. The four-crank engine without tandem-cylinders is the normal type for large marine engines, whether triple or quadruple, and is rarely departed from unless the power to be developed is so great as to make more than four cylinders desirable.

Slide-valves placed between the cylinders, with reversing gear worked from eccentrics on the shaft, are used in nearly all marine engines, the valves being generally of the piston form for the high-pressure and intermediate-pressure stages and of the flat form for the low-pressure stage (usually double-ported). The slide-valve serves, when "notched up" by the link-motion or other reversing gear, to give such variations of cut-off as are required, without the use of a separate expansion valve. The cylinders of large marine engines, like those of large land engines, have an internal *liner* with a space between it and the outer wall, which is made steam-tight and may be used as a jacket.

The air-pump is often driven by a rocking lever from one of the cross-heads of the main engine, but with large engines it is frequently an independent auxiliary. This is also the case with the circulating, bilge, and feed-pumps. In many cases the circulating pump is of the centrifugal type.

In the last large ships of the Royal Navy to be fitted with reciprocating engines, before the turbine came into general use, the speed had increased until the number of revolutions per minute ranged from 120 to 145, with a piston speed of 1000 feet per minute. In destroyers the engines were designed to run at 400 revolutions per minute with a piston speed of about 1200 feet per minute. In the engines both of the larger and smaller ships forced lubrication was introduced in the crank-pins, eccentrics and main bearings.

Surface condensation was introduced in marine practice by S. Hall in 1831, but was not brought into general use until much later. Previous to this it had been necessary, in order to avoid the accumulation of too dense brine in the boiler, to blow off a portion of the brine at short intervals and replace it by sea water. By the adoption of surface-condensers it became possible to use the same feed-water over and over again.

310. Locomotives. The ordinary locomotive consists of a pair of direct-acting horizontal or nearly horizontal engines, fixed in a rigid frame under the front end of a boiler of the type described in § 277, and coupled to the same shaft by cranks at right angles, each with a single slide-valve worked by a link-motion, or by a form of radial gear. The engine is non-condensing, and the exhaust steam, delivered at the base of the funnel through a blast-pipe, serves to induce a draught of air through the furnace. The demand for power in a locomotive changes frequently and quickly with changes of gradient in the line; and this method of forcing the draught has the great practical convenience that it makes the production of steam respond automatically to variations in the rate at which steam is taken by the engine. In some instances a portion of the exhaust steam, amounting to about one-fifth of the whole, is diverted to heat the feed-water. In tank engines the feed-water is carried in tanks on the engine itself; in other engines it is carried behind in a tender.

On the shaft which the engine drives are a pair of driving-wheels, which furnish tractive force by their frictional adhesion to the rails. A single pair of driving-wheels being insufficient to furnish as much tractive force as is required, two, three, or more pairs are used, and these are connected together by *coupling-rods* outside the wheels, which cause the work to be divided between as many of the wheels as are coupled in this way, and enable a greater fraction of the whole weight of the locomotive to be utilized in furnishing the frictional adhesion. Thus we have what are called "four-coupled," "six-coupled," "eight-coupled," and even "ten-coupled" locomotives. For express traffic the four-coupled and six-coupled arrangements are the most usual, with driving-wheels up to 6 ft. 9 ins. in diameter: for goods traffic with its heavier and slower trains the engines are more generally six- or eight-coupled with smaller driving-wheels. In *inside-cylinder* engines the cylinders are placed side by side within the frame of the engine, and their connecting-rods work on cranks in the driving shaft. In *outside-cylinder* engines the cylinders are spread apart far enough to lie outside the frame of the engine, and to work on crank-pins on the outsides of the driving-wheels, thereby dispensing with the cranked axle. In some locomotives there is a combination of two outside cylinders with either one or two inside the frames. The use, in this way, of three or four cylinders not only augments the power but promotes smoothness of running, and spares the bridges and other parts of the permanent way a great part of the "hammer-blow" or pulsating force that is caused by dynamic effects of revolving balance weights (§ 262).

The leading wheels are rarely used as drivers; it is preferred to have under the front of the engine two or four smaller wheels which do not form part of the driving system. These are carried in a *bogie*, which is a small truck supporting the front end of the frame on a swivel-pin or equivalent device which allows the bogie to turn, so that it may adapt itself to curves in the line, and thus obviate the grinding of tyres and danger of derailment which would be caused by using a long rigid wheel-base. The bogie appears to have been of English origin[1]; it was first brought into general use in America, but it is now a common feature of locomotives everywhere. Instead of a four-wheeled bogie, a single pair of leading wheels is also used, carried by a *pony* truck, which has a swing-bolster pivoted by a radius bar about a point some distance behind the axis of the wheels. Any such support must provide lateral as well as radial freedom, both being required if the wheel-base is to be properly accommodated to the curve. Another method of getting lateral and radial freedom was used by F. W. Webb, who set the leading axle in a box curved to the arc of a circle, with freedom to slide laterally for a short distance, under the control of springs[2]. On a long wheel-base the trailing wheels are also mounted so that they have lateral freedom.

Locomotives are often classified by stating successively the numbers of leading-wheels, driving-wheels, and trailing-wheels. Thus a locomotive of what is called the "Atlantic" type, in which there is a four-wheeled bogie in front, then coupled driving-wheels on two axles, and then an axle with a pair of trailing-wheels, is described as a 4-4-2. Similarly a 4-6-0 means an engine with a bogie, three coupled driving-axles, and no trailer; and a 4-6-2, which is called the "Pacific" type, differs from it in having a pair of trailing-wheels added. Engines for goods and mineral traffic are often 2-8-0 or 2-8-2, there being four coupled driving-axles. We shall have other examples of this nomenclature later.

Slide-valves of the piston type are commonly used instead of the flat D-shaped slide-valve which in the early days of the locomotive

[1] *Min. Proc. Inst. C. E.* vol. LIII, p. 50.
[2] *Proc. Inst. Mech. Eng.* 1883.

was universal. The Stephenson and other forms of link-motion requiring two eccentrics have given place generally to the Walschaerts valve-gear. This gear, which was for long a feature of continental practice, is now very usual in British and American locomotives also.

The outside-cylinder type is universal in America with piston-valves on the tops of the cylinders. Outside-cylinder engines are very general on the Continent, and increasingly common in England, owing to the difficulty of finding room within the frame for the larger cylinders which are required in consequence of the need of augmented power. Engines are now set to draw much heavier loads than was formerly the case, and often at higher speeds. Their boilers have been enlarged and their weight greatly increased, but this growth necessarily takes place subject to the limitations that are imposed not only by the rail gauge and the permissible load on each axle, but (what is more serious) by the "loading gauge," which depends on the dimensions of tunnels, etc.

Among British locomotives conspicuous in 1926 are Mr Gresley's three-cylinder "Pacifics" (4-6-2) on the L.N.E. Railway. Two of the cylinders are outside and one between the frames. All three have a diameter of 20 inches and a stroke of 26 inches, and drive the midmost pair of the six-coupled wheels, the diameter of which is 6 ft. 8 ins. The lateral dimensions and height of these engines are as great as the loading gauge will allow. They develop close on 2000 horse-power, with a tractive effort of about 30,000 lb.; the adhesive weight on each pair of drivers is 20 tons.

A slightly higher horse-power and tractive effort is found in the Great Western Railway 4-4-0 express engines of the "Castle" class. They have four cylinders, each 16 inches in diameter with a stroke of 26 inches, and work at a pressure of 225 lb.; the diameter of the drivers is 6 ft. 8½ ins. Two of the cylinders are inside and drive the foremost of the coupled axles; the other two are outside and drive the rear axle.

Both of these are non-compound. On the L.M.S. system, besides locomotives of other types, Sir Henry Fowler continues to build compound three-cylinder engines with one high-pressure cylinder of 19¾ inches diameter between the frames and two low-pressure cylinders of 21¾ inches outside, the stroke for all being 26 inches. The diameter of the drivers is 6 ft. 9 ins. For traffic of another class he also builds 0-10-0 engines where all the weight (nearly

74 tons) is taken by the coupled drivers, with four cylinders (non-compound), each $16\frac{3}{4}$ inches in diameter by 28 inches stroke.

In France locomotives are generally compound. Express engines of the Nord railway are six-coupled, often 4-6-4, with four cylinders arranged as De Glehn compounds, namely, two outside high-pressure driving the middle pair of coupled wheels, and two inside low-pressure driving cranks on the axle of the leading pair.

Among more powerful American types there are 2-10-0 engines built by the Baldwin Locomotive Works for goods traffic on the Pennsylvania lines, with two (outside) cylinders $30\frac{1}{2}$ inches in diameter by 32 inches stroke, working at a pressure of 250 lb., exerting a tractive force of 87,000 lb .and carrying on the driving-wheels a weight of 157 tons.

Few American locomotives are now built with compounded cylinders, the introduction of superheating and feed-water heating having been found to secure much the same advantages. But exceptionally powerful compound locomotives of the Mallet articulated type built some years ago are still doing good service. In the Mallet articulated engines the driving-wheels are coupled in two separate groups, the foremost group being carried by a truck which permits of a little sideways and angular movement with respect to the boiler, just as in a bogie, thus making possible a very long wheel-base. The boiler extends forward beyond the main frame, which carries the other group of driving-wheels, and its projecting end rests on the truck in front. The set of driving wheels in the truck are driven by a pair of low-pressure cylinders which the truck carries, and a pair of high-pressure cylinders fixed to the main frame drive the other set. All four cylinders are of the outside pattern[1]. This arrangement, originally introduced in France, has been largely used on American railroads, in forms such as 0-8-8-0, 2-8-8-2, and 2-10-10-2. Certain 2-10-10-2 Mallet engines built in America about 1918 may claim to be the largest locomotives in the world. They have two high-pressure cylinders 30 inches in diameter by 32 inches stroke and two low-pressure 48 inches in diameter by 32 inches stroke. The twenty drivers, with a diameter of 56 inches, carry a total weight of 617,000 lb., or more than $27\frac{1}{2}$ tons on each axle. The tractive effort is

[1] See Mr Mallet's paper on compound articulated locomotives, *Proc. Inst. Mech. Eng.*, July 1914.

147,000 lb., but this may be increased to 176,000 for starting, by admitting boiler steam to the low-pressure cylinders.

The Garratt locomotive is another articulated design, with two groups of coupled drivers, each group with its own cylinders carried on a separate truck. The boiler is supported on a beam between the two trucks, its weight borne by them through a pin joint at either end. A 2-8-8-2 Garratt engine[1] with six cylinders (non-compound), three on each truck, was introduced in 1925 on the L.N.E. Railway for dealing with coal traffic on a heavy gradient. The weight is about 18 tons or each coupled axle, and the tractive effort is nearly 73,000 lb.

[1] Designed by Mr Gresley and built by Beyer Peacock and Co. See *The Locomotive*, July 1925. An account of another 2-8-8-2 Garratt engine is given in the same journal, of date Feb. 1926.

CHAPTER XVI

AIR-ENGINES

311. Air-engines with external or internal combustion. The term air-engine may be used in a restricted sense to denote an engine in which the working substance is atmospheric air, or it may be extended to apply to any heat-engine which employs a gaseous working substance, as distinguished from a vapour which becomes condensed during some part of the cycle of operations. In this more extended sense the term would include engines in which the working substance is the mixed gas resulting from the combustion of fuel, whether gaseous, liquid or solid, within the engine itself. In other words, it would include gas-engines and oil-engines. These will be more particularly considered in the next chapter, but some of the remarks which follow, relating to air-engines in general, apply to them as well as to engines using atmospheric air.

When air alone forms the working substance, it receives heat from an external furnace by conduction through the walls of a containing vessel, just as the working substance in the steam-engine takes in heat through the shell of the boiler. An engine supplied with heat in this way may be called an *external-combustion* engine, to distinguish it from the very important class of engines in which the combustion which supplies heat occurs within a closed chamber containing the working substance. These last are *internal-combustion* engines.

By far the most common kind of internal-combustion engines are those in which the heat constituting the supply is generated within the cylinder itself by the combustion there of inflammable gas or vapour along with air. Often the two are mixed together before ignition, so that the combustion takes place in an explosive manner. But besides these there have been a few engines in which the combustible is solid fuel burning in a vessel apart from the cylinder, and the working substance takes up heat by coming into direct contact with the burning fuel. The air which maintains the combustion is itself the working substance. Mixed with products of combustion it passes into the cylinder to do work there; and so far as the action in the cylinder is concerned the engine may operate in the same manner as if the working gas were heated by conduction from an external furnace. Engines of this class will be briefly noticed in the present chapter, along with air-engines of the kind that are externally fired.

312. Advantage in respect of temperature range. Compared with engines using saturated steam, engines using air or other gases have the advantage that the temperature and the pressure of the working substance are independent of one another. When the working substance is saturated water-vapour, the upper limit of temperature is comparatively low in consequence of the high pressure with which high temperature is, in such cases, necessarily associated. But in an air- or gas-engine it becomes possible to use a much higher upper limit, and if the lower limit is not correspondingly raised an increase of thermodynamic efficiency results. It is true that the upper limit of temperature may be raised in the case of steam, by superheating; but even then a steam-engine continues to take in the greater part of its supply of heat at the comparatively low temperature at which the water is vaporized, and the direct thermodynamic advantage of the superheat is consequently small.

So long as external combustion is used, there must still be some considerable drop in temperature, of an irreversible and therefore wasteful kind, between the temperature in the furnace and the temperature at which the working air receives its heat, since without this no sufficiently rapid transfer of heat from one to the other could occur. Internal-combustion engines have the advantage that the temperature which is produced in the combustion is itself the upper limit in the thermodynamic cycle. This gives them, from the thermodynamic point of view, a great superiority, for the temperature reached in combustion is immensely higher than any temperature that it is practicable to give to a working substance so long as the supply of heat has to pass through the wall of a containing vessel capable of resisting pressure. When the heat has to pass through iron a temperature exceeding that of a low red heat, say 600° C. or 700° C., is out of the question, and in any actual air-engines of this class the maximum has been much lower. But in the cylinder of a gas- or oil-engine the temperature reached in the explosion is very usually 1600° C. and may approach 2000° C. Such temperatures are practicable within the cylinder because the heat that produces them does not have to pass through the metal, and the metal itself may be kept cool by means of a water-jacket so that it forms a safe container of the excessively hot gas.

313. Air-engine using Carnot's cycle. A theoretically simple and thermodynamically perfect form of external-combustion air-engine would be one following Carnot's cycle, in which heat is received while the air is at the highest temperature T_1, the air meanwhile expanding isothermally. After this the supply of heat is stopped, and the air is allowed to expand adiabatically until its temperature falls to the lower extreme T_2. At this it is compressed isothermally, giving out heat, and finally the cycle is completed by adiabatic compression, which restores the initial high temperature T_1. The indicator diagram for this cycle has been sketched in fig. 13, § 43. Practically, this action would be attended by the serious drawback that the volume to be swept through by the piston would be very great in relation to the work done. The inclination of adiabatic to isothermal curves for a gas is slight, and hence the area of the diagram, or the effective work done per revolution, is small in comparison with the two quantities of which it is the difference, namely the work done by the substance during the forward stroke and the work spent upon it during the backward stroke. The mean effective pressure would be very small in comparison with the greatest pressure for which provision would have to be made in designing the parts for strength. An air-engine using Carnot's cycle would consequently be excessively bulky and heavy, and also mechanically inefficient.

314. External combustion air-engine with regenerator: Stirling's air-engine. This objection is lessened when the use of a regenerator (§ 52) is substituted for the adiabatic steps of the Carnot cycle. In Stirling's engine, where the regenerator was first used, the working substance was cooled from the upper limit T_1 to the lower limit T_2 by passing in one direction through a regenerator, which stored the heat it extracted from the gas in such a way that when the gas was passed through the

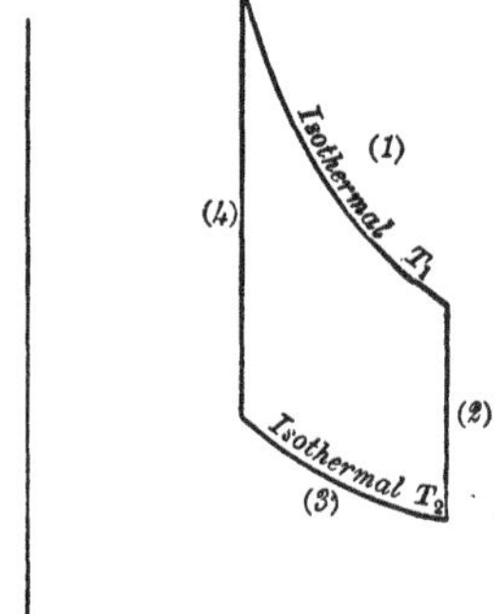

Fig. 262. Ideal Indicator diagram of Air-engine with Regenerator (Stirling).

regenerator in the opposite direction the heat was again taken up and the temperature consequently rose from T_2 to T_1. The cycle of operations has been described in § 53, and an ideal indicator diagram sketched there is reproduced in fig. 262.

Several forms of engine were designed by Stirling in which the action approximated to the cycle of that figure. The characteristic parts of one of them are shown in section in fig. 263. A is the

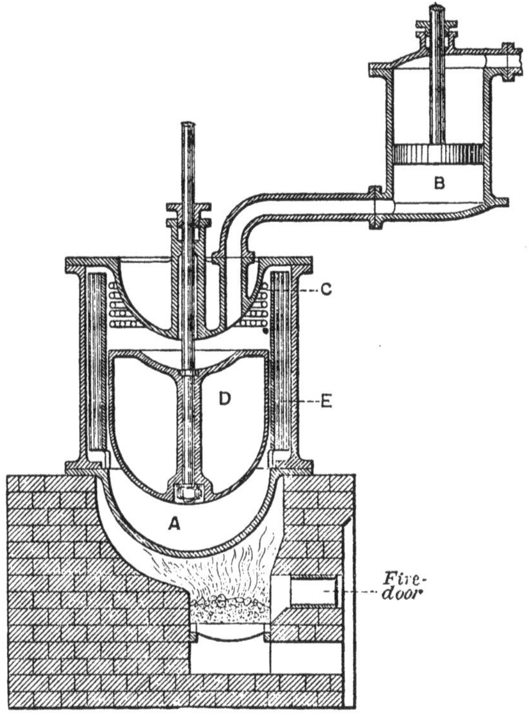

FIG. 263. Stirling's Air-engine.

heater, a closed iron vessel containing air, externally heated by a furnace beneath it. A pipe from the top of A leads to the working cylinder B. At the top of A is a cooler C, consisting of pipes through which cold water circulates. In A there is a displacer plunger D, which is driven by the engine; when this is raised the air in A is in the lower part of the vessel and is consequently taking in heat from the furnace, whereas when D is lowered the air in A is transferred from the lower to the upper part and is thereby brought into contact with the cooler. On its way from the bottom to the top of A, or from the top to the bottom, the air must pass through an annular lining of wire-gauze E. This is the regenerator, and the air in passing up through it becomes cooled, and in passing down again through it becomes heated. At the beginning of the cycle D is at or near its highest position. The air is then receiving heat at temperature T_1, and is expanding isothermally; this is the first stage in the diagram. Then the plunger D descends. The air is driven through the regenerator, where it deposits heat, and its temperature on emerging at the top is T_2. Next, the working piston makes its down-stroke; this compresses the air isothermally, the heat produced by compression being taken up by the cooler C. Finally the plunger is raised, and the working air again passes down through the regenerator, taking up the heat it left there, and rising in temperature to T_1. In the actual engine the working cylinder was double-acting, another heating vessel, precisely like A, being connected with the cylinder B above the piston.

The actual forms in which Stirling's engine was used are described in two patents by R. & J. Stirling (1827 and 1840[1]). An important feature in them was that the air was compressed by means of a pump which formed an additional organ of the engine, so that its average pressure was kept much above that of the atmosphere. The pressure was commonly as much as 150 lb. per sq. inch with the air cold. Stirling's cycle is theoretically perfect whatever be the density of the working air, and the use of compression affects the theoretical thermodynamic efficiency only if the ratio of adiabatic expansion and compression be altered. But it gives a higher mechanical efficiency, and also, what is of special importance, it increases the amount of power developed by an engine of given size. To see this it is sufficient to consider that with compressed air a greater amount of heat is dealt with in each stroke of the engine, and therefore a greater amount of work is done. Practically the use of compressed air also increases the thermodynamic efficiency by reducing the ratio of the heat wasted by external conduction and radiation to the whole heat.

A double-acting Stirling engine of 50 I.H.P., used in 1843 at the Dundee foundry, appears to have realized an efficiency of 0·3.

[1] The 1827 patent is reproduced in Fleeming Jenkin's Lecture on Gas and Caloric Engines, *Inst. Civ. Eng.*, Heat Lectures, 1883–84. See also *Min. Proc. Inst. C. E.* 1845 and 1854.

Notwithstanding very inadequate means of heating the air it consumed only 1·7 lb. of coal per I.H.P.-hour[1]. This engine remained at work for three years, but was finally abandoned on account of the failure of the heating vessels, though the highest temperature reached by the air was only about 343° C. In one form of the engine as described in Stirling's patent the regenerator was a separate vessel; in another the plunger D was itself constructed to serve as regenerator by filling it with wire-gauze and leaving holes at top and bottom for the passage of the air through it.

315. Ericsson's air-engine. Another mode of using the regenerator was introduced in America by Ericsson, in an engine which also failed, partly because the heating surfaces became burnt, and partly because their area was insufficient. In Ericsson's engine, which was tried on a considerable scale in the steamship "Caloric," the temperature of the working substance was changed by passing through the regenerator while the pressure remained constant. Cold air was compressed by a pump into a receiver, from which it passed through a regenerator into the working cylinder. In so passing it absorbed heat from the regenerator and expanded. The air in the cylinder was then allowed to expand further by taking in heat from a furnace under the cylinder until its pressure fell to near that of the atmosphere. The cycle was completed by the discharge of the air through the regenerator. The indicator diagram approximates to a form bounded by two isothermals and two lines of constant pressure[2].

Compared with Stirling's engine Ericsson's had the practical disadvantage that the mean pressure of the working substance was much lower, the lower limit of the pressure range being the pressure of the atmosphere. This required the engine to have an excessive bulk relatively to its power. With four working cylinders, each 14 feet in diameter, and a stroke of six feet, it developed only about 300 indicated horse-power, the mean effective pressure being only a little over 2 lb. per sq. inch. The consumption of coal was reported to be 1·87 lb. per I.H.P.-hour, but a large part of the indicated power must have been expended in overcoming the friction of the engine itself.

[1] See Rankine's *Steam-Engine*, p. 367. The consumption per brake H.P. was much greater.

[2] For a diagram of Ericsson's engine see *Proc. Inst. Mech. Eng.* 1873.

316. Modern air-engines of the Stirling type. Externally-heated air-engines are now employed only for very small powers—from a fraction of 1 H.P. up to about 3 H.P. With the coming of the internal-combustion engine their development ceased to have interest or importance. Powerful engines of the type would be scarcely practicable, on account not only of their excessive bulk

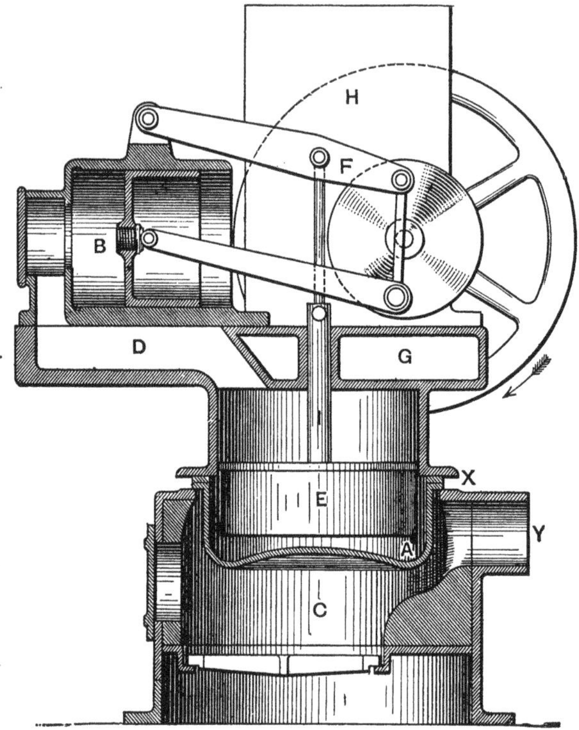

FIG. 264. Robinson's form of Stirling Engine.

but also of the difficulty which would be experienced in heating large quantities of air. By keeping the working substance highly compressed, giving it a mean density much in excess of that of atmospheric air, the bulk of the working cylinder and displacer might be reduced, but the difficulty would remain of getting enough heating surface and of preserving the heater from being burnt through its exposure to oxygen at a high temperature. The small externally-fired air-engines that are now manufactured resemble the original Stirling engine very closely in the main features of their action, and comprise essentially the same organs.

One of these modern Stirling engines is a small domestic motor designed by H. Robinson (fig. 264). It has no compressing pump and the mean pressure of the working air is equal to that of the atmosphere. The range of pressure is slight—so slight indeed that no packing is needed in the piston or other working parts—and the engine develops only a fraction of one horse-power. It has been used for such purposes as pumping water for the supply of country houses, or driving laboratory appliances.

A is the heater and displacer cylinder; B is the working cylinder, which communicates with A by a passage D. A is heated externally by a small coke fire at C or by a gas flame from a Bunsen burner. The displacer E takes its motion from a rocking lever F connected by a short link to the crank-pin, and is about 90° in phase ahead of the working piston. In the figure the displacer is at the bottom of its stroke and the piston has still half the back stroke to perform. The displacer E is itself the regenerator, its construction being such that the air passes up and down through it as in one of the original Stirling forms. On the top of the displacer cylinder is a water-vessel G, which is the cooler, and this is kept in communication with the circulating water-tank H. The account which has already been given of the Stirling cycle will serve as a description of the action in this engine. A conspicuous feature is that there are neither valves, packing, nor glands; but the absence of compression, which makes this possible, limits the efficiency of the engine as well as its power.

A somewhat larger engine of the Stirling type, called the Rider engine, is shown in fig. 265. It follows substantially but not exactly the Stirling cycle. A and B are two cylinders, open at the top, with plunger pistons C and D, which are connected to cranks nearly at right angles. Between the two is the regenerator H. Round the lower part of C is the cooler E, a jacket through which cold water is kept circulating. Under the lower part of B is the furnace F which heats the air contained in the space G below the plunger D. In the position shown in the figure, D is rising, and C is just beginning to rise. Nearly all the working air has been compressed into G and is expanding as it takes in heat, doing work against the plunger D and also against C. By the time D reaches the top of its stroke C is about half-way up: air is passing rapidly through the regenerator from G to the space under C, and is cooled first by the regenerator and then by the water-jacket E. As D comes down this transfer of the air continues and the pressure

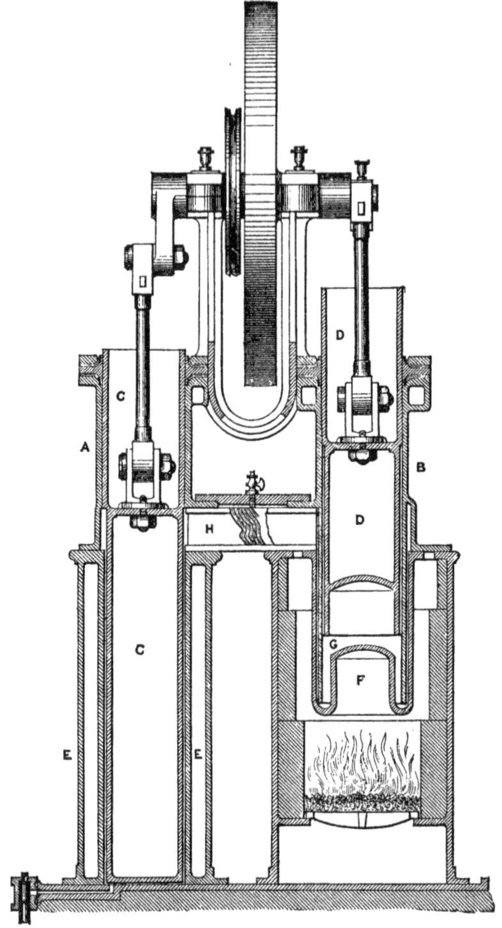

Fig. 265. Rider's Air-engine.

falls. Then C follows, compressing the air beneath it while the cooler E absorbs the heat, and finally forcing all the working air back through the regenerator into the heater, when the cycle begins again. The maximum pressure reached during the cycle is about 20 lb. per sq. inch.

The action is of course continuous, but we may broadly distinguish the following four stages:

(1) The air, previously compressed to a small volume (in G), takes in heat at its highest temperature and expands, doing work on D and subsequently to some extent on C.

(2) After this expansion it is transferred through the regenerator to the cold cylinder A, storing heat in the regenerator and losing pressure. During this process little work is done on or by the air, since the actions on the plungers nearly balance. In other words, the volume does not materially change.

(3) The air which is now in A, expanded to large bulk and at a low temperature, is compressed by the descent of C and gives out heat to the cooler E. During this process work is done upon the air by the fly-wheel.

(4) The compressed air is transferred through the regenerator to G, rising in temperature and pressure. In this process, again, little work is done by or upon the air.

317. Internal-combustion air-engines. The difficulty already referred to of getting heat into and out of a gaseous working substance is a fundamental objection which has prevented the external-combustion air-engine from coming into use on any large scale. The activity of a heating surface is vastly greater when the substance that is being heated is changing its state from liquid to vapour than when the substance is already a gas. And similarly a gas that is being cooled by conduction through a surface will part with its heat far less readily than a vapour which is being condensed in the process.

This objection applies when the air-engine is of the external-combustion type, but so far as the heating process is concerned it is removed by causing the combustion to occur within the engine itself. So far as cooling is concerned the difficulty also disappears when the substance which is to reject heat is expelled to the atmosphere instead of being used over again after cooling. Hence it has been possible for the internal-combustion engine to attain a much greater efficiency.

318. Internal-combustion engines using solid fuel. The earliest practical example of the internal-combustion engine (if we leave guns out of account) appears to have been the hot-air engine of Sir George Cayley[1], of which Wenham's[2], Buckett's[3] and Bénier's engines are more recent forms. In these engines coal or coke is burnt under pressure in a closed chamber, to which the fuel is fed through a species of air-lock. Air for combustion is supplied by a compressing pump, and the engine is governed by means of a distributing valve which supplies a greater or less proportion of the air below the fire as the engine runs slow or fast. The products of combustion, the volume of which is increased by their rise in temperature, pass into a working cylinder, raising the piston. When a certain fraction of the stroke is over the supply of hot gas is stopped, and the gases in the cylinder expand, doing more work and becoming reduced in temperature. During the return stroke they are discharged into the atmosphere, and the pump takes in a fresh supply of air. Fig. 266 is a diagram section of the Buckett engine. A is the working piston, the form of which is such as to protect the tight sliding surface (at the top) from contact with the hot gases; B is the compressing pump, and C is the valve by which the governor regulates the rate at which fuel is consumed by admitting more or less of the air under the grate through the channel F. D is the air-lock and hopper through which fuel is supplied, and E is the exhaust-valve through which the products of combustion are finally expelled. In a reported trial of an engine indicating 20 horse-power the number of pounds of fuel burnt per hour was 1·8 per I.H.P. and 2·5 per B.H.P.

In an engine such as this it was practicable to let the working substance reach a higher temperature than could be reached with external firing, and in fact in Wenham's engine it was about 600° C. But a limit was still imposed by the fact that the hot gas had not only to be contained in iron vessels, under pressure, but had to go through passages and valves between the combustion chamber and the cylinder. Hence in practice the highest temperature in the working cycle was necessarily much lower than it is in a gas-engine or oil-engine, where no transfer of the working substance takes place after it is heated by internal combustion, until it has cooled itself by doing work.

In engines of this class the degree to which the action is thermodynamically efficient depends very largely on the amount of cooling the gases undergo by adiabatic or nearly adiabatic ex-

[1] *Nicholson's Art Journal*, 1807. See also *Min. Proc. Inst. C. E.* vol. IX.

[2] *Proc. Inst. Mech. Eng.* 1873.

[3] Fleeming Jenkin, *loc. cit.*

pansion under the working piston. Without a large ratio of expansion the thermodynamic advantage of a high initial temperature is lost. In any kind of internal-combustion engine the gases have to be discharged at atmospheric pressure, and consequently a large ratio of expansion is possible only when there is much initial compression. Compression is therefore an essential condition, without which a heat-engine of this type cannot be

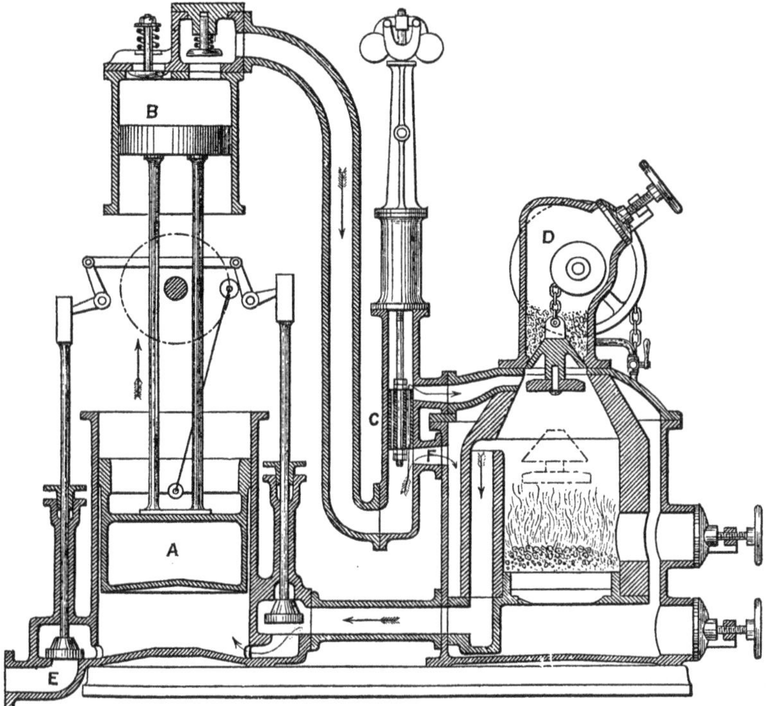

FIG. 266. Buckett's Internal-Combustion Air-engine.

made efficient. It is also, as has already been pointed out, an essential feature in any air-engine which is to develop a fair amount of power without excessive bulk. When we come to discuss the theoretical action of gas- and oil-engines we shall see how dependent the thermodynamic efficiency is on the amount of compression which the working substance undergoes before ignition takes place.

319. Advantages of gaseous or liquid fuel for internal combustion. Internal-combustion engines using solid fuel have been but little used, and that only for small powers. But the use of liquid and gaseous fuel has given the internal-combustion engine a position of great and constantly increasing importance. The fundamental difference is that gas or oil used as fuel for internal combustion leaves no ash—all the products are gaseous—and further, the process of combustion may go on intermittently with a separate admission and ignition of the fuel, in suitably regulated quantity, for each working stroke. For these two reasons combustion may take place in the working cylinder: there is no need for it to proceed in a chamber separated from the cylinder by valves, as in Buckett's or any other solid-fuel engine.

This has two conspicuous effects. It makes automatic action practicable, by dispensing with a furnace; it also greatly increases the fraction of the potential energy of the fuel which the engine may convert into work. Incidentally, it reduces the weight, making it possible to obtain a moderate number of horse-power with a much less heavy mechanism.

Compared with a steam plant, the internal-combustion engine using gaseous or liquid fuel is, or may be, thermodynamically more efficient, much more independent of attendance, and (except in very large sizes) lighter in relation to the power developed.

These advantages have led to a rapid development in the design of such engines and in the applications to which they are put. The gas-engine, at first using ordinary illuminating gas as fuel, established its position as a convenient prime-mover for domestic purposes and in small workshops or factories. Its uses are too familiar to require comment. The petrol engine revolutionized road traffic and made flying possible. More recently the internal-combustion engine burning heavy oil has entered into serious competition with steam in cases where economy of fuel mainly determines the choice. For ship propulsion, especially, it has made a notable advance. About half the tonnage built in 1925 was engined with oil motors of the Diesel type, and these have been applied to some vessels of almost the largest size. For example, the twin screws of the motor ship "Asturias," of 22,000 tons, put into service in 1926, are driven by a pair of oil-engines each indicating 10,000 horse-power. Even this large output has already been exceeded by an engine which is designed to give 15,000 brake horse-power in a single unit.

The theory and development of internal-combustion engines using gaseous or liquid fuel will be dealt with in the next chapter.

CHAPTER XVII

GAS-ENGINES AND OIL-ENGINES

320. Early gas-engines. Lenoir. In this chapter we have to deal with internal-combustion engines in which the fuel is a combustible gas or liquid. The first successful engines of this class were supplied with fuel in the form of a gas. Gas-engines in this sense have become an important source of power, but a much larger part is now played by engines which use fuel that is liquid at ordinary temperatures. These may be collectively called oil-engines: they include engines taking a readily vaporizable spirit such as petrol or benzol, which becomes a vapour or very fine spray in the carburettor while it is being drawn into the cylinder along with the necessary supply of air, or a light oil such as paraffin (kerosene) which is usually vaporized by contact with hot metal and mixed with air before compression and ignition, or a heavy oil of the kind that serves for injection into the cylinder of a Diesel engine where the air is already compressed and highly heated by compression. It was not, however, until the gas-engine had demonstrated the practicability and advantage of internal combustion that the various forms of liquid-fuel engine were evolved. The subject is now one of such magnitude that only a few of its more salient points can be touched on here.

The first gas-engine to be brought into practical use was Lenoir's. It dates from 1860, but before that several inventors had described engines none of which got beyond the experimental stage. In 1820 the Rev. W. Cecil read a paper before the Cambridge Philosophical Society giving details of experiments made with an engine of his own design driven by the explosion of a mixture of hydrogen and air. It is probable that this was the earliest gas-engine to be actually worked. A patent by W. Barnett (1838) makes the important proposal to use compression of the explosive mixture: it also describes a method of ignition which was afterwards applied for a long time in other engines with success. Another patent by Barsanti and Matteucci, in 1857, describes a free-piston engine of a type which afterwards took practical form in the hands of Otto and Langen. But none of these, or other, early proposals bore immediate fruit, and it was not until Lenoir brought out a design in which various difficulties were overcome that the gas-engine entered on its practical career[1].

Lenoir's engine was of great simplicity. There was no compression, and in this respect the action was seriously defective from the thermodynamic point of view, and was a retrograde step after Barnett's proposal. During the early part of the stroke air and gas, in proportions suitable for combustion, were drawn into the cylinder. At about half-stroke the inlet valve closed, and the mixture was immediately exploded by an electric spark. The heated products of combustion then did work on the piston by expanding during the remainder of the forward stroke, and were expelled during the back stroke. The engine was double-acting, and the cylinder was prevented from becoming excessively heated by a casing through which water was kept circulating. This water-jacket is a feature that is retained in most gas-engines and oil-engines. When they are large it is indispensable as a means of keeping the working surfaces of the piston and cylinder cool enough to permit of lubrication. In Lenoir's engine every stroke was active, two explosions taking place per revolution, one on each side of the piston.

An indicator diagram from a Lenoir engine is shown in fig. 267. After the explosion the line falls, partly from expansion, and partly from the cooling action of the cylinder walls. In consequence of there being no compression of the explosive mixture before the combustion takes place the amount of subsequent expansion is very small: the final volume is less than double the volume after explosion. Hence the amount of cooling through expansion in this engine was small, and the efficiency low; the waste heat discharged in the products of combustion as they leave the cylinder, or taken up by the water-jacket by conduction through the walls, formed a very large part of the whole heat generated. Only some 4 per cent. of the energy of the gas was converted into work, whereas in a good modern gas-engine over 30 per cent. is converted. Notwithstanding what we should now consider its excessive consumption of gas the Lenoir engine found considerable application between 1860 and 1870. Hugon's engine, introduced five years

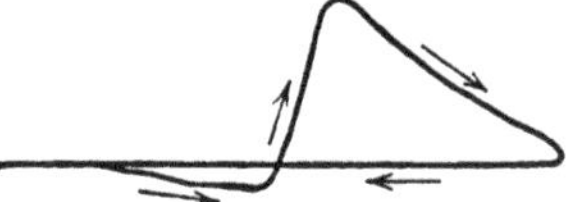

FIG. 267. Lenoir engine diagram.

[1] For an account of the early history of the gas-engine see Sir Dugald Clerk's book on *The Gas, Petrol, and Oil Engine*, vol. I, 1909.

later than Lenoir's, was very similar. A novel feature in it was the injection of a jet of cold water to keep the cylinder from becoming too hot. These non-compressive engines are now obsolete; the last to survive was a small engine designed by Bischoff, the mechanical simplicity of which gave it an advantage in cases where but little power was required.

321. Otto and Langen's atmospheric gas-engine. In 1866 Otto and Langen gave working shape to suggestions made as early as 1857 by Barsanti and Matteucci, by introducing a free-piston engine which, as to economy of gas, was distinctly superior to engines of the Lenoir type[1]. Like them it did not use compression. The explosion occurred early in the stroke, in a vertical cylinder, under a piston which was free to rise without doing work on the engine-shaft. The piston rose with great velocity, so that the expansion was much more nearly adiabatic than in earlier engines, and the ratio of expansion was greater, for the momentum of the piston carried it so far that the pressure of the gas fell below that of the atmosphere. After the piston had reached the top of its range the gases became further cooled by giving up heat to the walls of the cylinder, and, their pressure being below that of the atmosphere, the piston came down, this time in gear with the shaft, and doing work upon it. The burnt gases were discharged during the last part of the down-stroke. A coupling like the free-wheel gear of a bicycle allowed the piston to be automatically thrown out of gear with the shaft when rising, and into gear when descending. This "atmospheric" gas-engine came into somewhat extensive use in spite of its noisy and spasmodic action. It achieved an indicated thermal efficiency of about 16 per cent. in favourable cases, and about 11 per cent. on the brake. After a few years it was displaced by a greatly improved type, in which the direct action of Lenoir's engine was restored, but the gases were compressed before ignition.

322. The four-stroke cycle of Beau de Rochas and Otto. The advantage of compressing the explosive mixture before igniting it, in order to make the subsequent expansion large, was recognized by several of the early inventors. It was discussed in a remarkable pamphlet published by Beau de Rochas in 1862. He pointed out that compression might be carried to any extent

[1] *Proc. Inst. Mech. Eng.* 1875.

short of that which would cause the mixed gas to explode in consequence of its heating effect. He further suggested a means of compressing the explosive mixture without using a separate compressing pump. His plan was to have the following four operations take place, on one side of the working piston, during four successive strokes or two revolutions of the crank-shaft.

(1) Drawing in the charge of gas and air during one whole stroke of the piston.

(2) Compression during the return stroke (into a comparatively large clearance space behind the piston).

(3) Ignition at the dead-point, followed by expansion during the third stroke.

(4) Discharge of the burnt gases from the cylinder during the fourth and last stroke.

This was the earliest account of the "four-stroke" cycle of operations which is now used in most gas-engines. Beau de Rochas further pointed out that, besides compression, high speed and small cylinder surface were conditions to be aimed at as favourable to economy. Extremely valuable as were the suggestions contained in his pamphlet they were for a long time unproductive. It was not till 1876 that Otto, who with Langen had been engaged in the manufacture of the "atmospheric" engine, introduced the highly successful gas-engine in which this action is carried out. The new engine was called the Otto "silent" engine to distinguish it from its noisy predecessor, the engine of Otto and Langen. It was the first gas-engine to come widely into use, and it formed the model to which, after the expiry of Otto's patent, other gas-engines were indebted for the chief features of their action. The manufacture of the Otto engine in England by Messrs Crossley led to its rapid introduction. At first the only fuel available was illuminating coal-gas, and the engine was used for rather small powers only, in which the cost of gas fuel, though greater than that of the fuel for a steam-engine of corresponding power, was more than balanced by the greater convenience and saving in attendants' wages. For a good many years nearly all gas-engines were of less than 20 horse-power. In 1881 Emerson Dowson applied a cheap form of gaseous fuel, obtained by the partial combustion of anthracite in a "producer" of his own design, to the driving of a gas-engine. Since then the application of producer gas has been effected in many ways, and in 1895 it was shown by Thwaite that gas-engines might be driven by the so-called waste gases of the blast-furnace. The use of cheap gas gave an impetus to the design of large gas-engines, which came to be built in sizes generating hundreds and even thousands of horse-power.

Nearly all gas-engines of small or moderate power, and many of the largest, use the four-stroke cycle of Beau de Rochas and Otto. For brevity these are often spoken of as *four-cycle* engines. The alternative is a *two-stroke cycle* introduced by Dugald Clerk which will be referred to later (§ 326). Engines using it are sometimes called *two-cycle* engines. Most petrol motors use the four-stroke cycle. In Diesel engines both the four-stroke and two-stroke types are common.

323. The four-stroke cycle engine. In the original Otto engine, and in most other small examples of the four-stroke type, and also in some large ones, the cylinder is single acting, with a trunk piston, and the explosive mixture is compressed before ignition into a clearance space at the back end of the cylinder. The volume of the clearance depends on the amount of compression which is desired. In the early forms of Otto engine it was generally more than half the volume through which the piston moved, but a much larger amount of compression is now usual. High compression makes for efficiency, but when it is too high there is a risk of premature ignition: the permissible amount depends on the character of the fuel.

To complete the action requires two revolutions of the crank-shaft. During the first forward stroke of the cycle gas and air are drawn in by the piston. During the first back stroke the mixture is compressed into the clearance space. It is then ignited as the crank approaches the dead-point, and the second forward stroke, which is the only working stroke in the cycle, is performed under the pressure of the heated products of combustion. During the second back stroke the products are discharged into the atmosphere through an exhaust-valve, with the exception of so much as remains in the clearance space, which (except when special means are taken to remove it) is allowed to dilute the explosive mixture in the next cycle. The cylinder is kept cool enough to admit of lubrication, by means of a jacket through which a continuous circulation of water is kept up. The admission and exhaust-valves are worked by cams on a lay-shaft which is geared to run at half the speed of the crank-shaft, so that the several events take place once in two revolutions of the engine.

324. Ignition in gas-engines. In early forms of the Otto engine the ignition of the compressed gases was effected by carrying a flame, through a narrow port in a slide-valve, from a gas jet that was kept burning outside to the mixture within. The slide-valve was worked from the half-speed lay-shaft. To prevent the gases from blowing back through the valve when the explosion took place the slide was arranged so that the port in it which served as a vehicle for the flame had passed under a cover which shut it off from the atmosphere, before it reached the fixed port in the cylinder cover through which the flame passed in to ignite the contents of the cylinder. This mode of igniting the gases is now obsolete. The device which took its place was an ignition tube, namely, a small closed tube of metal or porcelain maintained at a bright red heat by a flame playing on its outside surface. A portion of the explosive mixture was allowed to enter the tube from the cylinder at the time when it should be fired. In most cases the ignition tube was used in conjunction with a "timing valve" which determined the instant at which explosion should occur by being opened to allow a portion of the compressed explosive mixture to enter the ignition tube, but in some gas-engines the "timing valve" was dispensed with and the ignition tube was always in free communication with the cylinder, the instant of firing being determined only by the compression of some of the explosive mixture into the tube. The ignition tube has been superseded by electrical ignition. For this purpose a "sparking plug" is provided, with an insulated terminal from which a spark or stream of sparks is discharged at the desired instant by the use of a battery and induction coil or a high-tension magneto machine. An alternative, now little used, is to produce a low-tension spark by the mechanical breaking of a contact within the cylinder, in a circuit having a considerable amount of self-induction.

In Diesel engines no igniting device is required: the fuel is not injected until the air which fills the clearance space has been so much heated by compression that the oil catches fire as it is sprayed in. In what are called "semi-Diesel" engines the air is not so highly compressed, but a detached (and uncooled) part of the clearance space forms a "hot bulb" into which the fuel is injected, and in that the compressed air acquires a sufficiently high temperature to make the fuel ignite as it enters.

325. Governing of gas-engines. The speed of a gas-engine is usually regulated by a centrifugal governor, and in small engines this often acts by cutting off the supply of gas when the speed exceeds a certain limit, making the engine miss one or more explosions. The governor determines whether the gas valve shall or shall not be opened, by means of a "hit and miss" arrangement of the kind briefly referred to in § 235. A cam fixed on the side-shaft, so that it makes one revolution for every two revolutions of the engine, opens the gas-admission valve by acting on a lever through an intermediate roller. This intermediate roller is carried by an arm which is caused to move sideways by the governor, in such a manner that when the speed exceeds a certain value the roller is removed and consequently the cam fails to act on the lever, and the admission valve remains closed. In some instances a stepped cam is used, giving admission to various amounts of gas corresponding to various positions of the centrifugal governor, with the effect that the explosive mixture is weakened when the speed rises. The tendency, however, to miss fire with weak mixtures is an objection to this method of regulating, and more generally the gas is freely admitted when the speed is below the limit, and completely cut off when the limit is passed. In some small gas-engines the inertia of a reciprocating piece is used instead of the inertia of revolving pieces to determine the admission or non-admission of gas. When the speed exceeds the assigned limit the acceleration of the oscillating piece becomes sufficient to displace it in such a way that the gas-admission valve misses the stroke.

For large gas-engines, and also for petrol motors, the usual method of governing is to control the quantity of explosive mixture taken in per cycle, without greatly altering the proportion in it of fuel to air. For this purpose both gas and air may be throttled, or the mixed gas may enter through an admission valve of the mushroom or lift type, on which the governor acts so as to alter the time during which the valve is opened.

With a Diesel engine, governing is readily effected by controlling the period during which oil is injected at the beginning of the stroke.

326. The Clerk two-stroke cycle. Sir Dugald Clerk, who has done much for the development of the gas-engine both as inventor and investigator, introduced in 1881 an engine in which the explosion occurs once in each forward stroke of the piston instead of once in each alternate forward stroke as in the Otto or Beau de Rochas cycle. The Clerk cycle has therefore the advantage of reducing the proportion of idle strokes by two-thirds. His two-cycle engine requires a pump or displacer, and is in this respect less simple than the Otto four-cycle engine, in which the working cylinder itself acts as pump. The Clerk engine, though it met with some favour as a competitor of the Otto during the life-time of

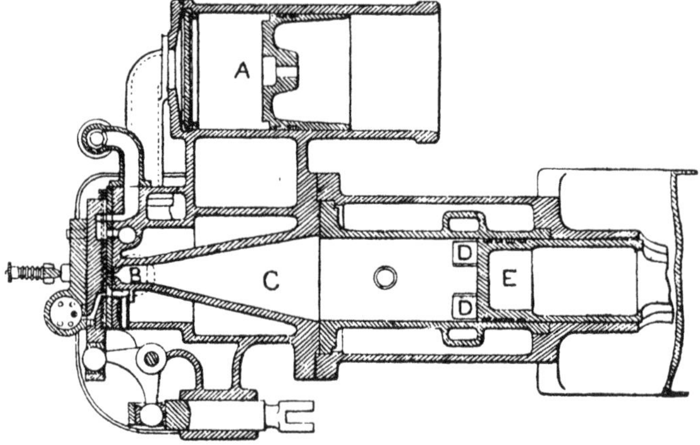

FIG. 268. Clerk engine of 1881. Horizontal section through cylinder and pump.

Otto's patent, fell after that out of use and remained in disuse so long as gas-engines of small power only were built; but in the modern development of oil-engines and of large gas-engines the Clerk cycle has been revived, and is now adopted in some of the most important forms, including several of the Diesel engines which serve for the propulsion of motor ships. Many engineers consider that it will take a still more prominent place in the future of internal combustion.

In the Clerk engine of 1881 (fig. 268) the gas and air were inhaled by an auxiliary piston in a separate cylinder *A*, forming the pump, from which they were delivered at a low pressure (about 4 lb. per sq. inch) to the main cylinder just after the main piston had completed its working stroke. They entered at *B*, passing through a trumpet mouth or large cone *C* forming the back end of the cylinder, which had the effect of removing the kinetic energy of the stream, and hence of allowing the fresh charge to enter without intermingling much with the products of combustion already in the cylinder. The fresh charge drove the products of combustion in front of it, causing these to be expelled at exhaust-ports *D* in the wall of the cylinder close to the front end of the stroke. The piston *E*, returning, closed these exhaust-ports and compressed the fresh mixture, which was ignited as usual when its compression was completed, that is to say when the piston passed its dead-point at the back end. The indicator diagram for the Clerk cycle is almost identical with that given by the Otto engine; the chief difference is that there is an almost sudden drop some way before the end of the stroke, when the exhaust-ports begin to be uncovered by the piston's advance.

327. Scavenging. In the ordinary Otto cycle the clearance space is left, at the end of the exhaust stroke, full of products of combustion, which mix with the incoming charge. To avoid this some of the early gas-engines working on the Otto cycle had a pair of idle "scavenging" strokes added to the four strokes necessary for the cycle, with the object of clearing the burnt gas out more completely before the next admission of explosive mixture. After the usual exhaust stroke the two scavenging strokes took place: in the first of them air was drawn in atmospheric pressure and in the return stroke it was expelled, with the result that the clearance space was left full of nearly pure air instead of products of combustion. There was, however, no sufficient advantage in this to compensate for the drawback of having additional idle strokes, and the use of such strokes was soon given up. The presence of some burnt gas in the mixture does not prevent ignition nor interfere with the completeness of combustion. A considerable amount of scavenging can be secured by making use of the momentum of the stream of escaping exhaust gases to draw in fresh air, by opening the air-admission valve before the exhaust-valve is closed. The clearance spaces are now so small, in consequence of the high compression used in modern engines, that the volume of burnt gas they contain is relatively unimportant. The principal advantage in sweeping them out with fresh air before the charge enters is to reduce the risk of premature ignition.

The word scavenging is applied broadly, in relation to internal-combustion engines of all types, to describe any action in which the products of combustion are driven out by the admission of fresh air.

328. Ideal action with combustion at constant volume. The action of a gas-engine is complicated by conduction and radiation, especially by radiation, between the working gas and the walls of the containing vessel, and also by the fact that the process of explosive combustion is not instantaneous, but takes an appreciable time to be completed. It is convenient however to consider an ideal action in which (1) there is no exchange of heat between the gas and the walls, and (2) all the heat of combustion is generated at a particular instant, namely, when the volume is constant at the end of the compression stroke, before expansion begins. Such an ideal action affords a useful standard for comparison with the performance of a real engine.

Consider then an ideal engine in which there is no transfer of heat between the working substance and the metal, and in which the combustion occurs only while the piston is at the dead-point. The indicator diagram of such an ideal engine, working on the four-stroke cycle, would take the form shown in fig. 269. *OM* is the volume of the clearance, which is the volume occupied by the mixture during combustion, and *MN* is the volume swept through by the piston. *AB* is the admission at atmospheric pressure; *BC* is the compression, which by assumption is adiabatic; *CD* is the rise of pressure caused by the explosion; *DE* is the expansion, also adiabatic, which constitutes the working stroke. At *E* the exhaust-valve opens, with the result that part of the gas at once escapes and the pressure falls to that of the atmosphere. *BA* represents the exhaust stroke by which the cylinder is (except for the clearance) emptied of gas preparatory to receiving a fresh charge in the first stroke of the next cycle.

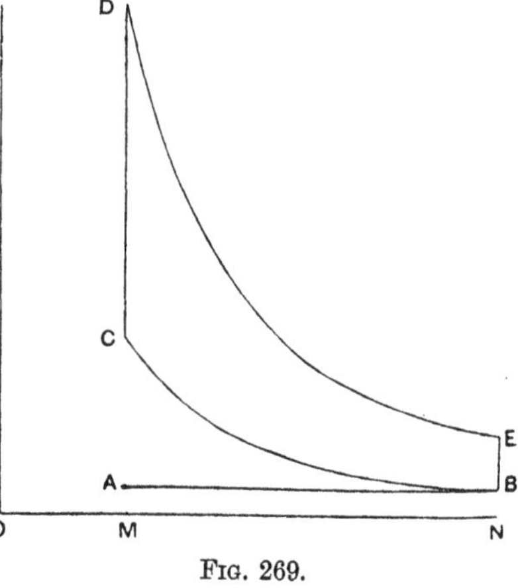

Fig. 269.

If the engine were one having a two-stroke cycle, the lines *AB* and *BA* would be omitted from the indicator diagram for the working cylinder, which would then consist simply of the figure *BCDEB*.

From *C* to *D* (in either case) the whole energy due to the chemical reaction goes to heat the mixture, for by hypothesis none is lost to the walls. The rise of pressure from *C* to *D* can be readily calculated when the rise of temperature resulting from this accession of heat is known, provided we also know what is the change in specific volume due to the change in chemical constitution brought about by the explosion. With the mixture used in gas-engines there is very little change in specific volume: that is to say the burnt products, if brought to the same pressure and temperature, would fill very nearly the same volume as they filled before chemical union took place. With mixtures of coal gas and air the specific volume is reduced after explosion by two or three per cent. in ordinary cases. With some explosive vapours the specific volume is slightly increased. The changes being in any case small, it is convenient in considering an ideal engine to ignore them, and to treat the working substance as if it were a gas whose specific volume does not alter. Further, the largest constituent is air, and that of the burnt charge is nitrogen, and the specific heat of nitrogen is, for equal volumes, the same as that of air. Hence for the purpose of obtaining a simple standard with which real engines may be compared, a practice is adopted of treating the working substance as if it were air, to which between *C* and *D* there is imparted a definite quantity of heat, which may be calculated when we know the composition of the explosive mixture and the heats of combustion of its various constituents. Knowing the temperature at *C*, we could calculate the rise of temperature and consequently the rise of pressure at *D*, if the average specific heat (at constant volume) between *C* and *D* were known.

When this calculation is based on the assumption that the same value may be assigned to the specific heat as we know it to have at low temperatures, the calculated rise of temperature turns out to be much larger than is ever found in a gas-engine, and the difference is too great to be explained by the loss of heat which, in a real engine, takes place by radiation and conduction to the cylinder walls while combustion is going on.

At one time it was supposed that this difference was to be accounted for by incompleteness in the combustion itself. But it is now known that the specific heat of a gas increases considerably at high temperatures, and it is largely for this reason that the temperature and consequently the pressure reached after explosion falls short of the value which might otherwise be expected.

329. Ideal cycle with constant specific heat. Air standard. It is nevertheless instructive to study an ideal cycle in which the working substance is supposed to have a constant specific heat, the action taking place in the manner assumed in fig. 269. This was done by a Committee of the Institution of Civil Engineers, who on these lines determined what is generally called the *air standard* for comparison with the performance of actual internal-combustion engines[1]. The assumptions made are that there is no transfer of heat between gas and metal, that there is complete combustion at the dead-point, and that the specific heat is constant. Let T_0 be the absolute temperature at which the working mixture is taken in, T_1 the temperature to which it is compressed, T_2 the temperature after explosion, and T_3 the temperature after expansion. Fig. 270 shows the cycle, with the stages numbered to correspond with these suffixes. Let K_v be the specific heat at constant volume, which is treated as constant for the purposes of the calculation. Then the heat taken in, namely, the heat generated in the explosion, is $K_v (T_2 - T_1)$. The heat rejected is $K_v (T_3 - T_0)$, for it makes no difference whether the products of combustion are cooled on release to the atmosphere or kept in the cylinder after expansion and cooled there to atmospheric temperature, at constant volume, before release. Hence the efficiency is

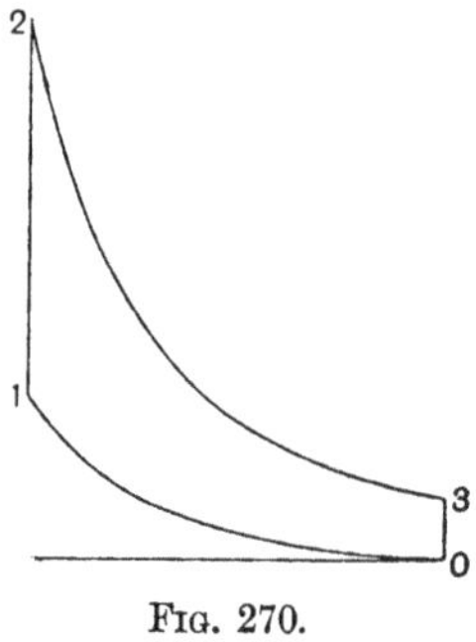

FIG. 270.

$$\frac{K_v (T_2 - T_1) - K_v (T_3 - T_0)}{K_v (T_2 - T_1)} \text{ or } 1 - \frac{T_3 - T_0}{T_2 - T_1}.$$

Writing r for the ratio of compression, which is also the ratio of expansion, we have by § 41

$$\frac{T_0}{T_1} = \left(\frac{1}{r}\right)^{\gamma-1} \text{ and } \frac{T_3}{T_2} = \left(\frac{1}{r}\right)^{\gamma-1}.$$

[1] Report of a Committee on the Efficiency of Internal Combustion Engines, *Min. Proc. Inst. C. E.* vols. CLXII and CLXIII.

Hence also
$$\frac{T_3 - T_0}{T_2 - T_1} = \frac{T_0}{T_1} = \left(\frac{1}{r}\right)^{\gamma-1}.$$

The efficiency of this ideal cycle, that is to say the air-standard efficiency, may accordingly be written
$$1 - \left(\frac{1}{r}\right)^{\gamma-1}.$$

This expression is practically important as showing the beneficial influence of compression. With increased amounts of compression it shows that the efficiency of the air standard increases as follows, taking γ to be 1·4:

Ratio of Compression	Air-standard Efficiency
1	0
2	0·242
3	0·356
4	0·426
5	0·475
6	0·512
7	0·541
8	0·565
10	0·602
15	0·661
20	0·698

It will be seen from these numbers and from the curve (fig. 271)

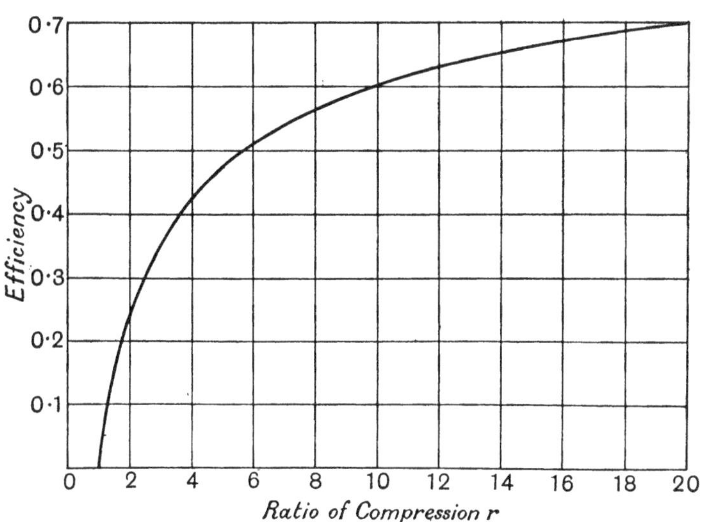

FIG. 271. Efficiency of "air standard."

that there is at first a very rapid gain of efficiency with compression, but that when the compression is high the thermodynamic advantage of increasing it becomes comparatively slight.

The actual efficiencies that are found in trials of engines of the constant-volume class are considerably lower than these air-standard figures, owing to heat losses and to the combustion not being instantaneous, as well as to the fact that the specific heat is not constant but increases at high temperatures. All these departures from the air-standard ideal conspire to lower the average temperature at which the working substance receives—that is to say develops—its heat. When account is taken of the variation in specific heat theoretical values of the efficiency may be calculated for various ratios of compression: these fall short of the air-standard efficiencies by about 20 per cent. but preserve a nearly constant ratio to them when the compression is altered throughout the usual range[1].

In very favourable cases the measured efficiency, in trials of engines working with combustion at constant volume, is as high as 0·37. This corresponds to about 68 per cent. of the air standard, or to about 83 per cent. of the theoretical standard that is set when account is taken of the changes in specific heat.

With all engines of this class there is a practical limit to the amount of compression: it must not be so great as to cause pre-ignition by unduly raising the temperature of the mixture during the compression stroke, nor so great as to give to the explosion, when it does occur, the peculiarly sudden character known as detonation[2] (§ 337). The highest useful compression ratio differs with different kinds of fuel. With coal gas the ratio used in practice is about $6\frac{1}{2}$ or 7. With producer or blast-furnace gas it is more nearly 7. With petrol vapour it is ordinarily between 4 and 5. The limit arises from the compression of air and fuel together, in a mixed state. When air alone is compressed, as in the Diesel engine, and the fuel is injected only when combustion is intended to occur, the same considerations do not hold; compression may then be usefully carried much further.

[1] A method of making such calculations will be found in the author's *Thermodynamics for Engineers*, §§ 164–167.

[2] See Ricardo on *The Internal Combustion Engine*, vol. II, chap. II.

330. Air standard for constant-pressure type. The idea of an air standard may be applied not only to the constant-volume type of internal-combustion engine, to which ordinary gas-engines, petrol engines, and most oil-engines approximately conform, but also to a type in which the pressure of the working substance does not change while combustion is taking place. Suppose that the air is separately compressed into the clearance space before any fuel is admitted and that fuel is then forced in, burning as it enters, while the piston begins its forward movement. By suitably regulating the rate of admission of the fuel the pressure may be kept constant till the combustion is completed.

Here the heat is supplied at constant pressure. We may further imagine the rejection of heat also to occur at constant pressure, if we suppose that the products of combustion are expanded adiabatically down to atmospheric pressure before they are discharged. The ideal indicator diagram would then take the form sketched in fig. 272. Under these conditions (which are not realized in practice) we should have an engine of constant-pressure type rejecting as well as receiving heat at constant pressure. Its air-standard efficiency is readily expressed in a form corresponding to that found for an engine of constant-volume type. We are concerned here with the specific heat at constant pressure, K_p. Treating it as constant, the heat taken in is $K_p(T_2 - T_1)$, the heat rejected is $K_p(T_3 - T_0)$, and the efficiency is

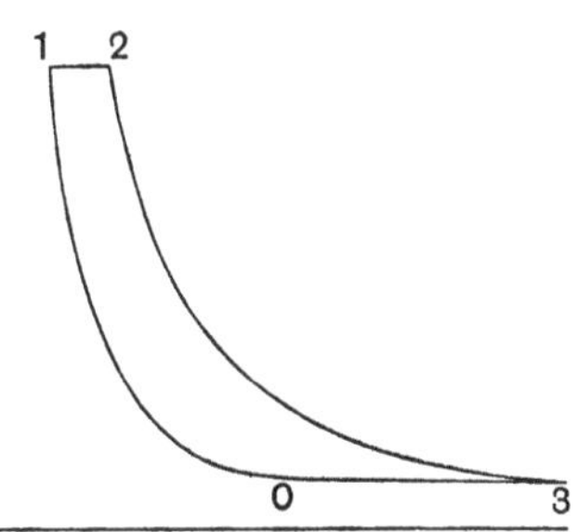

FIG. 272. Constant-pressure type.

$$\frac{K_p(T_2 - T_1) - K_p(T_3 - T_0)}{K_p(T_2 - T_1)} \text{ or } 1 - \frac{T_3 - T_0}{T_2 - T_1}.$$

The ratio of adiabatic expansion is equal to the ratio r of adiabatic compression, and $T_0/T_1 = T_3/T_2$. Hence the air-standard efficiency is given by the same expression as before[1], namely

$$1 - \left(\frac{1}{r}\right)^{\gamma-1}.$$

[1] It is interesting to note that this same expression applies to *three* ideal types of engine:

(1) The constant-volume type, in which heat is received and rejected only at constant volume.

It follows that for equal ratios of compression there would be no thermodynamic advantage in substituting a constant-pressure type of engine for the constant-volume type. But by avoiding any admixture of the fuel with the air before compression it becomes practicable to compress much more strongly, with a consequent gain of efficiency.

The Diesel engine approaches the constant-pressure type as regards its process of combustion, for that generally occurs in such a way as to maintain a nearly uniform pressure behind the piston while the oil is being injected and burnt. But the complete expansion of fig. 272 is not practicable: release takes place when the volume is equal to that at which compression began, and accordingly the long toe of the diagram, from 0 to 3 in the figure, is lost. Notwithstanding this drawback, the Diesel cycle, owing to the very high compression which it allows, gives in practice a higher efficiency than is reached with engines of the constant-volume type.

In all practical forms of internal-combustion engine expansion is incomplete: to make it complete would require the expansion stroke to be much longer than the compression stroke. When release occurs the products of combustion are discharged at a high temperature and with much unutilized internal energy. Attempts have been made to avoid this loss, notably by Atkinson who devised forms of engine with a relatively long expansion stroke. The advantage in respect of thermal efficiency was considerable, but the complication of such engines stood in the way of their success, and in modern forms the thermodynamic advantage of high expansion is sacrificed to mechanical simplicity.

331. Indicator diagrams of the actual engine. In all actual gas-engines combustion takes place at constant, or nearly constant volume. An indicator diagram from an early example of the Otto engine working in its normal manner is given in fig. 273, where

(2) The constant-pressure type, in which heat is received and rejected only at constant pressure.

(3) The constant-temperature type (Carnot's engine of § 43) in which heat is received and rejected only at constant temperature. For the efficiency of that engine is $1-\frac{T_2}{T_1}$, and $\frac{T_2}{T_1}=\left(\frac{1}{r}\right)^{\gamma-1}$, when r is used to denote the ratio of *adiabatic* compression (not of isothermal compression, as in § 43).

AB is the admission stroke, BC the compression, CDE the working stroke, and EA the exhaust. There is a rapid rise of pressure on explosion, so rapid that the volume has not very materially altered when the highest pressure is reached; the specific heat at constant volume may therefore be used without serious error in calculating the amount of heat which this rise accounts for. By the time of maximum pressure, at D, combustion is nearly complete and some heat has been lost to the walls by radiation and conduction. In the expansion curve, from D to E there is a continued loss of heat to

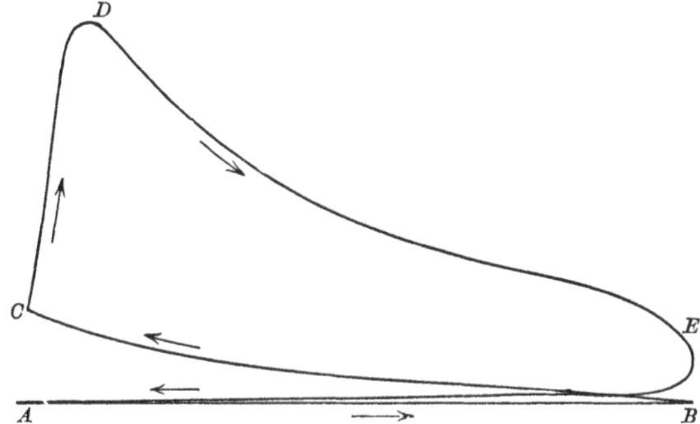

FIG. 273. Indicator diagram from an Otto gas-engine.

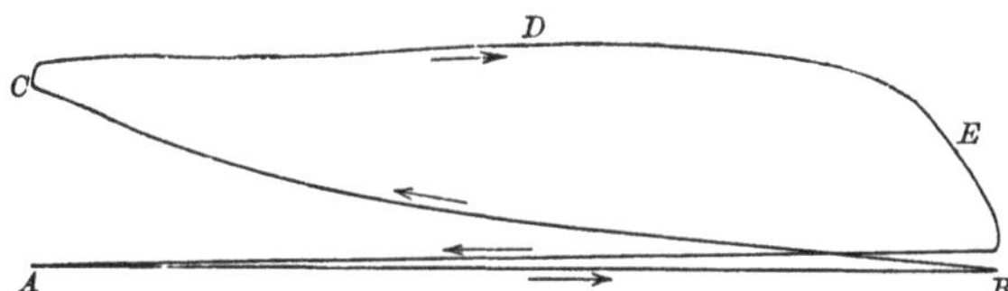

FIG. 273 *a*. Otto engine diagram with weak explosive mixture.

the walls; there may also be some slight further generation of heat within the gas. The high specific heat during expansion, resulting from the fact that the absolute temperature is then ranging from something like 2000° C. to 1000° C., gives the index γ for adiabatic expansion a value nearer 1·3 than 1·4, and so tends to keep the curve up notwithstanding the losses of heat.

With a very weak mixture (fig. 273 *a*) the spread of the ignition may be so slow that the mixture is burning during nearly the whole of the working stroke, with the effect (in this example) that the pressure is kept nearly constant up to release. In such abnormal cases the gases discharged to the exhaust may contain some unburnt fuel.

In the engine which gave the indicator diagram of fig. 273 the compression was less than is now usual: the volume of the clearance was about 40 per cent. of the volume swept through by the piston, and hence the compression ratio was 3·5. With the compression

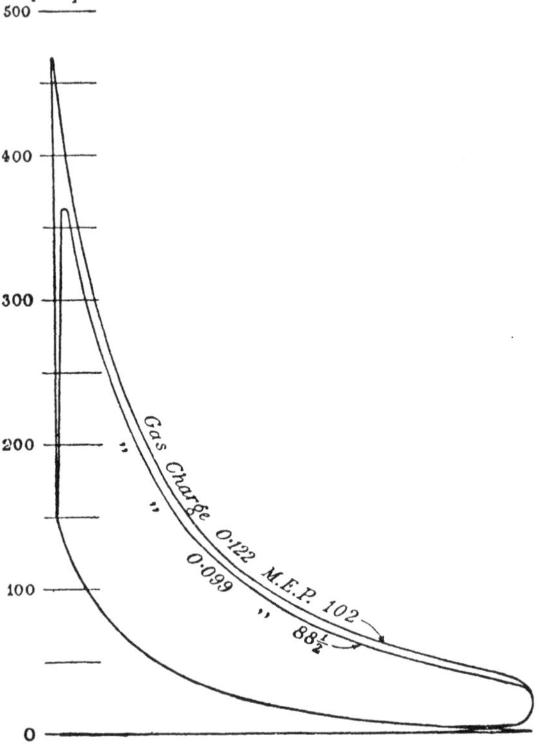

FIG. 274. Indicator diagrams from a Crossley Otto engine, 1908 (Hopkinson).

ratios of 6 or more that are common in modern gas-engines, the mixture is compressed to 150 or 200 lb. per sq. inch before explosion. For a normal mixture the combustion under these conditions is rapid and the maximum pressure approaches 500 lb. per sq. inch. Two typical indicator diagrams from one of Hopkinson's papers[1] are given in fig. 274 for a 40 horse-power Crossley engine, with a compression ratio of 6·37, using coal gas in charges of two grades

[1] B. Hopkinson, *Proc. Inst. Mech. Eng.* April 1908; *Scientific Papers*, p. 267.

of richness. In one case the mean effective pressure was 102 lb., in the other 88½ lb. The thermal efficiency was about 0·33 for the stronger mixture and 0·35 for the weaker. With a weaker mixture still it became 0·37. From these and other experiments it has been established that, within the range of mixture strengths which will ignite regularly and burn completely, the weaker mixtures give a somewhat higher efficiency.

The indicator diagram of a constant-volume internal-combustion engine serves not only to measure the work done but to trace the changes of temperature throughout the cycle, provided the temperature at one point is independently ascertained, and provided also there is no leakage of the working substance. The gas itself serves as a thermometer, since T varies as PV. Then if we know the "suction temperature," which means the temperature when compression begins, it is easy to find from the observed relation of pressure to volume what the temperature is at any other point, such as the end of compression, the point of highest pressure and the end of expansion. We must, however, also know what is the amount of the "chemical contraction," or change of specific volume which the substance undergoes as a result of combustion, apart from the change of temperature. In this example it was about 3 per cent. The gas mixture had a temperature of 100° C. With these data the diagram shows that the absolute temperature at the end of compression was 657° C.; at the highest point it was close on 2000°; and at release it was nearly 1300°. Hopkinson estimates that the indicated work represented 33 per cent. of the heat of combustion, that there was a loss by radiation and conduction of 28 per cent., and that 39 per cent. was discharged in the burnt gases. Most of the 28 per cent. went into the water-jacket: it is a loss that has to be incurred in order that the cylinder and piston may be kept cool enough for lubrication.

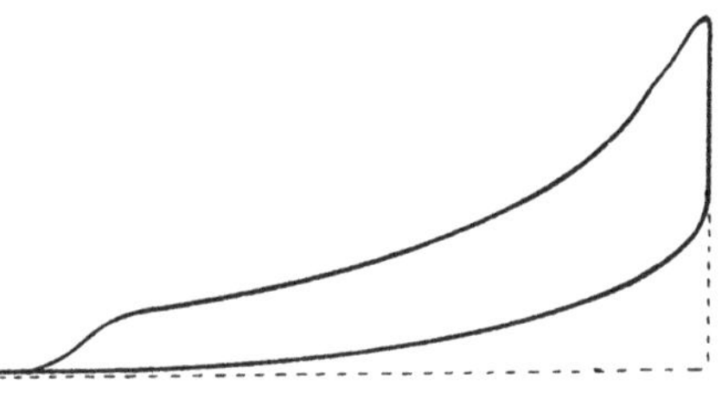

FIG. 275. Indicator diagram of gas-engine working on Clerk's cycle.

Fig. 275 is a characteristic diagram of a gas-engine working on the Clerk cycle, with exhaust-ports in the cylinder which are uncovered by the piston as it approaches the end of its stroke.

332. The process of explosive combustion. The explosion of gaseous mixtures in relation to the action of internal-combustion engines has formed the subject of a series of Reports by a Committee of the British Association[1] which describe investigations and discuss points of theory. Much light has been thrown on what happens in the cylinder of a gas-engine by experiments in which gas mixtures have been exploded in closed vessels of constant volume, with devices for registering the rise of pressure in relation to the time from the instant of firing, and also the progressive changes of temperature at various points within the vessel. Among noteworthy experiments of this class are those of Hopkinson[2], whose scientific study of gas-engine problems has done much to elucidate the subject.

A fundamental distinction must be drawn between the process of explosive combustion in a mixture initially at rest, and the process in a mixture which has turbulent motion. The existence of turbulent motion causes ignition to spread from the point of origin in a very different manner and may result in a much quicker combustion of the whole. Turbulence is in fact essential to the proper rapid action of high-frequency petrol and other motors; without it an efficient combustion of the charge could not occur in the available time. This is made clear by studying the action in non-turbulent mixtures.

Let an explosive mixture, homogeneous and at rest to begin with, be ignited at any point, say by a spark. A flame spreads like a wave from the point of ignition, travelling at a rate which depends on the pressure, so that each portion of the mixture ignites in turn, the most distant portions last. When the initial pressure is that of the atmosphere, the flame may travel at the rate of only about five feet per second to begin with, even in a rich mixture such as one of gas to nine of air. The rate depends on the richness of the mixture as well as on the pressure: in a weak mixture it takes much longer for the ignition to spread through the whole volume. When the initial pressure of the mixture is high the ignition flame travels much faster.

The gas first ignited, close to the sparking plug, burns at nearly

[1] Reports of the Committee on Gaseous Explosions, *Brit. Assoc. Reps.* 1908–1916. The Seventh Report (1914) contains a useful summary.

[2] B. Hopkinson, *Proc. Roy. Soc.* 1906 and 1910; *Scientific Papers*, pp. 367–422.

constant pressure, being surrounded by a large elastic cushion of unignited gas, and its combustion is practically completed before the pressure has risen. But as the spread of the flame brings more of the gas into action, the pressure rises, and the portion which was first burnt becomes compressed. This compression is nearly adiabatic: its effect is to raise the temperature of that portion much above the temperature to which it was brought by combustion, and above the temperature which is reached in combustion by the outlying parts of the gas, which are compressed before they become ignited.

In Hopkinson's experiments a mixture of nine parts of gas to one of air was fired at atmospheric pressure in a cylindrical vessel with a capacity of about 6 cub. ft. It was ignited by an electric spark at the centre, and developed a maximum pressure of about 80 lb. per sq. inch, which was reached a quarter of a second after firing. The temperature was observed near the centre and at other points. Immediately on ignition the temperature at the centre rose very rapidly to 1225° C., while the pressure remained nearly constant. In the later stages of the explosion, when the burnt gas at the centre was being adiabatically compressed, its temperature rose above the melting-point of platinum, probably to 1900°. This is a much higher temperature than was reached in the outlying portions, which were first compressed adiabatically and then heated by combustion. When the maximum pressure was reached, the mean temperature inferred from it was 1600°. Hopkinson concluded that even in a vessel impervious to heat, the portion of the mixture first fired would be hotter than the outlying portions by about 500°, when the combustion of the whole was practically complete.

The view was formerly held that even after any portion of the gas had become ignited a considerable time elapsed before the chemical reaction in that portion was complete, and the name "after-burning" was used to describe the supposed slow continuance of the process. According to this theory, it was believed that after the flame had spread throughout the whole mixture a large fraction —perhaps 40 per cent.—of the whole heat due to the chemical action had still to be evolved and that its evolution went on slowly, after the temperature and pressure had begun to fall through the cooling action of the walls, or, in a gas-engine, through both that and the expansion of the gas. The phenomena which seemed to support this view are now explained as being due to variation of specific heat; and it is generally accepted that little or no after-burning in this sense occurs, any continued combustion of the mixture as a whole being mainly or entirely due to delay in the ignition of all portions of the gas. Practically the full evolution of heat in each portion of the gas takes place almost at once when the flame reaches that portion, but there is necessarily some delay before the whole charge is ignited, and there may be considerable delay in completing the ignition of those portions which are in proximity to the walls, especially when the mixture is weak.

In explosion experiments with weak mixtures the spread of the flame is much slower, so slow indeed that it may be largely affected by convection currents set up by the ignition of the gas nearest to the spark. The gas in the upper part of the vessel may be completely ignited while the lower part of the vessel is still full of unburnt gas. By stirring the contents of the vessel, so that the gases are in motion when the spark passes, a much more rapid combustion of the whole can be secured.

333. Effect of turbulence. The effect of turbulence in assisting rapid inflammation of the whole charge is felt, though in a less degree, in strong mixtures as well as in weak mixtures. It was demonstrated in some of Hopkinson's closed-vessel experiments by using a revolving fan to stir the mixture before the spark was passed. Clerk has pointed out its importance as a factor in the working of an internal-combustion engine. When a fresh charge is drawn in and compressed the gases are still in more or less violent motion at the moment of ignition. This has the great advantage that the igniting flame quickly reaches all portions of the charge, and combustion is completed at an early stage of the expansion stroke. Clerk observed that when the explosive charge in a gas-engine was not fired after the first compression, but was fired after three successive compressions, so that the turbulence set up on its entry had time in part to subside, the process of combustion was generally prolonged, with the result of giving a flat diagram and a wasteful action. In a high-speed engine the whole expansion stroke may take only one-fiftieth of a second, or less, and the explosion is over in a small fraction of that time; this would be impossible were it not for the effect of turbulence in causing the flame to spread quickly from the sparking plug to the most distant portions of the charge.

In the design of high-speed engines it is important to make this distance no longer than is necessary, and also to avoid giving the clearance space a form that would tend to damp out the turbulence which the charge has acquired while it was being inhaled[1].

The detonation which may occur when certain combustible mixtures are too highly compressed (see § 337) is a type of sudden action entirely distinct from the rapid inflammation and consequently rapid combustion which turbulence assists. It is in fact less likely to be met with when there is turbulence than when the mixture is initially at rest.

334. Gaseous fuels. The following kinds of gaseous fuel are available for use in gas-engines.

Ordinary *coal gas*, as manufactured for public supply, serves in many small engines where the convenience of the supply compensates for its comparatively high cost as an engine fuel. It is obtained by distilling the volatile constituents of coal in retorts and purifying the gases that are given off by extracting tar, ammonia, etc. A typical analysis by Bunsen and Roscoe, quoted by Clerk, is as follows:—

Constituent	Parts by volume	Volumes of oxygen required for combustion	Volumes of resulting products in gaseous state
Hydrogen, H_2	45·58	22·79	45·58
Methane, CH_4	34·9	69·8	104·7
Carbon monoxide, CO ...	6·64	3·32	6·64
Ethylene, C_2H_4	4·08	12·24	16·32
Tetrylene, C_4H_8	2·38	14·28	19·04
Sulphuretted Hydrogen, H_2S	0·29	0·43	0·58
Nitrogen, N_2	2·46	—	2·46
Carbon dioxide, CO_2 ...	3·67	—	3·67
	100·00	122·86	198·99

The calorific value of average coal gas per cub. foot is about 333 units[2]. Each cubic foot of gas requires rather more than

[1] Cf. Ricardo, *loc. cit.* vol. II, p. 90 *et seq.*

[2] The unit used in these statements is the pound-calory or pound-degree-centigrade, which is 1·8 times the British Thermal Unit.

6 cub. ft. of air for complete combustion, and when used in a gas-engine the gas is mixed with from 8 to 11 times its volume of air. Under favourable conditions a gas-engine will use 15 to 16 cub. ft. of coal gas per hour per brake horse-power.

Natural gas issues from coal and oil-bearing strata in Pennsylvania, West Virginia, and other places. It consists largely of methane, sometimes to the extent of 93 per cent. by volume, along with a little hydrogen, ethane (C_2H_6), and inert constituents.

Coke-oven gas, which is distilled from coal in the process of manufacturing coke, forms an engine fuel of comparatively high power, though short of that of the coal gas distilled in retorts. It contains about 63 per cent. of hydrogen, 23 of methane, and 5 of carbon monoxide, and has a calorific value of about 230 units. The large proportion of hydrogen makes it necessary to use a lower compression ratio than would be safe with ordinary coal gas.

Water gas is made by passing steam through incandescent carbon. To maintain the carbon in a state of incandescence air is blown through it alternately with the steam, the alternation taking place at intervals of a few minutes. In many cases water gas is enriched for illuminating purposes by adding the gaseous products of decomposition of a certain amount of oil. Without enrichment the main constituents in water gas are 48 to 50 per cent. of hydrogen, and 36 to 40 per cent. of carbon monoxide, the remainder being inert. Its calorific value is about 160 or 170 units per cub. foot.

Producer gas is obtained, in its simplest form, by blowing air through coke or other carbon fuel, part of which is in a state of incandescence. The gas producer is essentially a device in which fuel is burnt with an insufficient supply of air. The combustion is therefore incomplete; an excess of carbon is taken up by the oxygen of the air, forming carbon monoxide, which is the effective constituent of the resulting fuel. Rather less than 30 per cent. of the whole heat of combustion of the carbon is generated in the producer, leaving fully 70 per cent. to be generated in the subsequent combustion of the gaseous fuel. In some producers coal is used instead of coke; consequently the gaseous fuel, while still consisting largely of carbon monoxide, contains also some hydrogen and methane which add considerably to the calorific value. Small producers generally use coke or anthracite to escape the difficulty which arises when tar is one of the products. Producer gas made from coke has a calorific value of about 40 units per cub. foot, and that made from coal about 55 units.

In most large gas producers the air is forced in under a low pressure. But for small or moderate powers it is very common to have a self-contained plant consisting of engine and producer combined, in which the engine sucks in air through the producer, thereby generating gas as it is wanted. *Suction producer* plants of this kind are made in large numbers by various firms, in sizes ranging from less that five up to several hundred horse-power. They furnish an economical and compact means of obtaining power, requiring but little attention, and using only about 1 lb. of anthracite or $1\frac{1}{4}$ lb. of coke per brake horse-power-hour. They can be adapted to utilize peat, wood refuse, saw-dust, husks and other combustible matter.

An improved kind of producer gas, applied as a gas-engine fuel by Emerson Dowson, is got by causing a certain proportion of steam to mix with the air passing through the incandescent coke or coal; the gas thereby obtained is intermediate in composition between water gas and the gas given by a simple air-fed producer. Many forms of producer are used which embody this principle and yield a mixed type of cheap gaseous fuel, very suitable for use in gas-engines, with carbon monoxide and hydrogen as its chief active constituents. The Mond producer is a large scale type, in which the process is carried out with great economy owing partly to the scrubbing arrangements by which the ammonia in the gas is recovered, to form ammonium sulphate for fertilizing purposes, and to the very complete regenerative devices which are employed to save what would otherwise be waste heat. In generating Mond gas cheap bituminous slack serves for fuel and the proportion of steam is large, about $2\frac{1}{2}$ lb. per lb. of fuel. Mond gas contains some 24 per cent. of hydrogen, 2 of methane, and 16 of carbon monoxide. Its calorific value is about 80 to 85 units per cub. foot. In other examples of mixed producer gas the proportion of hydrogen is as a rule lower, averaging something like 15 to 18 per cent., and the carbon monoxide is higher, about 20 to 25 per cent. The calorific value generally ranges from 80 to 90 units.

Blast furnace gas. A blast furnace acts like a gas producer of the air-fed type, yielding a gas which may contain some 28 per cent. of carbon monoxide, with more or less hydrogen and methane, according as the furnace burns coal or coke. The calorific value is in some cases only about 40 units, but may be higher. Notwithstanding its low calorific value blast furnace gas works well in gas-engines, and is in fact the fuel used in many engines of the largest size. When properly freed from dust a mixture of this gas with air may be highly compressed without risk of premature ignition. In some large power installations the gas furnished by blast furnaces is enriched by an auxiliary supply coming from coke-ovens or special producers.

335. Liquid fuels. Liquid fuel for internal-combustion engines is chiefly obtained from crude petroleum by fractional distillation. The more volatile constituents, which boil off at temperatures below about 150° C., form the readily vaporizable motor spirit familiar as petrol (or "gasoline"). Next comes the intermediate type of oil commonly called paraffin or kerosene, which is used in lamps and in some oil-engines of a non-Diesel type, and after it the heavy oils which serve as Diesel-engine fuel, or for burning in boiler furnaces. The whole residue, after the motor spirit has been distilled off, may be used in Diesel and other "heavy oil" engines; but on economic grounds it is preferable to separate, by continued distillation, the engine-fuel oil from still heavier constituents which may form lubricants and serve other industrial purposes. These oils are also obtained, in comparatively small quantities, by distillation from shale.

Petroleum is a mixture of many hydrocarbons, in at least eight definite series, each series being a succession of compounds which form a regular progression as to the number of atoms of hydrogen and carbon that make up the molecule. In each series the lower members (that is those with comparatively few atoms in the molecule) have lower boiling-points, lower density, and less viscosity. Thus in the series which chemists call the *paraffins* the first member to be liquid at ordinary temperatures is pentane, C_5H_{12}, the boiling-point of which, at atmospheric pressure, is 37° C.; this is followed by hexane, C_6H_{14}, with a boiling-point of 69°, heptane, C_7H_{16}, with a boiling-point of 98°, and so on, the general formula for the series being C_nH_{2n+2}. Another important series is the *naphthenes*, with the formula C_nH_{2n}. The name *aromatics* is applied to a group of series one of which is the benzenes, C_nH_{2n+6}, a series which includes benzene, C_6H_6, toluene, C_7H_8, and xylene, C_8H_{10}.

The composition of petrol differs widely from sample to sample, depending on the district of origin of the petroleum from which it has been distilled. It consists essentially of a mixture, in various proportions, of the lower members of the series which have been named. Something like 60 per cent. of paraffins, 30 per cent. of naphthenes, and 10 per cent. of aromatics is not unusual. Hexane is often a prominent constituent. The presence of aromatics is helpful in allowing a comparatively high compression ratio to be used. The net (or lower) calorific value of petrol varies from about 10,250 to 10,500 calories per lb. For complete combustion it requires $14\frac{1}{2}$ to 15 times its weight of air. Heavy oils, which are made up of higher members of the various series, have a somewhat lower calorific value, usually ranging from 9800 to 10,100 calories per lb.; 10,000 may be taken as a typical figure.

Liquid fuels for engines are also obtained by distilling coal tar. Commercial benzol, which is produced in this way, contains some 70 or 75 per cent. of benzene, with 20 per cent. or more of toluene and a small quantity of other hydrocarbons. It has the advantage of taking a high compression ratio without pre-ignition or detonation.

Still another practicable fuel for engines is alcohol, which may one day become specially important because it is (as Brame remarks) the only known medium by which the heat energy of the sun can be made to supply liquid fuel within a reasonable time. When the natural stores of coal and oil are exhausted, alcohol, produced by distillation or fermentation of vegetable derivatives, may still be available. The net calorific value of ethyl alcohol ($C_2H_5.OH$) is about 6300 calories per lb., and that of methyl alcohol ($CH_3.OH$) about 5000.

336. Cracking of oils. In the distillation of petroleum it is found that when the heavier oils are allowed to fall on a hot surface, or are heated under pressure, a change may take place in their molecular constitution, the heavy molecules breaking up into lighter ones, so that the proportion of oil with a lower boiling-point is increased. This change is known as *cracking*. Cracking may be employed commercially in the preparation of liquid fuels, as a means of getting a greater yield of light constituents; on the other hand, it may occur accidentally within the combustion chamber of an engine, when incompletely vaporized drops fall on a hot metal surface, and may in that case be the cause of premature ignition, by producing a constituent of the mixture which is more susceptible to the combined effect of temperature and pressure. Such cracking may also cause carbon to be deposited. These effects make it undesirable to vaporize the fuel of an oil-engine by bringing it into contact with hot surfaces while still in the liquid state.

337. Detonation. Detonation is the nearly simultaneous explosion of all parts of an explosive substance by means of a compression wave, as distinguished from an explosion in which inflammation spreads from one part to another in a manner like that of slow combustion. In the explosion of certain vapour mixtures it is found that detonation may occur, apparently as the latest stage of an ignition process which may have begun slowly. It occurs through the development of a compression wave whose pressure heats the still unburnt portions of the charge to a temperature at which they suddenly combine, without waiting to be reached by the igniting flame. Detonation or "pinking" may be distinguished from normal explosion by its sharper noise. It is to be avoided in the working of engines, not only because it may produce damaging stresses, but because it is apt to cause local heating which may result in the pre-ignition of a subsequent charge. The tendency to detonate, more than anything else, determines the highest useful compression ratio for any particular kind of petrol or other vapour mixture[1].

It has been found that by adding a small quantity of certain substances (called "dopes") to the fuel the tendency to detonate can be much reduced, and the advantage of a higher compression ratio thereby made practicable. A remarkably effective dope for this purpose—as was discovered by Midgley in 1922—is lead ethide or tetra-ethyl-lead, $Pb(C_2H_5)_4$. When a quarter per cent. by volume of this substance is added to any of the usual brands of petrol, the compression ratio may safely be increased by nearly 40 per cent., with a gain of 14 per cent. in power for the same fuel consumption. Other substances such as nickel carbonyl have a more or less similar action.

[1] See Tizard on the Causes of Detonation, *Proc. N.E. Coast Inst. of Eng.* May 1921; Tizard and Pye on the Ignition of Gases by sudden compression, *Phil. Mag.* July 1922; Report of the Empire Motor Fuels Committee, *Inst. of Automobile Engineers*, Part 1, 1924.

To explain the large protective effect that may be produced by a small quantity of a suitable dope, Callendar[1] points out that when petrol is inhaled from a carburettor the spray contains minute drops from which the more volatile constituents evaporate, leaving nuclei of microscopic size. During the compression of the mixture these nuclear drops are likely to be foci of ignition; for as the pressure rises condensation of vapour occurs on them, giving out heat. This is because the entropy of saturated vapour in the lower members of the paraffin series increases with rise of temperature and pressure[2]. The nuclear drops are also more energetic than the surrounding gas in absorbing radiation. The dope apparently acts by coating them with a more stable substance which checks their tendency to serve as centres of ignition, and since they form only a small part of the whole vapour, it is intelligible that a very little dope should suffice to produce this effect.

The tendency to detonate is reduced when there is active turbulence, and when the combustion chamber is so designed as to avoid any long distance in the travel of inflammation from the spark plug.

338. Examples of gas-engines using the four-stroke cycle. Nearly all small gas-engines, up to say 150 horse-power, have a single cylinder, generally horizontal, with single action, and follow the four-stroke cycle. Fig. 276, which is taken from one of Clerk's papers, shows such an engine. It illustrates the usual features: the trunk piston, running with a mean piston speed of 800 or 900 ft. per minute; the water-jacket which surrounds the whole of the cylinder barrel as well as the combustion chamber, and has in it a constant circulation of cooling water; the mushroom valves opening inwards into the clearance space for admission and exhaust, which are worked by levers from cams on a half-speed side-shaft. In most modern designs the admission-valve is set not as in this example but opposite to the exhaust-valve, opening vertically downwards, so that the axes of both valve-spindles are in one straight line, which facilitates removal for cleaning or grinding; the gas-valve is usually carried on the same spindle as the admission valve. This arrangement will be seen in later illustrations.

[1] Callendar on Dopes and Detonation, *Engineering*, April-May 1926.

[2] The saturation line in the entropy-temperature diagram slopes to the right as it rises until it comes near to the critical point, instead of to the left, as in steam. Compare *Phil. Mag.* June 1920, p. 633.

In some engines a plan, due to Messrs Crossley, is used in which the admission-valve is opened a little in advance of the gas-valve, and before the exhaust-valve is closed. Then the momentum of the gases which are being discharged in the exhaust-pipe draws air in through the admission-valve, with the result of scavenging the combustion chamber before the next charge enters.

In the design of engines for much larger powers two alternative plans are followed, to increase cylinder dimensions and to multiply the number of cylinders. Large cylinders were at first found to present formidable difficulties from over-heating; these have been met by avoiding excessive compression and by arranging for

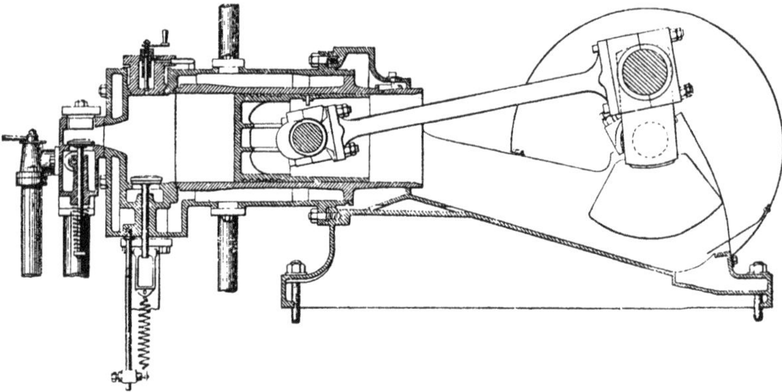

FIG. 276. "National" gas-engine.

effective water-cooling not only of the barrel and cylinder end but of the piston and valve seats. Up to a moderate size pistons are kept sufficiently cool by conduction to the cylinder walls; when that limit is passed the piston itself must be water-cooled. With these precautions many engines have been built which develop 1000 H.P. or more per cylinder, in a few cases as much as 2000 H.P., the fuel being usually blast-furnace or coke-oven gas[1].

Most English makers, however, have preferred to produce engines of considerable power by combining in one unit a number of cylinders of comparatively moderate size, thereby escaping, for

[1] Burstall, in his paper on the Gas-Engine in *Trans. of the First World Power Conf.* 1924, describes a Belgian example where 8000 H.P. is developed in four double-acting cylinders. Other papers in the same volume give particulars of large gas-engines at work in Germany and elsewhere. In many cases the heat of the exhaust gases is utilized to produce steam for use in a turbine.

the most part, the need to water-cool the piston. Thus the Premier Company, retaining the horizontal single-acting form and the four-stroke cycle, groups four or more cylinders together, with their valves operated by a lay-shaft behind the group. In fig. 277, which is a section through one of the cylinders, the lay-shaft is seen to the left, with the cams and levers by which it operates the admission-valve at the top of each clearance space, and the exhaust-valve at the bottom. At the back end of each cylinder there are two openings, one for the sparking plug, and the other for admitting air under pressure to start the engine.

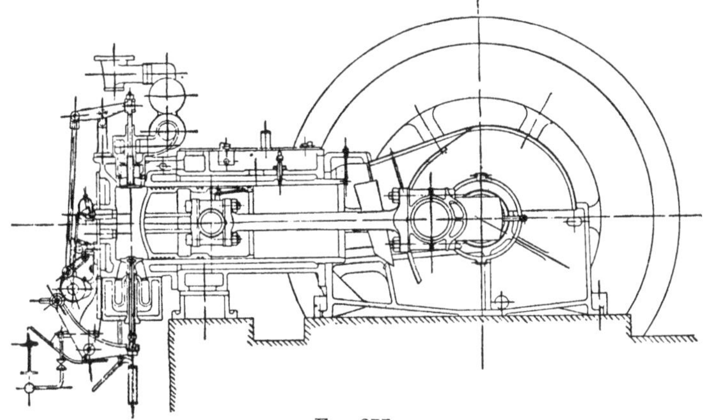

FIG. 277.

The National Company build a successful multi-cylinder gas-engine in which the cylinders are grouped vertically in tandem pairs, one pair over each crank. Each piston is single-acting, the top being the operative side. The upper and lower pistons of each pair are bolted together, and the lower part of the upper cylinder is enclosed so that the air in it is compressed by the upper piston on each downstroke, forming a cushion which promotes smoothness of running. The four-stroke cycle is used: above each piston the clearance space extends out sideways to form a pocket in which the valves are placed, the inlet-valve directly above the exhaust-valve. The gas-valve opens a little later and closes a little earlier than the inlet-valve, so that the first and last portions of the charge to enter consist of air. The pistons, which do not exceed 22 or 23 inches in diameter, are not water-cooled. The stroke is 2 feet and the normal speed 200 revolutions per minute. Four such pairs make up a unit developing about 1000 H.P.; six pairs, giving 1500 H.P., form a convenient unit in large installations.

These are single-acting engines. In an interesting double-acting design by the Premier Company four vertical cylinders are grouped to make two tandem pairs standing in one vertical plane perpendicular to the crank-shaft, which passes between the pairs. The two piston-rods drive a single crank through one triangular connecting-rod. The engine follows the four-stroke cycle, and the pistons are water-cooled.

A more usual double-acting engine, working on the four-stroke cycle, has two tandem horizontal cylinders with a slipper-block between and with the connecting-rod and crank at one end. Big engines of this kind, producing more than 1000 H.P. in each of the two cylinders, are built by the Maschinenfabrik Augsburg Nürnberg (commonly styled M.A.N.) and other makers, and are much used on the Continent in iron and steel works, often for driving blowers. They generally use blast-furnace gas as fuel. The general arrangement of such an engine, as built by Messrs Galloways, is illustrated in fig. 278. Fig. 279 gives a section through one of the cylinders with its piston and valves, and shows the arrangements for ample water-cooling of the piston (by means of a hollow piston-rod), also of the exhaust-valve seats and housings and the cylinder ends and stuffing-boxes, as well as the barrel. The valves are worked from a cam-shaft at the side, level with the piston-rod, the exhaust and inlet-valve at each end receiving their motion from the same cam. These engines, built in sizes up to 55 inch diameter and 55 inch stroke, develop fully 2500 H.P. in the tandem pair.

339. Large gas-engines using the two-stroke cycle. The two-stroke or Clerk cycle, though much used both in small and large oil-engines, is rarely found in any but the larger sorts of gas-engines. It allows exhaust-valves to be dispensed with, release taking place by the piston's uncovering a ring of exhaust-ports in the cylinder wall. In a single-acting engine these ports are placed near the end of the cylinder: in a double-acting engine they are placed at the middle and the piston is made long enough to keep them covered until the end of one or the other stroke is approached, as in the uniflow steam-engine (§ 302), the cylinder itself having a correspondingly greater length.

The Körting engine (fig. 280) is an important example of a

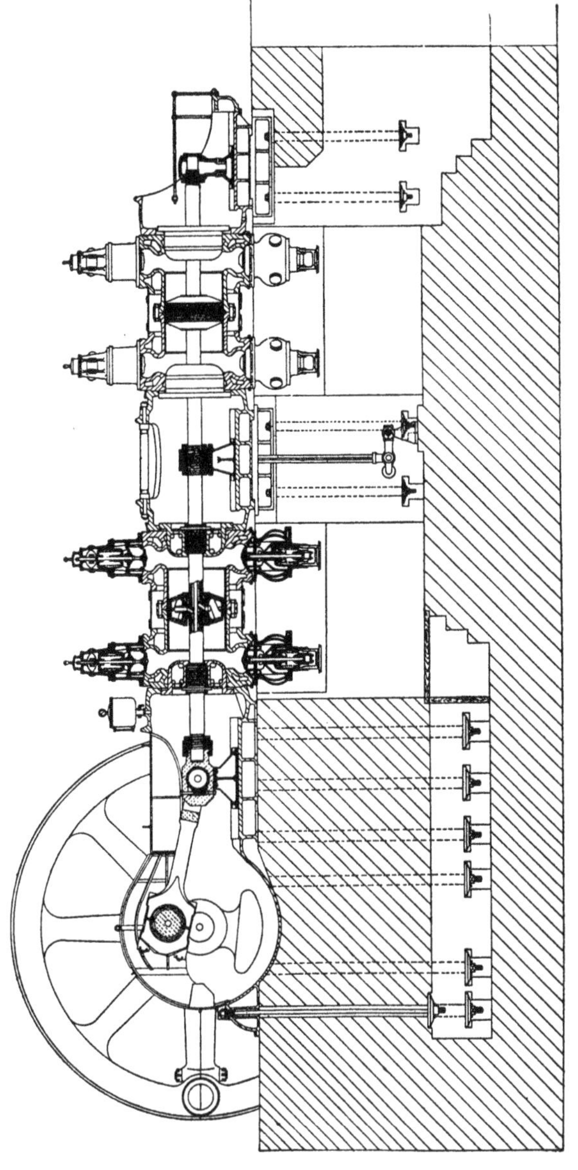

FIG. 278. Tandem double-acting gas-engine (Galloways).

double-acting gas-engine which works on the two-stroke cycle. There are two admission-valves *W* and *T*, one for each end of the

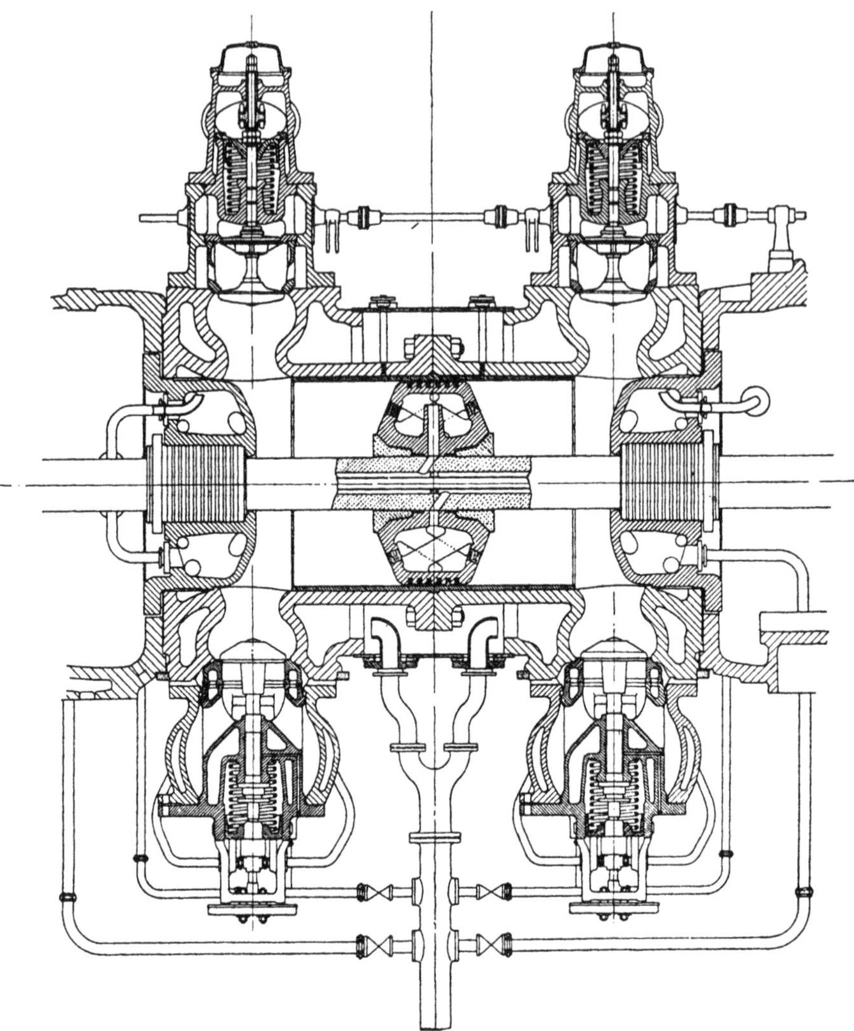

FIG. 279. Section of cylinder, piston, and valves.

cylinder. From a pump at the side, which is not shown, air, at a pressure slightly higher than that of the atmosphere, is supplied to the ducts *V* and *S*, and gas to *U* and *R*. The pumps are arranged so that only air enters the cylinder when the exhaust-ports begin to be uncovered, and gas follows when they are closing, with the effect that scavenging by air alone precedes the admission of a fresh charge. Engines of the Körting type have been built to develop 2000 H.P. in one cylinder.

Another arrangement for a two-stroke cycle has two pistons working in the same cylinder, alternately moving towards and away from one another, and each of them uncovering cylinder ports as it approaches the outer end of its travel. This is called the *opposed piston* type: its earliest practical form was in the Oechelhäuser gas-engine, which has been the parent of others for use both with gas and with oil. Here the cylinder is a mere tube with open ends, taking no longitudinal stress but only the radial pressure

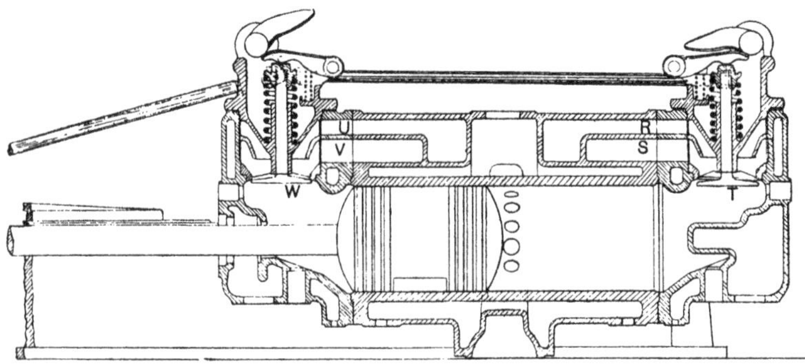

FIG. 280. Körting gas-engine.

of the gases within, and it has no inlet or exhaust-valves, for the belt of cylinder ports near one end serves for admission and the belt near the other end for exhaust. One of the pistons has a short connecting-rod leading to the crank: the other has an outlying cross-head beyond the cylinder and is connected through a pair of long rods to a pair of cranks set at 180° to the first. When the pistons are at the dead-point, in their most distant position, both belts of ports are uncovered: scavenging air (supplied by a pump) enters through one and drives the burnt gases out through the other, and this is followed by the working charge before the ports close. As the pistons approach one another the charge is compressed into the space between them. At the inner dead-points this space is a minimum, giving the desired degree of compression. The charge is then fired and the working stroke follows as the pistons recede until they again uncover the ports. This type of opposed-piston engine with three cranks for each cylinder has been adapted for use with oil as fuel and applied by Messrs Doxford of Sunderland in driving ships. For use with oil, air only is admitted at the inlet belt of cylinder ports before the compression stroke begins, and the distance between the two pistons is reduced so that the compression will make the air hot enough to ignite the oil, Diesel fashion, when it is injected under high pressure just as the expansion stroke is about to begin. The oil is injected under high pressure through a nozzle in the side of the cylinder, midway between the pistons. The engine has the merit of mechanical simplicity, no valves, no cylinder heads, and a balance of internal forces which relieves the main bearings of stress; its drawback is the heavy mechanism of long rods, cross-head, and double cranks, for connecting the more distant piston.

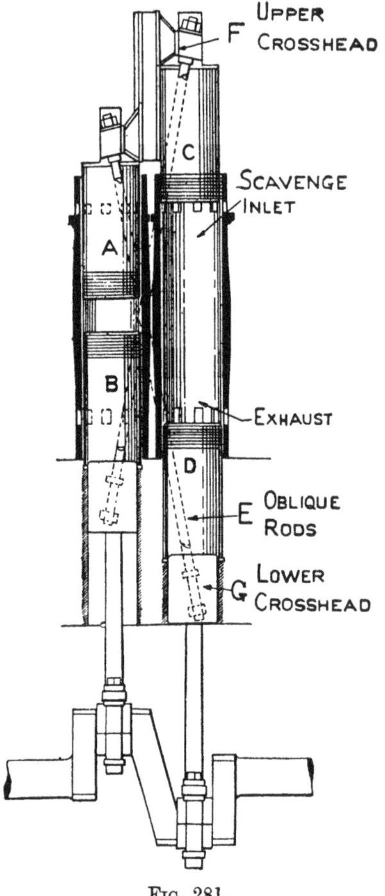

Fig. 281.
Fullagar's twin-cylinder engine.

An ingenious modification of it by Fullagar considerably reduces this objection. In Fullagar's opposed-piston engine[1] (fig. 281) the cylinders are grouped in twin vertical pairs, each over one crank, and the crank connects not only with the lower piston of the cylinder that is set directly over it but also with the upper piston of the adjoining cylinder, these pistons being joined by a pair of slightly inclined rods outside the cylinders. Thus, in the figure, the pistons *B* and *C* are

[1] See *Trans. N.E. Coast Inst. of Eng.* July 1914, for description of the Fullagar gas-engine, with tests by B. Hopkinson: also Hopkinson's *Collected Papers*, p. 349.

both connected to the left-hand crank, and D and A both to the right-hand crank. Upper and lower cross-heads F and G are provided, working on guides which take the small component of lateral thrust due to the inclined rods. At the outer end of the travel two belts of ports are uncovered in each cylinder, one serving for exhaust and the other for inlet, as in the engine just described. The exhaust-ports are made slightly longer than the others so that they open a little in advance, and thus the cylinder is completely scavenged by air that enters under a small pressure through the belt at the other end. The air-pump is formed by enclosing the upper cross-head. The general effect of Fullagar's device is to obtain the complete action of an opposed-piston engine with no more than one crank per cylinder. Originally introduced as a gas-engine, it is now made as an oil-engine for land purposes by the English Electric Company, and for ship propulsion by Messrs Cammell Laird & Co.

Imagine a cylinder with two opposed pistons to be lengthened beyond the admission and exhaust-ports, then doubled on itself and the ends joined, to form a pair of parallel cylinders enclosed at top and bottom as in fig. 282. This gives the "Duplex" engine of Mr Chorlton, a valveless double-acting engine which works on the two-stroke cycle. The exhaust-ports are at the middle of one cylinder and the admission-ports at the middle of the other. The pistons move up and down together, uncovering both sets of ports when they are at the bottom of their stroke, and again when they are at the top; there is complete scavenging by exhaust from the ports of one cylinder with simultaneous admission of air by the ports of the other. The curved passages at top and bottom serve as combustion chambers in successive strokes. In Mr Chorlton's design the crank of the piston that opens the exhaust-ports is set in advance of the other crank by about 15°, and the exhaust-ports are slightly longer, with the result that they open sooner than the admission-ports and also close a very little sooner.

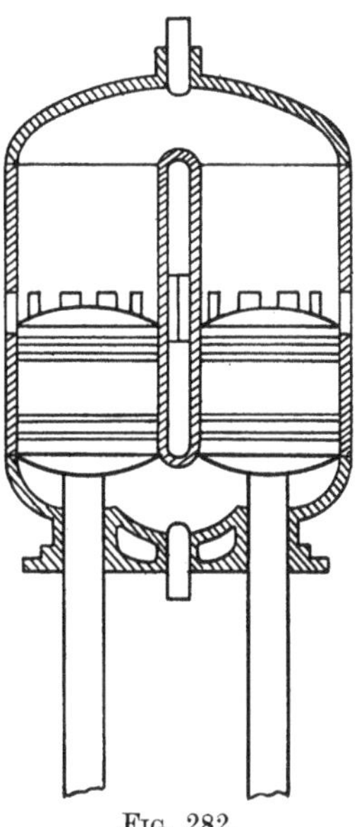

Fig. 282.
Twin duplex gas-engine.

340. The Humphrey internal-combustion pump. An interesting new departure was made in 1909 by Mr Humphrey's invention[1] of an internal-combustion pump of high efficiency in which nearly all the usual working parts of a pumping-engine are dispensed with. Explosion takes place in a chamber which is closed by a column of the water or other fluid to be pumped: the effect is to set the column oscillating, with the result that a part of it is discharged at a high level, or against a pressure, and the successive movements of the column complete the cycle of operations by first discharging the burnt gases, then inhaling a fresh charge, and then compressing it in preparation for the next explosion. What is substantially the Otto cycle (or, in a modified form of the apparatus, the Clerk cycle) is accomplished in an extraordinarily simple manner without the use of a piston, crank, or fly-wheel, and the effective return of work done in relation to the quantity of fuel burnt compares favourably with the best results obtainable in other ways. This is not only because of the absence of frictional mechanical losses, but because expansion in the working "stroke" of the column is carried so far as to bring the pressure down to a value approximately that of the atmosphere. The device escapes the usual mechanical limitation of a piston engine, which makes the ratio of expansion no greater than the ratio of compression.

Fig. 283 shows one form of the apparatus designed to pump water from a low level source E to an outlet F some 30 feet higher. The two are connected by a long discharge pipe D, the water in which forms the oscillating column. This pipe terminates, on the left, in the working chamber A, which has an inlet-valve B for the admission of the explosive charge, and an exhaust-valve C, both opening inwards. The exhaust-valve C is at the foot of a short pipe which projects some way into the chamber. It falls open by its own weight when a pawl H is released, which takes place when the valve G opens to admit water from the supply tank into the oscillating column. Imagine a charge in A to be already compressed and to be fired by the sparking plug K. Water is driven down in A and some is discharged at F. The momentum of the moving column carries the level in A down so far that the pressure of the expanding gases falls to about that of the atmosphere; the valve G opens and water is taken in from E. This also opens the exhaust-valve C, which remains open until the column in the return

[1] H. A. Humphrey, *Proc. Inst. Mech. Eng.* December 1909.

movement hits the valve C and closes it, at the same time locking it in place by means of the pawl. The residue of burnt products left at the top of the combustion chamber becomes compressed until the backward motion of the column stops and it again begins to move down in A. During this downward movement the inlet-valve opens and a fresh charge is drawn in. In the next backward movement this is compressed and it is fired, just as the pressure

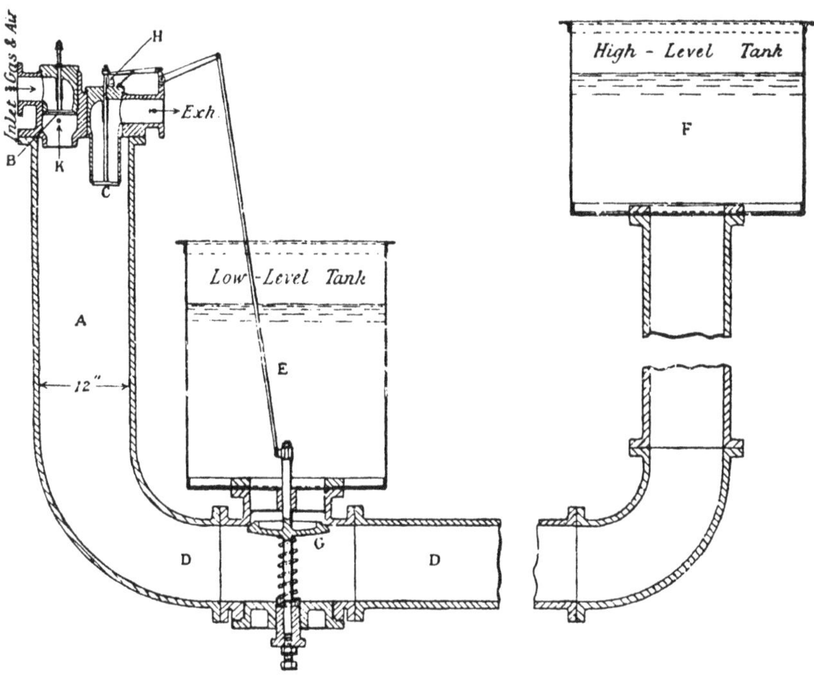

FIG. 283.

passes its maximum, by means of a firing device operated by changes of pressure. This completes the cycle, which is the cycle of Otto with the useful modification that the expansion stroke is longer than the inhaling and compression strokes. The column of water is maintained in a state of oscillation such that the alternate movements, to the same side, differ in range. A scavenging effect is obtained by letting the exhaust-valve open while the pressure within the chamber is still falling, so that air is drawn in. Thus when the column returns, discharging the burnt gases, it leaves a mixture of air and burnt gases in the clearance space. By having a valve separate from the exhaust-valve and at a higher level to admit air for scavenging, the residue in the clearance space is made to consist almost wholly of air. To start the pump from rest a charge of compressed air is introduced until the water is lowered somewhat beyond the usual charge level. The exhaust-valve is then forced open: the water consequently rises with momentum enough to close the exhaust-valve, compress the residue

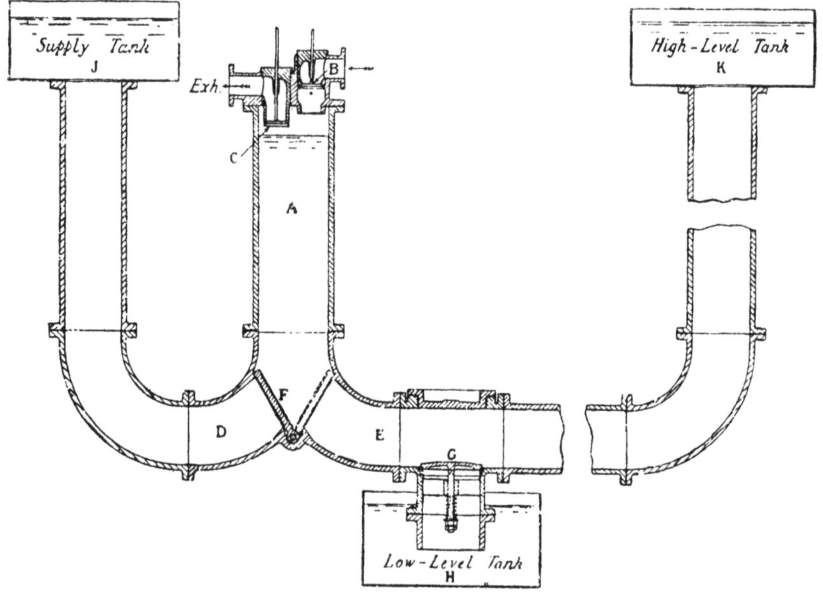

Fig. 284.

in the clearance space, fall again drawing in the charge, and then rise compressing it, after which ignition takes place.

A form of the Humphrey pump operating on what is in effect the Clerk cycle is shown in fig. 284. There the combustion chamber A ends in a pipe with two branches D and E, one or other of which is closed by a rocking valve F. The branch D leads to an auxiliary supply tank J, and the branch E to the rising main into which water can be drawn from the low-level supply tank H for discharge at the high level K. Starting from the condition sketched, with a compressed charge at the top of A and the valve F in the position which closes D, the charge is ignited, giving an impulse to the main water column AEK. When the expansion has caused the pressure to become atmospheric the exhaust-valve opens and the higher pressure in D throws the valve F over to the position shown by dotted lines. Water then comes down from J to discharge the burnt gases, and at the same time the main water column in E completes its movement towards the high-level discharge, taking in more water from the low-level supply tank H through the self-acting valve G. The auxiliary water column JDA closes the exhaust-valve C and compresses the residual gas into the cushion space. This stops and reverses its movement, and as it oscillates towards J a fresh combustible charge is drawn in. By the time this has happened, and the auxiliary column is beginning to return compressing the charge, the main water column in E returns, throwing F over to the initial position and completing the compression of the charge, after which the cycle repeats itself. A dash-pot connected to the valve F makes it move over somewhat gradually, and by regulating this control it may be arranged that the backward movement of the main column keeps up the water-supply of the auxiliary tank J, so that the net result of the whole action is simply a transfer of water from H to K. In this form of pump there is an explosion for every complete oscillation of the main column. The auxiliary column in D performs what is substantially the function of the displacing pump in the Clerk cycle.

341. The gas turbine. Attempts to drive a turbine directly by internal combustion have been numerous, but have not so far led to any valuable result. In Holtzwarth's gas turbine, which has been the subject of experiment on a considerable scale[1], a series of combustion chambers, grouped round an axis which is a prolongation of the turbine shaft, are successively charged and ignited, and the products of combustion from each in turn pass into a turbine where they are expanded in stages, just as steam is in a steam turbine. The chambers are kept from excessive heating by water-jackets, and in this way the temperature of the products, on their admission to the turbine nozzles, need not be impracticably high. But a serious obstacle to success is the need of compressing the air of the charge before combustion. A fairly high initial compression is, as we have seen, an essential condition of efficiency. In a piston engine the compression is readily effected by the piston

[1] See Lymn, *Trans. of the First World Power Conference*, 1924, vol. II, p. 931.

itself: an internal-combustion turbine provides no mechanism for it. A rotary blower is inefficient for high pressures, and the addition of a large reciprocating pump for the air would take away the mechanical simplicity which it is the object of the machine to secure. The only prospective advantage of the turbine would be that which comes from expansion to atmospheric pressure or near it. In such experiments as have been made this by no means compensates for the loss of efficiency that is due to an initially low degree of compression, and the net return of work is comparatively poor.

342. Classification of oil-engines. In the general classification of engines to which fuel is supplied in a liquid state a broad distinction is to be drawn between (1) those which compress a combustible mixture, consisting of air and fuel, either vaporized or in the form of a mist of fine particles, and (2) those which compress air alone and then inject the fuel into it at or near the dead-point. To the former class belong ordinary petrol motors, employing a carburettor to charge the air with inflammable matter before compression, and also some forms of engine using less readily vaporized oil: these may for brevity be called *mixture* engines. To the latter class, which may be called *injection* engines, belong Diesel engines and nearly all modern engines using heavy oil.

Among injection engines, besides those that employ the Diesel principle of compressing the air so much before the oil is injected that it burns as it enters, there are others often called semi-Diesel or hot-bulb engines where some assistance towards ignition is given, generally by having the fuel injected into a part of the compression space somewhat separated from the cylinder, and kept hotter by not being included among the parts that are cooled by water circulation. The temperature of the hot bulb or partially separate chamber is accordingly high enough, once the engine is running, to make the fuel ignite on injection, without so much compression of the air as the normal Diesel action requires. Further, in some engines of the Diesel class the oil is injected by means of a small quantity of still more highly compressed air; in others a high pressure pump forces a spray of oil in without this assistance. These modes of action are described respectively as air-injection and airless or "solid" injection.

Oil-engines, whether of the mixture or injection type, are (like gas-engines) classified as working on either the four-stroke cycle or the two-stroke cycle, and also as being either single-acting or double-acting. For small and moderate powers single action is much the more common; it is used also in many large engines, but there the desire to reduce weight and bulk is a strong incentive towards double-action. The four-stroke cycle is very common in oil-engines of all powers and using all grades of fuel, but the two-stroke cycle now takes an important place, especially in small cheap engines and also in marine and other engines of the largest class.

343. Petrol-engines. The carburettor. Engines using petrol or other easily vaporizable spirit are of the mixture class, inhaling the fuel along with the air by means of a carburettor, which is a device for causing the air, on its way to the admission-valve, to take up a due proportion of fuel. The fuel is taken in partly as true vapour but mainly as a mist of very fine liquid particles—a condition in which it is said to be "atomized." A true vapour is made up of molecules separated from one another; in the so-called atomized state each particle is an aggregate of very many adhering molecules, which may of course be converted into vapour by heating.

For engines such as those of motor cars, carburettors have to adapt the supply of oil automatically to large and frequent changes of speed. They act by aspirating the oil from the open end of a small tube which conveys it from a cistern where the level of the oil is kept constant by means of a float controlling an inlet-valve. The oil-tube is tapered to a fine opening and delivers into a wide pipe through which the air passes on its way to the engine, but at a constricted part of it where, in consequence of the constriction, the velocity of the air is relatively high and the pressure much less than atmospheric. Thus a jet of liquid particles is drawn out of the tube and travels on mixed with the air. But this arrangement, by itself, if adjusted to supply the right amount of fuel at high speed, would give too little at low speed, for increased velocity in the air current would suck in a more than proportionally increased amount of oil. To compensate for this and so secure a nearly uniform strength of mixture, a second oil-tube is added to which oil is supplied from the cistern at a constant rate, irrespective of the amount of air suction. This is the method of the Zenith carburettor, the action of which is shown diagrammatically in fig. 285. *A* is the cistern with its float which keeps the oil-level constant by acting on a needle-valve at the bottom, *B* is the main oil jet-pipe with it's pin-hole opening in the constricted part of the air-pipe or "choke-tube" *C*. The second oil-pipe *D* (which ends in an outer tube concentric with *B*) is fed by a small constant flow from the cistern into the intermediate well *E*. The two supplies together give the charge a nearly uniform richness over a very wide range of engine speeds. A third supply is obtained at starting through the pipe *F*, for when the engine is at rest the oil-level rises in the well *E* so that it is drawn up through *F* immediately the engine suction begins. In normal running, no oil enters through *F*. In very slow running the oil rises in *E* sufficiently to provide a reserve for acceleration when the throttle-valve *G* is opened.

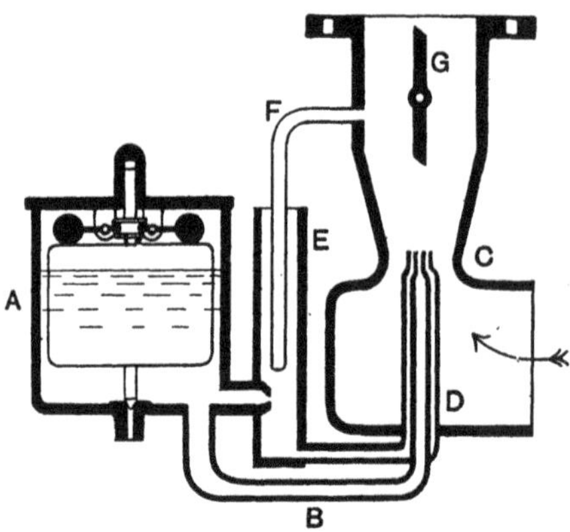

FIG. 285. Diagram showing principle of Zenith Carburettor.

The most usual form of engine for motor cars is a group of four vertical cylinders, water-cooled, with a "radiator" for dissipating to the atmosphere the heat taken up by the circulating water. The cylinders are single-acting and work on the four-stroke cycle, with all four cranks in the same plane, the two inner in the same phase and the two outer cranks at 180° from them, giving complete dynamic balance of couples and of primary forces (§ 264). Often six cylinders are used, with cranks at 120°, the two middle cranks being in the same phase, also the two end cranks, and the second and fifth.

Most of these engines have admission and exhaust-valves of the mushroom or poppet kind; a few of them, such as the Daimler-Knight engine, use sleeve-valve action, that is to say the lining of the cylinder forms a separate tube which is caused to slide relatively to the cylinder head and the fixed barrel, with the effect that this relative movement opens and closes cylinder ports at appropriate phases of the cycle.

In racing cars such engines may run at speeds of 4000 or 4500 revolutions per minute. Under favourable conditions their consumption of petrol is as low as 0·45 pint per brake horse-power-hour, corresponding to a conversion of about 31 per cent. of the thermal energy of the fuel into work on the brake.

For æro-engines there are, besides this "straight-line" arrangement of cylinders, many others. Among these is a V grouping of two inclined sets of cylinders, with twice as many cylinders as cranks, eight cylinders being grouped in two sets of four, or twelve cylinders in two sets of six, with each crank driven by two pistons, one in each set. In another type three sets of cylinders are combined, one set vertical and two inclined, with three pistons driving each crank. Another has a Maltese-cross arrangement of four cylinders per crank. In other types a radial grouping, usually of nine cylinders, is adopted, symmetrically spaced around the axis and working on one crank. Some of these radial engines are called "rotary" because the cylinders form a revolving system and the crank is fixed. In rotary engines and most of the fixed radial forms, as well as some other light engines, there is no water-cooling, but the cylinders carry a number of flat projecting ribs or fins which expose a much enlarged surface to the cooling action of the outside air. By this means, and by the use of very high piston speeds, high mean effective pressures, alloys of low specific gravity, machined forgings for cylinders, and the removal of all superfluous matter from fixed as well as moving parts, the weight of æro-engines, in relation to the power developed, has been reduced to 3 or in some cases even to 2 lb. per horse-power[1]. The weight of the lightest multi-cylinder water-cooled æro-engines is only slightly higher.

A familiar use of air-cooling is found in the numerous engines of small powers with one or two cylinders that are designed for motor cycles and motor boats, where lightness and cheapness are matters of primary importance.

344. Use of the two-stroke cycle. Many small petrol and other oil-engines work on the two-stroke cycle without requiring a separate displacement pump, by having the crank-case enclosed so that the under side of the piston serves to pump in the displacing air.

[1] For details of an air-cooled radial non-rotary engine, the British Jupiter, whose weight is barely 2 lb. per horse-power, see *Engineering*, Sept. 1925.

A closed crank-case acting as an air-pump is open to several objections, one being that lubricating oil is apt to be taken in as part of the charge. Engines of this type may not claim to be highly efficient in respect of fuel consumption, but they have great merits in simplicity, absence of valves, and low first cost. Many designs of this type are sold, some for petrol and others for paraffin, and it is not unusual to arrange that the same engine will take both fuels, petrol at starting and then paraffin when the combustion has become sufficiently hot to vaporize the paraffin spray.

An example of the class of enclosed crank-chamber engines by Messrs Petters of Yeoville is shown in fig. 286. Here the enclosing sides of the crank-chamber have flap valves which allow air to enter as the piston rises: then on the down stroke this air is lightly compressed and is ready to pass into the cylinder through scavenging ports on the right which are uncovered as the piston approaches its lower dead-point. The exhaust-ports on the left, being slightly higher, open before the air admission-ports. The top of the piston is so shaped that the entering air sweeps upward on the right, and scavenges the cylinder by driving the residue of burnt gases down on the left to the exhaust-ports. When the fuel is petrol the air is carburetted on its way from the crank-chamber to the admission-ports. When the fuel is paraffin or a heavier oil it is injected into a hot bulb or vaporizer at the top in the manner here illustrated, generally by being pumped in after the air of the fresh charge has already been compressed by the rise of the piston. The figure also shows, on the left of the hot bulb, a device for enabling the engine to start cold on heavy oil without preliminary use of petrol. This is a cartridge holder screwed into the bulb before starting, in which there is a paper tube with a slow-burning preparation to be ignited before insertion. The burning cartridge serves as igniter in the first strokes of the engine, and by the time it is consumed the holder and the bulb are hot enough to maintain the action. An alternative method of preliminary ignition is to use an electrically heated conducting strip in the bulb, or to apply a blow lamp or other external heater to the bulb for a short time before the engine starts.

In another form of two-stroke cycle oil-engine the crank-chamber is not itself used as a pump, but a cross-head made in the form of a piston is interposed between the crank and the main piston, and is enclosed so that it serves to pump in the scavenging air.

345. Hot bulb oil-engines. The devices for ignition which have been referred to above apply generally to hot-bulb engines,

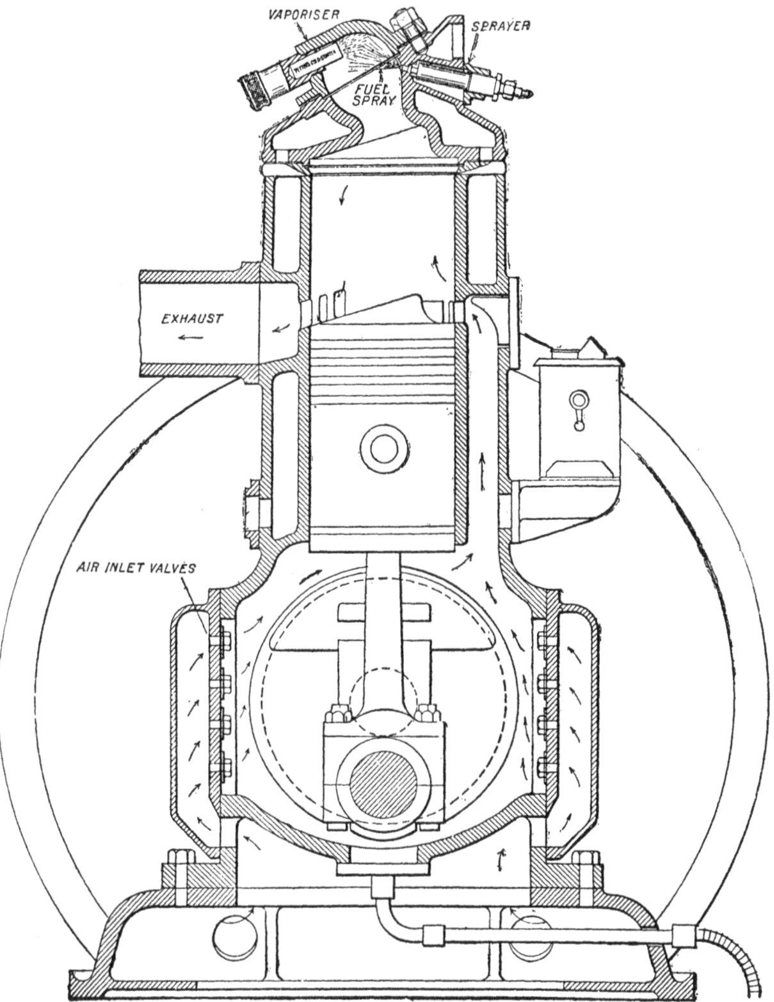

FIG. 286. Enclosed crank-chamber oil-engine working on two-stroke cycle (Petters).

whether working on the two-stroke or the four-stroke cycle. Early oil-engines made use of the exhaust gases to vaporize the oil completely before admission: it was then drawn in with air and treated as a gaseous charge. Later, about 1894, Messrs Hornsby introduced what was called the Hornsby-Ackroyd engine, in which the oil was injected by a small fuel-pump into a hot bulb while the engine cylinder was taking in its charge of air. The hot bulb formed an extension of the clearance space at the end of the cylinder, with a somewhat narrow neck between the two. The oil injected into it was vaporized by contact with the hot metal; its vapour remained not fully mixed with the air until near the end of compression, when enough air reached the hot bulb to allow ignition to take place at the temperature which was then reached. The amount of compression, as well as the form of the bulb, was so arranged that explosive combustion occurred at or close to the dead-point, as in a gas-engine, but without requiring a spark or other igniting agent. Before starting the engine the bulb was heated by applying a lamp. This engine, the invention of Mr Ackroyd Stuart, was very successful and is important as the precursor of the large class of modern hot-bulb engines using heavy oil. Most of these, however, compress the air more highly and do not spray the oil into the bulb until the compression is more or less complete[1].

Heavy oil-engines admitting the oil only when compression is complete, working with a compression of 300 lb. per sq. inch or thereabouts, with a hot bulb or its equivalent to assist ignition, are often called semi-Diesel engines, to distinguish them from the pure Diesel type in which the compression is so high (generally about 500 lb. per sq. inch) that the air is made hot enough to ignite the oil, as it enters, apart from any contact with metal other than the comparatively cool cylinder walls. In a hot-bulb or semi-Diesel engine the combustion generally coincides so closely with the dead-point that the indicator diagram is "peaky," resembling that given by an engine whose charge burns at nearly constant volume. In the true Diesel engine the admission of fuel is often more gradual, giving during the first part of the stroke a nearly level line which approximates to combustion at constant pressure.

An example of a semi-Diesel engine working on the four-stroke cycle is given in fig. 287, which shows one of Messrs Crossley's engines using heavy oil. The hot bulb, which is more or less spheroidal

[1] For an account of oil-engine development and description of certain modern forms, see F. H. Livens, *Proc. Inst. Mech. Eng.* July 1920.

in shape and projects from the end of the cylinder, has the air admission-valve above and the exhaust-valve below. The hand-

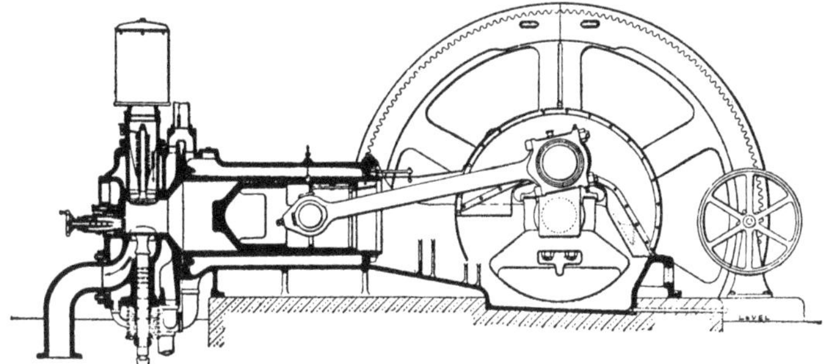

FIG. 287. Crossley engine for heavy oil.

worked valve seen at the back is to admit compressed air for starting. The oil-injection nozzle is at the side and is not seen in the figure. It appears however in fig. 288, which shows the bulb and cylinder end in horizontal section and illustrates an important feature in the working of the engine, namely, the device for promoting turbulence. For this purpose the piston has a contracted extension which nearly, but not quite, fills the neck of the bulb towards the end of the compression stroke, with the result that air coming from the cylinder acquires a violent eddying motion as it is pressed into the bulb, just before the oil spray is injected. This secures rapid inflammation of the whole charge, but apparently leaves the central portion of the air in the bulb so comparatively undisturbed during compression that it becomes hot enough, even when the engine starts cold, to ignite the charge, although the

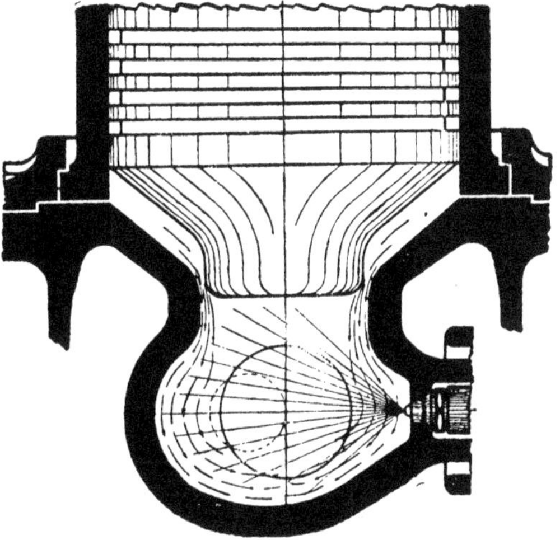

FIG. 288.

compression pressure is only 300 lb. per sq. inch. An engine of this class developing about 150 H.P. is reported to have consumed on trial 0·42 lb. of fuel oil per brake horse-power-hour, which means that 33 per cent. of the thermal energy of the fuel was converted into work.

346. Forms of Diesel engines. The idea of compressing the air separately, and taking advantage of the high temperature thereby produced as a means of igniting the fuel on its admission, is the fundamental feature of the engine introduced by R. Diesel about 1895, which in many different forms has now become an immensely important source of motive power, capable in practice of converting into effective work a larger fraction of the potential energy of the fuel than any other thermodynamic engine. Originally the Diesel engine worked on the four-stroke cycle, and this action is still preferred by many makers. There is however an increasing tendency to adopt the two-stroke cycle. The problem of scavenging is comparatively simple when, as in the Diesel engine, there is no mixture of fuel with the scavenging charge.

For the injection of fuel into the clearance space which forms the combustion chamber, Diesel's method was to provide an auxiliary supply of more highly compressed air which, mixing with the oil in a rapid passage through holes in small baffle plates placed in the injection nozzle, dragged it in as a fine spray which was scattered from the nozzle in the form of a spreading flame. For a time this method of injection was generally used, and it is still retained by the makers of some conspicuously successful Diesel engines; but it is now becoming recognized that the simpler method of solid injection, where the oil is forced in under high pressure without previous admixture with air, separates the particles sufficiently well to secure their efficient combustion, when a suitable form of spraying nozzle is used. Airless injection is therefore often preferred: it allows the designer to dispense with a wasteful feature, the high pressure air-pump, generally working in three stages, which brought the air for injection to a pressure much above that of the air in the clearance space.

Air compressed to a more moderate pressure is however a necessary element of all large Diesel engines to provide for starting and (in marine use) for manœuvring.

The four-stroke cycle and air injection are retained by the

Mirrlees firm, who were the first to take up the manufacture under Diesel's patents in Great Britain and have built many such engines, especially for land purpose. These features are also retained by Messrs Burmeister and Wain of Copenhagen, whose installation on board the "Selandia" in 1912 was the first step in the application of the internal-combustion engine to the driving of big ships. In the hands of Messrs Harland and Wolff as well as other builders the B. and W. engines, as they are called, have been very successfully applied to marine use and have (up to 1926) played a more considerable part than any others in the development of motor ships. Originally single-acting, they are now made double-acting in large sizes, the lower end of the cylinder being provided for this purpose with an enlargement at the side to make room for valves and give a space for combustion more or less clear of the piston-rod. At the top end, as in four-stroke cycle Diesel engines generally, the oil injection nozzle has a central position in the cylinder cover, with mushroom valves on either side for admission and exhaust. The twin-screw motor ship "Asturias," which made its first voyage in 1926, is driven by two B. and W engines, each with eight cylinders of 33 inches bore and 59 inches stroke, running at 115 revolutions per minute and developing together 15,000 brake horse-power[1].

The Werkspoor engine, which is also four-cycle with air injection, arranges for double-action by having an outlying combustion chamber at the lower end, with lower compression there than at the top where the pressure reaches the usual Diesel value of about 500 lb. The engine is started with single-action, fuel being admitted to the top end only, and the exhaust gases are diverted to heat the lower chamber so that it may presently take up its work. Thus at the top end there is pure Diesel action, at the bottom end semi-Diesel.

Messrs Vickers, who have had much experience in the design of Diesel engines for submarines, use airless injection, spraying the oil directly into the cylinder under a pump pressure of about 4000 lb. per sq. inch. They retain the four-stroke cycle, as do several other well-known makers.

The Tosi or Beardmore-Tosi engine[2] is a four-cycle form with

[1] For a full account of the Burmeister and Wain marine Diesel engine see Blache, *Trans. Inst. Eng. and Ship., Scotland*, April 1925.

[2] See First Report of the Marine Oil-Engine Trials Committee, *Proc. Inst. Mech. Eng.* Nov. 1924.

air injection and with the distinguishing feature that the same mushroom valve on the cylinder head serves for both suction and exhaust: outside it there is a two-way "director" valve which connects the passage alternately to the atmosphere and to the discharge pipe.

The two-stroke cycle is adopted by Messrs Sulzer in many Diesel engines for land and marine service, with (as is usual in this action) cylinder ports for air-admission and exhaust. A feature in their design is that there are auxiliary admission-ports above those through which the scavenging air enters: these remain open for a little after the exhaust-ports are covered, and so cause a small amount of supercharging by making the compression start from a pressure somewhat higher than that of the atmosphere.

Messrs Doxford, as was mentioned in § 339, build a two-stroke cycle opposed-piston Diesel engine on the lines of the Oechelhäuser gas-engine, with three cranks per cylinder. Tests of this engine, which uses airless injection, show it to have a high thermal efficiency. Messrs Cammell Laird also employ the two-stroke cycle, adapting the Fullagar design of opposed pistons in which each pair of twin cylinders require only two cranks (§ 339)[1]. An ingenious two-stroke engine with double action is made by the North-British Diesel Engine Co. where the cylinders, as in the Doxford and Fullagar engines, are tubes without fixed covers and consequently take no longitudinal stress. Their ends, in the North-British engine, are closed by cylindrical plugs (over which they can slide) which are fixed to the framework, and a small reciprocating motion is given to the cylinders over the plugs. This serves to open and close the admission or scavenge ports. The movement of a long central piston closes and opens the exhaust-ports. There is no piston-rod, but a gap between the upper and lower portions of the cylinder allows a cross-head to project from the middle of the piston, and from that the motion is taken to the crank[2].

In a few Diesel engines which use solid or airless injection the oil is first delivered into a small "ante-chamber" in the cylinder head, where a little of it is burnt, and the pressure thereby generated serves to project the remainder violently into the cylinder

[1] For trials of these two types, see Third and Fourth Reports of the Marine Oil-Engine Trials Committee, *Proc. Inst. Mech. Eng.* Jan. and May 1926.

[2] See Maclagan, *Trans. Inst. Eng. and Ship., Scotland*, April 1924.

through a nozzle tapering to a fine opening that connects the two. The ante-chamber is not a hot bulb, for it as well as the cylinder is cooled by a water-jacket, but the pressure both in it and in the cylinder is high enough to cause ignition. Its function is to secure more perfect pulverization of the entering fuel by adding to the final pressure of injection. An example of the ante-chamber device is found in the Worthington engine[1].

The Still engine, already mentioned in § 87, adds to the thermal efficiency of the Diesel type by using the heat of the circulating water and the exhaust gases to generate steam which acts on the underside of the piston. As an oil-engine it is single-acting and follows the two-stroke cycle, with cylinder ports for air-admission and exhaust and with a piston-head shaped more or less like that in fig. 286 to facilitate scavenging. An account of it, with results of tests, will be found in the Second Report of the Marine Oil-Engine Trials Committee[2]. In some of these tests the quantity of fuel oil consumed was only 0·35 or 0·36 lb. per brake horse-power-hour.

These brief notices of a few leading types will serve to show that the marine Diesel engine is, at the time of writing, in a tentative stage of its evolution. Unlike the marine steam-engine, its type is not yet fixed: its introduction to marine service is too recent for that. Many different forms are being tried, and while several have proved themselves to be efficient and practicable it cannot be said that any one form has established itself as the best. In respect of thermodynamic efficiency no very clear distinction can be drawn. A consumption of about 0·4 lb. of fuel oil per brake horse-power-hour is seen to be easily attainable, in more forms than one, which corresponds to a conversion of 35 per cent. of the net thermal energy of the fuel into useful effect. Considerations of simplicity in construction and working, of weight, cost, maintenance, of handiness in manœuvring, and above all of reliability in continuous service, are much more likely to determine the choice than any such small differences in thermodynamic efficiency as have hitherto been apparent. On general grounds it may be expected that the adoption of the two-stroke cycle, perhaps with double action, will tend to reduce what is at present a serious drawback to the marine Diesel engine, namely its great weight in relation to

[1] See also Nägel on Diesel Engines, *Trans. of the First World Power Conf.* 4291, vol. III, p. 32.

[2] *Proc. Inst. Mech. Eng.* March 1925.

its power. The comparatively low speed of revolution, the very high pressures which the structure has to stand, compared with the mean effective pressure, and the difficulty of cooling in large engines, all make for heavy weight. The weight of most Diesel engines is about 400 lb. per brake horse-power—a figure in sharp contrast to that which has been reached in small quick-running internal-combustion engines of other types. It is possible that the future development of oil-engines for motor ships may be on other lines; that, as was the case with steam turbines, some form of reducing gear will be resorted to—perhaps an electrical form—to reconcile the necessarily slow speed of a propeller shaft with the use of smaller, faster, and lighter engines.

APPENDIX

TABLES OF THE PROPERTIES OF STEAM

Table A. Properties of Saturated Steam, in relation to the Temperature.

A*. Properties of Water at Saturation Pressure.

B. Properties of Saturated Steam, in relation to the Pressure.

C. Volume of Steam in any dry state.

D. Total Heat of Steam in any dry state.

E. Entropy of Steam in any dry state.

F. Specific Heat, at constant pressure, of Steam in any dry state.

In these tables most of the figures are taken, with the author's and publishers' permission, from *The Callendar Steam Tables*[1], in which more complete data will be found. For higher pressures *The Enlarged Callendar Steam Tables* (1924) should be consulted.

In calculating his tables Callendar used a "characteristic equation" which he devised to express the relation between the pressure, volume, and temperature of steam—whether saturated or superheated—throughout the range of values that are common in steam-engine practice. This equation agrees well with the results of observation within a wide but still a limited range. It is in the form

$$V = \frac{RT}{P} - c + b.$$

If steam were a perfect gas, we should have $V = \frac{RT}{P}$, as in § 34. The terms b and c are introduced to express the actual deviations from the ideal volume it would occupy as a perfect gas.

Because the molecules are not simply mathematical points, but have some size, the volume of the gas is increased by the quantity b, which Callendar treats as a constant, and calls the "co-volume." The quantity c expresses the reduction of volume that arises from a temporary linking of the molecules during their encounters. It is not a constant but depends on the temperature, becoming less as

[1] Edward Arnold and Co. 1915.

the temperature rises. Callendar calls it the "co-aggregation volume," and expresses it in the form $c = \frac{C}{T^n}$, where C is a constant. For steam the index n is taken as $\frac{10}{3}$.

When Callendar's equation is applied to steam, the volume being expressed in cubic feet per pound, and the pressure in pounds per cub. foot, R is 154·17 (foot-pounds), C is 157520000, and b is 0·01602. The equation accordingly becomes

$$V = \frac{154{\cdot}17T}{P} - \frac{157520000}{T^{\frac{10}{3}}} + 0{\cdot}01602.$$

This applies to steam in any dry state, whether saturated or superheated, within the usual range. It does not of course give the volume of a wet mixture. The volumes in Tables A, B, and C are calculated from it.

As an example, apply it to find the volume of steam at a pressure of 160 lb. per sq. inch, superheated to 270° C. Here $P = 160 \times 144$, and $T = 563{\cdot}1$, making $V = 3{\cdot}6340 - 0{\cdot}1205 + 0{\cdot}0160 = 3{\cdot}5295$ cub. ft., which agrees with the figure in Table C. The volumes given in that table are calculated for various temperatures ranging from 400° down to the temperature of saturation, and below it. The volumes below the temperature of saturation refer to water-vapour in a supercooled state, such as is temporarily set up by sudden expansion in the absence of nuclei on which condensation may occur (§ 134).

To obtain numerical values for the other properties of steam Callendar uses certain experimental data, and assumes that, within the range to which his equations apply, the specific heat at constant pressure of steam in a very rarefied condition, namely, when the pressure is indefinitely low, may be treated as independent of temperature. He then deduces, by methods which are fully described in the author's *Thermodynamics for Engineers*, Chapter VIII, the following formulas for the calculation of the tables:—

For the total heat of steam in any dry state,

$$I = 0{\cdot}47719T - \frac{(\frac{13}{3}c - 0{\cdot}016)\,P}{1400} + 464.$$

For the internal energy in any dry state,

$$E = \frac{10}{3}\,\frac{P\,(V - 0{\cdot}016)}{1400} + 464.$$

For the entropy in any dry state,

$$\phi = 1{\cdot}09876 \log_{10} T - 0{\cdot}25356 \log_{10} P - 0{\cdot}002381 \frac{cP}{T} - 0{\cdot}21964.$$

The following expression, which applies only to saturated steam, connects the temperature with the saturation pressure p_s in pounds per sq. inch:—

$$\log_{10} p_s - \frac{0{\cdot}4057\,(c - 0{\cdot}016)\,p_s}{T} = 21{\cdot}07449 - \frac{2903{\cdot}39}{T} - 4{\cdot}71734 \log_{10} T.$$

To determine the volume, total heat, and entropy of water at saturation pressure (V_w, I_w and ϕ_w), Callendar makes certain further assumptions which we need not go into here[1]. He deduces the formula

$$I_w = 0{\cdot}99666\,t + (I_s - 0{\cdot}99666t)\,\frac{V_w}{V_s};$$

and, since $L = I_s - I_w$,

$$L = (I_s - 0{\cdot}99666t)\left(1 - \frac{V_w}{V_s}\right).$$

Here t is the temperature on the Centigrade scale, or $T - 273{\cdot}1$. In applying these formulas V_w/V_s may be taken as sensibly equal to $0{\cdot}00004p_s$.

Also $$\phi_w = 0{\cdot}99666 \log_e \frac{T}{273{\cdot}1} + \frac{V_w L}{T\,(V_s - V_w)}.$$

The values of ϕ_w and ϕ_s are related by the equation

$$\phi_s = \phi_w + \frac{L}{T}.$$

The Phase-function G. In addition to these various quantities, the tables give values of a function G which is defined by the equation

$$G = T\phi - I.$$

This function has the important property that it remains constant during a process of evaporation or condensation at constant pressure. In other words, it does not change when the substance changes its physical state or "phase" without alteration of temperature, as in melting or freezing, or in passing from liquid to vapour or *vice versa*. Consider for example a unit quantity of wet steam which is becoming drier by the evaporation of some of its water, under constant pressure and therefore at constant temperature T. Let a small quantity of the water in the mixture be evaporated by the taking in of a corresponding quantity of

[1] See *Thermodynamics for Engineers*, Art. 210.

heat δI. The entropy of the mixture consequently changes by a quantity $\delta\phi$ such that $T\delta\phi = \delta I$. Since $G = T\phi - I$,

$$\delta G = T\delta\phi - \delta I = 0.$$

That is to say, there is no change of G in any step of the process when water is passing into steam. G for dry saturated steam (G_s) has the same value as for water at the same pressure and temperature (G_w), or for an equilibrium mixture of steam and water in any proportions. It is convenient to have a name for the function G, and the fact that it remains constant in any substance during a change of phase has led the writer to call it the *phase-function*.

A useful application of the phase-function G is to determine the total heat of wet steam when the entropy of the mixture is known. For this purpose we may write, for the total heat per pound of the wet mixture,

$$I = T\phi - G.$$

Thus let steam expand adiabatically from an initial condition in which the total heat was I, and the entropy ϕ, until its temperature has fallen to some value T_2, at which it is wet. The total heat has then become

$$I_2 = T_2\phi_1 - G_2,$$

and by looking up G_2 in the tables (for the temperature T_2, or the corresponding pressure) I_2 is readily evaluated. Since the expansion is adiabatic, the entropy is still ϕ_1. The heat-drop is

$$I_1 - I_2 = I_1 - T_2\phi_1 + G_2.$$

This is the most direct method of calculating the heat-drop in adiabatic expansion. As an example, take the case considered in § 94, where steam supplied at 180 lb. per sq. inch, and superheated to 250° C. (so that I_1 was 702·95, and ϕ_1 was 1·6188), was adiabatically expanded to a pressure of 1 lb. per sq. inch, making T_2 311·84. From Table B, G after expansion is 2·61. Hence the heat-drop is

$$702{\cdot}95 - 311{\cdot}84 \times 1{\cdot}6188 + 2{\cdot}61 = 200{\cdot}7.$$

The use in this connexion of the phase-function G is only a matter of convenience: the procedure of § 94 gives the heat-drop readily enough though not quite so quickly.

Specific Heat at constant pressure. It may be shown (see *Thermodynamics for Engineers*, Art. 206) that in a gas which satisfies Callendar's characteristic equation the specific heat at constant pressure is given at any temperature T and pressure P by the formula

$$K_p = \frac{n(n+1)cP}{T} + K_p',$$

where K_p' is the value at zero pressure, assumed by Callendar to be independent of temperature. Table F gives numerical values of K_p throughout the range usual in practice.

Cooling effect of throttling. It is also easy to show that the cooling effect of throttling, when expressed as the amount of heat which would have to be supplied to prevent the temperature of the gas from falling when its pressure falls by unit amount as a result of throttling, is

$$(n + 1)\, c = b.$$

Hence observations of the cooling effect when dry steam is throttled serve to check the accuracy of the constants in this equation. It was, in fact, on the results of throttling experiments that Callendar mainly relied in settling the values of n and c.

Adiabatic expansion of steam in the dry state. In applying his equation to steam Callendar assigns to the constant n a value such that

$$nR = K_v',$$

where K_v' is the specific heat at constant volume when the pressure is indefinitely reduced, a quantity which, like K_p', is independent of the temperature. In this condition of extreme tenuity steam behaves as a perfect gas, so that (§ 36)

$$K_p' - K_v' = R,$$

and consequently $\quad K_p' = (n + 1)\, R.$

The numerical values for steam, as used by Callendar in the calculation of his tables, are (in calories) $K_v' = 0{\cdot}36707$; $K_p' = 0{\cdot}47719$, R being 0·11012, which is the equivalent in calories of the 154·17 foot-pounds given above.

When superheated steam expands adiabatically ϕ of course remains constant, and this requires that, *so long as none of the steam condenses,* $\dfrac{P}{T^{n+1}}$ or $\dfrac{P}{T^{\frac{13}{3}}}$ should remain constant, and also that $\dfrac{P\,(V - b)}{T}$ should remain constant (see *Thermodynamics for Engineers*, Art. 209). It follows that in such expansion the further relations hold:—

$$T\,(V - b)^{0{\cdot}3} = \text{constant; and } P\,(V - b)^{1{\cdot}3} = \text{constant}.$$

These results hold only so long as the expanding steam remains in the homogeneous state of a gas, whether superheated, saturated, or supercooled. They cease to be true when part of it liquefies.

When the steam is initially superheated they apply down to the temperature of saturation, and may continue to apply below it if (as in a steam jet) the steam becomes temporarily supersaturated without condensation (§ 134). The quantity b is so small compared with V at ordinary pressures that in dealing with the adiabatic expansion of steam which remains dry it usually suffices to take $TV^{0\cdot3}$ and $PV^{1\cdot3}$ as constant.

TABLE A. *Properties of Saturated Steam.*

Temp. Cent. t	Pressure, pounds per sq. inch p_s	Volume, cub. ft. per lb. V_s	Total Heat, lb.-calories per lb. I_s	Entropy, per lb. ϕ_s	Latent Heat, lb.-calories per lb. L	Internal energy, lb.-calories per lb. E_s
0	0·0892	3275·9	594·27	2·17602	594·27	564·21
10	0·1788	1693·8	599·01	2·11649	589·03	567·85
20	0·3399	922·19	603·72	2·06221	583·78	571·48
30	0·6162	525·81	608·40	2·01247	578·49	575·07
40	1·0703	312·45	613·04	1·96688	573·15	578·64
50	1·7888	192·72	617·63	1·92490	567·75	582·17
60	2·8873	122·91	622·16	1·88621	562·29	585·66
70	4·5156	80·804	626·60	1·85039	556·72	589·07
80	6·8627	54·596	630·95	1·81712	551·05	592·41
90	10·161	37·815	635·19	1·78619	545·25	595·67
100	14·689	26·789	639·30	1·75732	539·30	598·83
110	20·777	19·370	643·26	1·73027	533·17	601·86
120	28·808	14·271	647·07	1·70485	526·85	604·78
130	39·213	10·696	650·72	1·68092	520·32	607·58
140	52·482	8·1431	654·19	1·65831	513·57	610·23
150	69·150	6·2895	657·47	1·63689	506·56	612·73
160	89·800	4·9232	660·55	1·61657	499·29	615·08
170	115·06	3·9015	663·44	1·59724	491·75	617·27
180	145·59	3·1275	666·14	1·57884	483·93	619·30
190	182·08	2·5339	668·65	1·56128	475·82	621·19
200	225·24	2·0738	670·96	1·54453	467·41	622·91
210	275·78	1·7134	673·09	1·52851	458·69	624·48
220	334·38	1·4285	675·06	1·51326	449·69	625·93
230	401·89	1·2007	676·87	1·49868	440·38	627·23
240	478·74	1·0178	678·55	1·48480	430·81	628·43
250	565·63	0·8695	680·12	1·47161	420·96	629·53

TABLE A*. *Properties of Water at Saturation Pressure.*

Temp. Cent. t	Pressure, pounds per sq. inch $p_w = p_s$	Volume, cub. ft. per lb. V_w	Total Heat, lb.-calories per lb. I_w	Entropy, per lb. ϕ_w	Phase-function G, lb.-calories per lb. $G_w = G_s$
0	0·0892	0·01602	0	0	0
10	0·1788	0·01603	9·98	0·03585	0·181
20	0·3399	0·01605	19·94	0·07046	0·714
30	0·6162	0·01609	29·91	0·10393	1·58
40	1·0703	0·01614	39·89	0·13631	2·78
50	1·7888	0·01621	49·88	0·16770	4·30
60	2·8873	0·01629	59·87	0·19815	6·13
70	4·5156	0·01638	69·88	0·22774	8·26
80	6·8627	0·01648	79·90	0·25652	10·68
90	10·161	0·01659	89·94	0·28454	13·38
100	14·689	0·01671	100·00	0·31186	16·36
110	20·777	0·01684	110·09	0·33853	19·60
120	28·808	0·01698	120·22	0·36460	23·10
130	39·213	0·01713	130·40	0·39011	26·86
140	52·482	0·01729	140·62	0·41511	30·86
150	69·150	0·01746	150·91	0·43963	35·10
160	89·800	0·01765	161·26	0·46373	39·58
170	115·06	0·01785	171·69	0·48743	44·29
180	145·59	0·01807	182·21	0·51078	49·22
190	182·08	0·01831	192·83	0·53381	54·38
200	225·24	0·01856	203·55	0·55654	59·75
210	275·78	0·01885	214·40	0·57904	65·33
220	334·38	0·01914	225·37	0·60128	71·12
230	401·89	0·01946	236·49	0·62332	77·11
240	478·74	0·01980	247·74	0·64517	83·30
250	565·63	0·02016	259·16	0·66687	89·68

TABLE B. *Properties of Saturated Steam.*

Pressure, pounds per sq. inch p	Temp. Cent. t	Volume, cub. ft. per lb. V_s	Total Heat, lb.-calories per lb. I_s	Entropy, per lb. ϕ_s	Latent Heat, lb.-calories per lb. L	Phase-function G, lb.-calories per lb. $G_s=G_w$
0·1	1·59	2940	595·03	2·1662	593·44	0·005
0·2	11·69	1524	599·81	2·1068	588·15	0·246
0·3	17·99	1038	602·77	2·0727	584·83	0·58
0·4	22·66	790·7	604·97	2·0482	582·38	0·91
0·5	26·41	650·5	606·73	2·0299	580·40	1·23
1	38·74	333·1	612·46	1·9724	573·83	2·61
2	52·27	173·5	618·67	1·9159	566·52	4·69
3	60·83	118·6	622·53	1·8833	561·83	6·30
4	67·23	90·54	625·38	1·8600	558·28	7·64
5	72·38	73·44	627·64	1·8422	555·38	8·81
6	76·72	61·91	629·52	1·8277	552·92	9·86
7	80·49	53·59	631·15	1·8156	550·76	10·81
8	83·84	47·30	632·57	1·8049	548·82	11·69
9	86·84	42·36	633·85	1·7956	547·09	12·50
10	89·58	38·39	635·01	1·7874	545·50	13·26
12	94·44	32·37	637·02	1·7731	542·61	14·67
14	98·66	28·02	638·77	1·7611	540·12	15·94
16	102·41	24·73	640·26	1·7506	537·83	17·12
18	105·79	22·16	641·60	1·7414	535·75	18·20
20	108·87	20·08	642·82	1·7333	533·87	19·22
22	111·71	18·37	643·92	1·7258	532·09	20·18
24	114·34	16·93	644·93	1·7189	530·44	21·09
26	116·80	15·71	645·85	1·7126	528·88	21·95
28	119·11	14·66	646·74	1·7069	527·42	22·78
30	121·28	13·74	647·54	1·7016	526·02	23·56
32	123·35	12·94	648·30	1·6966	524·67	24·33
34	125·31	12·22	649·02	1·6919	523·40	25·07
36	127·17	11·59	649·69	1·6874	522·17	25·77
38	128·96	11·02	650·34	1·6831	521·00	26·45
40	130·67	10·50	650·95	1·6792	519·87	27·12
42	132·31	10·03	651·53	1·6754	518·77	27·76
44	133·89	9·603	652·08	1·6719	517·71	28·40
46	135·41	9·212	652·61	1·6685	516·68	29·00
48	136·88	8·853	653·12	1·6651	515·69	29·59

TABLE B (*continued*). *Properties of Saturated Steam.*

Pressure, pounds per sq. inch p	Temp. Cent. t	Volume, cub. ft. per lb. V_s	Total Heat, lb.-calories per lb. I_s	Entropy, per lb. ϕ_s	Latent Heat, lb.-calories per lb. L	Phase-function G, lb.-calories per lb. $G_s = G_w$
50	138·30	8·520	653·60	1·6620	514·71	30·16
60	144·79	7·184	655·77	1·6479	510·22	32·85
70	150·46	6·218	657·61	1·6359	506·23	35·30
80	155·52	5·487	659·20	1·6256	502·59	37·54
90	160·09	4·913	660·59	1·6165	499·24	39·62
100	164·28	4·451	661·82	1·6082	496·11	41·58
110	168·15	4·070	662·93	1·6007	493·18	43·40
120	171·75	3·751	663·92	1·5938	490·40	45·13
130	175·13	3·479	664·83	1·5875	487·76	46·78
140	178·31	3·245	665·69	1·5818	485·27	48·37
150	181·31	3·041	666·49	1·5765	482·90	49·89
160	184·16	2·862	667·22	1·5715	480·61	51·34
170	186·88	2·703	667·90	1·5666	478·40	52·75
180	189·48	2·562	668·53	1·5620	476·26	54·10
190	191·97	2·435	669·13	1·5577	474·20	55·42
200	194·35	2·320	669·69	1·5538	472·21	56·69
210	196·66	2·216	670·20	1·5502	470·26	57·94
220	198·87	2·120	670·70	1·5465	468·38	59·13
230	201·02	2·034	671·19	1·5429	466·55	60·31
240	203·09	1·954	671·64	1·5395	464·76	61·45
250	205·10	1·880	672·07	1·5362	463·00	62·58
260	207·04	1·811	672·48	1·5332	461·30	63·66
270	208·93	1·748	672·88	1·5303	459·65	64·72
280	210·77	1·689	673·25	1·5274	458·02	65·77
290	212·57	1·634	673·61	1·5246	456·41	66·79
300	214·32	1·583	673·96	1·5219	454·84	67·80
350	222·45	1·368	675·52	1·5096	447·44	72·57
400	229·75	1·206	676·84	1·4991	440·63	76·96
450	236·42	1·079	677·97	1·4897	434·28	81·06
500	242·57	0·977	678·97	1·4814	428·31	84·92

TABLE C. *Volume, in cubic feet per lb.,*

Temp. Cent.	Pressure in pounds per sq. inch						
	20	**40**	**60**	**80**	**100**	**120**	**140**
400	35·988	17·973	11·967	8·9648	7·1632	6·0009	5·1043
350	33·295	16·617	11·058	8·2785	6·6107	5·4989	4·7048
300	30·594	15·254	10·141	7·5848	6·0509	5·0284	4·2980
290	30·052	14·981	9·9569	7·4449	5·9378	4·9330	4·2153
280	29·510	14·706	9·7718	7·3045	5·8241	4·8372	4·1323
270	28·967	14·431	9·5863	7·1636	5·7100	4·7409	4·0487
260	28·425	14·156	9·4002	7·0221	5·5953	4·6441	3·9646
250	27·881	13·880	9·2134	6·8799	5·4798	4·5465	3·8798
240	27·337	13·603	9·0260	6·7370	5·3637	4·4481	3·7942
230	26·791	13·326	8·8376	6·5933	5·2467	4·3490	3·7078
220	26·246	13·048	8·6483	6·4486	5·1289	4·2490	3·6205
210	25·699	12·768	8·4582	6·3031	5·0101	4·1481	3·5324
200	25·150	12·488	8·2668	6·1564	4·8901	4·0459	3·4430
190	24·601	12·206	8·0743	6·0085	4·7690	3·9427	3·3524
180	24·050	11·923	7·8805	5·8592	4·6465	3·8380	3·2606
170	23·497	11·638	7·6850	5·7083	4·5224	3·7317	3·1670
160	22·944	11·352	7·4878	5·5558	4·3966	3·6238	3·0718
150	22·388	11·063	7·2886	5·4012	4·2687	3·5138	2·9745
140	21·829	10·773	7·0871	5·2443	4·1386	3·4016	2·8751
130	21·268	10·479	6·8832	5·0850	4·0061	3·2869	2·7731
120	20·705	10·183	6·6762	4·9227	3·8706	3·1791	2·6681
110	20·138	9·8840	6·4661	4·7572	3·7318	3·0482	2·5599
100	19·567	9·5809	6·2522	4·5878	3·5892	2·9235	2·4480

of Steam in any dry state

Temp. Cent.	Pressure in pounds per sq. inch						
	160	**180**	**200**	**250**	**300**	**350**	**400**
400	4·4609	3·9605	3·5601	2·8395	2·3591	2·0159	1·7586
350	4·1091	3·6459	3·2753	2·6081	2·1635	1·8458	1·6075
300	3·7501	3·3240	2·9831	2·3696	1·9605	1·6683	1·4492
290	3·6771	3·2585	2·9235	2·3206	1·9187	1·6317	1·4164
280	3·6036	3·1924	2·8634	2·2712	1·8765	1·5945	1·3830
270	3·5295	3·1258	2·8027	2·2212	1·8337	1·5568	1·3491
260	3·4550	3·0587	2·7416	2·1709	1·7894	1·5186	1·3148
250	3·3797	2·9908	2·6797	2·1196	1·7463	1·4796	1·2796
240	3·3037	2·9222	2·6171	2·0677	1·7015	1·4399	1·2437
230	3·2269	2·8529	2·5536	2·0149	1·6559	1·3994	1·2071
220	3·1492	2·7826	2·4893	1·9614	1·6094	1·3580	1·1695
210	3·0706	2·7114	2·4241	1·9068	1·5620	1·3158	1·1310
200	2·9908	2·6390	2·3576	1·8511	1·5135	1·2723	1·0914
190	2·9097	2·5655	2·2900	1·7941	1·4637	1·2276	1·0505
180	2·8274	2·4906	2·2210	1·7360	1·4126	1·1816	1·0083
170	2·7434	2·4140	2·1504	1·6760	1·3598	1·1339	0·9645
160	2·6578	2·3358	2·0782	1·6145	1·3054	1·0846	0·9190
150	2·5701	2·2555	2·0039	1·5508	1·2489	1·0332	0·8714
140	2·4802	2·1731	1·9273	1·4851	1·1902	0·9796	0·8217
130	2·3878	2·0881	1·8483	1·4167	1·1290	0·9235	0·7694
120	2·2924	2·0001	1·7663	1·3454	1·0649	0·8645	0·7141
110	2·1937	1·9089	1·6810	1·2708	0·9975	0·8022	0·6557
100	2·0913	1·8140	1·5920	1·1926	0·9263	0·7461	0·5934

TABLE D. *Total Heat I, in lb.-calories per lb.,*

Temp. Cent.	Pressure in pounds per sq. inch						
	20	40	60	80	100	120	140
400	784·70	784·20	783·71	783·22	782·73	782·23	781·74
350	760·68	760·04	759·39	758·74	758·10	757·45	756·81
300	736·61	735·74	734·88	734·01	733·15	732·28	731·42
290	731·78	730·86	729·94	729·02	728·10	727·18	726·26
280	726·95	725·97	724·99	724·02	723·04	722·06	721·08
270	722·11	721·07	720·03	718·99	717·95	716·91	715·87
260	717·27	716·16	715·06	713·95	712·84	711·73	710·62
250	712·43	711·24	710·06	708·87	707·69	706·51	705·32
240	707·57	706·31	705·04	703·78	702·51	701·25	699·99
230	702·71	701·36	700·01	698·66	697·30	695·95	694·60
220	697·85	696·40	694·95	693·50	692·05	690·60	689·16
210	692·97	691·42	689·86	688·31	686·76	685·20	683·65
200	688·08	686·42	684·75	683·08	681·41	679·74	678·08
190	683·19	681·39	679·60	677·81	676·01	674·22	672·43
180	678·28	676·35	674·41	672·48	670·55	668·62	666·69
170	673·35	671·27	669·19	667·10	665·02	662·94	660·85
160	668·41	666·16	663·91	661·66	659·41	657·16	654·91
150	663·46	661·02	658·59	656·15	653·71	651·28	648·84
140	658·48	655·84	653·20	650·56	647·92	645·28	642·64
130	653·48	650·61	647·75	644·88	642·01	639·14	636·27
120	648·46	645·34	642·21	639·09	635·97	632·85	629·73
110	643·40	640·00	636·59	633·19	629·79	626·38	622·98
100	638·31	634·59	630·87	627·15	623·43	619·71	615·99

of Steam in any dry state.

Temp. Cent.	Pressure in pounds per sq. inch						
	160	**180**	**200**	**250**	**300**	**350**	**400**
400	781·25	780·76	780·26	779·03	777·80	776·57	775·33
350	756·16	755·51	754·87	753·25	751·63	750·02	748·40
300	730·55	729·69	728·82	726·66	724·50	722·34	720·17
290	725·34	724·43	723·51	721·21	718·91	716·61	714·31
280	720·11	719·13	718·15	715·71	713·26	710·82	708·37
270	714·83	713·79	712·75	710·14	707·54	704·94	702·34
260	709·51	708·40	707·29	704·52	701·75	698·98	696·21
250	704·14	702·95	701·77	698·81	695·85	692·89	689·93
240	698·72	697·46	696·19	693·03	689·87	686·70	683·54
230	693·24	691·89	690·54	687·16	683·77	680·39	677·01
220	687·71	686·26	684·81	681·19	677·56	673·94	670·32
210	682·10	680·54	678·99	675·10	671·22	667·34	663·45
200	676·41	674·74	673·07	668·90	664·73	660·56	656·39
190	670·63	668·84	667·04	662·56	658·08	653·59	649·11
180	664·76	662·83	660·90	656·07	651·24	646·41	641·58
170	658·77	656·69	654·60	649·40	644·19	638·98	633·77
160	652·66	650·41	648·16	642·53	636·91	631·28	625·66
150	646·41	643·97	641·53	635·44	629·35	623·26	617·17
140	640·00	637·36	634·72	628·12	621·52	614·92	608·32
130	633·41	630·54	627·67	620·50	613·33	606·16	598·99
120	626·61	623·48	620·36	612·56	604·75	596·95	589·15
110	619·57	616·17	612·77	604·26	595·75	587·24	578·73
100	612·27	608·55	604·83	595·52	586·22	576·92	567·62

TABLE E. *Entropy ϕ of Steam*

Temp. Cent.	Pressure in pounds per sq. inch						
	20	40	60	80	100	120	140
400	2·0100	1·9331	1·8878	1·8556	1·8304	1·8097	1·7921
350	1·9729	1·8958	1·8503	1·8178	1·7924	1·7714	1·7536
300	1·9326	1·8551	1·8092	1·7764	1·7506	1·7293	1·7111
290	1·9242	1·8465	1·8006	1·7676	1·7417	1·7203	1·7021
280	1·9155	1·8378	1·7917	1·7586	1·7327	1·7112	1·6928
270	1·9067	1·8288	1·7826	1·7494	1·7234	1·7018	1·6833
260	1·8977	1·8197	1·7734	1·7401	1·7139	1·6921	1·6735
250	1·8885	1·8104	1·7639	1·7305	1·7041	1·6822	1·6635
240	1·8791	1·8009	1·7543	1·7206	1·6941	1·6721	1·6532
230	1·8696	1·7911	1·7443	1·7105	1·6839	1·6617	1·6426
220	1·8598	1·7812	1·7342	1·7002	1·6733	1·6509	1·6316
210	1·8498	1·7709	1·7238	1·6896	1·6625	1·6398	1·6204
200	1·8396	1·7605	1·7131	1·6786	1·6513	1·6284	1·6087
190	1·8291	1·7497	1·7021	1·6673	1·6397	1·6166	1·5966
180	1·8184	1·7387	1·6907	1·6557	1·6278	1·6044	1·5841
170	1·8074	1·7274	1·6791	1·6437	1·6155	1·5917	1·5711
160	1·7961	1·7157	1·6670	1·6313	1·6027	1·5785	1·5575
150	1·7845	1·7037	1·6546	1·6184	1·5894	1·5648	1·5433
140	1·7726	1·6913	1·6417	1·6050	1·5755	1·5504	1·5285
130	1·7604	1·6785	1·6283	1·5911	1·5610	1·5354	1·5129
120	1·7478	1·6652	1·6144	1·5766	1·5458	1·5196	1·4964
110	1·7347	1·6515	1·5999	1·5613	1·5299	1·5029	1·4790
100	1·7213	1·6372	1·5848	1·5454	1·5131	1·4852	1·4605

in any dry state.

Temp. Cent.	Pressure in pounds per sq. inch						
	160	180	200	250	300	350	400
400	1·7768	1·7633	1·7511	1·7250	1·7034	1·6849	1·6687
350	1·7381	1·7243	1·7118	1·6852	1·6630	1·6439	1·6271
300	1·6952	1·6810	1·6682	1·6406	1·6175	1·5976	1·5798
290	1·6861	1·6718	1·6539	1·6311	1·6077	1·5873	1·5696
280	1·6767	1·6623	1·6493	1·6212	1·5976	1·5771	1·5589
270	1·6670	1·6525	1·6394	1·6111	1·5872	1·5664	1·5479
260	1·6572	1·6425	1·6293	1·6006	1·5764	1·5553	1·5365
250	1·6470	1·6322	1·6188	1·5898	1·5652	1·5438	1·5246
240	1·6365	1·6216	1·6081	1·5786	1·5537	1·5319	1·5123
230	1·6257	1·6106	1·5969	1·5671	1·5417	1·5194	1·4995
220	1·6146	1·5993	1·5854	1·5551	1·5292	1·5065	1·4860
210	1·6031	1·5876	1·5735	1·5426	1·5162	1·4929	1·4719
200	1·5912	1·5755	1·5611	1·5296	1·5026	1·4788	1·4571
190	1·5789	1·5629	1·5482	1·5161	1·4884	1·4639	1·4416
180	1·5661	1·5497	1·5348	1·5019	1·4735	1·4482	1·4251
170	1·5527	1·5360	1·5208	1·4870	1·4577	1·4316	1·4077
160	1·5387	1·5217	1·5060	1·4713	1·4411	1·4140	1·3891
150	1·5241	1·5067	1·4906	1·4548	1·4235	1·3953	1·3693
140	1·5088	1·4908	1·4743	1·4373	1·4047	1·3753	1·3482
130	1·4926	1·4741	1·4570	1·4186	1·3847	1·3539	1·3253
120	1·4755	1·4564	1·4386	1·3986	1·3631	1·3307	1·3006
110	1·4574	1·4375	1·4190	1·3772	1·3399	1·3056	1·2737
100	1·4381	1·4174	1·3980	1·3541	1·3147	1·2783	1·2443

TABLE F. *Specific Heat, at constant pressure, of Steam in any dry state.*

lb.-calories per lb.

Temp. Cent.	Pressure in pounds per sq. inch											
	20	40	60	80	100	120	140	160	180	200	300	400
400	·480	·482	·485	·488	·490	·492	·495	·498	·501	·503	·516	·528
380	·480	·483	·486	·489	·492	·495	·498	·501	·504	·507	·522	·536
360	·481	·484	·487	·491	·494	·498	·501	·504	·508	·511	·528	·545
340	·481	·485	·489	·493	·497	·501	·505	·508	·512	·516	·535	·555
320	·482	·486	·491	·495	·500	·504	·509	·513	·518	·522	·545	·568
300	·482	·488	·493	·498	·504	·509	·514	·519	·524	·529	·556	·582
280	·483	·490	·496	·502	·508	·514	·520	·526	·532	·538	·569	·599
260	·484	·492	·499	·506	·513	·520	·527	·535	·542	·549	·585	·620
240	·486	·494	·503	·510	·519	·528	·536	·545	·553	·562	·604	·646
220	·487	·497	·507	·517	·527	·538	·548	·557	·568	·578	·627	·678
200	·489	·501	·513	·515	·537	·549	·561	·573	·585	·598	·657	
180	·492	·506	·521	·535	·549	·564	·579	·593	·608	·622		
160	·494	·512	·530	·548	·565	·583	·601					
140	·499	·520	·542	·564								
120	·504	·530										
100	·511											

INDEX

www.ingramcontent.com/pod-product-compliance
Ingram Content Group UK Ltd.
Pitfield, Milton Keynes, MK11 3LW, UK
UKHW042207080726
473066UK00007B/285

* 9 7 8 1 1 0 7 6 1 5 6 3 2 *